THE STORY OF ALCHEMY
AND EARLY CHEMISTRY

(The Story of Early Chemistry)

BY

JOHN MAXSON STILLMAN

The more I try to understand chemistry, the more I am convinced that the methods, achievements and aims of the science can be realized only by him who has followed the gradual development of the chemical ideas.
M. M. Pattison Muir

DOVER PUBLICATIONS, INC.
NEW YORK

Published in Canada by General Publishing Company, Ltd., 30 Lesmill Road, Don Mills, Toronto, Ontario.
Published in the United Kingdom by Constable and Company, Ltd., 10 Orange Street, London WC.2.

This Dover edition, first published in 1960, is an unabridged and unaltered republication of the work originally published in 1924 under the title *The Story of Early Chemistry*.

International Standard Book Number: 0-486-20628-9
Library of Congress Catalog Card Number: 60-3183

Manufactured in the United States of America
Dover Publications, Inc.
180 Varick Street
New York, N. Y. 10014

PREFACE

The pioneers in the history of chemistry, J. F. Gmelin, Thomas Thomson, Ferdinand Hoefer and Herman Kopp, devoted much able and serious labor to the early developments in the growth of the science. Later historians of the science have laid the emphasis upon the more modern development, and have depended largely upon the earlier histories for their summaries of early chemists.

In the mean time, however, much serious attention has been given to ancient and medieval writers by certain modern scholars, and their conclusions have altered, in important respects, the story of the growth of chemical knowledge and speculation. Such investigators are M. Berthelot, Pierre Duhem, Edmund von Lippmann, B. Haureau, John Ferguson, Otto Lagercrantz, Karl Sudhoff, F. Dieterici, and many others.

The desirability and importance of a re-writing of the history of early chemistry was brought home to the present writer during the fifteen years in which he conducted an advanced class in the history of chemistry at this University. Retirement as emeritus in 1917 offered the opportunity for time and study; and the fortunate presence in the library of the University and in the Medical Department, of an unusual collection of early books, journals, and proceedings of scientific academies, encouraged the ambition. The large private library of Mr. Herbert C. Hoover, relating to early chemistry, metallurgy and mining, made freely accessible to the writer, added importantly to the resources of valuable works.

The endeavor has been to tell the story of the development of chemical knowledge and science, from the earliest times to the close of the eighteenth century, in a connected and systematic way, not as a condensed encyclopedia, but rather by placing the emphasis upon such discoveries and speculations as have made a decided impress on the growth of the science. Thus the names of many chemists are missing which occur in the earlier histories. None, however, of real significance in the growth of chemical science is intentionally omitted.

For the benefit of critical readers of this book, the author has thought it desirable to append a bibliography of the principal works consulted in its preparation, not including journals or proceedings of standard societies. In general, it has seemed advisable to translate into English the many quotations from ancient and modern languages, with such references as would enable those interested to verify the accuracy of the translators.

The author also takes this opportunity of acknowledging his indebtedness to many friends and colleagues for friendly assistance, in particular to the President of the University, Dr. Ray Lyman Wilbur, to Librarian George C. Clark, and to the Department of Chemistry for many needed facilities; to Dr. Wm. F. Snow (A. B. Chemistry, '96), for the generous donation of a fund used for supplying stenographic assistance. Also especial acknowledgment is due to Professor B. O. Foster, of the Department of Classical Literature in this University, for his cordial aid in translating and revising many translations from ancient or medieval Latin; and to my colleagues of the Department of Chemistry: Professors S. W. Young, E. C. Franklin, and R. E. Swain, for their generous assistance in reading the manuscript in progress and for their many valued suggestions in connection therewith.

Stanford University
California

CONTENTS

FOREWORD

Shortly before the first proofs of the *"Story of Early Chemistry"* were received from the publishers, the author, John Maxson Stillman, passed quietly away at his home at Stanford University after only a few hours of acute illness. On this account it has seemed desirable that the book should be prefaced by a brief sketch of the life and character of its creator, and I have gladly undertaken this labor, with the hope that more than thirty years of close association as colleague and friend may have reasonably qualified me for the task.

Professor Stillman was born at New York on April 14, 1852. His early years were spent at Sacramento, California, and later at San Francisco. In 1874 he was graduated from the University of California, to which, after two years of study at Würzburg and Strassburg, he returned as instructor in chemistry. In 1885 his Alma Mater granted him the degree of Doctor of Philosophy, and later in his life, in 1916, conferred upon him the degree of Doctor of Laws. After some years of service at the University of California, he accepted a position as superintendent and chief chemist at the Boston Sugar Refinery, a position which he held for ten years, when he resigned to become the first executive head of the Chemistry Department of the newly founded Leland Stanford Junior University. He assumed the responsibilities of the new position in January, 1891, and was continuously active until 1917, when he retired as *professor emeritus*. He died on December 13, 1923.

Professor Stillman was a man of broad and diverse interests and activities, and whenever he undertook a thing, it was with fine a enthusiasm and great energy. He had, first of all, a profound respect for sound scholarship, and this not only led him to equip himself as thoroughly as possible in matters of learning, but it also became a living influence upon those with whom he came into contact, an influence which awakened in others aspirations for self-improvement and carried with it a realization of the value of knowing things well. He had a keen eye for the beautiful in art, and his collector's instinct brought him many fine books and etchings, and particularly a large collection of Japanese prints and carvings, all of which he loved solely because of their aesthetic appeal to him. In social affairs and usages, he had a keen and discriminating taste, which, together with an unembarrassed social manner, made him a charming host and a gracious presiding officer at social gatherings, where he was master of a genial humor that put everyone at ease, and though his ready wit and *repartée* sometimes grazed the skin slightly, they never punctured it, and above all never humiliated.

Stillman's participation in all things having to do with the day's work was always very active and very effective, and he was frequently called upon to do more than his share. But he never stinted himself in the response. A vigorous honesty with himself and an unusually keen instinct in divining the possible and probable results of an administrative policy, combined with a fine idealism and a high sense of duty and responsibility to his superiors in administration made his counsel and executive skill invaluable during the formative period of the young university. He gained much pleasure from this general administrative work, and was by nature well constituted for it, being able, on occasion, to enforce an unpopular ruling with so much of diplomacy as to arouse a minimum of antagonism. He was for several years vice-president of the University, and often acting president. Here, within the limitations of his power he always stood for sanely progressive policies, and a goodly store of worldly wisdom told him when to fight, when to bide his time, and when, if necessary, to yield.

As a department executive, where his authority was almost autocratic, his attitude was always forbearing, kindly, conciliating and helpful, but he was nevertheless a jealous guardian of his rights and prerogatives. His willingness to freely discuss questions of policy, to listen patiently to opposing views, and his always unruffled, dignified and gentlemanly bearing were largely responsible for an almost ideal atmosphere in the faculty of which he was the chief. It was never a pleasure to him to make a showing of his power, although if necessary he never shrank from it, but he preferred always, even though it took time, to settle things by peaceful methods.

There was something in Stillman's art as a teacher that almost invariably commanded the respect, admiration and devotion of his pupils. It was not merely that he lectured well, and taught well in the laboratory; nor was it merely that he was painstaking, patient and generous to a fault of his time and energy. That he had a strong, inborn instinct for teaching and took great delight in fathoming the workings of immature and even slow minds is quite true, and that he was invariably affable and courteous is equally true, but all these things do not quite explain the high esteem in which he was held by so many of his students. If it is to be explained at all, I think it was due to a fine power that was his, of keenly discerning the deeper spiritual characteristics and mental traits of each of those with whom he came into contact, and thus of subtly distinguishing between individuals and meeting each on his own ground. I doubt if any serious student ever had cause to feel that he was just a specimen of the *Genus*

Studiosus consulting the professor, but rather that he was *himself* going to talk over his work or his affairs with a good friend who was better informed and wiser than he. But whatever the explanation, the relation was always a most admirable one, based on mutual respect and friendliness.

It was out of Professor Stillman's labors as a teacher that *"The Story of Early Chemistry"* was born and grew to what it is. For much of his life he had given increasing attention to the history of chemistry, and for many years taught the subject to small classes. Gradually covering new ground and extending his knowledge of the field, he finally gained a breadth of view which he felt might justify some contributions to the literature of the subject. These began with a number of shorter articles, namely:

1912. *Basil Valentine, a 17th Century Hoax.* (Popular Science Monthly)

1915. *The Dawn of Modern Chemistry.* (Popular Science Monthly)

1917. *Contributions of Paracelsus to Medical Science and Practice.* (The Monist)

1918. *Chemistry in Medicine in the Fifteenth Century.* (Scientific Monthly)

1919. *Paracelsus as a Reformer in Medicine.* (The Monist)

      Paracelsus as a Theological Writer. (The Open Court)

      Paracelsus as a Chemist and Reformer in Chemistry. (The Monist)

      The Character and Ethics of Paracelsus. (The Open Court)

1923. *Petrus Bonus and Supposed Chemical Forgeries.* (Scientific Monthly)

In 1920 *"Theophrastus von Hohenheim, called Paracelsus"* came from the press and finally, during the later years of the *emeritus* professorship, although there were still many other demands on his time and energy, he completed *"The Story of Early Chemistry."* In this book he planned to develope in parallel from the earliest known beginnings the history, on the one hand of the chemical arts, on the other hand of chemical thought and theory, concluding the work with the downfall of the phlogiston theory. He aimed at a book that should be found readable by those whose knowledge of the science was not profound, as well as by those professional chemists who find little time to delve into such matters for themselves.

STEWART W. YOUNG

STANFORD UNIVERSITY
  CALIFORNIA

THE STORY OF
EARLY CHEMISTRY

CHAPTER I

THE PRACTICAL CHEMISTRY OF THE ANCIENTS

The beginnings of the arts we call chemical are lost to us in the buried civilizations that have left no records sufficiently decipherable to afford us definite knowledge, but so far as remains and records of the oldest civilizations exist, they give evidence of the great antiquity of many chemical arts.

These earliest evidences are naturally those that relate to the practical arts rather than to the natural philosophy or speculations which the practical workers of those times used to explain or interpret the facts as known to them. These theories and speculations, if indeed they were recorded at all, were in the form of records which were peculiarly liable to destruction from the elements.

The human mind is so constituted that it finds a need to attempt to account for observed phenomena, so that theory and practice are inseparable. The natural curiosity we entertain to know what, for example, the earliest natural philosophers thought about the nature and changes of substances finds little satisfaction until a time when written records exist, as in Greece in the fifth or sixth century before Christ, or in India at very early dates.

Our knowledge of the very earliest developments of chemical arts is dependent upon the discovery of products of these arts which have been preserved under circumstances which permit reasonably reliable estimates of their origin and approximate age.

Such products are, for example, articles of metal, pottery, glass, cements or mortars, pigments and dyed materials. Analyses of such articles give much valuable information as to the development of certain arts at various periods. Remains of tools, factories or furnaces, etc., also furnish information at times.

Thus M. Berthelot analyzed a small votive figure from the excavations at Tello in Ancient Chaldea, and found it to consist of nearly pure copper. The age of this figure is variously estimated at from 3000 to 4000 B. C. A small metal cylinder from Egypt of a period estimated at about 4000 B. C. was also of copper. Thus the mining and metallurgy of copper is at least 5,000 years old, and as to how much older, evidence from dependable chronology may be lacking.

It appears from evidence from many localities that copper was in use for a long time before bronze came into use. The readiness with which bronze can be cast and its greater hardness for articles of use afforded manifest advantages when once known. Bronzes of copper and tin seem also to have been of great antiquity. Somewhat later lead was utilized, and much later we find zinc entering into their composition.

Angelo Mosso[1] analyzed metal from the statue of Pepi, dating from the sixth dynasty (estimated about 2500 B. C.), and found it to consist of copper with 6.56 per cent tin; while a bronze plate of the same period contained 9 per cent tin. A metal plate attributed to the first dynasty (3400 B. C.), contained 3.75 per cent tin.

Rathjen and Schulz[2] analyzed various articles of Egyptian origin of periods from about 3500 B. C. to 350 B. C.

---

[1] *The Dawn of Mediterranean Civilization*, London, 1910.
[2] *Beiträge aus der Geschichte der Chemie.* Edited by P. Diergart, Leipzig and Vienna, 1909, p. 212–213.

The earliest of these, a chisel-shaped tool of about 3500 B. C., was of pure copper (99.9 per cent). So also a small figure of about 1300 B. C. was of pure copper. Some fifteen other articles, dating from 1900 B. C. to 350 B. C., were of copper alloyed with tin, ranging from 3 to 14 per cent tin, or with tin and lead, the lead ranging from small quantities, probably unintentional, up to 25 per cent. One figure, of 700 B. C., was of copper with 1.72 per cent arsenic. All of these bronzes contained small quantities of iron, and often small quantities of nickel, cobalt and arsenic, probably unintentional constituents.

Bronzes of copper and tin were found by Schliemann in the ruins of Troy, Tyrins and Mykenae, indicating origins of as early as 2000 B. C.

Layard[3] gives the composition of bronze articles found in the ruins of Nimroud which show fairly uniform composition of the alloy of copper and tin.

BRONZE FOUND IN NIMROUD

Object	Per Cent Copper	Per Cent Tin
Bowl.	89.51	10.63
Hook.	89.85	9.78
Figure of a bull.	88.37	11.33
Bell.	84.79	14.10

Berthelot also found that the most ancient articles of Egyptian origin are of copper without addition of other metals. Bronzes of copper and tin he finds as early as the sixth dynasty. Indeed, in a weathered metal fragment from a tomb of the third dynasty, according to Masperot, he found a very considerable admixture of tin, the quantity being sufficient to serve as rather conclusive evidence

[3] Layard, *Nineveh and Babylon*, 1859, p. 571.

as far back as some 3000 years B. C. tin bronzes were made.[4]

Berthelot also found that articles of metal from ancient Chaldea, dating from 1000 B. C. to 3000 B. C., were composed of copper alone, while a statuette of about 2600 B. C. was of copper and lead in the ratio of about one of lead to four of copper; while another article of Chaldean origin of similar antiquity was of copper and tin with about 12 per cent tin.[5]

The ancient use of tin in bronze is established by many such data from many localities.

That the ancients recognized tin itself as a distinct metal is not, however, to be inferred. It is quite probable that the tinstone (oxide) was used directly in the furnaces, not previously reduced and added as a metal, because, so far as can be inferred, alloys which were manufactured by the ancients were generally made by mixing the ores in the furnace, not by melting together the metals themselves. The Greeks named tin "Kassiteros," though probably this name includes the ores as well as the metal.

There has been much speculation as to the sources whence the ancient Egyptians obtained the tin for their bronzes. No nearby sources have been discovered. Geologic evidence is to the effect that tin occurred in Persia, and it may have been from this region that the earliest supplies came. It is also possible that sources of tinstone from farther south on the African continent may have been drawn upon, but any evidence to that effect is also lacking.

The Greek name "kassiteros" is allied to the more ancient names for tin among Assyrians, Acadians and Babylonians (kazazatira, ikkasduru, kastira).[6] The Sumerians in Southern Babylonia (Shinar), evidently possessed a knowledge of tin as a constituent of bronze as early as about 3000 B. C., and it is not impossible that this region was the earliest source of tin for Egypt and the Mediter-

---

4 Berthelot, *Archéologie et Histoire des Sciences*, p. 6 *ff.*
5 Berthelot, *op. cit.*, p. 75 *ff.*
6 Von Lippmann, *Entstehung und Ausbreitung der Alchemie*, pp. 578, 579.

ranean countries.[7] Just when the sources of tin in Britain
became available to the ancient world about the Mediterranean is difficult to determine. References in ancient
authors, however, make evident that certainly by the fifth
century B. C. tin was received from that region. The
price of the metal was lowered and the uses of bronze much
expanded by the opening up to trade of the rich deposits
of the British Islands.

So late as the first century of our era, tin was called
by the Latins white lead (plumbum candidum or album),
as distinguished from our lead (plumbum nigrum). The
metal by itself seems not to have been used for making
articles of use or ornament, though its use for coating copper vessels to protect them from rust or corrosion in use
was known to Pliny and to Dioscorides. According to
Pliny, this art was supposed to have been introduced from
Gaul. Pliny says that white lead is naturally more dry,
while black lead is always moist; consequently, the white
without being mixed with another metal is of no use for
anything. This is a curious attempt to explain physical
properties on the basis of the Aristotelian theory of the
elementary qualities of matter—moist, dry, hot and cold.

The word "stannum" (modern Latin for tin) is used
by Latin writers of later ancient periods not to designate
tin, but an alloy of lead and tin in varying proportions,
practically our pewter.

Lead, called by the Greeks "molybdos," by the Latins
"plumbum," by reason of the wide occurrence of its ores
and the readiness of its reduction, was known at a very
early period. It was used by the Babylonians in the form
of thin plates for engraving inscriptions, and by the Egyptians and other early civilized peoples for a variety of
purposes. We have already noted its use by the Egyptians
as a constituent of bronzes, a use which Pliny also records
in Roman times. The Egyptians called lead the mother of
metals, an idea which may have arisen from the frequent

---

[7] Von Lippmann, *op. cit.*, p. 552.

occurrence of silver in lead ores, leading to the belief that the silver grew from or was generated from lead. This idea in turn may have been the germ of the idea of the later alchemists that mercury instead of lead was the generator of other metals.

The metal iron and articles manufactured from iron were also known from very early times. The great perishability of iron as compared with the other useful metals known to the ancients makes difficult the settlement of the much disputed question as to whether copper or iron was first made use of. It seems, however, to have been known to the Egyptians as early as 2500–2900 B. C., and in Babylon also it was evidently known at a very ancient epoch. According to Von Lippmann, the earliest manufactured article of iron whose age is approximately established was found in the pyramid of Cheops (about 2500 B. C.), though earlier mention is found in Egyptian inscriptions. A lance head from a tomb of about 1800 B. C. is said to be the earliest known iron weapon of established age.[8]

The applicability of iron to the making of weapons would depend upon the time at which its more or less perfect conversion into steel was effected, a period which though several centuries before our era, yet probably was not as early as when good bronze weapons were in use. By about 1300 B. C., however, steel seems to have been used by the Egyptians.

Greece seems to have first received iron from Asia Minor about 1500 B. C., and to have used it on a large scale some three or four centuries later.[9]

That gold and silver were known and greedily sought by the most ancient of the civilized nations is too well known from the evidence of manufactured articles of the greatest antiquity to require confirmation here.

Gold articles are amongst the ornaments from the prehistoric stone age of Egypt; and in the earliest dynasties

[8] Von Lippmann, *op. cit.*, p. 610.
[9] Von Lippmann, *op. cit.*, p. 616.

of the historic period, the working of gold was evidently wide-spread, and the art well developed.

In other countries of the ancient world, gold appears to have been in use from their earliest civilizations for ornamental purposes.

Articles of gold from ancient sources vary much in purity as the frequent occurrence of notable quantities of silver in the native metal was in earlier times not recognized, nor were methods of separating silver from gold adequately developed. Gold, as obtained by the Egyptians, was often especially rich in silver, so that the color was notably light, and was considered by them as a different metal—a white gold or "asem." Beads and gold leaf of the twelfth dynasty (perhaps 2000 B. C.), analyzed by Berthelot, gave 82.94 per cent gold to 16.56 per cent silver, and 85.92 per cent gold to 13.78 per cent silver.

That silver should have been of later discovery, as it appears to have been in Egypt, is not surprising, considering that it does not occur free to any extent, but has to be recovered by chemical processes from its ores. In Egypt, therefore, from about 3000 to 1500 B. C., it seems to have been rare and more valued than gold.

Mercury (Greek—hydrargyros, liquid-silver; Latin— argentum vivum, live or quick silver) is stated to have been found in Egyptian tombs of from 1500 to 1600 B. C. Ancient Hindu and Chinese literature also gives evidence of their familiarity with it, but reliable data as to the period when it was first recorded are lacking, owing to the frequent revisions and additions to the ancient Hindu and Chinese authorities. In early times, mercury was not generally classified among the metals (which were, in fact, in no way very definitely characterized). From its Greek and Latin names, it may be inferred that its relation to silver was something of a problem in their theory.

The concept of a "metal" in the sense in which we use it—a distinct elementary substance of fixed and characteristic properties, chemical and physical—was never attained

by the ancients. The word itself originally meant the
mines, and was later interpreted to designate the products
of the mines. When Dioscorides, for instance, says that
quicksilver is found ἐν μετάλλοις[10] he does not mean that mercury is found in all metals—an alchemistical idea—but that
it is found native in the mines.[11]

Such groupings of substances as we may call attempts
to classify them were on the basis of their properties—
luster, malleability, stability; or of their applicability to
similar purposes, and naturally varied much at different
times and places.

P. C. Ray, in his *History of Hindu Chemistry,* quotes
from the Chakara, "gold and the five metals—silver, copper, lead, tin and iron."

According to Oppert[12] various Hindu classics give classi-
fications differing in many respects. Thus the Sukraniti-
sara gives gold, silver, copper, tin (and zinc), lead and iron.
The Bhavaprakasa names gold, silver, copper, tin, mercury,
lead and iron. The Danasagara gives gold, silver, bronze,
copper, lead, tin (and zinc), iron and brass. The Sukha-
boda classes gold, silver, brass, lead, copper, tin, iron,
bronze and the lodestone.

Latin and Greek writers of ancient epochs apparently do
not make any attempt to classify the metals as such. In
the early centuries of our era, however, there gradually
developed a mysticism among chemical writers due to Egyptian and Chaldean religious doctrines or magical ideas, and,
among these, there developed a fanciful relation of the
metals as such to the sun and the planets, and as a consequence there arose the notion that it was necessary to
confine the number of metals to seven. Thus, Olympiodorus, in the sixth century of our era, gives the following
as the metals and their relation to the planets:

---

[10] Dioscorides, V, 110.
[11] On the origin and development of the word metal. Cf. Strunz, Fr.,
Ueber die Vorgeschichte und Anfänge der Chemie, 1906, p. 31 *ff.*
[12] *Beiträge aus der Geschichte der Chemie,* p. 129 *ff.*

Gold...................................the Sun
Silver................................the Moon
Electrum............................Jupiter
Iron..................................Mars
Copper.............................Venus
Tin..................................Mercury
Lead...............................Saturn

When electrum, alloy of silver and gold, was rejected as not being a distinct substance, tin became attributed to Jupiter, and mercury was permitted to enter the mystic circle and was attributed to the planet Mercury. This classification served as a catalogue and definition of the so-called metals for many centuries, in fact, throughout the middle ages of Europe.

The ancients and the chemists of the medieval period had indeed no such rational basis as we have to-day for distinguishing certain substances as possessing constant and invariable proportions. When Pliny, for instance, speaks of several kinds of "aes" (copper, bronze or brass being included under that term), of two kinds of silver, etc., he is expressing an idea common to the thinkers of his time, that all substances might vary in properties according as the four so-called Aristotelian elements, fire, air, earth and water, entered in varying proportions into their constitution. Even so late as the sixteenth century, we find Paracelsus voicing the traditional belief when he says that there are many kinds of gold, just as there are many kinds of pears or of apples.

Not only were methods of quantitative analysis lacking, but there existed no hypothesis in their philosophy which could have suggested the possibility of such methods. For an understanding of the chemical ideas of ancient and medieval chemists, it is important that this fact be kept constantly in mind.

Of other common metals, it does not appear that the ancients had any distinct recognition. Zinc either was never obtained in the metallic state, or, if so, it was never distinguished from lead or tin. Its ores were used in the manufacture of brass, and the term "cadmia" seems to have been applied to such ores as well as to the oxide of zinc obtained as crusts or dust from the brass furnaces. The use of zinc ores as raw material in the manufacture of brass cannot be definitely traced beyond the first or second century before Christ.

A passage quoted from a work ascribed to Aristotle, περὶ θαυμασίων ἀκουσμάτων (Latin, *De Mirabilibus Auscultationibus*), has been by Kopp and later writers adduced as an indication of an earlier origin for brass from copper and zinc. The passage says that "it is said that Mossynoican bronze (χαλκός) is very brilliant and light colored, not because it has tin added to it, but because an earth occurring there is fused with it."

The passage, to be sure, would not be very conclusive even if authentic, though a fair question might be raised. The work in question, however, seems to give very scanty evidence in support of the claims that it originated with Aristotle, for it contains among other evidences of a later origin, a reference to the Pantheon at Athens built by Hadrian, which fact locates its authorship at a period as late as the first century A. D. when brass from zinc was in frequent use.[13]

Aurichalcum (Greek ὀρείχαλκος), meaning a gold-colored bronze, is applied by Latin writers of that time, Strabo, Pliny and others, to the alloys of copper and zinc which we call brass. The same Greek word was used by Homer and other earlier writers, but there is no evidence that the "golden bronze" of their times contained zinc. The question as to what the writers of the period from Homer to Aristotle meant by the gold bronzes has been much debated, but the writer knows of no specimens of bronzes of their period which contain zinc as a constituent, except in such very small and insignificant quantities that they are

---

[13] Cf. Wilhelm von Christ, *Geschichte der Griechischen Literatur*, 5th ed., Munich, 1908–1913, Th. I, p. 686.

evidently accidental constituents of their ores.[14] The name zinc first appears in Paracelsus in the sixteenth century, "Zincken," and it is characterized by him as a bastard metal.

The metal antimony seems not to have been recognized by the ancients, though the sulphide of antimony, called "stimmi" or "stibi," was known and used by them for blackening the eyebrows and for medicinal purposes, as was also the crude oxide obtained by roasting the native sulphide. Yet small ornamental articles discovered in an ancient necropolis of Transcaucasia (Redkin-Lager) were analyzed by M. Virchow[15] and found to consist of almost pure antimony; and M. Berthelot found the cylindrical spout of a vessel from the ruins at Tello, estimated to be of a period of between 3000 and 4000 B. C., to consist of practically pure antimony. In this connection, it is interesting to note that both Dioscorides and Pliny, in describing the preparation of medicines by roasting the sulphide, note that, if the process is not conducted with care, the substance changes into lead. It is therefore probable that the metal when obtained was not distinguished as other than a kind of lead.

The art of glass making is of very ancient origin with the Egyptians, as is evident from the glass jars, figures and ornaments discovered in the tombs. Paintings on the tombs of the early dynasties have been interpreted by earlier archæologists as descriptive of the process of glassblowing.[16] Flinders-Petrie, the eminent archæologist, considers these illustrations, however, as representing smiths blowing their fires by means of reeds tipped with clay. This interpretation, though not universally accepted, is held by many modern critics, and there is certainly no evidence existing in the form of blown-glass vessels of such early

---

[14] Cf. Paul Diergart, *Journal für Praktische Chemie, Neue Folge,* Vols. 61, 66, 67; *Zeitschift für Angewandte Chemie,* 1903. Cf. also J. A. Phillips, "Metals and Alloys Known to the Ancients," *Journal of the Chemical Society,* Vol. 4, p. 252 *ff.*

[15] *Verhandlungen der Berliner Gesellschaft, für Anthropologie,* 1884.

[16] Cf. Sir Gardner Wilkinson, *Ancient Egyptians,* 3d ed., 1847, III, p. 89.

dates. Glass-blowing is apparently of Egyptian origin, but of a date approximately at the beginning of our era.

The remains of glass furnaces discovered by Flinders-Petrie at Tel-El-Amarna (Eighteenth dynasty, about 1400 B. C.) illustrate the manufacture of rods, beads and jars or other figures, formed apparently by covering clay cores with glass and later removing the cores. Egyptian glass articles—beads, jars, figures, mosaics—were of colored glass, often beautifully patterned. Transparent and color-less glass seems not to have been manufactured until the centuries approaching the beginning of the Christian era.

Glass manufacture in India was also of ancient origin, but definite data are difficult to ascertain. So also Chinese glass manufacture is doubtless many centuries old, but satisfactory chronological data are difficult to obtain.

Schliemann discovered glass beads in the mines of Ti-ryns, and notes that lead was present in considerable quan-tities in certain specimens.

From analyses of ancient Egyptian and Roman glass articles, it is shown that generally the glass from these sources was a soda-lime glass with rather high soda con-tent as compared with modern soda-lime glass.

The analyses of Egyptian and Roman glass on the next page illustrate the general character of their composition.[17] Potash from wood ashes does not appear to have been used by either the Egyptians or Romans in ancient times, native sodium carbonate being found in arid districts of Egypt. The given analyses do not differ from those of some soda-lime glasses of modern times, though the better mod-ern grades show somewhat higher silica, higher lime and lower soda content, yielding a glass more resistant to weather and acids than were the glasses above described.

Lead was used in glass from very ancient times. Ber-thelot[18] analyzed a vase of the Fourth dynasty in Egypt which contained about one quarter lead.

---

[17] Muspratt, *Chemie* (*4te Auflage*), 1888–1905, III, 1366.
[18] Berthelot, *op. cit.*, p. 17.

ANALYSES OF EGYPTIAN AND ROMAN GLASS	SiO_2	Na_2O	CaO	Fe_2O_3	Al_2O_3	MnO	MgO	SO_3
Analyzed by Benrath								
Egyptian glass rod, colorless.	72.30	20.83	5.17	.51	1.19	...	...	...
Egyptian disk used in games.	70.58	20.70	6.54	.99	1.19	...	...	...
Egyptian disk, bottle green..	71.15	18.76	8.56	.25	.84	.44	...	...
Analyzed by Schüler								
Egyptian glass rod, brown...	65.90	22.33	8.42	.94	1.44	.94	...	...
Analyzed by Benrath								
Roman bottle..............	70.16	17.47	8.38	1.24	2.25	...	...	...
Roman ampulla, greenish....	68.10	20.53	6.51	1.09	1.30	1.98	.49	...
Roman ampulla, green......	67.96	22.39	5.12	.68	1.86	1.67	...	.32
Roman urn................	70.32	21.95	3.04	1.92	1.61	.87	.29	...
Analyzed by Schüler								
Roman urn................	70.58	18.86	8.00	.53	1.80	.48	...	...
Roman tear bottle..........	71.45	16.62	6.14	1.02	2.55	.17	...	...
Analyzed by Sigwart								
Roman glass from tomb.....	64.25	23.22	7.54	3.52		...	1.44	...

Pottery, its manufacture and decoration, is an industry of prehistoric antiquity, and the application of glazes and enamels is a work of the most ancient origin in the earliest civilizations in Egypt, India, China, and Asia Minor. So also the beginnings of the art of weaving and of the art of dyeing are lost in antiquity. Mummy cloths of varying degrees of fineness, still evidencing the dyer's skill, are preserved in many museums. Some of the finest are of the period of 3000 B. C. or earlier. The invention of the royal purple, which appears to be of Cretan origin, was perhaps as early as 1600 B. C.

From the painted walls of tombs, temples and other structures which have been protected from exposure to weather, and from the decorated surfaces of pottery, chemical analysis often is able to give us knowledge of the materials used for such purposes. Such data also serve at times to assist in the interpretation of the often unclear or incomplete descriptions given by extant ancient writers.

Thus pigments from the tomb of Perneb, which was presented to the Metropolitan Museum of New York City in 1913, were examined by Maximilian Toch.[19] The date of the structure is estimated at 2650 B. C. A red pigment

19 *Journal of Industrial and Engineering Chemistry*, 1918, X, p. 118.

proved to be the red oxide of iron, hæmatite; a yellow consisted of clay containing iron or a yellow ochre; a blue color was a finely powdered glass; and a pale blue was a copper carbonate, probably azurite; greens were malachite; black was charcoal or boneblack; gray, a limestone mixed with charcoal; and a quantity of pigment remaining in a paint pot used in the decoration, contained a mixture of hæmatite with limestone and clay.

Pigments of Greek origin, dating from 1500 to 500 B. C., examined by A. O. Rhousopoulos[20] showed red pigments to be cinnabar, and iron oxide; a black pigment was the black oxide of manganese; blues were due to copper or to mixtures of copper and iron; whites were carbonate and phosphate of lime.

Rammelsberg[21] analyzed a blue powder used as a pigment in an ancient Egyptian tomb, and found it to consist of silica 70.50 per cent; lime 8.53 per cent; copper oxide 13.00 per cent; ferric oxide 3.71 per cent; magnesium oxide 4.18 per cent.

The analysis of a dark blue glass bead found in an Egyptian tomb reported by Lepsius (loc. cit), as analyzed by Clemm and Jahn, gave 2.86 per cent cobalt oxide, while a bead of lighter blue contained 0.95 per cent cobalt oxide.

Sir Gardner Wilkinson brought samples of pigments from the walls of Thebes which were examined by Dr. Ure. A green pigment, not dissolved by hydrochloric acid, became a brilliant blue color when it was so treated, a small quantity of yellow ochre being dissolved out. The blue residue was a powdered blue glass, which on analysis showed copper and iron as its coloring constituents. A blue pigment was a similar glass unmixed with any ochre. A red pigment was mainly iron oxide with some alumina, "a red earthy bole." A black pigment consisted of boneblack mixed with a little gum. A white pigment was a

---

20 P. Diergart, *Beiträge aus der Geschichte der Chemie, Zum Gedächtniss von G. W. A. Kahlbaum*, p. 172 ff.

21 Quoted by Lepsius, *Abhandlungen der Akadamie der Wissenschaften zu Berlin*, 1871, p. 63.

practically pure chalk; and a yellow was a yellow ochre.[22]

Pigments of a later Roman period from the baths of Titus (first century A. D.) were examined by Sir Humphrey Davy. Red colors he found to consist of cinnabar, red lead (minium) and red ochre (ferric oxide). Yellows were yellow ochre and chalk mixed with some red lead or with litharge. Green was due to copper carbonates which for lighter shades were mixed with chalk. A blue pigment was a blue glass (a copper silicate) mixed with chalk. Blacks and browns were of carbon or of black oxide of manganese, sometimes mixed with iron oxide. A sample of pigment of pale rose color in a broken pottery jar was found to owe its tint to some organic dye.

Davy found that he could reproduce the blue glass above mentioned by fusing together fifteen parts of sodium carbonate, twenty parts powdered flint and three of copper filings. This is of interest in connection with a statement of Vitruvius to which reference will be made later.

The foregoing examples will serve to illustrate the character of the evidence furnished by chemical analysis of surviving samples of the products of early chemical industries.

It is, after all, comparatively a narrow range of products of chemical arts that, through their analytical examination, can give us evidence as to the materials and, inferentially sometimes, as to the processes in use before any literary remains from ancient times are to be found. From such few ancient writings as touch upon the arts and manufactures in comprehensible detail, and which have survived the destruction of time, we may learn much that is more specific regarding the chemical knowledge of the ancients.

Of such writings as deal more or less with subjects involving the chemical arts, those of most importance are certain works of Theophrastus of Eresus (about 372–288

[22] Wilkinson, *Ancient Egyptians*, III, p. 301 *ff*.

B. C.), Vitruvius, a Roman architect of the first century
B. C., Dioscorides Pedanus, a Greek physician of the first
century A. D., and the Elder Pliny, also of the first century
A. D. Some brief allusions are contained also in the writings of Plato (died 347 B. C.), Aristotle (384–322 B. C.),
Diodorus Siculus (about the first century B. C.), and
Strabo, the geographer, though Dioscorides and Pliny have
incorporated in their later writings the important facts
of these writers.

While the great Greek philosopher, Aristotle, contributed
to a dominating degree toward the development of the
theory of matter and its changes, and exerted a great influence upon the chemical philosophy of the Middle Ages
as well as of the ancients, his writings contribute little
of information as to the chemical knowledge of his time.
He refers to some of the substances used for pigments
such as ochre, minium, and sandarach. He states that
from the crude iron from the smelting furnaces a more useful product is obtained by re-fusing several times, whereby
a slag separates and the iron becomes tougher or more
malleable. He states that sea water is made fresh and fit
for drinking by percolation through clay, though he does
not explain the basis of his belief. Aristotle's writings
speak of a wax vessel as used for this purification, but
Diels and von Lippmann have shown that the fact alluded
to was doubtless originally described by Democritus and
his word κεράμινος (clay) probably was changed by the careless of some copyist to κήρινος (wax), and this is account-
able for Aristotle's error and for similar errors by his
commentators.[23] His references, however, are more casual
than descriptive.

Plato also has some allusions to facts of chemical interest,
though his interest in such matters lay rather in the
theories of the structure of matter in general than in facts
of a practical character. He considers gold as consisting

---

[23] Cf. E. von Lippmann, *Abhandlungen und Vorträge zur Geschichte der Naturwissenschaften*, 1913, II, pp. 98, 99, 162–167.

of particles which were homogeneous in character, differing only in size. This might be considered as an approach to the modern concept of an element, but Plato, like Aristotle, accepted the theory of the four elements constituting all other substances. He refers to the work of the artisans in the separation of foreign matter, earth and stones from gold, thus leaving the gold associated only with silver or copper (χαλκός) and sometimes iron. From these, it is separated only by repeated fusions until the pure gold remains behind. He speaks of the formation of the rusts of copper and of iron, interpreting these changes as caused by the loss of some of their elementary earth. It is quite possible that this notion of Plato's backed by his great authority may have contributed to the idea long prevalent among the early chemists that what we call oxidation was accompanied by a loss of something from the substance burned.

Plato mentions white lead, sulphur, oreichalcos (golden bronze), and other common substances obtained by chemical processes.[24]

Plato and Aristotle in their voluminous writings on many subjects evidence a knowledge of the common properties of metals and other substances, but nowhere do they give any indication of knowledge other than such as was common among all well-informed men.

Theophrastus of Eresus was a philosopher of importance in the history of the natural philosophy of the ancients, but he also wrote some works upon subjects more or less closely related to certain chemical facts. These are: his brief work upon rocks or minerals, περὶ τῶν λίθων; a treatise upon plants, περὶ φυτῶν ἱστορίας; and a fragment "Upon Odors." As the earliest author whose works have come down to us dealing more or less circumstantially with certain phases of chemistry, his data are of particular interest.

---

[24] Cf. E. von Lippmann, "Chemisches und Physikalisches aus Plato," Journal für Praktische Chemie, Neue Folge, 76, p. 513 ff.

In his work upon minerals or stones, Theophrastus[25] describes many natural minerals and products derived from them in ways clearly recognizable, though many others are so ill-defined as to be not now readily identified.

The ideas of Theophrastus as to the nature and origin of minerals were based upon the theories of his master, Aristotle, but in this surviving work he does not enter into theories of the origin. His treatise begins by stating that of things formed in the earth, some have their origin from waters, others from earth. Water is the basis of metals; earth of stones, whether precious or common. This early statement, brief as it is, is interesting as the ideas of the origin of metals and of minerals from earth, water, air and heat or fire dominated chemical philosophy for nearly two thousand years after its first promulgation by Plato.

Our cinnabar was known to Theophrastus under that name (κιννάβαρις). He states that it is found in Spain. Quicksilver (hydrargyros) can be obtained from cinnabar by rubbing it with vinegar in a copper vessel with a copper pestle. He also states that an artificial cinnabar (an imitation) is washed from the sands at Ephesus. This latter statement occurs also in later writers, though what it may mean, unless it is bright red hæmatite or red ochre, is hard to say. From Pliny's statement, referring to the above from Theophrastus, it appears that he considered it to be a red pigment used in painting ships, and it was probably essentially red oxide of iron.

"Chrysocolla" is applied by Theophrastus to malachite, the native copper carbonate, though other green-colored minerals may have been included. He states that chryso-colla and smaragd are thought by many to be the same thing. The latter term was used for the emerald, but also was manifestly applied to malachite. Both chrysocolla and smaragd are used for soldering gold, says Theophras-

---

[25] Theophrastus of Eresus περὶ τῶν λίθων, with English translation by John Hill, London, 1746.

tus. The carbonates of copper, verdigris and malachite were so used, while the smaragd (emerald) could not have had any such use. The ancients were evidently confused by the green-colored minerals and had difficulty in discriminating between them.

Cyanos was a blue gem of much value, and it has been identified as the stone called lapis lazuli, though Theophrastus also refers to another kind of cyanos, which has in it chrysocolla. This doubtless refers to our azurite, a hydrated copper carbonate, used by the ancients as a blue pigment, and known to the Latins as *armenus,* so named after the locality, Armenia, from which it was largely obtained.

Among red and yellow earths used in pigments, Theophrastus mentions *miltos* "found sometimes in iron mines." Pliny mentions this same substance under the name of *rubrica,* used for painting ships. "The Greeks," says Pliny, "call this red earth *miltos.*" This *miltos* may be essentially the same substance that Theophrastus elsewhere calls an artificial or imitation cinnabar. The yellow ochre, mentioned by Theophrastus, was doubtless clay containing ferric hydroxide which we have previously seen from analysis of pigments from ancient buildings to have been largely used for yellow paint. Theophrastus says that, if it is heated, it yields a purple color. The "purple" of the ancients comprised a wide range of tints from red to brown, as well as our purple, and the change of the yellow to red or brown red by heating the yellow ochre is what occurs in the baking of bricks from yellow clays. The synopis of Theophrastus was a red ochre.

Orpiment and realgar were known to Theophrastus under the names of "arrhenikon" (or "arsenikon," whence later was derived our "arsenic"), and "sandarach" respectively. Cerussa (our white lead), used as a pigment, and externally in medicine, was obtained by submitting lead to the action of the fumes of vinegar in closed vessels for ten days, after which time the "rust" was scraped off, and the process repeated. The material so obtained was powdered, boiled for a long time with water, and allowed to settle out. The common so-called Dutch process for the manufacture of white lead is then at least as old as Theophrastus, and the directions for the preparation as given by him are frequently repeated by later writers in almost the same terms for many centuries.

In a similar manner is prepared "ios" (our verdigris). Copper is placed over the lees of wine and the rust which forms is removed.

The *magnetis lithos* of Theophrastus (Latin, magnes), was a term which was applied to a variety of substances, and produced great confusion in ancient writings. Theophrastus names it among stones that may be easily cut or engraved, and describes it as a stone of elegant appearance, and much admired. It bears a resemblance to silver, though really a stone of an entirely different kind. In Pliny's time, the word was used to designate several distinct substances. More often Pliny means the loadstone or magnetic iron oxide, over whose mysterious attractive power for iron he rhapsodizes. He states, however, that there are several kinds of magnes—red, black, blue (the best), "and the most inferior of all are those from Magnesia in Asia. They are white, have no attractive influence on iron, and resemble pumice in appearance." A black magnes which is "female," and has no attraction for iron, is in all probability manganese dioxide (pyrolusite), known to have have been used by the ancients as a pigment and in glass making.

Pliny, in describing the manufacture of glass, states that it is made from soda (nitrum), sand, and magnes, "from the belief that it attracts the liquid of glass as it does iron" [26] It is evident that Pliny is here confused as to the substance used, but whether the magnes here mentioned was the black magnes (black oxide of manganese), or the white magnes (possibly a calcium carbonate or sulphate or

---

[26] Pliny, Book XXXVI, Chap. 66.

a magnesite), there is no means of knowing, as both these substances were used in glass previous to Pliny's time.

That Theophrastus knows the lodestone also is plain, though he merely alludes to it in passing, "Electron also is a stone. It is dug from the earth in Liguria, and has, like the before-mentioned,[27] a power of attraction. But the greatest and most evident attractive power is in that stone which attracts iron. But that is a scarce stone and found in but few places. It should however be ranked with these stones as it possesses the same quality."

The "haimatites" of Theophrastus, "seeming as if formed of concreted blood" and used as a pigment, was doubtless our haematite, formerly called "bloodstone." Pliny also says that in Ethiopia the "magnes called haimatites" is found, a stone of blood-red color, which when ground yields a pigment like that of blood.

The analyses previously quoted from Maximilian Toch would seem to show that hæmatite was used as a pigment by the Egyptians more than twenty-two hundred years before the time of Theophrastus.

The subject of glass-making was not particularly germane to the work of Theophrastus on stones, but there is a reference to it in connection with a statement that some earths may be melted by heat and become harder on cooling. He says that if glass is made, as some say, from glass-sand (velitis), that this also takes place by a compacting. "But most peculiar is that [glass] which is mixed with copper, for in addition to the melting and mixing, it has the additional property of causing a beautiful difference in color." This is apparently the first reference in literature to the use of copper in coloring glass.[28]

---

[27] The before-mentioned stone was a legendary stone produced from the urine of the lynx and which, from ancient references to it, was possibly the gem now called hyacinth. Theophrastus calls it the lyncurium (λυγκούριον).

[28] Theophrastus, περὶ τῶν λίθων, LXXXIV.

In the English translation of Theophrastus (περὶ τῶν λίθων) John Hill assumes that the original manuscript probably contained the word chalcites instead of chalkos, that is, flint, instead of copper. The assumption seems to be without authority, and the resulting interpretation less reasonable. *Op.cit.*, pp. 117–119 and footnote.

"Plaster of Paris" was familiar to Theophrastus.

"The stone from which gypsum is made by burning is like alabaster. Its toughness and heat when moistened is very wonderful. They prepare it for use by reducing it to powder and then pouring water on it and stirring and mixing well with wooden tools, for they cannot do this by hand because of the heat. They prepare it in this manner immediately before using, for in a very little while it becomes hard and not in condition to be used."

He mentions its strength as a cement for walls, and its use for whitewash and making images. It seems, he says, to have the heat and tenacity of lime and the viscous earths (clays?), but possesses these qualities in a higher degree than either.

It will be noted that the term "gypsum" is used by Theophrastus, as indeed by later ancient writers, to indicate the dehydrated sulphate of lime (plaster of Paris), rather than the mineral (gypsum) from which it is obtained, though he elsewhere alludes somewhat vaguely to certain natural earths under that name.

In his work, περὶ φυτῶν ἱστορίας (or *Enquiry into Plants*),[29] Theophrastus catalogues a large number of plants with discussions of their habitat, products, and uses for food, medicine and other purposes. There are comparatively few references to products or processes that are distinctly applications of chemistry, but there are a few of interest.

The "burning" of charcoal by the method still much used of submitting wood to partial combustion in earth-covered mounds is mentioned. The recovery of pitch from resinous trees was either by making incisions in the living tree and collecting the pitch which accumulated, or by a process somewhat similar to the charcoal burning, a process interestingly described by Theophrastus as follows:

"Having prepared a level piece of ground, which they make like a threshing floor with a slope for the pitch to run

---

[29] Edition used is Theophrastus of Eresus, *Enquiry into Plants*, and minor works on odours and weather signs, Greek and English text, Sir Arthur Hort, London and New York, 1916.

toward the middle, and having made it smooth, they clean the logs and place them in an arrangement like that of the charcoal burners, except that there is no pit, but the billets of wood are set upright against one another, so that the pile goes on growing in height, according to the number used. And they say that the erection is complete when the pile is one hundred eighty cubits[30] in circumference and fifty or at most sixty in height, or again when it is a hundred in height, if the wood happens to be rich in pitch. Having then thus arranged the pile and having covered it with timber, they throw on earth and completely cover it, so that the fire may not by any means show through, for if this happens, the pitch is ruined. Then they kindle the pile where the passage is left, and then, having filled that part up, too, with timber and piled on earth, they mount a ladder and watch wherever they see the smoke pushing its way out, and keep piling on earth, so that the fire may not even show itself. And a conduit is prepared for the pitch right through the pile, so that it may flow into a hole about fifteen cubits off, and the pitch as it flows out is now cool to the touch. The pile burns for nearly two days and nights. On the second day before sunset, it has burnt itself out and has fallen in; for this occurs if the pitch is no longer flowing. All this time, they keep watch and do not go to rest, in case the fire should come through; and they offer sacrifices and keep holiday, praying that the pitch may be abundant and good. Such is the manner in which the people of Macedonia make pitch by fire.''[31]

In the treatise ''Concerning Odors,'' Theophrastus enters into a considerable discussion of the nature, causes and sources of odors in general, and then describes the making of perfumes and unguents, with a rather full account of the various spices and odors, and of the oils used as vehicles for retaining the perfumes. While the catalogue of these is of no special interest here, the methods of extraction and preservation of the odoriferous materials are pertinent. In the first place, it is of interest to note

---

[30] The cubit varied in ancient times according to locality and period, from about seventeen to twenty-six inches.

[31] Theophrastus, *Enquiry into Plants*. Hort's Translation, II. pp. 229–233.

that no process of distillation was used. The odor-bearing materials were often used in dried and powdered form, and many different substances were often mixed in these powders. "In fact," says Theophrastus, "powders are better the more ingredients they have."

For unguents or ointments, the perfumes were extracted by subjecting the materials to treatment with warm or hot oils, which dissolved and preserved the essential oils which imparted perfume. The oils so employed were numerous. Benoil (balanos) was considered one of the best because it possessed no odor of its own, and because of its superior keeping qualities. Olive oil, sesame oil and the oil of bitter almonds were also used, the last named because of its own pleasant odor. The perfume-bearing plants or parts of plants used were very numerous, some of those most familiar to us being frankincense, myrrh, cassia, cinnamon, sweet-marjoram, cardamon, sweet-flag, thyme, myrtle, iris, rose, lily, and many others. Pliny, who evidently drew directly or indirectly largely from Theophrastus, treats extensively of unguents and of their uses and abuses in his time. He gives an illustration of the complexity of some of these mixtures. Not all of the substances mentioned are identifiable at present.

"A 'regal' unguent, so-called because it was first composed for the Parthian kings, was composed" he says, "of myrobalanus, costus, amomum, cinnamon, comacum, cardamum, spikenard, marum, myrrh, cassia, storax, ladanum, opobalsamum, Syrian calamus and Syrian sweet-rush, oenanthe, malobathrum, serichatum, cyprus, aspralathus, panax, saffron, cypirus, sweet marjoram, lotus, honey and wine. Not one of the ingredients in the compound is produced either in Italy, that conqueror of the world, or indeed in all Europe, with the exception of the iris, which grows in Illyricum, and the nard which is to be found in Gaul; as to the wine, the rose, the leaves of myrrh, and the olive oil, they are possessed by pretty nearly all countries in common." [32]

---

[32] Pliny, Book XIII, Chap. 2, Translation from Bohn's ed., III, p. 166.

In connection with the making of unguents, Theophrastus gives us the first notice in literature of the application of the principle of the water-bath.

"But in all cases, the cooking, whether to produce the astringent quality or to impart the proper odor, is done in vessels standing in water and not in contact with the fire, the reason being that the heating must be gentle, and there would be considerable waste if these were in actual contact with the flame, and further the perfume would smell of burning." [33]

A Greek philosopher and writer of about 400 B. C., Democritus of Abdera, was held in high esteem by writers on natural science and arts of the period of the Roman Empire, and by the early chemists or alchemists. Unfortunately, none of his writings have come down to us except in the form of citations or abstracts by later writers. His ideas upon the nature of matter, transmitted in this way, find their place in the history of ancient chemical philosophy. If we were to trust statements of Pliny and other writers of about that period, Democritus wrote treatises upon plants, and upon magic. Synesius in the fourth century A. D. states that he wrote four books on the colors, or tinctures, on gold and silver, on gems and on purple dyes. However, the authenticity of the contributions of Democritus of Adbera to chemistry or chemical ideas is much complicated by the fact that at a period probably a little earlier than the beginning of our era, a writer assumed the name of Democritus, who was a devotee of magic and mysticism, a pioneer among the early Greek alchemists. It seems very probable that many of the writings quoted by Pliny and accessible to Pliny were by the pseudo- Democritus. Pliny indeed has a passage which suggests the probability of such a confusion,[34] when he says that it was Democritus who sought the works of Dardanus in the tomb of that personage, and his own were composed in accordance with the doctrines there found.

---

[33] Theophrastus, *op. cit.*, p. 347.
[34] Pliny, Book XXX, Chap. 2.

"All the particulars there found are so utterly incredible, so utterly revolting, that those even who admire Democritus in other respects are strong in their denial that these works were really written by him. Their denial, however, is in vain; for it was he, beyond all doubt, who had the greatest share in fascinating men's minds with these attractive chimeras. There is also a marvelous coincidence in the fact that the two arts, medicine, I mean, and magic, were developed simultaneously; medicine by the writings of Hippocrates, and magic by the work of Democritus, about the period of the Peloponnesian War which was waged in Greece in the year of the City of Rome 300" (about 450 B.C.).

In the light of modern criticism of scholars of early chemistry, we may be justified in disagreeing with Pliny that the magical and mystical writings attributed to Democritus of Abdera are by that same philosopher whose notions of the atomic structure of matter and of other problems of natural forces have given him a place in the history of chemical theory.

The allusions to Democritus by Vitruvius, writing a century or more before Pliny, seem to apply to the real Democritus. Vitruvius says he wrote several works on the nature of things. Seneca attributes to him the invention of the reverbatory furnace, and the art of imitating natural gems, particularly the emerald, though it is probable that here also the real Democritus is confused with the pseudo-Democritus.

It is not improbable that more than one writer wrote under the name of Democritus, and that works of an alchemical character were written at a later period than the works on magic which Pliny alludes to, but even the latest period to which they can be ascribed is somewhere near the beginnings of our era.[35]

At any rate, we may safely assume that whatever is assigned to Democritus that is related to the practical arts of chemistry, is attributable to the pseudo-Democritus and belongs, in so far as it has significance, to the earliest

35 Cf. Berthelot, *Les Origines de l'Alchimie*, Paris, 1885, p. 145 *ff.*

literature of alchemy. We shall later have occasion to consider this literature.

In the first century B. C., as nearly as the internal evidence of his writings establishes their date, a Roman architect, Vitruvius, wrote the work through which he is known, *Ten Books on Architecture*.[36]

In the discussion of the materials used in various structures, and of pigments and colors used in their decorations, he often furnishes more specific information than is contained in earlier Greek or Latin writers. Pliny mentions him among his authorities and apparently cites him at times quite literally. It is also quite evident that Vitruvius does not always depend upon knowledge gained by personal observation or experience, but himself depends upon previous writers. In particular, it is evident that while he is familiar with the use of pigments, he is often dependent upon previous writers for his accounts of their sources and methods of preparation. He was, in other words, in no sense a practical chemist of the period. Nevertheless his contributions to our knowledge of the chemical arts of the time are valuable.

Bricks were used by the ancients both as sun-dried and as baked or burned bricks. Of the sun-dried bricks, Vitruvius says they should not be made of sandy or pebbly earth, for they are then too heavy and fall to pieces in the wall. The straw does not hold them together on account of the roughness of the material. They should be made of white, chalky or red earth, being then durable, not heavy to work with, and easily laid. They should be made in the spring or autumn, so that they will not dry out too quickly and crack; and they should not be used for two years after making. In Utica, he says, it was against the law to use them before five years. Sea sand is bad for mixing with the earth (terra) because it renders the bricks slow in drying, and a salty efflorescence is caused on the walls.

---

[36] Works consulted are: *Vitruvius, The Ten Books on Architecture*, translated by M. H. Morgan, London, 1914; *Vitruvii de Architectura Libri Decem*, edition of Valentinus Rose and Hermann Müeller-Strübing, Lipsiae, 1867.

Burned bricks, he says, are used for topping walls and for laying floors (tiles).

Lime for mortars or cements should be burned from stone which whether hard or soft is at least white. Lime from close-grained stone of the harder sort is best for structural parts, while lime from porous stone is adapted to stucco work.

After slaking, he directs to mix three parts of pit sand to one of lime, but if river sand or sea sand is used, then to mix two parts to one of lime, but to use with this a third part of burned brick pounded fine and sifted.

His explanation of the loss of weight in lime burning is characteristic of the idea prevalent in his time. "When lime is burned, the elements water and heat are ejected, hence the stone loses weight, though the bulk remains the same." He is here referring to the Aristotelian theory that all substances are composed of the four elements—water or moisture, fire or heat, air, and earth. The stone loses about one third of its weight in burning, says Vitruvius, which is fairly close to actual results, since perfectly pure limestone burned to a pure calcium oxide would lose forty-four per cent of its weight, a limit never reached in practice. When lime is to be used in stucco work, he specifies that it should be slaked a long time before using, otherwise crude bits are left and the stucco blisters and the smooth surface is spoiled.

The natural cement now known as Pozzuolan is clearly described by Vitruvius:

"There is also a kind of powder which from natural causes produces astonishing results. It is found in the neighborhood of Baiae and in the country belonging to the towns around Mount Vesuvius. This substance, when mixed with lime and rubble, not only lends strength to the structures of other kinds, but even when piers are built of it in the sea, they set hard under water."

The hydraulic character of Pozzuolan was therefore clearly recognized as well as intelligently applied by the early Roman builders.

Gypsum, he says, should not be used in stucco, because it sets too rapidly and thus interferes with even drying. Vitruvius, like Theophrastus, uses the word gypsum not in the sense of the native mineral, but rather to indicate what we call the plaster of paris which is produced by its "burning."

The Egyptians, Greeks and Romans used many colored pigments for the decorations of buildings both externally and internally, and they were very much concerned with their properties, especially their durability so that it is, therefore, natural that Vitruvius should devote considerable attention to their description. Many of these had been previously described by Theophrastus, and probably by other writers whose works are lost to us. Thus yellow ochre and red (iron) earths from various localities, the red ochre from Synopis, orpiment ("auripigmentum which in Greek is called arsenikon") our realgar—"sandarach," mentioned by Vitruvius—have been described by Theophrastus. With reference to sandarach, however, Vitruvius states that the sandarach obtained by heating white lead (cerussa) is more serviceable than that dug from the mines, thus evidencing a failure to distinguish clearly any essential difference between the native sulphide of arsenic or realgar, and the red lead obtained by igniting white lead.

The term "minium," as used by Vitruvius, denotes the red sulphide of mercury or cinnabar.

"Minium [he says], is an ore. During the digging, it sheds tears of quicksilver which the miners collect and save. The masses of ore as taken from the mine are so full of moisture that they are thrown into a furnace or oven in the laboratory to dry, and the fumes that are driven off from them by the heat of the fire, settle down on the floor of the oven and are found to be quicksilver (argentum vivum). When the lumps of ore are taken out, the drops which remain are so small that they cannot be gathered up, but they are swept into a vessel of water, and there they run together and combine into one."

Four pints of quicksilver, says Vitruvius, will be found to weigh one hundred pounds.[37]

Neither silver nor gold can be properly gilded, says Vitruvius, without the use of quicksilver. When gold has been woven into a garment, and it becomes worn out, the cloth may be burned and the ashes thrown into water and quicksilver added. The quicksilver attracts all bits of gold and makes it combine with itself. The water is poured off, and the quicksilver squeezed through a cloth (pannum). The gold brought together by squeezing is retained, while the liquid quicksilver passes through. The recovery of gold by amalgamation is thus of ancient origin.

Pliny, a hundred years later, gives this process in much the same terms, but in place of the cloth (pannum), says "skins that have been well tawed." It may well be that Vitruvius may have originally written "pellem" instead of "pannum," and some later copyist may have ignorantly or inadvertently changed the word.

It is interesting to note that neither Vitruvius nor Pliny mentions the further necessary step of driving off by heat the mercury from the amalgam which is separated from the liquid mercury by the process they describe. Though this necessarily was done, they may have been uninformed upon that detail.

It may be recalled that Theophrastus uses the word "cinnabar" as we use it to-day, while Vitruvius uses the word "minium" to denote our cinnabar. There was much confusion in the writings of the ancients due to their difficulty in recognizing fundamental differences in many of the substances used as red pigments. So Vitruvius, still discussing his minium, explains that when used in

---

[37] This is Morgan's translation. Vitruvius says: "id autem cum sit quattuor sextariorum mensurae cum expendunter invenietur esse pondo centum." Vitruvius, VII, 8.

Assuming the sextarius to be 34.4 cubic inches, and the pondo centum to be 495,000 grains (Encycl. Brit. article, "Weights and Measures"), the specific gravity of mercury would be from the data of Vitruvius 14.2 as against present value of 13.59, a fair approximation. The value of the libra or pound varied more or less at different times. The value above given may not have been exactly the one used by Vitruvius.

decorating open apartments where the bright rays of the sun and moon can penetrate, it is spoiled by contact with them, loses the strength of its color and turns black. Among others, the secretary, Faberius, who wished to have his house in the Aventine furnished in elegant style, applied minium to all the walls of the peristyle; but after thirty days, they turned to an ugly and mottled color. He, therefore, made a contract to have other colors applied instead of minium. Vitruvius explains how this change of color may be prevented by covering the surface of the wall after painting with wax applied hot and rubbed down. It is quite evident that the wall in question was not colored by cinnabar, which does not so blacken by exposure, but was probably covered by red lead.

Vitruvius gives a test for detecting adulterations or substitutions for minium by heating a sample upon an iron plate until the plate is red hot. When the heat makes the color change and turn black, remove the plate from the fire, and if the minium returns to its former color, it is unadulterated; if it remains black, it is adulterated.

Both the red sulphide of mercury and the red lead have this property, and the test above given would not distinguish between them, but would give evidence of adulteration of either by many possible additions.

Vitruvius knows of the formation of a red substance obtained by heating white lead, but calls it a kind of sandarach, not minium.

The red coloring matter gave much confusion to the ancient writers generally. The term "cinnabar" (κιννάβαρις) was used to indicate the blood-red resin, dragon's blood, and by Theophrastus for our cinnabar. The term, "minium" was used by later writers for our cinnabar, but often also for red lead, and evidently the users did not know how to distinguish between them.

Dioscorides (first century A. D.), speaking of cinnabar says, "Some incorrectly think that cinnabar is the same as minium (ἄμμιον), for minium, from a certain stone in

Spain, is mixed with silver-sand. Elsewhere it is not known. When heated in the furnace, it turns to a brilliant flame-like color. The vapor it gives off is suffocating. It is used by painters." This description leaves room for doubt as to whether red lead or cinnabar is referred to. But the real "cinnabar," he goes on to explain, is the red resin, Dragon's blood.

Pliny uses the word minium to denote our cinnabar. In describing "rubrica," a red iron pigment, he says,

"The Greeks call this red earth miltos, and give to minium the name of cinnabar, and hence the error caused by the two meanings of the same word, this being properly the name given to the thick matter which issues from the dragon when crushed beneath the weight of the dying elephant [dragon's blood]. Indeed this last is the only color which in painting gives a proper representation of blood. This cinnabar, too, is extremely useful as an ingredient in antidotes and various medicaments. But, by Hercules, our physicians, because minium also has the name of cinnabaris, use it as a substitute for the other and so employ a poison."

Red lead, obtained by heating white lead, Pliny calls a spurious kind of sandarach.

The above is a typical illustration of many confused notions of the ancients due to the fact that they possessed no knowledge of the elementary constituents of substances. The criteria for classification and nomenclature were based upon superficial phenomena, or upon the sources or the applicability of the substances to particular purposes. So long as the concept prevailed that all substances consisted of variable quantities of the four Aristotelian elements, and that their properties were determined by the proportion of these elements, it was not possible for them to conceive of the possibility of a method of analysis based upon elementary compositions of bodies as understood in modern times.

The realization that substances are made up of definite masses of elementary substances, and that these might be separated from one another by analytical methods so as to determine the chemical constitution of bodies, was to wait many centuries for development.

Chrysocolla, Vitruvius says, is a green pigment brought from Macedonia and dug up in the vicinity of copper mines. As with Theophrastus, this is doubtless our malachite. Vitruvius states that those who cannot use chrysocolla on account of its cost employ a blue color (coeruleum) mixed with the plant called lutum, and obtain a very vivid green. Pliny also states this fact, but adds that it gives a very inferior color.

This word "chrysocolla" of the ancients, which denotes malachite, was not confined to that mineral, as appears particularly from the extended description of Pliny. He mentions the substance dug from the mines in proximity to gold, but he also states that it is a liquid found in the shafts of mines—a slime hardened by the cold of winter till it has the hardness of pumice. The most valued is from copper mines, the next best from silver mines, and that from the gold mines is inferior. In the mines also an artificial chrysocolla is made by allowing water to percolate into the veins during the winter and spring, and evaporating these in July and August.

The goldsmiths make a chrysocolla of their own from the rust of Cyprian bronze (copper), urine and soda (nitrum). This they use for soldering gold. It will be recalled that the word "chrysocolla" means a solder or cement for gold. From Pliny's description, not only malachite but the evaporated residues from copper and iron vitriols produced by the weathering of sulphide ores, and carbonates of copper, verdigris, or mixtures of carbonate and acetate of copper more or less pure, all passed under the name of chrysocolla. In fact, anything which was green and would serve as a solder for gold, or could substitute for malachite as a pigment, might pass as chrysocolla.

Vitruvius, like Theophrastus, describes the formation

of verdigris (ios) by the action of vinegar on copper, or on "burned copper" (oxides of copper), or by hanging copper plates over vinegar, or burying the copper in old and sour lees of wine. "Coeruleum," a blue pigment, is described by Vitruvius as having been first made in Alexandria, afterwards at Pozzuoli.

"The method [he says], is strange enough. Sand and the flowers of nitrum are brayed together to a meal, and copper is grated by means of coarse files over the mixture. This is made into balls by rolling in the hands. The dried balls are put into an earthen jar and this into a furnace. When they have lost their properties through the intensity of the fire, they yield coeruleum."

As the "flowers of nitrum" were a superior grade of carbonate of sodium, the result of the treatment would be a blue glass, more or less soluble to be sure. It will be remembered that just such a glass was analyzed by Sir Humphrey Davy from the baths of Titus, and imitated by him through fusing powdered flint, soda and copper filings. We know also that the Egyptians, at least, also used some cobalt ore for giving blue colors to glass.

Pliny says there were formerly three kinds of coeruleum: the Egyptian, most esteemed of all; the Scythian, which is easily dissolved; and the Cyprian, which is now preferred as a color to the preceding; but Pliny sheds no new light on their nature or preparation.

Pliny also states that coeruleum is a kind of sand. It seems probable that besides the blue glass, native blue minerals were also used, as for instance the cyanos of Theophrastus and of Pliny, probably lapis lazuli, and azurite, the other kind of cyanos referred to by Theophrastus as containing chrysocolla.

Armenium, a blue pigment, merely alluded to by Vitruvius, is probably azurite, for Pliny says that armenium is a thinner color than coeruleum and very much cheaper.

Indicum, mentioned by Vitruvius and described by Pliny and Dioscorides as a production of India, being a slime which adheres to certain reeds there, is our indigo. When powdered, says Pliny, it is black in appearance, but when diluted in water, it yields a marvelous combination of purple and blue ("coeruleum"). Pliny says the proper test for indicum is to lay it on hot coals. If genuine, it produces a fine purple flame. This is an early application of the well-known volatilization of indigo by heat. It was frequently adulterated by staining pigeon's dung with indigo, or imitated by coloring certain earths or chalks with woad.

Usta (burnt ochre), used for coloring stucco surfaces, is said by Vitruvius to have been obtained by heating sil (yellow ochre) to a white heat and quenching in vinegar. Theophrastus also gives this preparation, though omitting the quenching with vinegar. It is hard to understand how quenching with vinegar could have had any value unless to dissolve out any chalk or limestone constituents which if present might dilute the color appreciably. Pliny, giving the same method for obtaining usta, states that it was first discovered accidentally by the burning of white lead. Here red lead is confused with ferric oxide; as we have previously seen, it has been confused with cinnabar and with realgar (sandarach).

The manufacture of white lead is described by Vitruvius as previously by Theophrastus, and as later by Dioscorides and Pliny. The process of making verdigris from copper is also given by Vitruvius as in Theophrastus and as later by Dioscorides and by Pliny. Theophrastus and Dioscorides name it ios. Vitruvius and Pliny call it æruca (bronze or copper rust).

The ostrum of Vitruvius, a beautiful and costly purple color, was obtained from certain marine shell fish. It varies in shade according to the regions where found, being black in the north, as Pontus and Gaul, red in the south as at Rhodes, and blue or violet in the intermediate regions. The shellfish are collected and broken with iron tools, and the purple fluid exudes. "On account of its saltness, it soon dries up unless honey is added to it." Large quantities of the shellfish were collected for a very small quantity of the dye.

This is a description of the color obtained from certain varieties of murex. As Tyre was one of the cities where it was prepared and used with skill, "Tyrian purple" became a name familiar to literature. Pliny gives a much more specific account of the varieties of "murex" and "purpura" used and the method of collecting the dye. He also tells of its use in dyeing wool, though from this account there is not much to be gained except that the dye was boiled down in vats to a relatively small volume after adding a certain quantity of salt, and that the wool, cleansed from grease, is soaked for some five hours in the boiling dye, being again soaked if the color is to be deeper. To produce the Tyrian hue, says Pliny, the wool is soaked in the uncooked juice first of the variety of shellfish called "pelagiæ," and afterwards in that of the "buccinum." The color is best when it resembles the color of clotted blood.

Black pigment described by Vitruvius was made from lampblack or charcoal. He describes in detail the method of manufacture of lamp black for this purpose.[38]

"A place is built like a *laconicum* [this structure he elsewhere describes as a circular chamber with domed ceiling, used for vapor baths], and nicely finished in marble smoothly polished. In front of it a small furnace is constructed with vents into the Laconicum and with a stokehole that can be very carefully closed to prevent the flames from escaping and being wasted. Resin is placed in the furnace. The force of the fire in burning compels it to give out soot into the *laconicum* through the vents and the soot sticks to the wall and curved vaulting. It is gathered from there and some of it is mixed and worked with gum for use in writing ink, while the rest is mixed with glue and used on walls by fresco painters."

A good black may also be obtained more simply by char-

---

[38] Vitruvius, *Morgan*, p. 218.

ring shavings and splinters of pitch pine and pounding them in a mortar with size [glue].

The lees of wine dried and similarly charred and ground with glue yield an excellent black and the better the wine from which it comes, the better the imitation, not only of the ordinary black, but even of indicum. By indicum in this connection Vitruvius doubtless refers to India ink or China ink, for Pliny also, in describing black pigments, after mentioning soot and lampblack and charcoals as above, says after Vitruvius that the black from wine lees, if the wine is of good quality, will bear comparison with that of indicum. He further states that indicum is a substance imported from India and that the composition of it is unknown to him.[39]

As both Vitruvius and Pliny have described under the same name indicum, the blue or purple indigo, this black indicum is doubtless India ink, known to have been made in China before our era. It is also probable that the ancients in Europe did not know whether the black and the blue indicum were of essentially different origin or not. As a matter of fact, the India ink also has lampblack as its base.

Pliny mentions other black pigments used for various purposes—bitumen for painting statues and protecting copper vessels;[40] burnt ivory (boneblack) and a black obtained by dyers; a black inflorescence which adheres to the brazen dye-pans (copper oxide). The sæpia also secretes a black liquid, but from this he says no color is prepared. That black oxide of manganese was used by the ancients as a pigment, we know from analyses already referred to, but no clearly recognizable reference to this substance has been identified in the ancient authors.

In the treating of water supplies and the conduction of water, Vitruvius touches upon items of chemical interest. Thus in digging wells, he emphasizes cautions to be ob-

---

[39] Pliny, Bohn ed., Book XXXV, Chap. 25.
[40] Pliny, Book XXXV, Chap. 51.

served, for sometimes sulphur and bitumen are present, or alum (a term covering a number of soluble astringent salts of different character), and sometimes, "currents of air, which coming up in a pregnant state through porous fissures to the places where wells are being dug, and finding men engaged in digging there, stop up the breath of life of their nostrils by the natural strength of the exhalation. So those who do not quickly escape from the spot are killed there. To guard against this, we must proceed as follows: Let down a lighted lamp, and if it keeps on burning, a man may make the descent without danger. But if the light is put out by the strength of the exhalation, then dig air shafts beside the well on the right and left. Thus the vapor will be carried off by the air shafts as through nostrils."

This is interesting as an early record of methods of recognition of the danger from carbon dioxide and a method for safeguarding the workers. Empirical knowledge of ventilation methods in mines was doubtless of very ancient origin, because of the mining experience of the ancients.

Vitruvius recommends that pipes of earthenware and not of lead be used for conducting water, for lead is harmful, because white lead is formed from it, and this is said to be hurtful. Hence if what is produced from it is harmful, no doubt the thing itself is not wholesome. This we can exemplify from the workers in lead smelters (ab artificibus plumbariis), since in them the natural color of the body is replaced by a deep pallor. For when lead is smelted in casting, the fumes from it settle upon their members and day after day burn out the virtues of the blood.

Lead poisoning was familiar to the ancient medical authorities, but the application of that knowledge in discouraging the use of lead pipes for water supplies on sanitary grounds is of very modern origin.

In the first century of our era, two works important for their records of early chemical knowledge were written. These are the treatise in five books on *Materia Medica* by Dioscorides Pedanus, a Greek physician, a work considered

by modern critics to have been completed about 75 to 80
A. D., and the *Historia Naturalis* of the Elder Pliny completed about 77 A. D. Both these works were received as
authorities and were extensively copied so that copies have
come down to us that may be considered reasonably free
from additions or interpolations of later dates. They
both, in so far at least as facts pertaining to chemistry are
concerned, depend upon previous authors, and there is a
decided similarity in their descriptions, so much so that
H. Kopp in his early history of chemistry considered that
Pliny copied from Dioscorides. It may now be safely assumed, however, that neither of the two writers was cognizant of the other's work, as their manuscripts were too
nearly contemporary, and it has been shown by M. Well-
mann[41] that the principal source from which they drew for
the subjects they treat in common was a work by Sextius
Niger, an author mentioned by both writers, and several
times specifically quoted by Pliny. He wrote in the early
part of the first century A. D., but his writings have not
been preserved to our day.

Dioscorides Pedanus was born at Anazarba in Cilicia
in Asia Minor. He apparently served as military physician
in the Roman campaigns in Asia Minor, and his work,
Materia Medica, was held in high repute, its influence extending in Asia Minor even to comparatively recent times.[42]

As the materia medica of the ancients included almost
everything conceivable in the vegetable, animal and mineral
kingdoms, the writings of Dioscorides include consideration
of many substances prepared by chemical arts, or serving as raw materials for chemical arts. His point of view
is that of the medicinal uses of substances, and there is
no reason to suppose that he personally had any experience

---

[41] *Hermes,* Vol. 24, p. 530 *ff.*
[42] For authorities on the Chemistry of Dioscorides cf. Kopp, *Geschichte der
Chemie,* 1843; Hoefer, *Histoire de la Chimie,* Paris, 1842.
E. von Lippmann, *Zeitschrift für Angewandte Chemie,* XVIII, p. 1209 *ff.*
Text of Dioscorides used by the author is Pedanii Discoridies Anazarbei *De
Materia Medica,* Edition of C. Sprengel, Leipzig, 1829. Greek text with
Latin translation.

with chemical operations. On the contrary, the evidence appears that he is depending upon some previous writers and notably apparent is his dependence upon the above-mentioned Sextius Niger. The range of subjects and the scope of his treatment of chemical subjects is necessarily limited by the pharmacological character of his book.

Caius Plinius Secundus, the "Elder Pliny," was born 23 A. D., and died in 79 A. D. at Stabiæ in the eruption of Vesuvius which overwhelmed Herculaneum and Pompeii. In early manhood, he was a cavalry officer; in later life he held the office of Procurator in Nearer Spain under the Emperor Nero. His official duties evidently left him much leisure for study, for he was said to have been a constant reader, and was himself a prolific author. His nephew, the "Younger Pliny," has listed the works of his uncle as follows:

The Use of the Javelin by Cavalry, a work in one book.

The Life of Q. Pomponius Secundus, in two books.

The Wars in Germany, in twenty books.

The Student, in three books.

On Difficulties in the Latin Language, in eight books.

Continuation of the History of Aufidius Bassus, in thirty-one books.

Natural History, in thirty-seven books.

Of all these writings, none has been preserved to our day except the last named, and that was completed about two years before his death, or about 77 A. D.

It might be inferred from the variety and extent of these writings that comprehensiveness rather than a high degree of scholarly accuracy would characterize the work of Pliny, and the evidence furnished by the *Natural History* bears out the justice of such an inference. An industrious student of Greek and Latin manuscripts by earlier writers, with a real enthusiasm for all facts pertaining to the phenomena of nature, he intended this latest product of his genius to be an encyclopedia of the facts, arts and sciences depending upon or related to natural phenomena. Thus the geography of his time, the productions of the various countries, descriptions of known plants, animals, minerals, materia medica, agriculture, mining, metallurgy, and the industries having to do with naturally occurring raw materials, were all germane to and more or less completely discussed in this work.

In the preparation of his work, he used apparently all accessible authorities, and he lists the names of over five hundred of them. Of these, a large proportion are not represented by works remaining to us.

Pliny supplements the data compiled from these authorities by the results of his own knowledge and observation.

His work is not merely a record of facts, but is also full of the legends, myths, and superstitions of the time, often indeed recorded with protests against their absurdity, but often also soberly accepted. This feature, however, is of much human interest in giving an understanding of ancient points of view on many subjects. Taken all in all, the *Natural History* of Pliny is an extremely valuable compendium of the knowledge of his time, and in scope and comprehensiveness it far exceeds any other work which has come down to us in the domain it covers.

The work of Pliny includes many subjects related to the chemical knowledge and industries of his time. But Pliny evidently had very little knowledge himself on such subjects and his accounts taken from other writers are frequently lacking in accuracy. Whether this inaccuracy was due to imperfect interpretation of his authorities, or to the fact that the earlier writers were themselves but imperfectly informed upon the subjects treated, it is not possible to say, though the latter is in all probability at least a contributing cause. It follows that many of the descriptions of technical operations as described leave much room for conjecture as to important details.

Gold is treated by Dioscorides not from the point of view of mining or metallurgy, but from certain properties pertaining to its use in medicine. He mentions that it is

capable of extremely fine subdivisions, and that, in the
form of thin flakes or leaf, it serves as an antidote for
quicksilver poisoning. This would appear to suggest carrying the idea of the formation of an amalgam into medical
practice, though that inference may not be in accordance
with any established facts.

Copper (χαλκός) characterized by its red color, yields by
ignition either by itself or after the addition of sulphur,
salt, or alum, a burned copper, a substance of astringent
properties used as an emetic. This burned copper is best
for medicine when it is red and gives a red powder when
ground. If it is black, it has been overburned. This seems
clearly to be a discrimination between the red cuprous oxide and the black cupric oxide. "Flowers of copper" obtained by pouring water on heated copper in the form of
red scales is doubtless also cuprous oxide. It is easily
powdered. It is sometimes adulterated by the addition of
copper filings, and this adulteration may be detected by
adding vinegar which with the genuine article gives ios
(verdigris). This ios is also obtained by hanging copper
plates over vinegar. This method is given by Theophrastus, it may be recalled.

Copper, burned copper and flowers of copper with vinegar also yield ios. It may be assumed that as between
verdigris (carbonate) and acetate of copper, no distinction
was made; ios of the Greeks and chrysocolla of the Latin
writers cover both. Also the method of obtaining ios by
rubbing copper and vinegar in a copper mortar is given by
Dioscorides as previously by Theophrastus. When Theophrastus speaks of chrysocolla, he refers to malachite or to
some other copper salts or mixtures of salts, vitriols, etc.

Chalcanthon (Latin *chalcanthum*) is evidently used by
Dioscorides to designate the sulphate of copper (blue vitriol), and also to include mixtures of sulphates of copper
and iron, or even the sulphate of iron itself (green vitriol).
The best, he says, is blue and transparent, and obtained
by evaporation to blue crystals, but also it is obtained as

exudations from ore bodies in the mines. To detect adulteration of ios by chalcanthon, he says to heat it on an iron shovel. If chalcanthon is present, it becomes red in color. Such a test, however, could only indicate the presence of iron, for verdigris so heated turns black, to be sure, but so also does copper sulphate, while if green vitriol (ferrous sulphate) is added in considerable proportion as would take place in adulteration of the green verdigris by green vitriol, the resulting substance after ignition is red or reddish brown. Dioscorides also says that while the best chalcanthon is blue, the boiled is not so good for medicine, but better for black colors. What this may mean may be inferred from Pliny's information that chalcanthum is atramentum sutorium, shoemakers' black. It is prepared in Spain from the water of wells or pits which contain it in solution. This water is boiled with an equal quantity of pure water and then poured into large wooden reservoirs. Across these reservoirs there are a number of immovable beams, to which cords are fastened and sunk into the water by means of stones; upon which cords a viscous sediment attaches itself in drops of a vitreous appearance, somewhat resembling a bunch of grapes. Upon being removed, it is dried for thirty days. It is of an azure color, and of a brilliant luster, and is often mistaken for glass. When dissolved, it forms the black dye that is used for coloring leather.

The value of chalcanthum in coloring leather black, doubtless in conjunction with tannin, would depend upon the iron present, and as both Dioscorides and Pliny refer to variations in color of different grades of chalcanthum, it is evident that both green and blue vitriol and mixtures of the two passed under that designation.

Pyrites is described by Dioscorides as a kind of stone from which copper is made. It resembles brass in color, and strikes sparks easily. There is no evidence that any discrimination was made by Dioscorides between iron and copper pyrites.

By cyanos, Dioscorides, as Theophrastus and Pliny, means our lapis lazuli, and by armenion our azurite, both being blue-colored minerals. Pliny refers to these also. Cyanos, he refers to as a kind of iaspis (jasper) of a blue color, and armenium as a mineral of blue color, thinner in color and cheaper than coeruleum.

Iron is obtained from misy, a yellow, gold-appearing, hard stone (pyrites?). Pliny vaguely describes under the same name a product formed by roasting a copper ore. According to Berthelot, the misy of Pliny is the product of a gentle oxidation of copper pyrites, a mixture of basic sulphates of iron and copper.[43]

Quicksilver (hydrargyros) is obtained in Spain, according to Dioscorides, from minium ($\check{a}\mu\mu\iota o\nu$) falsely called cinnabar. From this falsely called cinnabar, it is obtained by heating in an iron dish placed in an earthen vessel which is provided with a domed cover that is luted on with clay. The quicksilver collects in drops on the domed cover. This crude method of distillation is of interest as being one of the earliest notices of distillation as a method of separating a substance. Quicksilver, he says, is a violent poison when taken internally, perforating the intestines by its weight. The fumes given off in its smelting are also poisonous.

In the smelting of lead, there is produced a lead slag, yellow, vitreous and dense, and a spodos. Spodos with early writers was a general term for any condensed dust or ash—like the substance resulting from the condensation of volatilized products in the furnace. In this case, both the slag and the spodos were evidently more or less pure oxide of lead. Lead, heated to melting, with constant stirring, either by itself or after the addition of sulphur or of white lead, yields first a black powder (suboxide?), and then molybdaena (litharge). The molybdaena of Dioscorides and of Pliny usually means litharge, but sometimes also is used as synonymous with galena, the native sulphide.

---

[43] Berthelot, *Introduction à l'Etude de la Chimie*, pp. 14, 15.

Dioscorides notes[44] that litharge treated repeatedly with common salt and warm water gives a white product which is separated and used as medicine. This, according to Kopp, is the earliest reference to the formation of lead chloride.

White lead, its preparation, uses, and the fact that heated it gives a red substance resembling sandarach, are described by Dioscorides just as previously by Theophrastus.

Zinc, as previously stated, was not recognized by the ancients as a distinct substance. As its ores (calamine) were much used in the manufacture of brass, it is difficult to conceive that it was never obtained in the metallic state, owing to the readiness with which its ores are reduced. But if obtained, it is probable that it was not considered as other than a variety of lead or tin, not well adapted to the uses made of these metals.

Cadmia is described by Dioscorides as produced in the manufacture of brass (or bronze) in the form of crusts or cakes of varying color, particularly when too much "cadmia" has been used in the furnaces, meaning here too much of the native ore of zinc. Lighter forms of the same substance are pompholyx and spodos. Cadmia, pompholyx and spodos are used in medicine, when ground and washed.

Pompholyx was prepared for medical purposes by a special process of re-fusing the crude cadmia.[45] A furnace was placed on the first floor of a two-story structure, the furnace opening at the top into a settling chamber constituting the second story. The cadmia in small pieces was fed in at the top of the furnace together with charcoal, and a blast maintained by bellows. The fine dust settled on the walls and ceiling, white in color. A coarser dust, settling on the floor, was distinguished as spodos. The process was then that the crude zinc oxide was reduced by the heated charcoal, reoxidized to oxide, settling in the chamber—a refining process. Pliny also, speaking of cad-

---

[44] Dioscorides, V, 102.
[45] Dioscorides, V, 85.

mia, evidently includes both the ores and the oxide under that title. He says, "For as the stone itself from which brass is made is called cadmia, so necessary for the fusion and useless in medicine, so also it is found in the furnaces."

And again, discussing aes (here meaning bronze or brass), "It is also made from a stone containing aes (e lapide aeroso) which they call cadmia." As any definite knowledge of the composition of ores was lacking in Pliny's time, this statement may be interpreted as meaning that one of the raw materials from which brass was made was a stone called cadmia.

Tests for detecting adulteration of pompholyx, as given by Dioscorides, are to add vinegar which imparts to it a brassy odor, a color like pitch and a disagreeable taste, and to throw it upon glowing charcoal in which case it heats up giving an appearance like air. Doubtless the latter test depends upon the fact that if the zinc oxide is pure it is reduced by the charcoal, volatilizes and oxidizes again as a bluish white smoke. White substances usually used as substitutes or adulterants would behave differently.

Tin (kassiteros) is mentioned casually by Dioscorides, as used for covering vessels of copper, and as one of the substances which may be used for vessels to contain mercury without being attacked, a curious error.

Arsenikon and sandarach mean to Dioscorides, as to Theophrastus, respectively orpiment and realgar. The former, "yellow scales or plates," is used in medicine as a depilatory and a caustic. Heated alone, or with charcoal, it loses color and leaves a mass which cooled and powdered is a deadly poison (arsenious oxide). Curdled milk is said to be an antidote. Sandarach, red like cinnabar (dragon's blood, he means), behaves when heated like arsenikon, and in general has properties similar to that substance. He notes that it gives a sulphureous odor when roasted.

Stimmi (the native black sulphide of antimony) is used for staining the eyebrows. Heated with charcoal, it yields "lead." Pliny, who calls it stibi, states that it is prepared for medicinal use also by heating, covered with cow dung in a furnace, after which it is quenched with woman's milk and pounded with rain water in a mortar. The turbid liquid is poured off from time to time into a copper vessel and purified by soda (nitrum). The lees from it which are rejected are recognized by their being full of lead and falling to the bottom.[46] It is evident that the metallic antimony when thus reduced by roasting with charcoal or other organic matter was not distinguished from lead.

Quicklime (ἄσβεστος —unslaked) obtained by burning marble, or shells of marine shellfish, is described as being sharp, burning and caustic. Its activity is increased by long burning. It is slaked by standing overnight in water yielding a heavy white mass. Quicklime is capable of mixing with oil.

Gypsum (plaster of Paris), Dioscorides says, is poisonous taken internally, though it is added to wine of hellebore. He deprecates the use of gypsum in adding to wines, as such wines are injurious to the body and especially to the nerves. The custom of "plastering" wines by the use of calcium sulphate was evidently in use quite extensively. Pliny says it was added to correct acidity. The custom is still in vogue, particularly in the south of Europe, though controlled by law in many countries. Its value consists, not in correcting acidity, but in promoting clarification and improving the color and keeping qualities of the wine.[47]

It is interesting with reference to the above statements of Dioscorides, to note in S. P. Sadtler's *Industrial Organic Chemistry*, written 1800 years afterward, the statement that the practice of plastering "undoubtedly has an injurious effect upon the consumers of wine."

Common salt used as a condiment and as a preservative was known from time immemorial. In the time of Dioscorides and Pliny, it was described as derived from various

[46] Pliny, Book XXXIII, Chap. 35.
[47] Cf. Thorpe, *Dictionary of Applied Chemistry*, article on wines.

sources and in various commercial grades, depending on
the sources of its occurrence and the locality whence imported.   Rock salt from mountains, mined in blocks or
masses, sea salt, salt from evaporation of waters of saline
springs and lakes are discussed in much detail by Pliny
in particular, who catalogues also the many uses to which
it is put.   A flos salis, or flower of salt, seems to have been
a very fine flour of salt, perhaps obtained from the dried
foam of the sea beach.   Pliny's descriptions of various
kinds of salt suggest possibilities of other than common
salt, but do not characterize such in terms that render them
intelligible to us.   The "ammoniacal salt" of Dioscorides
is not, as was sometimes supposed, our sal ammoniac, but
was common salt from Egypt in the vicinity of the temple
of Ammon.   Pliny, discussing common salt and the places
where it is found, says:

"King Ptolemaeus discovered salt also in the vicinity of
Pelusium when he encamped there, a circumstance which
induced other persons to seek and discover it in the scorched
tracts that lie between Egypt and Arabia, beneath the
sands.   In the same manner, too, it has been found in the
thirsting deserts of Africa as far as the oracle of Hammon." [48]

Apicius says that sal ammoniacum should be roasted before using in the kitchen.   This would bar any interpretation as a salt of ammonium.   Arrian (second century A. D.)
mentions ammoniacal salt as essentially the same as common salt, but as used in sacrifices because it was considered
purer.[49]   Later writers for many centuries used the term
"ammoniacal salt" to indicate a preferred grade of common salt.   The application of the term in a modern sense
of ammonium chloride has not been traced in literature
earlier than to the works of perhaps the tenth or eleventh
century A. D., and was not in common use until as late as
the thirteenth century.[50]   Pliny also states that the dis-

---

[48] Pliny, Book XXXI, Chap. 39.
[49] Kopp, *Geschichte der Chemie*, III, p. 237.
[50] Kopp, *Geschichte der Chemie*, III, pp. 237, 238.

tricts of Cyrenaica are distinguished by the production of hammoniacum, a salt so called because of its being found beneath the sands there. He describes it as of unpleasant flavor, but highly useful in medicine. It occurs in long pieces not transparent.[51]

Under the designation of styptaria in Greek and alumen in Latin, Dioscorides and Pliny include a number of more or less soluble substances occurring in nature, or artificially prepared, and which are of a more or less well characterized styptic or astringent character. There are several varieties, liquid and solid, black and white. Every kind of alumen, says Pliny, is a liquid product exuding from the earth, the concretion of it commencing in winter and being completed by the summer sun. Liquid alumen, if genuine, should turn black when pomegranate juice is added. The solid alumen is pale and rough in appearance and turns black on application of nutgalls. Alumen is astringent and corrosive. White alumen is used in the dyeing of wool with bright colors. A kind of alumen called by the Greeks "schiston" splits into white filaments and is produced from the mineral chalcites from which bronze (aes) is produced.

From such information as this, it seems evident that any naturally occurring astringent salts were called alums, and that these may possibly have included our alum, though there is no certain evidence of the fact; but they certainly did include iron sulphate and mixed sulphates of iron and other metals, as the test for iron by nutgalls or pomegranate juice is quoted as a test of genuine character. Whether the white alum used as a mordant in dyeing wool bright colors was our alum or a white vitriol (zinc sulphate), there is no evidence to determine. Delafosse thinks that Pliny's alum was more commonly a double sulphate of iron and aluminum. Beckmann (*History of Inventions*) and Kopp do not believe that our alum was known to the an-

---

[51] Berthelot thinks that the description of the sal ammoniacum of Dioscorides and Pliny might sometimes apply also to sodium carbonate. *Introduction à l'Etude de la Chimie*, p. 237.

cients. Berthelot thinks that white liquid alum was probably a sulphate of aluminum more or less pure.[52]

The nitron of the Greeks and the nitrum of the Latin
writers, was carbonate of sodium, generally as obtained
from evaporation of alkaline waters in arid regions, and
with such natural impurities or admixtures as were inci-
dental to its occurrence. Early translators were often confused by the word, interpreting it as our niter, potassium
nitrate, but the evidence of ancient writers describing
properties and uses is very conclusive that the terms apply
to carbonate either of sodium or potassium. If niter itself was known to them, there is no certain evidence that
they distinguished it from the common alkali salts known
as nitrum.

Dioscorides states that nitron occurs as an exudation
from the earth, and from certain waters, particularly from
certain lakes in Egypt. It varies in color from whitish to
reddish. It is of fatty consistency or feel, caustic, and of
biting taste. Its activity is increased by heating. When
purified, it is white and dissolves in water. Pliny says it
is not changed by the action of fire. This would not apply
to the nitrate. Very similar to the natural nitron, says
Dioscorides, are the ashes obtained by burning plants (potassium carbonates mainly).

Pliny states that the lees of wine when dried will burn
without the addition of other fuel, and that the ashes so
produced have very much the nature of nitrum.[53]

The uses of nitrum, as given by Pliny, include its use in
glassmaking; in making bread; internally for colic pains;
and, when mixed with oil or by itself, for skin eruptions.
All these uses unmistakably indicate sodium carbonate. He
states also that it is destructive to vegetation, destroys the
shoes of the laborers, intensifies the green color of vege-

[52] Berthelot, op. cit., p. 237.
[53] The English translation (Bohn ed.) translates here "nitrum" by
"niter" and comments upon this by stating that "they are tartrates and
have no affinity at all with niter." (XIV, 26) Vol. III, p. 268. The same
misinterpretation occurs elsewhere in this translation, though the original
word "nitrum" is very frequently used instead of any attempted translation.

tables, and that it mixes with oil. Different commercial grades had special names, among which were foam of nitrum (spuma nitri) and flowers of salt (flos salis), though the properties of these appear to be the same in all essential particulars as nitrum. Dioscorides states that for certain medicinal uses, it is taken up with vinegar (sodium acetate). Vinegar is the only acid reagent distinctly recognized by the ancients. Dioscorides states that it is formed on standing, from wine—date wine, fig wine, and similar liquids "whose power is not sufficient to keep the sweetness of the original juice." He mentions its use in dissolving soda (nitron), plant ashes and iron rust for medicinal purposes, and its use for preparing white lead and verdigris. This latter use has already been alluded to.

Pliny also mentions its preparation from wine and figs. He states that poured upon rocks in considerable quantities, it has the effect of splitting them. This statement perhaps has its basis in the disintegrating effect which vinegar would have on rocks which are carbonates, or which contain carbonates, though Pliny has an exaggerated notion of its use in that way, a legendary idea shared by some other early writers.

He also says that poured upon earth, it foams, though here also he gives no indication of any knowledge that a limestone chalk or other carbonate rock is necessarily a condition for such effervescence.[54]

Pliny and Dioscorides give extensive catalogues of the applications of vinegar to a great variety of medicinal uses, both internally and externally.

Dioscorides also notes that numerous plants or parts of plants, as the bark, leaves and roots of the oak, nutgalls, sumac, etc., contain a substance sour and astringent, which is used in medicine and for tanning leather, and for coloring and darkening the hair. The tannin, which is the essential constituent, was not, however, more definitely identified. It will be recalled that in the form of juices or

---

[54] Pliny, Book XXIII, Chap. 27.

extracts, it was used in tests in which the black color formed with iron salts was the determining factor.

Starch (Greek, amylon, Latin, amylum), is said by Dioscorides and Pliny to be made from wheat, the best coming from Egypt and Crete. It was prepared by soaking the grain in water about five times until thoroughly softened, the water finally drawn off and the wheat trodden out. The starch thus separated is washed, sieved and dried in the sun on new bricks. It must be dried quickly as when wet it soon sours.

Oils and fats are discussed by Dioscorides, though little of interest is added to the earlier statements of Theophrastus. Fats of the bear, lion, panther, stag, elephant, camel, ass, fox and serpent, are mentioned on account of special virtues they are supposed to possess in healing. Dioscorides mentions, as Theophrastus had already done, that certain fats and resins are heated, not over free fire, but inclosed in tight vessels suspended over or set in a vessel of heated water—the principle of the water-bath. He notes the interesting fact that to prevent fats from becoming rancid, they were covered with honey.

Naphtha, occurring in Babylon, Dioscorides says, is white in color, though sometimes found black. Fire attacks it with great energy, so that it even seizes upon it from a distance.[55]

Bitumen (asphaltos), the best from Judaea, occurs in Phoenicia, Sidon, Babylon and Zacynthos. In Agrigentum in Sicily, it swims as a liquid on the surface of springs where it is used instead of oil for lamps and is falsely called oleum siculum (Sicilian oil), though it is a kind of liquid bitumen.[56] It will be remembered that the lamps of the ancients were open lamps, not with closed oil reservoirs.

Herodotus writing in the fifth century before Christ, describing the method of the building of the walls of Babylon, tells that for a cement for setting the bricks, they em-

[55] Dioscorides, I, 101.
[56] Dioscorides, I, 99.

ployed hot bitumen. The source of the bitumen was the Is, a small stream flowing into the Euphrates, eight days' journey from Babylon. Lumps of bitumen are found, he says, in great abundance in this river.

The same author refers to a well near Ardericca in Cissia, whither Darius had transported his Eretrian prisoners, from which they get produce of three kinds.

"For from this well they get bitumen, salt and oil, procuring it in the way that I will now describe. They draw with a swipe, and instead of a bucket, make use of the half of a wine skin; with this the man dips and after drawing pours the liquid into a reservoir wherefrom it passes into another and there takes three different forms. The salt and the bitumen forthwith collect and harden, while the oil is drawn off into casks. It is called by the Persians 'rhadinacé,' is black, and has an unpleasant smell."

This is probably the earliest unmistakable reference in literature to a petroleum industry.

In connection with the recovery of certain oils from tar and resin, Dioscorides describes a crude process of distillation. The vessel in which the heating takes place has flocks of loose wool in the throat or upper part above the boiling liquid, and the distilled oil condensing in this wool is obtained by removing and squeezing out the oil. Pliny also describes this method for obtaining turpentine oil from the resin. These contemporaneous records of Dioscorides and Pliny, both very probably borrowed from Sextius Niger, are of interest as the earliest records of a crude process of distillation as a method of isolating a chemical product. The recovery of quicksilver from the domed cover of the vessel in which "minium" (our cinnabar) was heated, as given by Dioscorides, and already noted, is of similar significance.

Glue is mentioned by Dioscorides as made from oxhides and a better quality from the stomachs of certain fishes found in the Black Sea (fish glue).

Sakkaron is described as a kind of solidified honey from India and Arabia Felix, similar to salt in consistency, and crushing like salt between the teeth. It is soluble in water. Pliny also, doubtless quoting from the same authority as does Dioscorides, says, "Arabia, too, produces saccharon, but that of India is the most esteemed. This substance is a kind of honey which collects in reeds, white like gum, and brittle to the teeth. The larger pieces are about the size of a filbert. It is only employed, however, as medicine." Von Lippmann, the authority on sugar and the history of sugar, does not think that this is an allusion to cane sugar, as there is no evidence that sugar entered into use in Europe for centuries later. Its production in India is, however, of great antiquity, and it is not impossible that the substance described was, in fact, cane sugar, which was really known to writers anterior to Pliny and Dioscorides, even if its importation from India had been discontinued at an early period. The brief and almost identical description by Dioscorides and Pliny would seem to show that they had no further knowledge of it than they obtained from the common source of their information.

Poisonous substances described by Dioscorides include conium, strychnia, colchicum, aconitum, the poppy, hellebore, and the mandragora. From the last named, a wine is made which produces so heavy, long continued and unconscious a sleep that physicians perform difficult operations by its use. Pliny also says that it is given before incisions or punctures are made in the body, in order to ensure insensibility to pain.

Dyestuffs of organic origin, known to Dioscorides, were numerous, among them being alkanna, madder, kermes, woad and indigo.

Ink (melanos) was made from lampblack or soot from burning resins, mixed with gum or glue. Dioscorides mentions that chalcanthum is added. This addition is difficult to understand. Chalcanthum as we have seen was a term including copper sulphate, ferrous sulphate and mixtures of the two. If a solution of nutgalls or other solution of tannin were used, the addition would be comprehensible as forming a black tannate of iron, but no such addition is mentioned.

Likewise we find that Pliny describes ink, which he calls atramentum, as a product made from lamp black and glue, but he also makes no mention whatsoever concerning the addition of chalcanthum.

Fermented liquors, wines, meads and beer were known in all countries from the most ancient times. Their use at the time of Dioscorides and Pliny was extensive and excessive. They naturally entered largely into medicine. It is worthy of note that Dioscorides ascribes injurious action to their continual use, and advises they be used only as occasional stimulants. The effect of new wine in accelerating the pulse may be avoided, he says, by adding water and boiling until this is again evaporated. That the reason for this lies in the elimination or reduction of the alcohol content was beyond the understanding of the time.

Beer, from grain, especially barley, he considers as especially deleterious, as it bloats, promotes obesity, attacks the kidneys through its diuretic properties, and irritates the nervous system and the brain.

Many contributions of Pliny to our knowledge of the chemistry of the ancients have been already mentioned in relation to subjects discussed by his predecessors or by his contemporary Dioscorides, but many subjects are treated by him which were not included in the works of these authors.

Of the chemistry of the metals, Pliny writes much more extensively and in much greater detail than do the other authors of his period or of earlier periods. In introducing the subject, he says:

"We are now about to speak of metals (metalla), of real riches, the standards of value of things, objects for which we diligently search within the earth in many ways, for in some places it is dug up for gold, silver, electrum, or copper, elsewhere for riches in gems and pigments, to decorate our fingers and our houses; elsewhere we rashly

seek iron more esteemed than gold amidst wars and car-
nage." [57]

It is not apparent that the word "metal" with Pliny
had any such definite meaning as we apply to it. Originally the word meant the mine itself, gradually extended to
the products of the mine, and probably in a more restricted
sense to the more valuable products of the mine, the precious and useful metals, without attempting to draw any
clear distinction between these and other mineral products.
According to Lepsius, quoted by M. Berthelot in his chapter on "Metals with the Egyptians," [58] the Egyptians distinguished in their inscriptions eight mineral products,
particularly precious, arranged in the following order:
gold, electrum, silver, lapis lazuli, emerald, copper (or
bronze), iron, lead. Here also the classification is of products of the mines with no distinction, such as we recognize,
of metals as such. It is probable that this also was the
understanding of those of Pliny's time. The various substances which Pliny writes of in the book beginning with
the above quotation, include many substances from the
mines which are not metals as well as the metals themselves, a fact which seems to confirm the indefiniteness of
the designation "metal" at this time.

Gold, its occurrence, mining, properties and uses are
treated at length by Pliny. Gold is obtained in the form
of grains found in running streams, the Tagus in Spain,
the Padus (Po) in Italy, the Hebrus in Thracia, the Pactolus in Asia, the Ganges in India, "and there is no gold
in a more perfect state than the gold so found."

A second method is by sinking shafts in the earth, or
seeking it amongst the débris of mountains. It is often
located by washing the surface outcrop of the veins. The
covering of earth which gives indication of gold is removed, a bed is constructed, and the earth washed, and
according to the residue, the richness of the vein is con-

---

[57] Pliny, Book XXXIII, *Proemium*.
[58] *Les Origines de l'Alchimie*, p. 211 *ff*.

jectured. Shafts being sunk, the gold is found running in veins. It is found adhering to the gritty crust of marble (quartz?) and interlaced with the particles of the rock. Wooden pillars are placed to prevent the earth from falling into the shafts. The ore extracted is crushed and washed, then heated by fire and powdered. The dust or scoria escaping from the furnace chimneys is again crushed and melted. The crucibles used for this are of a white earth similar in appearance to potter's clay, there being no other substance capable of withstanding the strong currents of air, the action of the fire and the intense heat of the melted metal.

The third method of obtaining gold, he says, "surpasses the labors of the giants." It consists in driving long galleries into the mountains, the miners working by the light of torches, many of them never seeing the light of day for many months together. Not infrequently clefts are suddenly formed, the earth sinks in and the workmen are crushed beneath the weight of the mountain above. Arches are left at intervals to support the galleries. Barriers of quart (silex) are sometimes met and penetrated by fire and vinegar.

This latter statement of Pliny's appears to be the repetition of a prevalent tradition, for Livy and Plutarch credit Hannibal with this method of splitting rocks during the passage through the Alps.[59] Building fires for the purpose of cracking and loosening rock was doubtless in use, for Diodorus Siculus mentions this fact also. It is also said by later writers that at a much later period the practice existed of heating the rocks with fires and then deluging them while hot with water. This practice may possibly be the basis of Pliny's evidently incorrect statement. It may be recalled that Pliny, speaking of vinegar, says that it has the power of splitting rocks, evidently referring to the process he here describes.

Such rock barriers are also broken, he explains, by the

---

[59] Cf. Pliny, Bohn's ed., IV, p. 480, footnote.

use of heavy iron-shod beams, weighing often a hundred and fifty librae (equivalent to approximately one hundred and seven pounds avoirdupois), and certain tough layers are attacked with hammers and wedges. The broken fragments are passed out from hand to hand to the mouth of the gallery. When operations are completed, beginning with the last they cut away the wooden pillars that support the roof. When symptoms of yielding are observed by sentinels stationed for the purpose, alarms are given, the workmen called from their labors, and the mountain is cleft asunder "hurling its débris to a distance with a crash, which it is impossible for the imagination to conceive."

Another great labor is that of bringing rivers from mountains of higher elevation by aqueducts and by cutting away the rocks, sometimes for a distance of a hundred miles, for washing this mass of mountain ruin. The water is received in reservoirs constructed with sluices, and released so as to wash the heavy débris to lower levels, where trenches or ditches are provided, in the bottom of which are layers of ulex, (a plant) "rough and prickly," for arresting and holding the gold that may be carried along. These plants are afterward dried and burned to recover any gold left in them. The gold obtained from these washings, he says, is very pure and often in large lumps, sometimes weighing ten libra or more.

The water and suspended earth finally arrive at the sea—a cause, says Pliny, which has greatly tended to extend the coasts of Spain by these encroachments upon the sea.

For comparison with the account of gold mining in Spain, there is an interesting account of Egyptian gold mining by a Greek writer of a century or more earlier than Pliny, Diodorus Siculus. He bases his knowledge, he tells us, not merely upon accounts given by Agacarthades and Artemidorus and some others "who have in their writings nearly followed the truth," but upon his own observations, "having sojourned in Egypt, associated with many of the priests and conferred with ambassadors and others from Ethiopia." His account is moulded, he claims, upon the agreement of all these sources.[60]

The mines are situated in the confines of Egypt and in Ethiopia, and in neighboring regions of Arabia. The gold occurs in white veins in the earth, which is there black in color. These white (quartz?) veins are followed in the mining. Multitudes of slaves, criminals, or captured prisoners of war—men, women and children, are employed in the work. They are chained and fettered, and are cruelly driven by barbarian soldiers. No rest is given them, not even if feeble or sick, but by blows they are kept at work till they drop dead in the midst of their insufferable labors.

The workers in the galleries carry lamps on their foreheads as the galleries are not otherwise lighted. The large masses of ore are broken out by picks and by loosening the rock by fires. Boys take the loosened lumps and carry them to the surface. Here men take them and break them into small pieces with iron mortars and pestles. These small pieces are taken by old men and women and ground to powder in hand mills placed in long rows. The fineness of grinding is determined by samples given the workers. The finely powdered ore is then taken by the masters of the work, placed upon slightly hollowed wooden boards or inclined planes, and skillfully washed with water to remove all earthy particles and leave the clean gold. This gold is then mixed with lead, salt, a little tin and barley bran, placed in an earthen pot, the cover luted on, and the pot heated in the furnace for five days and nights. When cooled, only refined gold remains, the other matter has disappeared and the gold diminished a little in weight.

From this description of the metallurgical operation, it would appear to be a process of cupellation, which would remove base metals, though not silver. The lead oxide

---

[60] Diodorus Siculus, *The Historical Library*, Translation of G. Booth, London, 1814, I, p. 157,

must have been volatilized and some slags formed adhering to the crucible. The cover could not have been closely luted if the process is otherwise correctly described.

As to whether or no the ancients had a method of separating the silver from the gold is not certain, though a more or less complete separation was perhaps made. Strabo states that such was accomplished in Spain by repeated heatings or fusions. Pliny says that gold is melted with twice its weight of salt and three times its weight of misy, and again melted with two parts of salt and one of a stone called schistos. This process, he says, leaves the gold pure and incorruptible. He does not mention this operation in connection with the separation for silver, however.

If we assume with Berthelot, that misy was partly oxidized pyrites, containing basic sulphates of iron and copper, and that schistos was a rock related to hæmatite or an alum schist, the operation would have some action in converting silver to chloride and the process would resemble the now obsolete cementation process of separating silver and gold. This process consisted in heating the alloy in granulated form with a "cement" consisting of two parts brick dust and one part salt in a porous earthen pot for thirty-six hours at a temperature below melting. The silver is converted into silver chloride and afterwards removed by washing.[61]

The customary tests for the purity of gold with the ancients were color, weight (specific gravity), and the streak made by rubbing the metal upon the touchstone, a black silicious stone. Pliny states that by this method the experts could tell to a scruple how much gold, silver or copper was present—"their accuracy being so marvelous that they are never mistaken." [62]

It is not improbable that the ancient metallurgists by their somewhat crude methods, succeeded in removing silver from its natural alloys with gold, at least to the extent

---

[61] Cf. T. K. Rose, *Metallurgy of Gold*, 5th ed., 1906, p. 397.
[62] Pliny, Book XXXIII, Chap. 43.

necessary to bring the gold to a degree of purity which satisfied the requirements of their tests of color, specific gravity and streak.

Gold ornaments, articles and coins of very early and established antiquity are not abundant, and chemical analyses of them are not numerously recorded. Such analyses as have been made show wide variations in the purity of the gold, from a pure gold to gold with very high silver content, and the proportions of the two vary much between these limits, just as they do in the native gold from placers or mines, so that the analyses do not afford satisfactory evidence as to ancient standards of purity nor as to the results of their methods of separation.

Concerning the properties of gold, Pliny emphasizes the facts that it is the only substance that suffers no loss by the action of fire, and that the oftener it passes through fire, the purer it becomes. He mentions its difficulty of fusion, and that it does not wear away by handling, other metals soiling the hands by the substance which rubs off. He notes its malleability and its capability of extreme subdivision, so that an ounce may be beaten into seven hundred and fifty leaves of more than four fingers in length by the same in breadth. It can also be spun and woven like wool. "I have myself seen Agrippina, the wife of the Emperor Claudius, on the occasion of a sham naval combat which he directed, seated by him attired in a military scarf made entirely of woven gold without any other material." [63] Gold also resists the corrosive action of salt and vinegar "things which obtain the mastery over all other substances." Gold forms no rust. Gold found as dust or in masses (nuggets) is in a state of perfection, but all other kinds of gold have to be purified by art.

That Pliny knew of the use of mercury for recovering gold from the ashes of textiles containing it, has been previously noted.

The use of gold leaf for gilding of metals or other mate-

---

[63] Pliny, Book XXXIII, Chap. 19.

rials is described by Pliny. Metals, particularly silver and bronze or copper, were gilded by applying a film of quicksilver to the metal surface after cleaning with a mixture of salt, vinegar and "alum," then laying on the gold leaf and heating to a high heat to expel the quicksilver. For gilding marble or other substances which "do not permit of being brought to a high heat," the white of egg was used to attach the gold leaf, and for gilding wood, a substance called "leucophoron." This substance seems from his description to be a mixture of earths, practically a red or yellow clay.

Pliny does not explicitly state that the metals are heated after gilding to expel the mercury from the amalgam, but when he states that the white of egg is used on marble and other substances which cannot be heated to high heat, the inference seems clear that such was the process.[64] Even in the case of copper, Pliny says the white of eggs was sometimes fraudulently used instead of mercury.[65] Gilded bronzes of ancient origin still in existence bear testimony that the ancient artisans knew how to do very good work in the gilding of metals.

The mining and metallurgy of silver are treated of by Pliny in a manner rather suggestive than clearly descriptive. He states that silver occurs in almost all provinces, but the richest mines are in Spain. It is never found except by sinking shafts, for it does not, like gold, give evidence of its presence by shining particles. The earth in which it occurs may be ash-colored or red. He mentions a mine in Spain where the mountain has been penetrated to the distance of fifteen hundred paces, and that laborers are kept busy in shifts baling water night and day. He says that exhalations from silver mines are dangerous, especially to dogs. Evidently carbon dioxide is the exhalation referred to, especially dangerous to dogs, because the heavy gas is more concentrated near the floors of the drifts.

---

[64] Pliny, Book XXXIII, Chap. 20; Book XXXV, Chap. 17.
[65] Pliny, Book XXXIII, Chap. 32.

Silver is not to be melted except with lead or galena, "a name given to the vein of lead that is mostly found running near the veins of silver ore." Submitted to the action of fire, part of the ore is precipitated as lead, while the silver is left floating on the surface like oil on water. Certainly not a very lucid description, but many such faulty descriptions illustrate Pliny's vagueness of knowledge of technical operations. By *scoria of silver*, Pliny generally means the oxide of lead obtained in the smelting of silver ores with lead, for he says the scoria (several varieties are named), are used like molybdaena (litharge) for making plasters to promote cicatrization of wounds. Scum of silver or foam of silver (spuma argenti) in several varieties of color or density was also evidently litharge. The scum of silver, he states, is obtained by melting the silver and allowing it to flow into a lower receptacle where it is lifted by iron spits or stirrers in the midst of the flame in order to make it lighter. The process he attempts to describe, can hardly be other than an operation to get rid by oxidation of any lead mixed with it, and the volatilization of the lead oxide formed.

Pliny describes a method used by the Egyptians for darkening the surface of silver vessels. The silver is mixed with two thirds of finest Cyprian aes, and a proportion of sulphur equal to that of the silver. This mixture is melted in an earthen vessel well luted with potter's earth. This custom, he adds, has now passed to our triumphal statues, the value of the silver being enhanced by deadening its brilliancy. Silver may also be blackened, he says, by the yolk of a hard-boiled egg, but this color is easily removed by the application of vinegar and chalk. That silver becomes stained by contact with mineral waters "and the salty exhalations from them," is doubtless an observation dependent upon the presence of hydrogen sulphide in some spring waters.

There are two kinds of silver, says Pliny—on placing a piece of it upon an iron shovel and heating it to a high heat, if the metal remains white, it is of the best kind; if it turns red, it is inferior; if black, it is worthless. This test is evidently to distinguish silver from white alloys made with intent to deceive—the red and black colors are perhaps the oxides of lead or of copper present in such alloys, though we must always remember that Pliny's descriptions are not always reliable in details.

Fraud, however, Pliny tells us, has invented a method of stultifying this test by immersing the shovel in urine, "the piece of silver absorbs it as it burns and so displays a fictitious whiteness." This addition of organic matter may be supposed by its reducing action to prevent the oxidation for a time and so interfere with the test as to impose on the unexpert.

Electrum (Egyptian-asem) was by the ancients considered as a distinct metal—just as silver and gold were distinct metals. It is supposed that it was first known to the Egyptians in the form of an alloy, either native, or as the product of the working of a naturally occurring ore. It was sufficiently different in appearance and weight from either gold or silver to receive a distinctive name. In Pliny's time, the word was also in use, though recognized as an alloy of gold and silver. In all gold, says Pliny, there is some silver, a tenth part in some, an eighth part in others. In one mine only, at Albucrara in Gelaecia, (Spain), the proportion of silver is only one thirty-sixth; hence this gold is more valuable than any other. "Whenever the proportion of silver exceeds one fifth, it does not resist on the anvil" (becomes brittle?). An "artificial" electrum, he says, is also made by mixing gold and silver.

Concerning quicksilver, Pliny adds little to what has been already stated. It is of interest, however, to note that he considers the native quicksilver as different from that obtained by heating "minium" (cinnabar). He calls the latter "hydrargyros," a substitute for the native argentum vivum. There are two methods of obtaining this substance, either by pounding "minium" with vinegar, with a mortar and pestle of bronze or copper (aes), or by putting "minium" into flat earthen pans, covered with a lid, and then enclosed in an iron pot well luted with potter's clay, and the latter heated by fire maintained with bellows. The vapor (condensed) is then removed, that is found adhering to the cover. This vapor is like silver in color and like water in fluidity.

Pliny notes the poisonous character of quicksilver, and that it even pierces vessels "by the agency of its malignant properties." All substances except gold float upon the surface of quicksilver.

Iron, its sources, varieties and uses are discussed quite at length by Pliny without contributing anything very specific as to its metallurgy or properties. He refers to the hardening of the metal by plunging it while hot into water, and states that the differences in value of various kinds of iron are due to some extent to the ores, but the main differences come from the quality of the water into which the heated metal is plunged. Smaller articles are often quenched in oil, as they become too brittle if water is used. Iron rust is spoken of and its uses in medicine described. Also the product of the action of vinegar upon iron rust [acetate] is said to be a remedy for erysipelas. For protecting iron structures from rusting, a coating of a mixture of white lead (cerussa), gypsum and tar was used.

Pliny, like all other ancient Latin writers, uses but one term "aes" to designate copper, bronzes, and brass. Nor is it to be concluded from anything he says that he realizes any fundamental difference between these substances. Greek writers used the term chalchos (χαλκός) in the same sense. That there are many different kinds of aes, he knows, distinguished by varying colors, malleability and especially by the locality where manufactured. The Corinthian aes was highly valued and apparently rare, as Pliny says there was a mania for collecting it. It existed in three varieties, white like silver, yellow like gold, and a third in which there is an equal mixture of aes. He states that formerly gold and silver were melted together with aes. Delian aes, much used for statues, and Aeginetan aes were also much valued. Cyprian aes, from the various references to it and its uses, was probably copper, pure or nearly pure. From chemical analyses of ancient bronzes, we have seen that the oldest are either alloys of copper and tin or pure copper; that later, lead often enters into their composition, and that in Pliny's time, zinc alloys (brasses) were in use, but except in a few special instances, Pliny gives no information that would permit the inference that he, or the authorities from whom he draws, has any knowledge of what constituted the differences between the alloys comprehended under the designation of chalchos or of aes, or that the difference in properties was caused by particular constituents.

M. Berthelot well expresses the fundamental ignorance of the ancient writers in matters of this kind.

"Let us insist upon this point, that neither the Greeks nor the ancient Romans have ever employed two distinct and specific names for copper and bronze, and that we should not look for two words among the ancient Orientals. The word "aes" was applied to copper and to its alloys with tin, lead and zinc. In order properly to understand the ancient texts, it is necessary to eliminate from our minds precise definitions acquired by the chemistry of our time; for elementary bodies have not, at first sight, any specific character which distinguishes them from compound bodies. Nobody in antiquity considered the red copper as an element which it was necessary to isolate before combining it with others. The ancients, I repeat, never conceived of alloys as we do by referring them to the association of two or three elementary metals, such as our copper, our tin, our lead, elementary metals which we melt together to obtain bronzes or brasses. But they operated chiefly upon the ores of these metals more or less pure, ores called cadmias or chalcites; they mixed these before the operation of manufacture and casting of the metal proper. Sometimes, though rarely, they mixed with these, alloys and

metals obtained from the first casting (jet). Every metal
and alloy, red or yellow, which was alterable by fire was
called χαλκός or aes; every white metal and alloy fusible
and alterable by fire was called originally lead. Later two
varieties were recognized, black lead, which comprised our
lead and more rarely antimony, and white lead, which comprised our tin and certain alloys of lead and of silver.'' [66]

When Pliny attempts to describe the ores of aes, he does
not understand that these substances contain the metal
which is to be made. He understands only raw materials
used in their making. Thus cadmia is mentioned. We know
that this cadmia was a zinc ore, but Pliny mentions it as
a source of aes, in the same way that he mentions the real
copper ore or ''chalcites.'' They were for him and his
times merely raw materials whose treatment in the furnace
resulted in the making of the product, a variety of aes.

Cyprian aes is itself of two kinds according to Pliny—
coronarium and regulare, both of them ductile. The former
can be made, he says, into thin leaves and is therefore probably copper itself. In other mines, they prepare the regulare and also the caldarium which breaks when hammered,
but all kinds if sufficiently melted and heated will become
malleable.

Pliny mentions the making of certain bronzes for special
purposes by adding to the bronze and melting with it, certain proportions of silver-lead, or of lead and silver-lead.[67]
This silver-lead he elsewhere[68] says is made of equal parts
of black lead and white lead, that is, lead and tin.

The compounds of copper, known to Pliny, are practically the same as already discussed, and his information
has been there referred to.

The plumbum candidum, (white lead), of Pliny is tin. He
states that it is more valuable than the ordinary or ''black''
lead, that there is a ''fabulous story of its having been
brought in boats of osiers covered with hides from islands

[66] Berthelot, *op. cit.*, pp. 230, 231.
[67] Pliny, Book XXXIV, Chap. 20.
[68] Pliny, Book XXXIV, Chap. 48.

in the Atlantic.'' The story was not entirely fabulous, for such were the coracles of the ancient Britons, and the Scilly Islands and Cornwall were ancient sources of tin, as they still are. Tin was also alleged to be obtained from Lusitania and Gallaecia in Spain, occurring as heavy pebbles in old river beds and collected by washing in connection with the gold occurring there. When melted in the furnace, they are converted into tin (plumbum candidum). It was in the Gallic provinces, says Pliny, that the method was discovered of coating articles of copper (aes) with tin so that they were scarcely distinguishable from silver. Tin was tested by pouring it when melted upon paper (charta), which then gives the appearance of being broken not by the heat, but by the weight. This test it would appear must have depended upon the low melting point of tin as compared with other white metals or alloys; thus when properly applied not burning or scorching the paper though breaking it. The paper was then made from the papyrus; hence the modern name.

The term "stannum," as used by Pliny, does not mean tin, but alloys of tin and lead, or silver and lead, alloys which were used instead of tin, probably in covering copper utensils, or for other purposes, as solder.

Lead, plumbum nigrum, its occurrence in connection with silver, its uses in making certain bronzes, for making lead water pipes, and in sheet form, are described by Pliny. Its oxide (Pb O) is described under the names of molybdaena, lithargyros, and galena, as the product of roasting lead in the air, and as produced in the furnaces where silver and gold are smelted. White lead (cerussa) and our red lead were also known and described by Pliny, much as by authorities already quoted.

Gypsum (plaster of paris), quicklime, cements, are discussed by Pliny, but little of interest added to information given by Theophrastus and Vitruvius.

Pigments are discussed in much detail by Pliny. Red pigments were minium (our cinnabar); cinnabaris, meaning dragon's blood, though he notes that the same term is sometimes applied to minium; sandarach (realgar), but including red lead (our minium) as a spurious kind of sandarach; rubrica, and sinopis, both evidently red oxides of iron or earths containing these; ochra obtained by burning rubrica;[69] usta, obtained by heating sil which is a yellow ocher reddened by heating as in burning bricks. Sandyx is a red color obtained by heating a mixture of equal proportions of rubrica and sandarach, a cheaper substitute for sandarach (realgar).

Yellow pigments were auripigmentum, our orpiment, the arsenikon of the Greeks; sil or Attic sil, a clay colored yellow by ferric hydroxide.

White pigments were paraetonium (from Egypt), the most unctuous of the white colors. "It is sea-foam, they say, solidified with slime and hence it is that minute shells are often found in it:" "melinum—the best from the isle of Melos," a white earth occurring in veins; cimolian earth, also used for scouring cloth and probably a white clay; eretria, white or sometimes ash-colored, an earth used as a pigment; cerussa or white lead. From Pliny's descriptions, it is difficult to guess whether any one of the white earths is a chalk or a clay, or possibly a magnesite or a meerschaum.

Green pigments were chrysocolla, malachite, or other basic carbonates of copper; and appianum, a green earth or chalk said to be a cheap and inferior color.

Blue materials used as pigments or dyes, were the lapis lazuli (ultramarine), azurite (armenium). Both of these sometimes were called caeruleum. Indicum was indigo imported from India. "Purpurissium" was the name given to a pigment made from chalk colored with a purple dye, but whether from murex, indigo or woad does not seem definitely stated.

Of sulphur, Pliny states, there are four kinds, but he

---

[69] Theophrastus says the opposite and Pliny may be, and probably is, in error.

makes no very intelligible characterization of their differences. "Live" sulphur (sulphur vivum), occurring in masses or blocks is the only kind used in medicine. The others are used respectively by fullers, for the fumigation of wool, and the preparation of lamp wicks (the latter evidently used for kindling as we use it in matches). Sulphur was also used in religious ceremonies, and for fumigating houses, and for fumigating (bleaching) cloth. The virtues of sulphur are to be perceived in certain hot mineral springs, and there is no substance that ignites more readily, "a proof that there is in it a great affinity for fire."

Bitumen, or asphalt, and naphtha are described much as Dioscorides describes them. "Maltha" is a product of similar character, will take fire and burn even upon water, and can be extinguished only by earth. The uses of bitumen were for medicines; for coating the inside of vessels of copper or brass for the purpose of protecting them from the action of fire; for staining bronze statues; as a cement instead of mortar for buildings, as in the walls of Babylon; for varnishing iron and the heads of nails to prevent their rusting.

It will be recalled that a crude form of distillation was described by Dioscorides and by Pliny where flocks of wool were used to condense the more volatile constituents of pitch or bitumen.

Pliny gives a description of a process a little more systematic for the recovery of tar from the "torch tree." The wood is chopped into small billets, placed in a furnace which is heated by fires lighted on every side. The first liquid that exudes flows like water into a reservoir made for its reception. In Syria, this substance is known as cedrium, and it possesses such remarkable power that in Egypt the bodies of the dead after being steeped in it are preserved from corruption. The liquid that follows is of thicker consistency and constitutes pitch properly so called. This is apparently a somewhat elaborated method of melting out the pitch from wood, similar to the process previously given by Theophrastus, but less crude and wasteful.

The industry of dyeing, a very important industry in ancient times, is rather slighted by Pliny. The reason for this is to be found in his own statement: "I should not have omitted to enlarge upon the art of dyeing, had I found that it had ever been looked upon as forming one of the liberal arts." There is room for doubt whether there existed any works sufficiently informing on this subject that Pliny might have used, for processes of this nature were in general rather carefully guarded secrets of the artizans who practiced them.

Nevertheless, there are some allusions in Pliny that pertain to the raw materials used. Thus are mentioned kermes, a species of coccus giving a red dye; anchusa, which is the alkanna or orcanet of more modern practice, used for imparting rich colors to wool; madder, of which alizarine is the color-giving constituent; besides indigo and murex purple to which allusion has already been made. Of the madder, he says large profits were made from it, and that it was used for dyeing wool and leather. Walnuts and seaweeds are also mentioned as dyestuffs. There is a reference in Pliny to the use of mordants as practiced by the Egyptians, which is interesting as showing that their methods were developed to a greater degree than might otherwise be supposed:

"In Egypt they dye clothing in a remarkable way. The white material is treated not with the colors, but with medicaments which absorb the colors. This done, the materials appear unchanged, but when immersed in a cauldron of boiling dye and immediately removed, they are colored. It is remarkable that though the dye in the cauldron is of one color only, the materials when taken out are of various colors according to the quality of the medicaments applied." And the colors so applied, Pliny says, will not wash out and the goods so treated are rendered more durable by the operation. The description, though lacking in specific detail, yet bears evidence to a considerable understanding by the dyers of the time of the influences of different mordants in modifying the colors fixed.

The manufacture of glass, as has been already stated, is of very ancient origin, thousands of years before the time of Pliny; and the period of its discovery as an art is too early to be determined with certainty. Pliny, however, repeats an ancient fable of its discovery by accident through merchants who moored their boats loaded with soda (nitrum) on the sands of a tidal river in Phoenicia in Syria. When preparing their meal on the sandy shore, they lacked stones to support their pots, and took instead lumps of soda from their cargo. When the fire became hot, they beheld transparent streams of an unknown substance flowing from the fire "and this, it is said, was the origin of glass." The story preserved in Pliny's record has been an often repeated tale in more modern literature. In the days of the Roman Empire, glass was extensively manufactured for ornaments, statues, imitation gems, and for drinking vessels.

Pliny says that glass is made not merely from sand and soda, but that later, magnes lapis began to be added, "from the idea that it attracts liquid glass as well as iron." Pliny here confuses the various minerals which passed under the name of magnes. It will be recalled that in speaking of the magnetis lithos of Theophrastus, facts were stated which might easily explain this confusion in the writing of an author who had no personal knowledge or understanding of the art. While magnes more often meant the lodestone or magnetic oxide of iron, it also included pyrolusite or black oxide of manganese, and a white magnes, which might have been a marble, a dolomite or lime sulphate. We know that manganese and lime are found in ancient glass articles, and whether the magnes here referred to was pyrolusite, or a limestone cannot be decided. Pliny also says that many other substances are used, shells and fossil sand and brilliant stones of various kinds.

It is melted, he says, by wood fuel, Cyprian aes being

added. The fusion takes place in contiguous furnaces, as with copper, and a dark mass of viscid appearance is the result. This mass is again subjected to fusion for the purpose of coloring it, after which it is blown into various forms, turned in the lathe or engraved like silver. Here again Pliny's account is confused. Copper was added for coloring blue, though Pliny has it added in the original fusion before coloring.

For imitating gems and semiprecious stones, Pliny says that glass was extensively used, and that it was with great difficulty that the imitations could be distinguished from the genuine. Obsidian, topaz, beryl, carbuncle, sapphire, jasper, opal, onyx, and emerald were thus imitated.

"Nay more than this, there are books, the authors of which I refrain from mentioning, which give instructions how to stain crystal (quartz) in such a way as to imitate emeralds and other transparent stones . . . and there are no frauds which bring greater profits.

"Still [says Pliny], the highest value is placed upon glass that is entirely colorless and transparent, as nearly as possible resembling crystal. For drinking vessels, glass has quite superseded the use of silver and gold, but it is unable to stand heat unless a cold liquid is first poured in. And yet, we find that globular glass vessels filled with water, when brought into the sun's rays become heated to such an extent as to cause articles of clothing to take fire." [70]

Pliny also states that some authors say that in India glass is made from broken crystal (quartz) and that in consequence there is none that can compare with it. It is well known that in China the glass industry was of very early development though the chronology of early arts of China is difficult to determine with exactness.

The manufacture of glass mosaics for decoration of buildings, still an important Italian art, Pliny describes as a recent invention, and as evidently not known when Agrippa constructed his baths, or he certainly would have

---

[70] "In tantum excandescunt ut vestes exurant," Book XXXVI, Chap. 67.

used them. Pavements of mosaic tiles were certainly of earlier invention and Pliny thinks they date from the time of Sylla.

Tests for distinguishing natural gems from their imitations in glass, as described by Pliny, depended upon observing them in sunlight, upon relative weights (specific gravities), the feeling of coolness in the mouth (conductivity), and differences in hardness, though the last named test was often not permitted by dealers, naturally enough.

Oils, wines and perfumes or unguents are treated at great length by Pliny, but few items of information germane to our subject are here contained which are not found in Theophrastus and Dioscorides. One observation of Pliny in connection with wines deserves attention. He says:[71] "There is now no known wine that ranks higher than the Falernian; it is the only one, too, among all the wines that takes fire on the application of a flame." This statement is very interesting if true, for no wine obtained by direct fermentation can contain a sufficiently high alcohol content to so take fire. If the alcohol content were sufficient for this, it must have been produced by some distillation process, by adding an alcohol produced by distillation, or in some other way increasing the alcohol content above that resulting from fermentation. Yet the history of distillation contains no evidence of any such process, and the method of "fortifying" wines by the addition of alcohol so far as we know dates from a much later period. Either then the makers of Falernian wine possessed a knowledge of this process which remained a secret with them, or the fact recorded by Pliny must be otherwise explained. It is conceivable that the addition of plaster of Paris might be carried so far as to increase the alcohol content to a high degree by removing water. Whether this could have taken place and the Falernian enjoyed its high reputation is a question not to be answered offhand. We know that the Falernian wine was a strong wine, for both

[71] Pliny, Book XIV, Chap. 8.

Pliny[72] and Dioscorides[73] in describing the medicinal properties of wines call attention to the unusally active influence of this wine in quickening the pulse. Pliny says indeed that no other wine so stimulates the venous system. It is conceivable also that as Falernian wine was a strong wine, that the phenomenon of taking fire may have been observed when the wine was heated and thus the vapor of alcohol was sufficiently concentrated to take fire when the flame was approached. That Pliny should make this statement unless some basis existed does not seem reasonable, but as to the correct interpretation of it, the field of conjecture lies open, for there is elsewhere apparently no record of facts related to the subject in ancient writings which help us to interpret Pliny's statement. In connection with this subject, it is interesting to note the previously cited method given by Dioscorides for reducing the stimulating effect of new wine by adding water and boiling it off again.

Water is a subject treated by Pliny very extensively from many points of view. Its physics, geophysics, the different kinds and sources of water, mineral springs of all kinds in a great number of localities are described, and there is much dealing with the marvelous, and current superstitions with respect to particular waters are accounted in great numbers. Some of his observations are pertinent to the scope of our inquiry. He states that some waters are impregnated with sulphur, some with "alum," some with salt, or soda or bitumen. Some deposit a thick crust on vessels when boiled. Such are not to be preferred for drinking water. Some waters have the property of petrifying twigs and branches of trees which are exposed to them. Bricks placed in certain waters "change to stone." In caverns in Mt. Corycus, the drops of water that trickle down from the stone harden to stone, and at Mieza the water petrifies as it hangs from the vaulted roof of the rocks. In

---

[72] Pliny, Book XXIII, Chap. 20.
[73] Dioscorides, V, Chap. 10.

other caverns, the water petrifies both as it hangs and after it falls, making columns.

The deposit made by the separation of dissolved carbonate of calcium from hard waters is thus interpreted by Pliny as an actual change of the water to the stone, a point of view entirely consistent with the theories of matter existent at the time. Thus Diodorus Siculus[74] says of crystal (quartz), "It is said that it is produced of the purest water congealed and hardened not by cold but by the power of the sun, so that it continues forever, and receives many shapes and colors, according as the spirits are exhaled."

As to the wholesomeness of water for drinking, he states that physicians consider running water more wholesome than stagnant or sluggish waters, that that water is best which has neither smell nor taste, that it is generally admitted that all water is more wholesome when it has been boiled, that well water is generally more wholesome than that from other sources, but only in the case of wells in which it is kept in agitation by repeated drawing and by percolating through the earth. Rain water, and water from melting snow or hail, were considered by not a few medical men as injurious for drink. Snow water and hail water, they explained, were injurious because all the refined parts had been expelled by agitation, which sounds like many other attempts to explain observed facts by hypotheses that do not explain.

Rain water, says Pliny, putrefies with great rapidity and keeps but badly on a voyage.

"It was the Emperor Nero's invention to boil water and then inclose it in glass vessels and cool it in snow; a method which insures all the enjoyment of a cold beverage without any of the inconveniences resulting from the use of snow."[75]

The danger in digging deep wells from "sulphureous" and "aluminous" effluvia which kill the well diggers, the test for danger by lowering a lighted lamp, and the digging

---

[74] Book II, Chap. IV. (Booth's Translation, I, p. 143).
[75] Pliny, Book XXXI, Chap. 23.

of wells to right and left for ventilation in case noxious vapors are present are described much as by Vitruvius, previously cited.

That the ancients had some definite idea of the sterilization of water by heat is indicated by Diodorus. He is discussing the possible source of the Nile and opposing the theory that it comes from the antipodes through the center of the earth, a theory advanced by certain natural philosophers puzzled by the fact that high water came during the rainless period in Egypt, and by the fact that the water was sweet, which suggested the influence of the hot regions. The waters of the Nile teemed with fish and were believed to breed mice and other animals: Diodorus says:

"As to causes alleged for the sweetness of the water, they are absurd, for if the water be boiled by the parching heat and thereupon become sweet, it would have no productive quality either for fish or other kinds of creatures and beasts, for all water whose nature is changed by fire is altogether incapable of breeding any living thing."

For the perfected philosophy of sterilization of water, the world was to wait for the results of the researches of Louis Pasteur.

The foregoing discussion is believed fairly well to illustrate the scope and character of the knowledge of practical chemistry possessed by the ancients, in so far as extant literature gives evidence.

It is obvious, however, that the practical chemists of the time, metallurgists, jewelers, glass makers, dyers, etc., must have possessed more specific and detailed knowledge than the authors whose works have come down to our times, but no contemporary records from their pens are preserved. From a somewhat later period, about the third century of our era, two very interesting original manuscripts have been preserved, which are of this character. These are trade manuals, so to speak, of chemists. These manuscripts are of much importance in the history of chemistry and deserve special consideration.

CHAPTER II

THE EARLIEST CHEMICAL MANUSCRIPTS

Though Egypt is generally recognized as the mother country of the chemical and alchemical arts, her monuments and literature have left little of early records to explain them to us. It is through Greek and Roman sources mainly that some of these ideas have been transmitted to us, but the character of these sources is not often such as to enable us to discriminate between the matter derived from Egyptian science and the confused interpretation or additions of the early Greek alchemists. At about 290 A.D. the Emperor Diocletian passed a decree compelling the destruction of all works upon alchemical arts and on gold and silver throughout the empire, so that it should not be possible for the makers of gold and silver to amass riches which might enable them to organize revolts against the empire. This decree resulted in the disappearance of a mass of literature which doubtless would have furnished us with much of interest in the early history of chemical arts and ideas.

By a fortunate chance, however, there have been saved to our times two important Egyptian works on chemical processes, the earliest original sources on such subjects. They were discovered at Thebes, and both formed part of a collection of Egyptian papyrus manuscripts written in Greek and collected in the early years of the nineteenth century by Johann d'Anastasy, vice consul of Sweden at Alexandria. The main part of this collection was sold in 1828 by the collector to the Netherlands government and was deposited in the University of Leyden. In 1885, C. Leemans completed the publication of a critical edition of the texts with Latin translation of a number of these manuscripts, and among these was one of the two works abovementioned. It is known as the Papyrus X of Leyden.

The eminent French chemist and student of the history of early chemistry, Marcelin Berthelot, subjected this work to critical analysis and published a translation into French with extensive notes and commentaries.[1]

On the basis of philological and paleographic evidence, its date is established as written about the end of the third century A. D. It is, however, manifestly a copy of a work previously written, as slight errors evidently due to a copyist are found. That the original is later than the first century A. D. is certain, as there are included in it extracts from the *Materia Medica* of Dioscorides. The work is a collection of chemical recipes and directions for making metallic alloys, imitations of gold, silver or electrum, dyeing and other related arts.

In 1913 at Upsala, Otto Lagercrantz published the Greek text with critical commentary and with translation into German of a similar Egyptian papyrus, the "Papyrus Graecus Holmiensis." This work like the Leyden manuscript is a collection of recipes for alloys, metal working, dyeing, imitations of precious stones and similar arts. Investigation developed that this manuscript also came from the Swedish vice consul at Alexandria, d'Anastasy, presented by him to the Swedish Academy of Antiquities of Stockholm, as in its records appeared a letter of thanks of date 1832. Here it slumbered apparently unnoticed until 1906 when it was transferred to the Victoria Museum at Upsala. Examination and comparison with the Leyden Papyrus made it evident that the new papyrus was not only contemporaneous, but in all probability was in part at least written by the same hand.

Both papyri were in remarkably well preserved condition. Both give internal evidence of having been copied

[1] *See Les Origines de l'Alchimie*, 1885; *Collection des Anciens Alchimistes Grecs*, 1887–1888; *Introduction a l'Etude de la Chimie*, 1889.

from other originals. Berthelot has suggested that the Papyrus X had been preserved in the mummy case of an Egyptian chemist, and Lagercrantz concurs in the opinion, and is convinced that the two works were the property of the same person, and that these copies were probably made as copies de luxe for the purpose of being entombed with their former owner in accordance with a common custom of placing in the tomb articles formerly owned or used by the deceased.

The two manuscripts taken together form an interesting collection of laboratory recipes of the kinds which Diocletian ordered destroyed and which apparently were very generally destroyed. The date ascribed to them is about the time of the decree of Diocletian, and it may be presumed that, in the mummy case, they escaped the execution of that decree.

The laboratory manuals from which these copies were made were written not for public information but for the guidance of the workers. The recipes themselves are often very detailed directions, but often also were mere hints or suggestions, sometimes elliptical to such an extent as to give no clear idea of the process as carried out.

The Leyden papyrus comprises about seventy-five recipes pertaining to the making of alloys, for soldering metals, for coloring the surfaces of metals, for testing the quality of or purity of metals, or for imitating the precious metals. There are fifteen recipes for writing in gold or silver or in imitation of gold and silver writing. There are eleven recipes for dyeing stuffs in purple or other colors. The last eleven paragraphs are extracts from the *Materia Medica* of Dioscorides, relating to the minerals or materials used in the processes involved.

Berthelot notes that the artisan who used these notes while a practical worker in metals, especially the metals used by the jewelers, seemed to be a stranger to the arts of enamels and of artificial gems. It is, therefore, of great interest to discover that the Stockholm papyrus supplements the Leyden recipes in this direction. The Stockholm manuscript contains in all about a hundred and fifty recipes. Of these, only nine deal with metals and alloys, while over sixty relate to dyeing and about seventy to the production of artificial gems. Some ten others deal with the whitening of off-color pearls or the making of artificial pearls.

There is considerable that is practically only a duplication of recipes contained in each of the manuscripts, and very similar recipes occur in both. The recipes in both are empirical with no evidences of any occult theories, nor any of that obscurity of language which is so characteristic of the later alchemists.

The parts dealing with the metals are largely concerned with producing passable imitations of gold, silver or electrum from cheaper materials, or with giving an external or superficial color of gold or silver to cheaper metal. There seems to be no self-deception in those matters. On the contrary, there are often claims that the product will answer the usual tests for genuine products, or that they will deceive even the artisans. The vocabulary of materials used is practically that of Dioscorides, with few changes in the meaning of such terms as are used by him, although at times the Latin equivalents of Vitruvius and Pliny have been employed.

There is little to be found in these manuscripts which suggests that there has been any advance in the practical arts as known in the times of Dioscorides and Pliny and which had been less specifically described by them, but the papyri, in the more definite and detailed directions they give, throw a very interesting light upon the somewhat limited fields of industrial chemistry, of which they treat.

Examples will best serve to illustrate the character of the recipes and of the knowledge of practical chemistry which underlies them. The following are from the Papyrus of Leyden, as found in the previously mentioned translation of Berthelot.

5. Manufacture of asem (electrum).[2]

Tin, 12 drachmas;[3] quicksilver, 4 drachmas; earth of Chios, 2 drachmas. To the melted tin add the powdered earth, then add the mercury, stir with an iron, and put it into use.

This, then, is a tin amalgam intended to give the appearance of asem or silver. The earth of Chios as described by Pliny appears to have been a white clay. Pliny says it was used by women as a cosmetic.

6.   The doubling (diplosis) of asem.

This is the way the doubling of asem is accomplished. Take refined copper (chalchos) 40 drachmas, asem 8 drachmas, button tin 40 drachmas. The copper is first melted and after two heatings the tin and finally the asem is added. When all is softened, remelt several times and cool by means of the preceding composition. (No. 5?)   Clean with coupholith (talc or selenite according to Berthelot). The tripling (triplosis) is effected by the same process, the weights being proportioned in conformity with what has been directed above.

This recipe would yield a pale yellow bronze containing mercury if, as seems probable, the preparation No. 5 is added.)

4.   Purification of tin.

Liquid pitch and bitumen, one part of each.   Throw it on and melt and stir.

Of *dry* pitch 20 drachmas, bitumen 12 drachmas.

This is manifestly a process of obtaining an unoxidized clean tin for further use.

---

[2] The numbers prefixed to the recipes are the serial numbers of the recipes in the manuscript in Berthelot's *Collection des Anciens Alchimistes Grecs. See* also Berthelot, *Archéologie*, Greek text and translation, p. 268 *ff*.

[3] The weights and measures used in these recipes are those which were current both in Egypt and Greece at the period, and though the values of the particular units, varied very considerably at different times and in different places, the following values given by Berendes in his translation of Dioscorides are probably not far from those attaching to the units used in these recipes.

Kotyle. .about	"	274 cubic centimeters
Chu.	"	3282 cubic centimeters
Obole.	"	.568 gram
Drachma.	"	3.411 grams
Stater.	"	6.822 grams
Alexandrian Mina.	"	546 grams

8. Manufacture of asem.

Take soft tin in small pieces, four times purified. Take of it four parts and three parts of pure white copper (or bronze, "chalchos"), and one part of asem. Melt and after casting, clean several times and make what you will with it. This will be asem of the first quality which will deceive even the artisans.

Copper was whitened by the ancients sometimes by alloying with arsenic. A recipe in this papyrus gives directions for this whitening of copper. No. 23.

16 & 17. Augmentation of gold.

To augment gold, take Thracian cadmia, make the mixture with the cadmia in crusts; or cadmia of Gaul,[4] misy and sinopian red, equal parts to that of gold. When the gold has been put into the furnace and has become of good color, throw in these two ingredients and removing [the gold] let it cool and the gold will be doubled.

Cadmia, it will be remembered, is the impure zinc oxide, containing sometimes lead and copper oxides, from the furnaces in which brass was smelted. Misy was the partly oxidized iron or copper pyrites, essentially basic sulphates of iron and copper. Synopian red was hæmatite. This mixture, assuming the reducing action of the fuel in the furnace, or of any other reducing agent not specified in the recipe would yield an alloy of gold and zinc, with some copper and perhaps some lead.

11. To make asem.

Carefully purify lead with pitch and bitumen, or tin as well; mix cadmia and litharge in equal parts with the lead. Stir till the mixture becomes solid. It can be used like natural asem.

Reduction in the furnace must here also be assumed. The soft white alloy so obtained must have been a cheap and poor substitute for electrum or silver.

31. Preparation of chrysocolla (solder for gold).

The solder for gold is prepared thus: Copper of Cyprus 4 parts,

---

[4] Thus Berthelot.   Von Lippmann translates γαλατικός as "Galatian," from Asia Minor.   Cf. Von Lippmann, *Entstehung und Ausbreitung der Alchemie*, p. 4.

asem 2 parts, gold 1 part. The copper is melted first, then the asem and finally the gold.

It will be recalled that the term "chrysocolla" was applied also to malachite, verdigris and copper acetate, all of these being used for soldering gold.

32.   To determine the purity of tin.
Having melted it, place paper (papyrus) underneath it and pour it out. If the paper is scorched the tin contains lead.

36.   To make asem black as obsidian.
Asem, 2 parts, lead, 4 parts. Place in an earthen vessel, throw on it a triple weight of native sulphur, and having put into the furnace, melt. After withdrawing from the furnace, beat and make what you will. If you wish to make figured objects of beaten or cast metal, polish and cut it. It does not rust.

This process yields a metallic mass blackened with sulphides of lead and silver, similar to the black silver bronze as described by Pliny.[5]

38.   To give objects of copper the appearance of gold, so that neither the feel, nor rubbing on the touchstone can detect it, to serve especially for a ring of fine appearance.
Here is the process. Gold and lead are reduced to a fine powder like flour, 2 parts lead to 1 of gold. When mixed, they are mixed with gum and the ring covered with this mixture and heated. The operation is repeated several times till the article has taken the color. It is difficult to detect because rubbing gives the mark (or "scratch") of a genuine article, and the heat consumes the lead but not the gold.

This is an interesting process of gold plating by using lead instead of mercury, the lead being oxidized and volatilized in the heating.

43.   Test for purity of gold.
Remelt and heat it. If pure, it keeps its color after heating, and remains like a coin. If it becomes whiter, it contains silver, if it becomes rough and hard, it contains copper and tin, if it softens and blackens it contains lead.

---

[5] Pliny, *supra*, p. 68.

56. To gild silver in a durable way.

Take quicksilver and gold leaf, making to the consistency of wax. Clean the vase with alum, and taking a little of the waxy material, spread it on the vase with the polisher and let it stand to fix. Do this five times. Take the vase with a linen cloth so that it be not soiled, and removing it from the coals, prepare ashes, smooth with the polisher and use it as a gold vase. It will stand the test for real gold.

The recipes for writing with letters of gold vary much according to the material upon which they were to be applied, as also with respect to their relative durability.

The following one was doubtless for decoration of articles which could be subjected to action of heat to expel mercury.

34. To write in letters of gold.

Take quicksilver, pour it into a suitable vase and add gold leaf. When the gold appears dissolved in the quicksilver, shake well, add a little gum, one grain for example, and letting it stand, write in letters of gold.

Other methods of manipulation for the preparation of gold amalgam appear in the manuscript, as for instance grinding the quicksilver and gold leaf in a mortar. One recipe directs drying and grinding the gold leaf to powder with gum, thus avoiding the use of quicksilver, but furnishing a writing which was evidently not so durable, and which could not be heated. Cheaper imitations of gold writing were also used as illustrated in the following.

58. Orpiment of gold color, 20 drachmas; powdered glass, 4 staters; or white of egg, 2 staters; white gum, 20 staters; safran   .   . After writing, let it dry and polish with a tooth. (An animal's tooth used by jewelers for polishing.)

In other recipes, the yellow or gold color is obtained by sulphur mixed with gum; the "bile of the tortoise," or of the calf, "very bitter," serves also for the color. These may be secret trade names for some substances of different character.

The processes of dyeing are treated much more fully in the Swedish papyrus than in the Leyden, and can better be discussed in connection with that work. It will suffice here to give one example, and in connection with it, one very similar in the Swedish papyrus, as illustrating the close connection between the two collections of recipes. From Papyrus X:

94. Preparation of purple.

Break in small pieces Phrygian Stone; bring to a boil and having immersed the wool leave it till it becomes cool, then throwing into the vessel one mina of algae,[6] boil and throw in the wool and letting cool, wash it in sea-water to the purple coloration. The Phrygian stone is roasted before breaking.

Berthelot considers the Phrygian stone probably to have been an alunite, or basic sulphate of aluminum and potassium. Pliny describes it as a porous stone resembling pumice which is saturated with wine and then calcined at red heat and quenched in sweet wine—the operation being three times repeated. Its only use is in dyeing cloths. If it were an alunite, this process, consisting essentially of roasting and lixiviating, would yield a solution of sulphate of aluminum valuable as a mordant.

The algae above-mentioned are manifestly the source of the dyestuff and as suggested by Berthelot were probably lichens such as were formerly much used and which yield the dyestuff called archil or orseille.

The recipe in the Swedish manuscript is as follows:[7]

Purple—Roast and boil Phrygian Stone. Let the wool stay in till cold. Then take it out; put into another vessel orseille[8] and amaranth, one mina of each, boil and let the wool cool in it.

It is pretty evident that the two recipes are practically the same, the one helps us to understand the other.

The Papyrus Holmiensis, contains but few recipes relating to the working of metals, and these are very similar

---

[6] This apparent duplication is in the text.

[7] Lagercrantz, *Papyrus Holmiensis*, p. 206.

[8] The Greek word φῦκος—sea-weed or algae—is interpreted by Lagercrantz and, as above noted, by Berthelot, as "orseille."

in form and content to some of those of the Papyrus X.[9] One peculiarity of the Swedish work, however, is worthy of note, namely, that recipes which are there given for imitation of silver (argyros), are essentially the same as those given for asem (electrum) in the Leyden Papyrus. This would seem to indicate that at the time of these papyri, from the point of view of these artisans, the two terms were more or less interchangeable, or that they used both terms loosely to indicate the white or nearly white alloys. It is of interest in this connection to note that in modern Greek the word "argyros" and "asemi" both mean silver.

The methods for whitening pearls are sometimes very simple. If they have a brownish tint as if smoked, it is directed to make a solution of honey in water, to add fig roots pounded fine, and to boil down the mixture. Spread it on the pearls and let it harden, then remove it and wipe off with a linen cloth. If the pearls are not yet white, repeat the process. Another method is to mordant or roughen the pearls by letting them stand in the "urine of a young boy," then covering them with "alum," and let what remains of the mordant dry. They are then put into an earthen vessel with "quicksilver" and "fresh bitch's milk." Everything was then heated together, the process being regulated. It was cautioned to apply the fuel externally and to maintain a gentle fire.

This recipe is rendered obscure by the use of the term "quicksilver" in an unusual sense. As suggested by Lippmann, it cannot be mercury, but was probably some finely divided substance of pearly or silvery character, calculated to give the pearly luster.

It is of course pure conjecture that it might have been the silvery particles from the scales of certain fishes, used in much more recent times in the making of artificial pearls,

---

[9] The work of Lagercrantz has been made the subject of a summary with critical commentary by Von Lippmann, the distinguished scholar of early chemical history, and corrections or emendations made by him have been considered where pertinent to this treatment of that work. Von Lippmann's papers are contained in the *Chemiker Zeitung* for 1913, Vol. 37. Cf. Lippmann, *Entstehung und Ausbreitung der Alchemie*, pp. 1–27.

and sometimes called "Oriental pearl essence," and which in suspension in water resembles quicksilver or silver in appearance. It might also have been mineral particles, mica or "glimmer."

The use of trade names for the purpose of concealing the character of the substance used where secrecy seemed desirable was not unknown at that period.

In one of the Egyptian papyri at Leyden, contemporaneous with those we are considering (Papyrus V), there is a passage which says:[10]

"Interpretation drawn from the sacred names, which the sacred writers employ for the purpose of putting at fault the curiosity of the vulgar. The plants and other things which they make use of for the images of the gods have been designated by them in such a way that for lack of understanding they perform a vain labor in following a false path. But we have drawn the interpretation of much of the description and hidden meanings."

The secret names in this manuscript which are placed with the real names are thirty-seven in number. They are such names as the later alchemists used extensively: "blood of the serpent," "blood of Hephaistos," "blood of Vesta," "seed of the lion," "seed of Hercules," "bone of the physician," etc.

It is very probable that the term "quicksilver" in the preceding recipe takes its name from a similarity in appearance rather than from the deliberate attempt to mystify, for these recipes are for the artisan himself, not for the public, but it is also possible that some special constituents of these recipes were intentionally so named as to avoid advertising unnecessarily the more valuable secrets of their business.

The "blood of the dragon" for the red resin of the pterocarpusdraco is doubtless a surviving remnant of the fanciful names used for mystification. The Swedish papyrus has a few other names of the same character, though in

---

[10] Berthelot, *Collection des Anciens Alchimistes Grecs,* Vol. I, p. 10.

general its vocabulary is plain and direct. Thus the Greek word for garlic σκόροδον is used to designate human feces, sometimes used in mordanting wool. The manuscript itself gives this translation.[11]

The term "blood of the dove" used in the papyrus, Von Lippmann has identified from other sources as meaning red lead or sometimes cinnabar.[12]

A curious method given for whitening a pearl is that of causing it to be swallowed by a cock, afterwards killing the cock and recovering the pearl, "when it will be found to be white."

The Swedish papyrus gives us what is apparently the earliest account of methods of making artificial pearls. One recipe is as follows:

Mordant or roughen crystal in the urine of a young boy and powdered alum, then dip it in "quicksilver" and woman's milk.

The word "crystal" often meant with the ancients quartz crystal, but it is very evident that with the authors of these notes the term was used in a more comprehensive sense to include other transparent or translucent stones. This use is very evident in the many recipes for imitation of precious stones, where the processes involve a degree of porosity or absorbent power towards colored solutions not possessed either by quartz crystal or by glass, while certain agates, micas, alabasters or other stones possess this property. In case of the above recipe, it is doubtful whether any such mordanting would in a reasonable time roughen the surface of real quartz crystal adequately. The "quicksilver" here mentioned is evidently the same substance of pearly luster previously referred to.

A more elaborate process for making artificial pearls is the following, suggesting the modern "Roman pearls."

"Take a stone easily pulverized, as glimmer, and pulverize it. Take gum tragacanth and soften it for ten days in

11 Lagercrantz, p. 185.
12 Von Lippmann, *Chemiker Zeitung*, 1913, p. 962.

cow's milk. When it is softened, dissolve it till it becomes thick like glue. Melt Tyrrhenian wax. Take also the white of an egg and "quicksilver." There must be two parts of "quicksilver" and three parts of stone, but of all other materials one part each. Mix (the stone and wax), and knead the mixture with the "quicksilver." Soften the paste in the solution of gum and the contents of the egg. Mix in this way the whole liquid with the paste. Then make the pearls which you wish according to pattern. The paste will soon be like stone. Make deep round impressions and bore them while moist. Let the pearls solidify and polish them well. Treated as they should be, they will excel the natural."

It may be remembered that Pliny speaks of the uses of glass for imitating precious stones, and that he also remarks that "there are books, the authors of which I refrain from mentioning, which give instructions how to stain crystal in such a way as to imitate emeralds and other transparent stones . . . and there are no frauds which bring greater profits."

It is just this art of staining "crystal" which is represented very fully in the Swedish papyrus. There is no reference to colored glass gems as manufactured by the glass workers. This manuscript gives us the detailed explanations which make Pliny's statement more intelligible.

The processes start with some stone presumably cut to form before coloring. The stone whether mica or so-called "crystal," or other stone, is either submitted after cleaning and mordanting to a color bath, whereby color is absorbed into the texture of the stone, or in some cases submitted only to a superficial stain or varnish. It is evident that some of these stains must have been more or less evanescent, depending upon vegetable dyes, while others may have been relatively permanent. It is not to be taken for granted that all the stones used were transparent or colorless before treatment, as many of the precious or semiprecious stones valued by the ancients were not transparent.

The substances used for cleaning and roughening the surface of the stone so as to facilitate the absorption or adherence of the color are various. "Alum," which doubtless comprised as with Dioscorides and Pliny salts of iron as well as of aluminum, is frequently used, although white alum is here often specifically mentioned. Urine is frequently used, its efficiency being doubtless due to the carbonate of ammonium formed on standing. Limewater, sodium carbonate, vinegar, and a solution of sulphur and lime (polysulphides of calcium) are other constituents of the mordanting solutions.

The stones thus prepared are then heated for a considerable time in color baths until the requisite coloring effect is obtained, when they are very carefully cooled to avoid cracking. The staining materials are of both mineral and vegetable origin—copper salts, especially acetate, for green as emerald; alkanna (orcanet) for red as garnet; indigo, used with resin, for "beryls." Pliny says the best beryls are of sea-green color, others are paler, amethystine or yellow. He says that, in India, they have a method of imitating precious stones, particularly beryls, by coloring crystal.[13] Armenian blue (azurite) dissolved in vinegar (yielding copper acetate), dragon's blood, cheledonium, orseille, the bile of the tortoise, or of the calf, or of the ox, are among the colors used, and there are others whose identity it is not easy to establish.

For the preparation of the verdigris, to be used for green stones, the directions are on the same line as described by Theophrastus, and later writers, but more specific.

A well-made sheet of Cyprian copper is cleansed with pumice and water, dried and lightly rubbed with a little oil. It is then hung in a cask over sharp vinegar in such manner that the vinegar does not touch it. The cask is carefully closed to avoid evaporation. If put in in the morning, the verdigris is carefully brushed off in the evening. When put in in the evening, it is brushed off the

---

[13] Pliny, Book XXXVII, Chap. 21.

following morning. The sheet is then returned to the cask, and the process continued till the copper is consumed. But each time that it is removed, a little oil is rubbed on the copper. The vinegar used in the process is rendered useless.

A clearer understanding of the art will be conveyed by a few typical recipes of this character.

To make a garnet.

Dissolve alkanna in oil. Add the "blood of the dove," fine Sinopian earth (essentially ferric hydroxide), and enough vinegar to keep the dye bath sufficiently fluid. Place mica (glimmer) in it, close the vessel and place it for ten days "under the dew." (?) If you wish it very clear wrap horsehair around it, tie it and hang it in the color bath.

It does not seem probable that in this recipe the "blood of the dove" is red lead or cinnabar as interpreted in other connections by Von Lippmann, for neither of these substances would be held in solution by vinegar. It is more probable that it is some vegetable red dye stuff. The value of the ferric acetate produced by the action of vinegar upon the Sinopian earth was perhaps that of fixing or rendering more permanent the color absorbed. Alkanna is the red dye from the roots of *anchusa tinctoria*.

To make an emerald, it is directed to take a stone called tabasis. This is interpreted by Lagercrantz as topaz, but as Von Lippmann suggests, it is more correctly translatable as the stone called tabaschir, an iridescent concretion of practically pure silica deposited in the joints of an Indian bamboo, and which from ancient times was endowed in popular belief with mystical medicinal properties. Its loose structure would permit of its absorbing colored stains, a property not belonging to the topaz. Whatever may have been the stone actually used, the process was as follows: The stone is soaked in liquid "alum" for three days. It is then placed in a solution of verdigris and vinegar (copper acetate), and gently heated for six hours. "Take it out and let it cool slowly, otherwise it will break."

Immerse the stone in oil for seven days. "The resulting stone will equal the natural."

Another recipe for emerald says to take iridescent Indian crystal and form stones from it. This very probably refers to the same stone as the one above-mentioned, and tends to confirm the interpretation of von Lippman. In this recipe the stone when shaped (cut) is immersed for three days in a paste made from alum schist, human feces and vinegar.

Then add vinegar to make the paste fluid, pour it out into a "foreign" pot (imported, and probably strong). Hang the stones in this in a basket so that they do not touch the bottom of the pot and boil gently over the coals. The pot must be covered and sealed with tallow. Blow with the bellows so that the fire may not be extinguished. Heat for two hours. Take then Macedonian chrysocolla and verdigris in equal parts, and the bile of a calf one-half part, and rub them together very fine. Pour on oil from unripe olives as measured by the eye. Then take wax and cover the stones and leave them in oil alone or with addition of Ricinus oil, put in a pot. Again hang the stones in a basket and heat for six hours. Again hang the stones on a horsehair and let them stay in the mixture overnight. Then take them out and you will find that they have become emeralds.

Though somewhat confused as to details, this recipe again depends evidently upon copper acetate for the green color, but uses olive oil as the medium for penetrating the pores or laminations of the stone.

Production of a beryl.

Mix black indikon (indigo—or India ink?) with resin and heat the crystal. If you let it cool in the mixture, it will become excellent beryl.

Preparation of chrysolith.

Heat "crystal," dip it into liquid pitch and cedar oil, and it will become chrysolith.

The chrysolith according to Pliny is a yellow or gold-colored stone. The above recipe merely covers the clear crystal with a yellow varnish.

The following is an interesting laboratory note:

Substitute for ricinus oil.

All crystal becomes dark by boiling in ricinus oil. Do not use, therefore, that material where it says "with ricinus oil," for the material is to be replaced. Use olive oil instead of ricinus oil.

The opening up of the texture of the stone so as to facilitate the absorption of color evidently was a matter of importance. Besides the slight corrosion by processes as above-mentioned, gentle and careful heating was evidently deemed useful. The following recipe is for that purpose:

Loosening up of stones.

Make sure that the stones are receptive and that the dense stones are loosened up. Insert (the stone) into a soft fig, lay it on the coals and the stone will be immediately changed.

The notes on dyeing form an important part of the Stockholm papyrus, and furnish more specific information as to methods and materials employed than any other source of information as to the dyeing processes in use in Egypt in ancient times.

The recipes are almost exclusively devoted to the dyeing of wool. The colors range from purples and reds to rose, yellow, green and blue, though the greater number of recipes have to do with purple. That term with the ancients, included deep red and even red brown as well as purples proper.

It is interesting to note that the purple from the murex, which is discussed at length, though not very clearly, by Pliny, is not used by these dyers. On the other hand, certain of their purples are characterized as successful imitations of the "Tyrian" or the "foreign" (imported) purple.

The processes described cover methods of cleansing the wool and freeing it from fats, various mordanting operations, and the dyeing proper. The dyeing was sometimes in two stages, a preliminary color being first given, and then modified by a second color bath.

For the cleaning of the wool, the customary reagents appear to have been ashes (alkaline carbonates), the water from potter's clay, probably a fine suspension of clay particles, these two substances being usually used together, or there was used a "soap-plant" (*struthion*), crushed and warmed with water. It is directed to put the wool into such an infusion, stir it around a little, take it out and dry. (Pliny describes a plant which he calls radicula "but called by the Greeks struthion." It furnishes a juice, he remarks, that is much used in washing wool, and that it is quite wonderful how greatly it contributes to the whiteness and softness of the wool.) Nitron (sodium carbonate), and clear lime-water, described as obtained by adding water to unslaked lime and after allowing it to stand until clear, pouring off the clear liquor, are other cleansing agents used.

The materials used as mordants are many. Alum, lime-water, milk of lime, ironrust and vinegar, alum and vinegar, nutgalls, solution of the roasted Phrygian stone, misy, copper and iron vitriols, blood-stone (hæmatite) and vinegar, the juice of unripe grapes, and the juice of pomegranates are among the common mordanting substances. The dyestuffs are numerous. For so-called purples were used alkanna (from *anchusa tinctoria*), safflower (*carthanus tinctorius*), komari (comarum palustre), orseille, woad, madder, kermes (a coccus from *quercus coccifera* of Southern Europe), hyacinth (?), mulberry juice, pomegranate blossoms, the root of the henbane (*hyoscyamus*), "krimnos," much used but not at present identified, and other materials. By the use of these singly or in combination and with different mordants, a wide range of colors was obtainable. By the use of some of these same dyes by different treatment, rose, scarlet and blue colors were obtained. A yellow color was produced by crushing together safflower blossoms and oxeye (*buphthalmum*), soaking in water, immersing the wool and drying.

A deep yellow was to be obtained by using gold-colored

litharge and quicklime in specified proportions, covering with water, stirring well and adding the wool with constant stirring. It may be presumed that this was a rather unsatisfactory process, and indeed the directions are followed in the manuscript by the remark that "the color changes after a time. If you add alkanna, the color is better." In another note elsewhere, we find a statement very pertinent to the above: "Lime ground with litharge gives many colors, nevertheless such that the wool does not retain them—first, milk white, then natural (wool-color), then deep (by dyeing in the cold)."

Hints for testing the quality of dyestuffs are given.

Woad should be heavy and dark blue if good, if light and whitish, it is not good. Syrian Kermes—crush those which are best colored and lightest, those which are black or spotted with white are bad. Rub up with soda and dissolve the fine colored.

Rub up the best colored madder and so make the test.

Purple colored and fast orseille is purple snail-colored, but the white spotted and the black is not good.

When you rub up very fine colored orseille, take and hold it in your hand. (A rough color test on the palm of the hand?)

Alum must be moist and very white, but that which contains saltness is not fit.

Of "flowers of copper" that fit for use should be either dark blue, a very green leek-color or in general possess a very fine color.[14]

It will be recalled that the chalcanthum of Pliny and Dioscorides was either blue vitriol, green vitriol or apparently more commonly a mixture of the two, obtained by the weathering of wet iron or copper-pyrites. The above specifications would appear to recognize these varieties of "flowers of copper." Some specimen recipes will perhaps convey a more adequate understanding of the processes employed.

---

[14] Flowers of copper ($\chi\alpha\lambda\kappa o\tilde{v}$ $\check{\alpha}\nu\theta os$), the flos aeris of Pliny, seems generally to be used for the copper oxide. In this manuscript, it seems, however, to be used as synonymus with chalcanthos, the blue or green vitriol. Otherwise, the above characterization would have no intelligible meaning.

A dye bath for three colors.

A color bath from which three colors can be obtained.
Crush and mix with water two thirds parts krimnos and one
part dyer's alum. Put the wool in and it will be scarlet red.
If it is to be leek green, add powdered sulphur with water. If it
is to be quince yellow, add soda and water.

Mordanting for Sicilian purple.

Put in the kettle eight chus of water, half a mina of alum, one
mina chalcanthum, one mina of washed wool. When it has boiled
two or three times, take out the wool, for if you leave it longer,
the purple will become red. Take the wool out and rinse it, and
it will be mordanted.

Mordanting and dyeing of genuine purple.

To the stater of wool, put in the vessel five oboles of alum, two
kotyls of water, boil and let it become lukewarm. Leave it till
early morning. Take it off and cool it. Then prepare a secondary
mordant by putting two kotyls of water and eight drachmas pomegranate blossoms in a vessel. Let it boil and add the wool. After
you have dipped the wool several times, lift it out. To the pomegranate blossom water, add about a ball of "alumed" orseille,
and color the wool as judged by the eye. If you wish the purple
to be dark, add a little chalcanthum and let the wool stand long
in it.

A recipe for mordanting for purple.

After the wool has been mordanted, take twenty drachmas of
good Sinopian earth, boil it in vinegar and add the wool. Add
two drachmas of chalcanthum. Lift the wool and place it in a
kettle of warm water and leave it one hour. Take the wool out
and rinse it.

This process is evidently a supplementary mordanting
with acetate of iron and copper or iron sulphate. The Sinopian earth was essentially ferric hydroxide or oxide.

Another recipe is essentially similar.

Reddle (ferric oxide) dissolved in vinegar produces purple.

This can only mean that this mordant gives purple color
to some dye which otherwise gives a plain red.

Dyeing of Tyrian or guaranteed excellent purple.
  7 drachmas alkanna.
  5 drachmas orpiment.
  1 ounce urine.
  5 drachmas unslaked lime.
  1 kotyl water.

The mordanting and dyeing seem to be here combined in one operation.

Dyeing scarlet.
  Take and mordant the wool with woad which blues it. Wash and dry it. Then take and crush kermes in water until dissolved. Then mix with it domestic orseille and boil. Put the wool in and it will become scarlet.

The foregoing will serve to illustrate the character and content of these two earliest known chemists' manuals. Written in the third century of our era, they nevertheless doubtless embody methods which had been without radical changes in vogue for centuries before, as many statements of Pliny and other writers of earlier date, while not so definite or specific, yet manifestly refer to just such processes.

To what extent these chemical arts originated in Egypt or to what extent they were dependent upon Asia Minor, Persia or perhaps India, it is difficult to determine, for we have no documentary evidence relating to these subjects, which is specific, of established antiquity and demonstrably free from later interpolations.

Traditions of ancient writers attribute some discoveries in these lines to India or Persia, or other Asiatic countries, but as to whether any of these countries contributed in any important way to the development of Egyptian chemical knowledge, or whether at some time these countries learned their arts from Egypt, we cannot safely determine from such tradition. It is quite certain that both in China and in India the chemistry of the metals and alloys, methods of dyeing and the use of certain chemicals in medicine were practiced at ancient periods, but their chronology is difficult to determine with certainty. In so far as western chemistry is concerned, it is generally admitted that the Greeks and Romans received their chemical arts mainly from Egypt.

In so far as concerns the processes described in the two manuscripts considered above, it will be observed that they are severely practical. In general they are easily comprehensible, expressed plainly in the language and common vocabulary of the time.

In treating of the making of gold, silver or electrum, there is no illusion as to any transmutation of the baser metals into precious metals. Their purpose is to produce an imitation that for practical purposes of the jeweler's trade will pass for the more expensive materials and yet will cost less.

The recipes in these manuscripts give evidence of a very considerable empirical chemical knowledge and the practices in the art of dyeing wool are rational and not essentially different from processes in vogue up to the time of the introduction of coal-tar colors or of better dyestuffs of vegetable origin than were known to the ancient world. They are entirely devoid of any evidences of mysticism or occultism which so characterize the writings of the later alchemists. There is no reference to the elements nor to any of the philosophical theories of matter, which were very generally entertained by earlier or contemporary authorities.

It is somewhat remarkable that these notes of an Egyptian artisan, assumed by Berthelot and Lagercrantz to belong to the priestly caste, because in Egypt such arts appear to have been strictly monopolized by them, should contain no traces of the mystery and secrecy with which they invested the practice of their science. The practice of magical arts, and the dependence upon superstitious observances were widely prevalent. But with however much mystery and secrecy these chemical workers may have invested their arts as concerned the uninitiated public, it is evident that there was little if any self-deception as to the nature of their processes.

Considering the character of their methods for "producing" gold and silver, and the claims they make as to their products standing the customary tests for genuineness, it is not difficult to understand why the Emperor Diocletian ordered the destruction of all such works as these under the fear that the standards of monetary value in the Empire might be threatened and that insurrections in the provinces might be financed by the production of artificial gold and silver; for at that time the means of distinguishing the purity of gold and silver, by weight or color or streak on the touchstone, were too imperfect to make sure of the possibility of detection of the fraudulent metal. At any rate, the risk was too great.

With the fourteen loose leaves which constituted the Stockholm papyrus, there was another leaf, not paged with the others, and which may or may not have been a part of the same lot as the two papyri. The writing on this unpaged leaf, though in uncial Greek like the others, is not by the same hand or hands, and the content is very different. All it contains is a magic formula or invocation which translated reads (according to Lagercrantz):

"Sun, Berbeloch, Chthotho, Miach, Sandum, Echnin, Zaguel, protect me while I make the composition. . . . And then annoint thyself and thou shalt observe the result with thine eyes."

The interpretation of this passage by Lagercrantz has been disputed by other philologists and the meaning according to Rubenstein[15] would be "Sun, Berbeloch, Chthotho, Miach, Sandum, Echnin, Zaguel, accept me who come before thee. Trust thyself [to the God], annoint thyself and thou shalt see him with thine eyes."

If, as is not certain, this leaf belonged with the other leaves and was part of the notes deposited in the mummy case of the former owner, the inference would lie near that

---

15 Cf. Lippmann, *op. cit.*, p. 600.

this owner probably belonged to the priesthood and that this was part of a ritual they were accustomed to use when about to perform certain experiments or operations. The words with which the invocation begins appear, according to Lagercrantz, much like the magic words which appear in magical papyri.

Berthelot, from the study of the Leyden papyrus and of other contemporaneous papyri of a nonchemical nature, concludes that the arts of magic and of these chemical arts were practiced by the same persons, though in both these manuscripts the text is free from magical or mystical content. If true, this fact would have very interesting bearing upon the mystical character of the works of later alchemists.

Pliny devotes considerable attention to magic and magicians, and though his historical data are not to be accepted as other than largely legendary, yet they doubtless well represent views prevalent in his times. Speaking of magic, he says:

"That it first originated in medicine, no one entertains a doubt; or that under the plausible guise of promoting health, it insinuated itself with mankind as a higher and more sacred branch of the medical art. Then in the next place, to promises the most seductive and the most flattering, it has added all the resources of religion, a subject upon which at the present day, man is still entirely in the dark.

"Last of all, to complete its universal sway, it has incorporated with itself the astrological art, there being no man who is not desirous to know his future destiny, or who is not ready to believe that this knowledge may with the greatest certainty be obtained by observing the face of the heavens. The senses of men being thus enthralled by a three-fold bond, the art of magic has attained an influence so mighty, that at the present day even, it holds sway throughout a great part of the world and rules the King of Kings in the East." [16]

Pliny attributes the origin of magic to Persia and par-

---

[16] Pliny, Book XXX, Chap. I, Bohn ed. "King of Kings" was the title of the Persian Kings.

ticularly to Zoroaster, supposed by him to have existed some "six thousand years before the death of Plato." Prominent writers on magic, according to Pliny were Osthanes, Pythagoras, Empedocles, Democritus and Plato, and "there is another sect also adepts in the magic art who derive their origin from Moses, Jannes, and Latopea, Jews by birth, but many thousand years posterior to Zoroaster."

Democritus is frequently cited by Pliny in connection with magical arts, and Democritus is a name high in authority with later alchemists. It is interesting to note that in the Stockholm papyrus, one recipe which seems to be a process for purifying copper by fusing with alum and salt is described as having been ascribed by Anaxilaus to Democritus.

Pliny apparently does not associate the arts of magic with the chemical arts, though a writer of a century later, Tertullian, affords evidence that such an association was present in legendary lore. Alluding to the legend of the angels who fell in love with mortal women and married them, and who were supposed to have taught magic arts to man, he says:

"They taught them the secret of worldly pleasures, they revealed to them gold and silver and their working, they taught them the art of dyeing cloths   .   .   .
They laid bare the secrets of the metals, they made known the virtues of plants and the power of magical incantations and described those singular doctrines which extend to the science of the stars." [17]

Among the papyrus manuscrpits in the Leyden collection, is one, Papyrus V, determined on paleographic basis to be of the same period as the manuscripts above-described. It also came from Thebes. This manuscript has been critically studied by Berthelot. It contains two chemical recipes of a character very similar to those in the other works.

One of them is a recipe for purifying gold by treatment with alum schist, salt and vinegar, vitriol and litharge. The

---

[17] Berthelot, *Les Origines de l'Alchimie*, p. 12.

other is a recipe for ink, which is composed of misy, chalcanthum, nutgalls, gum and some other substance designated by Z Z. Except for the last substance, it is a plain black ink, produced by tannate of iron in a solution of gum. The substance Z Z may, as Berthelot thinks, refer to the magical seven flowers and seven perfumes, the letter Z being used for the number 7. However this may be, the same manuscript which contains these two chemical recipes also contains magical formulæ, recipes for philters, incantations, divinations and dreams. It contains the names of Greek. and Egyptian divinities, and the names of Ostanes, Democritus, Moses, Abraham, Zoroaster and Pythagoras, traditional authorities among both magicians and alchemists.

While the recipes are like the others clear and practical, yet again it would seem probable that astrology, magic and the chemical arts were practiced by the same cult, probably the priestly caste of Egypt. And we may also reasonably infer that these operators, however willing they were to deceive others, were not self-deceived in the character of their work, nor confused in these operations to any considerable extent by metaphysical or mystical ideas.

The later chemistry, however, was the product of the influences of these practical chemical arts, combined with the mysticism of Asiatic or Egyptian origin, and the philosophy of the East and of Greece, respecting the nature of matter and the elements which impart to it its varying forms and properties.

The philosophy of the ancients as to the constitution of matter and the changes it undergoes, we will next consider.

CHAPTER III

THEORIES OF THE ANCIENTS ON MATTER
AND ITS CHANGES

From any evidences in the writings of the ancients having to do with chemical knowledge and arts, it would seem that their knowledge was empirical, little guided by theoretical concepts. Yet we are not therefore justified in assuming that theories were without influence, for experience teaches us that some sort of working hypothesis is a necessary accompaniment of progress in any experimental science.

Though the writers upon whose works we are mainly dependent for our knowledge of practical chemistry have little to say of the prevalent theories of matter, yet from other sources we know that speculations on such subjects have earnestly occupied the minds of men since the earliest period of recorded philosophy. Especially in the earliest records of India and of Greece are met serious efforts to account for the origin and changes of the material universe by consistent theories of the nature of matter and its changes.

These two nations developed the most consistent and logical theories, strangely parallel indeed in their development. Scholars are not agreed upon the question as to whether the development of the philosophy of nature in the two ancient civilizations has been entirely independent. Certain it is that, up to the present time, no historical evidence has been discovered which indicates any direct contact of Hindu and Greek thought, though it is not thereby rendered impossible nor even improbable that through Persian mediation Hindu concepts may have found their

way to Greek thinkers, if only in the form of imperfect and
incomplete suggestions. Scholars differ on this probability.
Thus Max Muller[1] says:

"It seems to me that until it can be proved historically
that the Greeks could freely converse with Indians in
Greek or in Sanskrit on metaphysical subjects or vice
versa, or until technical philosophical terms can be discovered in Sanskrit of Greek, or in Greek of Sanskrit
origin, it will be best to accept facts and to regard both
Greek and Indian philosophy as products of the intellectual soil of India and of Greece and derive from their
striking similarities this simple conviction only, that in
philosophy also there is a wealth of truth which forms the
common heirloom of all mankind."

Professor Richard Garbe[2] thinks:

"It is a question requiring the most careful treatment to
determine whether the doctrines of the Greek philosophers
. . . were really first derived from the Indian world of
thought, or whether they were first constructed independently of each other in both India and Greece, their resemblances being caused by the natural sameness of human
thought. For my part, I confess I am inclined toward the
first opinion without intending to pass an apodictic deci-
sion. . . . The historical possibility of the Grecian
world of thought being influenced by India through the
medium of Persia must unquestionably be granted, and
with it the possibility of the above-mentioned ideas being
transferred from India to Greece."

Professor Paul Deussen[3] who with Professor Garbe is
credited by Max Muller with having placed his name in
the front rank of Sanskrit scholars in Europe, is distinctly
of the judgment that the developments are independent, as
for instance, speaking of the Hindu theory of the five elements, he says:

"As in the Greek philosophy of Philalaos, Plato, and Aristotle, so also most Hindu thinkers distinguish five elements,

[1] *Six Systems of Indian Philosophy,* 1899.
[2] *The Philosophy of Ancient India,* 1897, p. 37, 38.
[3] *Allgemeine Geschichte der Philosophie.* Bd. I, 1906–1908.

ether, air, fire, water and earth. Any dependence of the
Greek upon the Hindu or of the Hindu upon the Greek is
not on that account to be thought of, for the reason that
the order of sequence is different in that with the Greeks
air comes between ether and fire. Further because in both
domains independently of each other, the simple observations of nature led to the consideration of the five states
of aggregation of matter—solid, liquid, gaseous, permanently elastic, and the imponderable as the five constituents
of the material world, to which correspond the five specific energies of the organs of sense. Finally in the Greek
as in the Hindu philosophy, the doctrine of the five elements gradually develops from simple conceptions.''

In the face of very striking resemblances in Hindu and
Greek theories of matter, and until historical evidence
shall be discovered, for which we may have long to wait,
scholars will doubtless continue to differ as to the independence or interdependence of the two philosophies according as the remarkable resemblances or the characteristic differences of the two developments impress them more
strongly.

Certain it is, however, that in so far as the two systems
influenced later European natural philosophy, Hindu influences were negligible and the theories of the Greek
philosophers, Leucippus, Democritus, Empedocles, Plato
and Aristotle were directly responsible for any logical
concepts of the changes of matter derived from ancient
times. If indebted at all to Hindu philosophy, it is only
through the Greeks. While entirely beyond the scope of
the present treatment to enter into a discussion of the
various Hindu schools of philosophy and the details of
their highly elaborated theories of matter, it will not be
out of place nor devoid of interest to give some notion of
the Hindu theory of matter in one of the more important
of its various theories.

The working hypothesis of modern physical sciences is
that observed phenomena follow the operation of natural
forces and are not produced by the intervention of super-

natural agencies. They are not due to magic, but are developed in accordance with the operation of laws and processes which are suitable subjects for the exercise of human reason or experiment.

This is not the point of view of primitive races who see supernatural causes in the common phenomena of life. So long as phenomena are satisfactorily explained by invoking the act of will or design by good or evil spirits, no scientific state of mind can exist.

That is one reason why it is of interest to trace the first clearly recorded efforts of man to account for things as they exist in the physical world without the direct act of a will or spirit, but by the operation of unalterable laws.

When, for instance, as in some primitive peoples, and as formerly with more civilized peoples, diseases were considered as sendings of gods or devils, for pennance, punishment, or mischief, there was no inducement for the development of scientific medicine, but charms, amulets, exorcisms and invocations were the proper and logical methods for cure. Only the realization that supernatural influences are not concerned made rational medical science a possibility.

It is these early endeavors to conceive how the world of matter has developed through physical means that interest us here. They are the first beginnings of the spirit of Science, though they are yet far from the method of modern science. Though the assumption of science is that no supernatural will or intervention interferes with cause and effect in nature, it does not follow that, therefore, it is atheistic, for the assumption that a creative intelligence works by natural laws in the natural world permits of the working hypothesis of science as well as does a purely materialistic theory.

The ancient philosophy of India is essentially religious, although Hindu philosophers generally endeavor to account for the material universe through the operation of forces inherent in the ultimate units or atoms of matter. The manifestation of a creative and directing will is also a fundamental concept. Thus it seems that the Hindu philosophy assumes that matter is indestructible and eternal, and motion, also, real or potential, is assumed as eternal. The premise that not any thing can come from nothing and that not any thing can become nothing seems early to have been accepted as a fundamental hypothesis. So also we find very early the idea of some primal substance from which all others are produced quite frequently accepted.

In the Hindu classics of an early period, there appears the notion that water is this primal matter. Thus in the Chandogya Upanishad, quoted by Deussen,[4]

"Only this water in solidified form are this earth, the atmosphere, the heavens, the mountains, plants and trees, wild animals, even to worms, flies and ants—they are all only this water in solidified state."

It is a curious coincidence that the earliest Greek philosopher whose speculations on matter have come down to us, Thales, also held that water is the primal matter, and even as late as the sixteenth century Van Helmont advanced a similar hypothesis on the basis of certain experiments.

Later still in Hindu writings appear references to three elements, fire, water and earth, then a fourth, air, appears. Finally, the number of the elements is accepted as five. The four elements, air, fire, earth, water, are recognizable by the senses, the fifth element, ether, being not recognizable by the senses, but a logical necessity for the manifestation of sound.

It is not possible to state whether the Hindu concepts of the four elements or of the five elements antedated the four elements of Empedocles or the five elements of Philalaos or Aristotle. This is largely because chronological data rarely enter into Hindu literature and the dates of the early classics are difficult to determine, as also the extent of changes and interpolation by later copyists.

---

[4] Deussen, *op. cit.*, I, 2, p. 172.

The atomistic theory of matter appears in well established and elaborated form in various systems of Hindu philosophy, differing in more or less essential characteristics in the various schools. The oldest of these systems which has come down to us in detailed character appears to be that of the *Vaiseshika*, attributed to Kanada, of whom little if anything in particular is positively known as to his life history. Whether or no the atomic theory of Kanada antedates the theory of Democritus, in Greece, is again uncertain. Professor Garbe's opinion is that beyond doubt the Indian theory is a long time after the theory of Leucippus and Democritus. L. Mabilleau,[5] on the other hand, considers the *Vaiseshika* system as several centuries earlier than Democritus. Reasons on both sides are apparently matters of inference rather than of demonstration. The atomistic theory of the *Vaiseshika* is too complex to be adequately presented here. Certain features of it are worthy of presentation for purposes of comparison with the development of the Greek theories.

This theory recognizes nine distinct entities constituting the universe. These are earth, water, fire, air (or wind), ether (akasa), time, space, soul, and "manas." The first four only are distinctly recognizable by the senses, while the fifth, akasa, though not directly recognizable by the senses, yet, as the medium of the transmission of sound, its existence is a necessary inference from data of sense. Time, space, and soul are not material, though existent. The "manas" is the medium through which impressions of sense are conveyed to the soul. The first four, therefore, correspond to the four elements of Empedocles; the fifth, ether, can be compared with little similarity to the ether of Aristotle. The first four elements are composed of atoms which are eternal, never created nor destroyed. Each of these four elements exists as atoms and also as aggregates of atoms. As atoms, they are imperishable.

---

[5] *Histoire de la Philosophie atomistique*, Paris, 1895: *Ouvrage couronné par l'Academie des Sciences morales et politiques.*

The elements which we see or feel are aggregates of atoms and as such are subject to change, but the atoms, which are invisible, do not change. The element earth possesses, as its specific quality, odor, but it also has taste, visibility (color), and may be felt. Water has for its distinguishing quality coolness; it does not possess odor, but visibility, and may also be apprehended by the sense of touch. When water has odor, it is due to the earth present in it, it is not pure. Fire (or light) has no taste nor odor, its specific quality is heat and it possesses visibility. Air without odor or visibility has for its characteristic quality feeling, but not hot like fire nor cold like water, but mild.[6]

Akasa, or ether, is assumed not to consist of atoms, but is infinite in extent, continuous and eternal. It cannot be apprehended by the senses, but is the carrier of sound. It is also described by certain authorities as all-pervasive, occupying the same space that is occupied by the various forms of matter, and therefore devoid of the property of impenetrability, characterizing the atoms of other elements. In this respect, it resembles the modern concept of the ether which conveys light. Deussen quotes from the Upanishad a passage which conveys an idea of akasa as the primal element from which the others were evolved.

"From the Atman (the universal soul or Brahma) arose akasa, from the akasa the wind (air), from wind fire, from fire water, from water earth. When this earth shall pass away, the reverse order of changes will take place, earth to water, water to fire, fire to air, air to akasa, akasa to Brahma."[7]

The atoms of the elements unite to form aggregates, first of two, then three of these double atoms. Thus the visible or tangible elements are formed and so compounds. While single atoms are eternal, aggregates of atoms are subject

---

6 Pliny writes in his *Natural History*, Book XV, Chap. 32: "It is a singular thing that three of the principal elements of nature—water, air and fire—should have neither taste nor smell, nor indeed any flavoring principle whatever."

7 Deussen, *op. cit.*, I, 3, p. 597.

to change, to birth and decay, which characterizes all the material things of the universe.

It would exceed the scope of this work to discuss further the complicated details of the atomic theories of the Hindus, or the variations existing in the different systems.[8] It is, however, pertinent here to emphasize that certain fundamental premises underlie these Hindu theories of matter.

These are; that matter is essentially eternal and indestructible; that matter in its essential constitution consists of a few elementary substances and that from these by combinations all the varied forms of matter in the universe, as well as all organisms have been evolved; that in these elementary particles or atoms, are inherent the properties which endow them with the possibilities of this development, and that this development is independent of any interference from supernatural sources, at least after the creative will has set in motion the process of development.

The Hindu philosophy is not atheistic, inasmuch as the great final source which set in motion the atoms, or which gave rise to ether, akasa, is Brahma or the impersonal soul or will of the universe. Through the soul (atman) which is not material, but yet an entity, the soul of the individual is linked to the universal soul. The atman is like ether and space unlimited and eternal, so that it does not travel from place to place like a material body, but is all pervading. The manas is the medium through which this omniscient and all-pervading atman is interpreted to the sense-impression of the individual.

Materialistic schools indeed evidently did exist in India, but they have left no literature and our knowledge of their existence seems to depend on arguments and criticisms by their opponents.

---

[8] Interesting descriptions may be found in Max Muller, *The Six Systems of Hindu Philosophy;* in R. Seal *The Positive Sciences of the Ancient Hindus;* and in Paul Deussen, *Allgemaine Geschichte der Philosophie,* 2te Auflage, Bd. I,

We shall when considering the theories of matter of the Greeks, have occasion to note how the ideas of the Hindus are in many respects, curiously paralleled, though the course of development is characteristically different.

Greek philosophy of nature, so far as its history has been traced, may be said to begin with Thales in the seventh century before Christ. The early philosophers, Thales, Anaximander, Anaximenes, Pythagoras, Parmenides, Heraclitus, Empedocles, Anaxagoras, Leucippus, Democritus, whose names are connected with the theories of matter and its changes, have left no original literary remains, except in scattered quoted fragments of more or less probable authenticity. For our knowledge of their theories, we are dependent upon later chroniclers and critics.

A more fortunate fate befell the writings of the later and greatest of Greek philosophers, Plato and Aristotle, whose works were so widely copied and so highly estimated that they have in large measure been preserved to our day.

To Plato and Aristotle are we mainly indebted for our knowledge of the physical theories of their predecessors, whose views they present apparently quite fully and fairly while subjecting them to the analysis and criticism of the agreeing or differing points of view of their own philosophic standpoint.

The Greek city of Miletus in Asia Minor furnished a little group of men who considered with seriousness the nature of causes and processes concerned in the development of the material universe. Only fragmentary knowledge of the nature of their speculations has come down to us, though from the brief accounts and references in later writers it is evident that they made an impression upon the thought of the time and contributed largely to the interest of other thinkers in the great problem. These Ionian philosophers are Thales (ca 624–545 B. C.), Anaximander (ca 611–546 B. C.) and Anaximenes who lived at about

550 B. C. Thales is credited in Greek tradition with having traveled in Egypt, with some inventions of theses in geometry, and with having predicted a certain eclipse of the sun. Plato relates that Thales, gazing upward to observe the stars, fell into a stream and was derided by a girl because in seeking what took place in the heavens, he overlooked what lay at his feet. Thus early, at least, was absentmindedness associated with the philosophic mind.

To Thales is credited the theory that the primal matter from which originated everything material is water, that water was the beginning and will be the end of all things. He is said also to have declared that everything is full of divinities. The lodestone has a soul because it attracts iron, and soul is defined by him as that which possesses the power of eternal motion.

From fragments of information such as these we may infer rather than positively know that Thales assumed that matter is eternal, that in the last analysis it is simple—one substance—and that it bears within itself certain inherent powers (souls or gods) by virtue of which the universe of matter is developed. This one simple substance he believes to be water, though why seems to be a matter of conjecture rather than of knowledge.

Anaximander appears to have accepted the same fundamental concept of the essential unity of matter, and of its eternal existence, as did Thales, but differs from his elder townsman in his views as to what that simple primal matter may be. Instead of water, he assumes a qualitatively undetermined primal matter, the *apeiron*. The *apeiron* is eternal and unlimited in extension. It is not any of the known elements; it is possessed of eternal motion, in consequence of which worlds are developed from it in space. As this world has so originated from the *apeiron*, so in time it will again be absorbed into it. There is something suggestive here of the akasa or ether of certain ancient Hindu concepts.[9]

---

[9] *See* ante, p. 110.

Anaximander's *apeiron* may be considered as something analagous to the akasa, though not yet had the idea of the other four elements come to the Greek mind.

Anaximenes, the youngest of the trio from Miletus, and reputed by tradition to be a pupil of Anaximander, follows his alleged master in the concept of a primal matter, unlimited in space and eternal in time which by its inherent energy of motion forms all other matter. Instead, however, of leaving this primal matter qualitatively undetermined, he sees in *air* this simple first substance from which all others are generated. Fire, he thought, was produced from air by a rarefaction process and other substances by condensation processes.

Heraclitus of Ephesus (about 490 B. C.), on the other hand, considered fire as the primeval element, but apparently viewed fire as also the moving and creative force of the universe, as a divinity indeed.[10]

The theories of the Ionian philosophers are not of the nature of scientific theories in the modern sense, for they were not intended as hypotheses to be tested by observation or experiment. They belong to the domain of speculation rather metaphysical than physical, but, let it be noted, very reasonable speculations such as the human mind must content itself with until more specific knowledge admits of scientific deduction. They were attempts to harmonize the evidence of the senses with the demands of human reason, without assuming the arbitrary acts of gods or devils as the causes of phenomena. In other words, they were attempts to account for the visible universe by process of natural law, rather than by supernatural agencies.

The school of philosophers which recognized Pythagoras (ca 570–490 B. C.), as its leader, attempted to reduce the theory of matter to a mathematical and geometrical basis. Pythagoras seems to have been primarily interested in

---

[10] Cf. Clemens Baeumker, *Der Problem der Materie in der Griechischen Philosophie*, Münster, 1890, pp. 19–33.

mathematics and astronomy. In the school of Pythagoras, it is said, the relation of numbers to the musical scale was first discovered. Many geometrical relations were first observed by Pythagoras or his followers. They seem indeed to have been so impressed with the power of numbers and of geometric forms that they endeavored to make these the basis of the physical universe, even, it is related, to the extent of holding that numbers and forms were the only realities. In so far, however, as can be judged by what we at present know, their efforts in this direction brought no constructive idea into the theories of matter and its changes. A Pythagorean follower, Philalaos (probably about 460 or 470 to 400 B.C.) is credited by a writer of a later century with the assumption that the five regular polyhedra determined the particles of the five elements. Thus the earth is made of cubes, water of ikosahedra, air of octahedra, fire of tetrahedra and ether of dodekahedra the most inclusive form of all. If this can correctly be credited to Philalaos instead of to some later Pythagorean, it is interesting as the earliest recorded acceptance in the Greek philosophy of nature of five elements including ether. This formulation is usually credited to Aristotle. Philalaos could easily have obtained the idea of the four elements from his contemporary, Empedocles, but not the fifth element, ether. Pythagoras himself was credited by later writers[11] with having studied magic and occult sciences in Egypt, Arabia and Persia. The Pythagoreans also held the theory of metempsychosis, and practiced mystical rites. In the absence of original writings of this school, it is uncertain what the exact nature of their theory of matter is, but it is evident that, fanciful and metaphysical as it is and in no tangible way connected with reasoning based upon observed phenomena, its tendency is rather confusing than promoting to clear thinking in physics. The historical importance of the Pythagorean

[11] As for instance by Diodorus Siculus, first century B. C., and Pliny the Elder.

concepts of matter lies in the fact that it strongly influenced the views of Plato, and through him emphasized
for many centuries a scholastic rather than scientific attitude toward physical problems.

Four natural philosophers of the fifth century, B.C. advanced ideas which were to leave a deep impress upon
theories of matter for many centuries.

Empedocles of Agrigentum in Sicily (ca 490–430) is
credited with the first announcement of the concept of the
four elements, earth, air, water and fire, as by their combinations forming all other substances in the universe.
Empedocles, like the Ionian thinkers, assumes that matter
is eternal and indestructible, but abandons the idea of the
unity of matter—the materia prima—he assumes for the
elements the attributes of immortality and therefore that
each of the four is through all changes unchangeable in
quantity. All other substances may perish, but they are
merely resolved into their constituent elements. The different properties of all substances which we perceive by our
senses are dependent on the different proportions in which
these elements are combined. As to the causes which
produce these combinations and separations, Empedocles
assumes specific attractions or repulsions which he typi-
fies as love and hate. It does not appear that he considers
these forces as intrinsic properties of the elements, but
rather as eternal forces acting upon them.

Many ideas attributed to Empedocles, as to the development of the universe, including living organisms, are fanciful and would seem to show that he was not a close or
logical reasoner, though we must remember that no writing
of Empedocles has come down to us, and we are dependent
only upon accounts of later authors for what we know of
his theories.

The formulation of the theory of the four elements
credited to Empedocles is however the first clear notion of
elements in a modern significance of the term which is found
in Greek or Western thought. It is namely a clearly ex-

pressed concept that the great variety of substances and bodies which we know are produced by the union of certain elementary units differing in their properties, but not themselves resolvable into simpler constituents.

Adopted with some important changes by Plato and Aristotle, the doctrine of the four elements became the generally accepted theory of matter until the rival doctrine of the three principles, the "tra prima" of Paracelsus, appeared, in the sixteenth century.

Anaxagoras, of Klazomenae in Asia Minor, (ca 500–427) considered the universe as consisting originally of infinite space filled homogeneously with a mixture of small particles, (seeds or as called by Aristotle homeomeria), of infinite variety and infinitely divisible. These particles may be considered as elementary particles of all known substances, air, gold, water, bone, flesh, etc. Upon this uniform but complex mixture acted an intelligence or a will, the "nous." By virtue of the "nous," the particles of like kind are brought together to form any substance which is produced, and when any substance is destroyed or perishes, these substances are again resolved by the nous into their constituent particles. The theory of Anaxagoras owes its historical interest to the abandonment of any attempt to account for the evolution of the material universe by physical properties of matter, and by frankly positing an external though perhaps impersonal intelligence as the organizing and directing force. It was Anaxagoras, says Mabilleau, who introduced the notion of an ordering and directing intelligence as the supreme cause of the universe which after him became the thought of the world and diverted the Greek spirit from the physical to the metaphysical.[12]

To Leucippus and Democritus the Greeks and the Western world are indebted for the first clearly defined atomic theory of matter. Leucippus was the teacher of Democritus,

---

[12] L. Mabilleau, *op. cit.* Cf. also citation from Plato's Phaedo, Trans. of Henry Cary, Everyman's Library, No. 456, p. 158.

and is credited with the origination of the theory, though little is known of him. He apparently wrote nothing and taught only verbally. The dates of his birth and death are unknown. Only from the better known data of his pupil, Democritus, is it inferred with reasonable probability that he was born about 500 or was contemporary with Empedocles and Anaxagoras.

Democritus of Abdera, in Thrace, (ca 460–370) was reputed to have traveled much in Egypt, Persia, Babylonia and even in Ethiopia and India, though these reports, while not improbable are not to be too easily credited. In a fragment of his own which has come down he alludes to a five years' residence in Egypt.[13] Many works by Democritus are named and cited by later writers, though only scattered fragments of not too certain authenticity are at present extant. It is again chiefly upon Aristotle and other commentators that we have to depend for our understanding of the atomic theory of Leucippus and Democritus, but Aristotle did not accept the atomic theory, though he enters quite at length into the analysis of the doctrines which he endeavors to refute. We may assume that in so far as that atomic theory has interest today, our information is fairly reliable.

The theory of the atomists starts again from the assumption that matter is eternal, and that nothing material can originate from nothing, nor can anything material pass into nothing.

They assume, however, that things material in the ultimate analysis consist of very minute but not infinitely small indivisible particles, atoms. These atoms are assumed by Democritus to be of the same kind or substance. qualitatively, but to differ in size, shape, position and presumably also in mass. The atoms exist in a vacuous space which separates them, and because of this space they are capable of movement. This concept of vacuous space was a troublesome idea for the ancient metaphysicians, for if it

---

[13] Cf. Deussen, *op. cit.*, 2, I, p. 137.

was vacuous space only, it was nonexistent, and how could we assume the existence of the nonexistent. The Eleatic school of philosophers (Parmenides and others) had assumed for this reason, that matter must be continuous in the universe. Aristotle later rejects the atomic theory partly at least because of this difficulty of conceiving a vacuum as existent. The atoms of Democritus are, however, capable of motion, and are indeed in ceaseless motion. As to the nature of this motion and the causes of atomic motions, Democritus is not very clear. Later atomists assumed that the cause was collisions as they were falling through space toward the center of the universe, or rising upward, but this concept cannot be traced to Leucippus and Democritus. Aristotle gives us to understand that they consider them to have been from eternity endowed with motion. From the motions of the atoms result their coming together to form combinations, or their separating to decompose substances. From such combinations of these atoms—essentially of the same substance, but varying in size, shape and position—arise all the changing phenomena of the material universe. They are all due to combinations and separations of atoms.

Since with Democritus these atoms are qualitatively the same, the four eternal elements of Empedocles have no fundamental significance. These also are caused by the combinations of the same atoms, and to his interpretation the four elements are merely more common or stable types of such aggregations, and to that extent only to be considered as different from the multitude of other substances.

Empedocles was tending toward the concept of an element as we define an element, Democritus toward the concept of an atom as we understand it, but there was apparently no thought of combining the ideas as we do when we speak of the atom of an element.

The atom of Democritus presents in its relation to the four elements, a certain analogy to the modern concept of the electron in its relation to the atoms of the elements, the elements being more stable aggregations than others, just as our atoms are often considered as relatively stable aggregations of electrons. But a world of experience and exact measurements lies between the metaphysical concepts of Democritus and the atomic theory of to-day.

Probably the concept of atomism could have gone little further than with Democritus so long as exact experimental means of questioning nature were not employed. The atomic theory of matter and indeed the effort to account for the phenomena of nature by physical causes were to lose in interest to the ancient philosophers through the influence of the two greatest philosophers of ancient times, Plato and Aristotle.

This was not because their theories of matter were more advanced than the ideas of Democritus or of Empedocles. Indeed, in a very essential particular, their views were less in line with scientific advance than their predecessors. For Plato and Aristotle were not so much concerned with accounting for phenomena by the operation of properties inherent eternally in matter as they were in interpreting the phenomena of nature as the expression of design, harmony and beauty, as the expression of a directing will and intelligence.

They abandoned the effort to account for physical phenomena by physical forces exclusively, and in this their logic differs from the modern scientific point of view.

It was by the weight of their great authority achieved by their importance in other lines of thought rather than by the merit of their theories of physical phenomena, that these two Greek thinkers acquired their dominion in the theories of matter which endured with increasing authority for nearly two thousand years.

Plato (427–347 B.C.), the great idealistic philosopher of Athens, and for some eight years the pupil of Socrates, contributed little of permanent influence in the specific doctrines of the nature of matter and its changes. Adopting something of the fanciful geometric concept of elementary matter from the Pythagorean school, along with the acceptance of the point of view of Anaxagoras that a directing intelligence was the cause of phenomena, he laid little stress on physical explanations of such phenomena.

His point of view in such matters is well illustrated in considerations which he puts into the mouth of Socrates in the Phaedo.

"Having once heard a person reading from a book written, as he said, by Anaxagoras, and which said that it is intelligence that sets in order and is the cause of all things, I was delighted with this cause, and it appeared to me in a manner to be well that intelligence should be the cause of all things, and I considered with myself, if this is so, that the regulating intelligence orders all things and disposes each in such a way as will be best for it. If any one, then, should desire to discover the cause of everything, in what way it is produced, or perishes, or exists, he must discover this respecting it, in what way it is best for it either to exist, or to suffer, or do anything else; from this mode of reasoning then, it is proper that a man should consider nothing else, both with respect to himself and others, than what is most excellent and best; and it necessarily follows that this same person must also know that which is worst, for that the knowledge of both of them is the same. Thus reasoning with myself, I was delighted to think I had found in Anaxagoras a preceptor who would instruct me in the causes of things, agreeably to my own mind and that he would inform me first whether the earth is flat or round, and when he had informed me would moreover explain the cause and necessity of its being so, arguing on the principle of the better and showing that it is better for it to be such as it is, . . . and if he should make all this clear to me, I was prepared no longer to require any other species of cause." [14]

This point of view is manifestly the antithesis of the standpoint of modern science. This point of view which dominates the views of Plato was shared also by his pupil,

---

[14] Plato's Phaedo, *op. cit.*, p. 177.

Aristotle, so that harmony, beauty, design, logical consistency came to be considered the criteria of the acceptability of theories rather than the data of observation or experiment.

Plato's concept of the nature of the universe is that of a duality, a material body and a soul or intelligence. His notion of matter is not easy to understand. It closely resembles that of Pythagoras, an indefinite something which does not differ demonstrably from space. When portions of this space are enclosed by bounding triangles or squares, the elements are formed differing according to the nature of these bounding surfaces and the resulting form of these elementary bodies. If the bounding surfaces are squares, then a cube results and the element earth is formed, because earth is the more stable or solid element and the cube is the most stable figure of all the regular polyhedra. If the bounding figures are such triangles that a tetrahedron results, fire is the element formed, because the sharpness of the points characterizes the penetrating power of fire. Air is formed of octahedra, water of icosahedra. Conceiving that some mathematical relation must exist between these and because a proportion is the most perfect of such relations, he forms the proportion:

Earth (cube): water (icosahedron):: water: air (octahedron)   :: air: fire (tetrahedron),

a strangely illogical use of mathematics, the absurdity of which has often been emphasized by critics. Manifestly this is all suggested by the Pythagorean concept of the geometric basis of matter. As rearrangements of these enclosing triangles might change the forms of the bodies, it was conceivable that elements might be changed one to another, except the cube which is the only figure bounded by squares, and square surfaces cannot bound other regular bodies except the cube.

This concept while accepting the four elements of Empedocles, yet introduces the idea of a possible change of one element to another. They are not, as with Empedocles, eternal as such, but functions of surfaces liable to re-arrangement. For the motive of any fundamental changes, we are to look to the directing intelligence, not to physical causes.

The four elements by their manifold combinations make up all the material universe. Water, thinks Plato, by heat is converted to vapor and eventually into air; by cooling, on the other hand, it is converted into snow or hail or ice; and under the earth, by heat or cold and pressure, it may be converted into rocks or stones.

The theories of Plato, as expressed principally in his Timaeus, while contributing little of permanent value to science, exerted a great influence upon ancient and medieval notions of matter and its changes, largely through the Neoplatonism of the Alexandrian school of Philosophy. The Timaeus will be considered more in detail in connection with the growth of the alchemical theories.

Aristotle (384–322 B.C.) of Stagyra in Macedonia, a pupil of Plato for some twenty years, developed a theory of matter which starts from Plato's fundamental concept of the reality of ideas and the less reality of material phenomena. Ideas are eternal, matter is subject to change. The study of the laws of nature with Aristotle was as with Plato the attempt to fathom the design of the universe, to show that it is for the common good and that its phenomena are in accord with the demand of the human mind for harmony and logical order. With Plato, he accepts the four elements of Empedocles, but rejects the Pythagorean idea of geometrical relations as accepted by Plato. He rejects the assumption of Empedocles of the eternal nature of the four elements, believing them capable of changing from one to another. He rejects the atomic theory of Democritus partly because he cannot conceive as logical the existence of a vacuum and hence the atoms with their inherent motion must be rejected. Matter, he holds to be continuous and to be indefinitely divisible, therefore again there can be no atoms whether in the Democritean or the

Pythagorean sense. He looks, as do Anaxagoras and Plato, to a world intelligence directing the development of the universe, and his efforts are not to trace physical effects to physical causes so much as to interpret relations so that they may seem intelligent, harmonious, logical.

The Aristotelian concept of the universe of matter is very elaborate. He assumes that the universe, including the heaven of stars, is spherical, that the earth is the center and that the universe revolves around this center. The universe is eternal in time but not indefinite in extent. Outside of the sphere of the universe there are no such things as space or time. It is spherical because that is the perfect form and representative of perfection, uniformity and eternity.

The four elements as such are subject to change. There must be something, however, back of these that is eternal and unchangeable. What this is, with Aristotle, it is not easy to understand. It is apparently not merely space as Plato seems to think, but something with at least latent power. It may be considered not as matter, for then it would be only another form of matter; perhaps the nearest interpretation is that it is the potentiality of matter.

The kinds of matter are five, an ether being added to the four elements of Empedocles and Plato. This ether is, however, not supposed to exist as a constituent of substances of this world, but to be the substance from which are formed the heavenly bodies and the sphere of the heavens in which these are set. This ether is eternal and unchangeable. Below the zone of the heavens lies the zone of fire, lightest of the four elements, and below this the air, and then water between the air and the earth which is the heaviest of the four. Characteristic motion is the property of the five elements. The most perfect motion is circular and this belongs to the ether, which has no tendency to approach the center of the universe nor to fly away from it, and the circular motion belongs to the eternal and unchangeable. All other motions may be resolved into combinations of circular and rectilinear motions, and to the four elements belong characteristic rectilinear motions, which would naturally be either toward the center of the universe, if intrinsically heavy, or away from the center if intrinsically light. Thus earth and water have motion toward the center or are heavy, while fire and air have motions away or are essentially light. This explains their existence in the relative positions they occupy in the four zones. The ether with its circular motion has no tendency either to approach or recede from the center of the universe and therefore is neither light nor heavy.

All natural things in this middle zone which we inhabit consist of mixtures of the four elements, in varying quantities. Thus the element water is not water as we know it, nor the element air the same as the air we feel. These are substances in which the real elements, water and air, predominate. Nor is the Aristotelian idea of combination of the elements the same as that held by Empedocles, nor by us at the present time. We conceive the various elements in a combination, however intimately combined, as still existent quantitatively unchanged, so that if we have the necessary power or skill, we may recover them unchanged in quantity from their combinations. Aristotle, however, considers these elements as combinations of certain qualities rather than as definite masses of unchangeable substances. The elements themselves may be converted into other elements by modifying the relations of their properties. Thus Aristotle considers water as an element possessing two qualities which constitute it water, viz., coldness and moisture (or liquidity). Air, as an element, is characterized by warmth and moistness; earth by coldness and dryness (or solidity); fire by warmth and dryness. So if water for instance can have its quality, coldness, converted to warmth, it would become the element air. The familiar phenomena of evaporation and boiling probably gave color to such an explanation.

These properties, cold, moist, dry, warm, are by Aristotle apparently considered as forces, which are pair-wise antagonistic forces, and if any one of them overcomes its opposite, the elements themselves are changed.

This curious notion of the nature of the elements and the fact that there are just four elements in the terrestrial zone of the universe, Aristotle arrives at somewhat in this way. The only absolute criterion of the existence of matter is the sense of touch. Sight and hearing are subjective phenomena dependent upon our senses, liable to errors of interpretation. The phenomena which affect the tactile sense may be analyzed into four elements, hot and cold, moist and dry. All other properties, color, odor, roughness, smoothness, he asserts are either nonessential or combinations of these four. From these four properties there may be made six pairs:

Cold and moist	Warm and dry
Warm and moist	Warm and cold
Cold and dry	Moist and dry

The last two pairs, however, are contradictory; the first four are the only possible combinations in matter, and these evidently constitute the four forms of elementary matter, and of these warm and dry characterize fire; cold and dry characterize earth; cold and moist characterize water; and warm and moist characterize air.

When these four elements combine to form the many substances that make up the material universe, their properties then blend into a composite in which the elements lose their identity. Aristotle makes it clear that he considers compound bodies homogeneous even in their smallest conceivable parts, so that the ultimate particle of flesh is still flesh. This is also the idea of Anaxagoras, already cited. To these simple substances of like particles Aristotle gives the name "homoiomere." It logically follows that the concept of the four elements of Aristotle differs fundamentally from that of Empedocles, for the smallest particle of a given substance would, by the theory of Empedocles, be ultimately divisible theoretically into elementary particles or atoms which are no longer the same substance as that from which they are separated. The four elements of Aristotle are manifestly not elements, either in the sense of Empedocles or in the modern sense of the definition.

The above is not a complete statement of the theory of matter of Aristotle, but will, it is hoped, give an idea of the elaborateness and complexity of the Aristotelian concept, and serve to illustrate how far removed was his method of developing the theory from the inductive methods of modern science. The concept of the four elements as qualitative factors in the constitution of other bodies, with their inherent forces of heat, cold, moist, dry, became accepted by later centuries as basic truth. His notion of a fifth element, variously interpreted, also held a place in the thought of later times, but his more complex notions of the nature of the elements and matter had little influence on the later development of natural philosophy.

The teleological point of view of Aristotle was in harmony with the doctrines of the great religions which dominated the thought of later centuries—Christianity, Mohammedanism, as well as of the older Hebrew theology —and this fact had much influence in maintaining the great authority of Aristotle into the period of the Renaissance. His influence on the development of physical science was probably on the whole rather retarding than stimulating, in that it tended to emphasize the interpretation of phenomena according to preconceived notions of fitness or design rather than by a rigid logic based on the determined facts or observed phenomena of nature. It emphasized the metaphysical rather than the physical considerations.

The Aristotelian theory of the elements according to which any element might be changed to another by changing one of its inherent qualities, hot, dry, etc., to its opposite, apparently helped to keep alive with the alchemists the hope of changing base metals into precious metals, a belief in the first instance dependent on failure to understand the nature of changes involved in processes employed for imitating gold and silver by cheaper alloys.

Indeed Aristotle himself seems to make a similar interpretation of changes, where speaking of making bronze, consisting of copper and tin, he states that the tin (or tin ore as kassiteros may have meant), vanishes almost entirely as if it were an immaterial condition of the resulting bronze, and escapes leaving behind with the copper a color only.

Aristotle marks the end of Greek influence upon the development of theories concerning the nature of matter and its changes. After his time, Greek philosophy spread in increasing circles, but in so far as the theories we are considering are concerned it lost rather than gained in interest and in clarity of thought.

The Stoics rejected the idealism of Plato and the teleological point of view of Aristotle, adopting a materialistic philosophy. Matter and nature they considered as eternal and even the soul was material. They however contributed nothing to constructive theories of matter or nature.

Epicurus (342–270 B.C.), revived the atomic theory of Democritus, though the efforts of his school to expound or develop it, appear not to have been very successful. Their theory is expounded very fully by the Latin poet Lucretius in his *De Rerum Natura*. Indeed it is said that it was this work that inspired Gassendi in the seventeenth century to revive the Democritan atomic theory as part of his campaign against the authority of the Aristotelian philosophy of nature.

The most notable feature of the Epicurean theory was an attempt to endow the atoms with a property which should account in the evolution of organic life and of man, for the accepted fact of free will. It attempts this by assuming in the atoms that their motions are due to gravity and therefore would be in parallel vertical lines, never colliding, except for the assumed fact that they have an inherent power of tending to swerve slightly, a beginning of volition. Lucretius thus interprets ideas:[15]

"When first bodies (atoms) are being carried downward straight through the void by their own weight, at times quite undetermined and at undetermined spots, they push a little from their path: yet only just so much as you could call a change of trend. But if they were not used to swerve, all things would fall downward through the deep void like drops of rain, nor could collision come to be nor a blow brought to pass for the first beginnings, so nature would never have brought aught to being. . . . Once again if every motion is always linked on and the new always arises from the old in order determined, nor by swerving do the first beginnings make a certain start of movement to break through the decrees of fate, so that cause may not follow cause from infinite time, whence comes this power of freedom for living things all over the earth, whence I ask is it wrested from fate, this power whereby we move forward, where our will leads each one of us, and swerve likewise in our motions neither at determined times nor in a determined direction or place, but just where our mind has carried us?"

Upon his contemporaries, the Epicurean atomic theory seems to have exerted little influence, and the same seems to be true of its revival by Lucretius. For writers of following centuries who are not philosophers seem to take no interest in the atomic theory, but follow Plato or Aristotle.

How generally the theories of Aristotle were accepted by the public at about the time of the beginning of the Christian era by Greek and Roman writers, is evidenced by allusions in prominent writings of the time on many subjects, though it must be admitted that the forms in which these ideas had been assimilated seem to have been empirical and elementary.

For instance, Diodorus of Sicily, Greek historian of the first century B.C. in describing the customs of the Egyp-

---

[15] Lucretius, *On the Nature of Things*, translation of Cyril Bailey, Oxford, 1910, p. 72 *ff.*

tians, attributes to them the personification of the elements with properties attributed to them which are apparently loosely formulated Aristotelian qualities. After mentioning Osiris and Isis as gods typifying the sun and moon, Diodorus says:

"They say that these gods in their natures do contribute much to the generation of all things, the one being of hot and active nature, the other moist and cold, but both having something of the air, and that by these all things are brought forth, and nourished, and therefore that every particular being in the universe is perfected and completed by the sun and moon whose qualities as before are five: a spirit of quickening efficacy, heat or fire, dryness or earth, moisture or water, and air, of which the world does consist as a man is made up of head, hands, feet and other parts. These five they reputed for gods; and the people of Egypt, who were the first that spoke articulately, gave names proper to their several natures according to the language they then spoke. They, therefore, called the spirit Jupiter which is such by interpretation because a quickening influence is derived from this into all living creatures. . . ."[16]

While the personification of the four elements as deities may well have been in Egyptian mythology earlier than Aristotle, yet the description of qualities of the elements are manifestly Aristotelian, though inadequately reproduced. That religious beliefs of oriental origin in which the elements are personified are older than Aristotle, and even than Empedocles, the earliest proponent of the four elements as constituents of matter, is evident because Herodotus (484–424 B.C.), a writer contemporaneous with Empedocles, in discussing the customs of the Persians, states that they make sacrifices to Jupiter "which is the name they give to the whole circuit of the firmament," and also to the sun, moon, to earth, fire, water and wind.

Strabo, the Greek writer on geography (ca 64 B.C to 20

---

[16] Diodorus Siculus, *Historical Library*, Book I, Chap. I, translation of G. Booth.

A.D.) in referring to the universe as a whole—this refers
to the zones of which earth is the central element, fire the
outermost and air and water the intermediate—says, "and
particularly in view of the hypothesis by which the four
bodies which of course we also call elements are made
spheres." [17]

The author of *Ten books on Architecture,* Vitruvius
Pollio, of the first century, B.C. alludes in places to the elements, as for instance in the following historical sketch:

"Among the seven sages, Thales of Miletus pronounced
for water as the primordial element in all things, Heraclitus
for fire, Euripedes, a pupil of Anaxagoras, and called by
the Athenians 'the philosopher of the stage,' for air and
earth. . . . But Pythagoras, Empedocles, Epicharnos
and other physicists and philosophers have set forth that
the primordial elements are four in number—air, fire, earth
and water—and that it is from their coherence to one
another under the moulding power of nature that the
qualities of things are produced according to different
classes." [18]

Again from Vitruvius:[19]

"For while all bodies are composed of the four elements,
that is, of heat, moisture, earth and air, yet there are mixtures according to natural temperament which make up the
natures of all the different animals of the world, each after
its kind.   Therefore, if one of these elements, heat, becomes
predominant in any body whatsoever, it destroys and dissolves all the others with its violence. . . ." Again,
"The reason why lime makes a solid structure on being
combined with water and sand seems to be this: that rocks
like all other substances are composed of the four elements.
Those which contain a larger proportion of air are soft,
of water, are tough from the moisture, of earth, hard, of
fire more brittle.   Therefore, if limestone without being
burned is merely pounded up small and then mixed with

---

[17] *Geography of Strabo,* translated by H. L. Jones, 1916, Vol. I, Book I,
3, 12, p. 205.   The suggestion of the translator that the above refers to the
Pythagorean concept of spherical atoms is far-fetched.
[18] Vitruvius, *op. cit.,* Book VIII, Introduction.
[19] *Op. cit.,* Book I, Chap. IV.

sand and so put into work, the mass does not solidify nor can it hold together. But if the stone is first thrown into the kiln, it loses its former property of solidity by exposure to the great heat of the fire and so with its strength burnt out and exhausted, it is left with its pores open and empty. Hence the moisture and air in the body of the stone being burned out and set free and only a residuum of heat being left lying in it, if the stone is then immersed in water, the moisture makes its way into the open pores, then the stone begins to get hot, and finally after it cools, the heat is rejected from the body of the lime.'' [20]

These attempts of Vitruvius to account for observed phenomena on the basis of an imperfectly comprehended Aristotelianism, would hardly have been approved by Aristotle himself. They serve to illustrate, however, how the fundamental ideas of matter of Aristotle were accepted as the basis upon which facts of experience must be explained if at all.

Pliny also in his *Natural History,* while he is not much concerned with this class of considerations, yet also evidently accepts the Aristotelian concepts as they had become conventionalized in his day.

''I do not find it doubted [he says], that there are four elements, the highest being fire, whence the eyes of so many shining stars, next that spirit which the Greeks and we call by the same name, air, that vital substance permeating all things and mixed in all, by the force of which, the earth and the fourth element, water, are balanced in the middle of space.'' [21]

In his theory of the development of the universe, Pliny follows the Stoics in discrediting the directing intelligence as adopted by Plato and Aristotle.

''The universe (mundus) and by whatever other name we please to call the heavens (coelum), by the vault of which all things are enclosed, is to be believed a divinity (numen)—eternal, without bounds, never created and

---

[20] Vitruvius, *op. cit.,* Book II, Chap. V.
[21] Pliny, *op. cit.,* Bohn ed., Book II, Chap. IV. Cf. also ante p. 110, footnote.

never to perish. To enquire what is beyond it, is no concern of man, nor can the mind of man form any conjecture respecting it. It is sacred, eternal and without bounds, all in all, indeed including everything in itself, infinite yet like what is finite; the most certain of all things, yet like what is uncertain. Externally and internally embracing all things in itself, it is the work of nature and itself is nature." [22]

These illustrations will serve to indicate very clearly how in about three centuries after the time of Aristotle, the Greeks and Latins had incorporated into the common thought of the period an apparently well conventionalized belief in the actual existence of the four elements with their characteristic qualities as constituting the great variety of substances making up the material universe. At this epoch, it does not appear, however, that there was any considerable question or serious dispute concerning the authoritativeness of these theories. Like insects in amber, those ideas derived from the natural philosophy formulated by Aristotle were preserved by custom and tradition until a time many centuries later, when the accumulated experimental data and new points of view which had been acquired invested the problems of the constitution of matter with fresh interest. It may be recalled that so late as the seventeenth century, Robert Boyle in writing his "Sceptical Chemist," considered the surviving faith in the four elements an object worthy the weight of his trenchant criticism.

In the domain which is covered by modern experimental sciences, the point of view of the ancients as compared with the present, is much the same as expressed by a student of the history of medical science, "The Greek process of reasoning was observation, speculation, deductive hypothesis; while the modern method is observation, experience, inductive conclusions." In medicine, Dr. Magnus points out that the Greek method of reasoning prevailed

---

[22] Pliny, *op. cit.*, Bohn ed., Book II, Chap. I.

from the sixth century B. C. to the nineteenth century A. D.[23]

For chemistry the same may be said, except that the modern point of view may be said to have been fairly well inaugurated by Robert Boyle in the seventeenth century A. D.

---

[23] Dr. Hugo Magnus, ''Der erkenntnis-theoretische Prozess in der vorhippokratischen Naturauffassung besonders bei Alkmaeon.'' In *Beiträge aus der Geschichte der Chemie*, herausgegeben von Paul Diergart, 1909, p. 59 *ff.*

CHAPTER IV

THE EARLY ALCHEMISTS

The chemistry of the ancients, as expressed by the writers from Theophrastus to Pliny and Dioscorides, was thoroughly practical. Their theories of the origin and changes of matter were based on their interpretation of the four elements as constituents of matter, principally as formulated by Plato and Aristotle.

There was no attempt at classification of phenomena or theories of chemistry in ancient times. There was no name to distinguish facts or ideas which we call chemical. The Greek word "Chemeia" first made its appearance in about the fourth century, A. D. and appears then to have been used to designate the arts of metal working particularly with reference to the supposed making of gold and silver from base metals. This supposed art does not seem to have been known to Pliny, nor does it appear that that art was known to other writers of his time. The two papyri from Thebes are the earliest manuscripts which give us any knowledge of the practices which seem to have given rise to the notion of transmutation of base metals into gold and silver, and these documents do not convey any idea that the practitioners were troubling themselves about any theories of transmutation. They were occupied in making alloys just as good, though very probably they knew no reason why their products under proper conditions might not turn out to be real gold or silver or electrum.

Other writers of about the same or of somewhat later date whose writings have been preserved to us in manuscripts in copies of about the eighth to eleventh centuries,

were however, wholly convinced of the reality of transmutation. The earliest allusions to the art call it the sacred
art, or the divine art, and the word "Chemeia" gradually
replaced these, and under later Arabian modification became "alchemeia," a word therefore of Greek origin with
the Arabic article prefixed. Primarily applied to the
processes supposed to be used for transmutation, the term
"alchemy" came ultimately to include the arts of chemistry
in general.

The origin of the word "Chemeia" has been the subject
of much discussion. Zosimos, an Alexandrian Greek alchemist of about the end of the third or the beginning of
the fourth century A. D. relates a myth which accounted in
his belief for the origin of the word. According to this
legend, the sacred or divine arts were revealed to man by
angels who fell from their high estate through their love
for mortal women. These secrets were revealed in the
book called Chemu, the book of Chemes or Chymes, whence
he says the art is called Chemeia. This Chemes is, however, not a historical personage and later scholars place
no credence in any basis for the legend. It is considered
probable that Chemeia was derived from the Greek word
χημΐ (Chemi) signifying black. Whether because of the black
soil of the Nile Valley, which gave to the Greeks the name
Chemi or Kemi for Egypt, or because of a "blackening"
which the early alchemists sometimes mention as a preliminary stage to the yellowing or whitening in the
"making" of gold or silver, is not certain.[1]

Certain it is that, by about the fourth century, the word
was used to designate the art of making the precious
metals from base metals, the actuality of which was the
common belief of the alchemists.

The actual basis for the belief in transmutation consisted in just such operations as we have seen illustrated
in the two Theban papyri. That these arts in Egypt were

---

[1] Cf. Hofmann in Ladenburg's, *Handwörterbuch der Chemie*, Bd. 2, Article
"Chemie," and especially V. Lippmann, *Entstehung und Ausbreitung der Al-
chemie-Herkunft des Namens Chemie*, p. 293 *ff.*

originally under control of the priesthood and by them were carefully guarded and surrounded with secrecy and mystery seems beyond question. The testimony of early writers and of legends and traditions point to Egypt as the source of the earliest notions on the sacred art. The legends and myths of early alchemy, however, give evidence also of influences from Persian, Chaldean and Hebrew sources as well as Egyptian and Greek.

All this points to Alexandria as the probable locality where the ancient alchemy took form and developed into a cult. When Alexander the Great conquered Egypt in 330 B.C. and his general Ptolemy became King of Egypt, the Greek city of Alexandria was founded, and soon became not only the most important city of Egypt, but through the foundation of schools and the accumulation of libraries became the acknowledged center of the intellectual world. The collection of manuscripts is estimated at from 400,000 to 500,000 works. Scholars from all parts of the then civilized world thronged there to take advantage of its books and its teachers. The culture which developed was a blending of Greek, Egyptian, Chaldean, Hebrew and Persian influences. Greek philosophy, Egyptian arts, Chaldean and Persian mysticism met and gave rise to strange combinations not always conducive to improvement upon the relative clarity of the Greek foundation.

As the power of Rome grew, Greek and Egyptian power declined. Egypt became a Roman province in 80 B.C. A fire, started, it is recorded, from ships burning in the harbor during Caesar's conquest of Alexandria, burned an important part of the collection of manuscripts of the Alexandria libraries. Under the Roman Empire, Alexandria, however, still exerted great influence and in the reign of Augustus was a metropolis second only to Rome itself, but in the succeeding centuries when Rome was suffering from internal disintegration and the Roman Empire was crumbling from successful barbarian invasions, Alexandrian culture also yielded to the general de-

moralization. In the third century, the conditions throughout the Empire were such as to justify the statement of a
competent critic—"In the tempest of anarchy during the
third century A.D. the civilization of the ancient world
suffered final collapse. The supremacy of mind and of
scientific knowledge won by the Greeks in the third century
B. C. yielded to the reign of ignorance and superstition in
these social disasters of the third century A.D." [2]

In the light of present knowledge, it was in the period of
the first to the third centuries that the mystical cult which
cultivated the fantastic ideas of that kind of chemical philosophy which later came to be called alchemy, first developed. The beginning seems to have been the development of a secret cult of Alexandrian mystics bound by oath
never to reveal to the uninitiated the mysterious knowledge
which they claimed to have. That the members of the cult
were originally of the Egyptian priesthood or foreign
scholars initiated by them, seems probable, for Egyptian
deities or mythological personages are prominent as
authorities in their writings. That the cult was of comparatively late development is evidenced by the prominence
of Persian, and Hebrew authorities which were also
frequently cited in their early writings. All this points
to the cosmopolitan influence of the Alexandrian schools,
the melting pots of Greek, Egyptian, Hebrew, Persian and
Chaldean philosophies, sciences, religions and superstitions. The universal sway of the Roman power and the
pax Romana had also the effect of spreading the various
cultures and national religions, but at the same time of
weakening their authority.

In the early centuries of our era, Rome and Athens contained temples of Egyptian Isis, and shrines to Mithra, the
Persian sun god, were frequent in Greek and Roman cities,
symptoms of a decline in the power of the ancient religions
in the centers of civilization under the Empire.

There was rising also the new and at first persecuted

[2] J. A. Breasted, *Ancient Times*, p. 674.

sect of Christians destined soon to supplant the old faiths.
Recognized and protected early in the fourth century under
the Emperor Constantine, the new sect as it gained influence waged war upon the schools of ancient pagan philosophies.   In 389 A.D. the Serapion of Alexandria was destroyed, and its library destroyed or scattered under an
edict of Theodosius calling for the destruction of all
pagan temples within the Empire, an order executed
with much severity and cruelty.   In the same year,
Zeno, Emperor of the East, closed the important school
at Edessa and its Nestorian teachers were banished,
finding refuge in Asia.   The Museum of Alexandria,
a real university, still maintained a precarious existence
until 415 when in riots incited by the Christians, the last
remnants of Alexandrian schools of philosophy and science
were swept away and the last notable teacher and philosopher of that school, Hypatia, fell a victim to the violence
of the mob.

The frequently repeated assertion that the library at
Alexandria was destroyed by Amru, the Arabian conqueror
in 640 A.D. is a story that lacks basis of truth.   The partial
destruction by fire during Caesar's siege, the ruin occasioned by disciplinary measures under Aurelian 273 A.D.,
the mandate of Diocletian ordering the destruction of all
books relating to the working of metals for fear of the
debasement of the currency, and the destruction of the
Serapion and the Museum above alluded to, had doubtless
left little to be destroyed.   Indeed the Arabs at that time
seem not to have been disposed to destroy but rather to
protect the remains of ancient science.   The story seems to
be based upon the narrative of an Arabian historian, Ibn
Khaldun, concerning the conquest of Persia.   The com-
manding general asked the Caliph Omar what was to be
done with a mass of books there found, and the Caliph is
reported to have answered "Throw them into the water.
If they contain anything of truth, we have received from
God a better guide.   If they contain falsehood, we are well

rid of them.[3] This story whether true or false has been by error transferred to Alexandria.

With the suppression of the schools of ancient science and philosophy and the banishment and scattering of their savants and disciples, scientific activities in the Christian countries became for many centuries dormant. The up-building of the doctrines and organization of the Christian Church dominated during the early middle ages the philosophy of life of civilized Europe and absorbed the attention of its scholars. The influence of the church was during that period not conducive to the advance of natural or physical science. Not indeed on account of any active hostility to natural science as such, but because of two fundamental points of view which under the influence of the early fathers as St. Adrian and St. Augustine dominated Christian thought. To the church of that day, this earthly life was only of importance as a discipline and preparation for the life after death. Only those things were worth while which were necessary preparation for the life to come and for the avoiding of the eternal torments of the unredeemed. What mattered, therefore, such trivial matters as the nature of the material universe and the laws and causes pertaining to it? In the second place, the neoplatonic philosophy of the late Alexandrian school which dominated whatever remained of the philosophy of nature itself tended to discourage the scientific inquiry into the physical causes of observed natural phenomena. This tendency was owing to the fact that this philosophy encouraged the belief in the mysterious and occult as complicating factors in the simplest and most ordinary events. When things mystical or miraculous might always be present in phenomena of the universe, there was little stimulus to study the operations of physical laws upon the continuity or invariability of which dependence could be placed.

Thus the study of nature from the scientific point of

---

[3] Cf. Friedrich Dannemann, *Die Naturwissenschaften in ihrer Entwicklung, etc.*, Leipzig, 1910, I, p. 223.

view was neglected. Even ancient Greek science with its early attempt at scientific reasoning became almost forgotten and its literature neglected. Speculations as to natural phenomena were largely confined to endeavors to harmonize observed facts with the Scriptures, or with their interpretation by church authorities, for the scholarship of Europe was largely absorbed in the problems of theology. Those arts were largely studied which were in harmony with the intellectual and emotional motives in religious life—logic, rhetoric, dialectics, grammar, etc.

Some attention was given to the study of arithmetic and geometry. Natural sciences—astronomy, botany and zoology—received some attention from a classificatory point of view, but the writings upon these subjects were curiously mingled with fabulous and mystical matter. Anything that may be considered as any material revival or continuation of the scientific interest in the study of nature in Christian Europe was to wait until the twelfth and thirteenth centuries.

It may well be that even the science as developed by the ancients, except in its practical applications, might have been lost to the world had its continuity not been maintained through other channels than the newly developing Christian civilization, so devoid of any scientific literature are these early centuries of Christian Europe.

The traditions of the ancient pagan schools and their literature were, however, preserved and cultivated especially by the Syrian scholars who took refuge in Persia, after the closing of the Alexandrian schools, and there founded and maintained schools modeled after the Alexandrian. By these scholars, the classical works of Plato, Aristotle, Galen, Dioscorides and others, and of some early chemical and alchemical writers, as the pseudo-Democritus and Zosimos, were preserved and translated into Syrian. Astronomy, astrology, medicine, alchemy, were among the subjects taught in their schools.

When the Mohammedan invasion of Asia Minor took

place, these Syrian scholars were patronized by the Caliphs, were employed in influential positions as physicians, astronomers, mathematicians, engineers, etc., and the Syrian manuscripts of Greek and Alexandrian authors were translated into Arabian. The early Mohammedan culture was more hospitable to these ancient sciences and philosophies than the early Christian, and thus Arabians became in medieval times the best trained scholars in mathematics, astronomy, medicine and chemistry. As the wave of Mohammedan conquest in the seventh and eighth centuries swept over Egypt and Morocco to Spain, Spain became the seat of a high degree of Mohammedan culture which endured until the final expulsion of the Moors in 1492 put an end to the Moslem rule in Western Europe. From Spain, however, the classical culture preserved by Syrian scholars and by them transmitted to Arab scholars, found its way to Europe, and Arabian mathematicians, physicians, alchemists, were held in high esteem as scientific experts. Arabian translations, elaborations and commentaries from ancient Greek and Greek-Egyptian authors received from Syrian versions and finally translated into Latin in the twelfth and thirteenth centuries, became the great authorities in natural science. So completely had the original Greek writings disappeared from sight in the middle ages of Europe that later centuries quite generally assumed that the Arabians were originators of very much that they had acquired and transmitted from original Greek and Alexandrian writers through Syrian and Arabic translations. Particularly was that true in the field of chemical knowledge, though modern research has made it clearer that the additions in that domain to the knowledge possessed by Alexandrian writers of the third and fourth centuries is of very subordinate significance. In the history of chemical science in Europe, Arabian influence is of importance because it was through this channel that interest in the science was again introduced to Latinized Europe. As previously noted, it was in Alexandria at about the beginning of our

era, so far as we can ascertain, that that phase of chemical
activity and speculation which we call alchemy originated.
The earliest alchemical writers whose writings have been
in part at least preserved to us were manifestly Alexandrian Greek-Egyptians. They wrote in Greek and their
writings contain allusions and traditions connecting with
the ancient Greek philosophy of nature, with Plato and
Aristotle, but also allusions and ideas related to Jewish,
Persian and Egyptian culture. In so far as these writings
contain references to the devices and methods of experimental chemistry, these early alchemists allude to just such
practical operations as we have seen in the Egyptian papyri
from Thebes, although they are rarely so definite and clear
as the latter descriptions and directions, and are mingled
with a confused mass of obscure allegorical narratives and
descriptions. These find their analogies in the fantastic
notions of the later Alexandrian neoplatonic philosophers
and related mystical cults belonging to the transition period
of the fall of the Egyptian and Greek culture and the
rise of the Christian philosophy with its mixture of traditions and ideas from many different ancient cults and
religions.

   Internal and external evidence are to the effect that the
phase of chemical activity and interest which so long held
the stage not only in Europe but in Arabia and Asia,
spreading even to India and China, had its origin in the
practices of the metal workers of Egypt and in the theories
of matter and its possible changes as developed in the neoplatonic school of natural philosophy.

   In so far as the neoplatonic philosophy as applied to
alchemy possessed a basis in ancient Greek philosophy, it
was based mainly upon Plato's conceptions as formulated
in his work entitled "Timaeus."

   This metaphysical physical science of Plato, imaginative
and fantastic in itself, became even less logical and more
fantastic by the elaborations and interpretations of the later
neoplatonists who "based their philosophy on revelations

of Deity and they found those in the religious traditions and rites of all nations."[4]

As the Timaeus of Plato appears to have furnished the more fundamental concepts which dominated the ideas of matter and its changes to the early and later alchemists, it will be of help in understanding some of these ideas if this work is explained in some detail.

In the form of dialogue, though substantially a monologue, Timaeus is represented as explaining to Socrates his formulation of the generation and development of the physical universe.

It will be remembered that the inductive method of modern science is not the method of Plato. The criteria which justify his conclusions are their reasonableness to the human mind. Ideas are the realities, the changing phenomena of the physical universe are but their transient images. Very illustrative of Plato's attitude in this respect is his discussion respecting the origin of the universe.[5]

"Now as to the whole heaven or order of the universe, . . . we must first ask concerning it the question which lies at the outset of every inquiry, whether did it exist eternally, having no beginning of generation, or has it come into being starting from some beginning? It has come into being, for it can be seen and felt and has body. And all such things are sensible and sensible things apprehensible by opinion with sensation belong as we saw to becoming and creation. We say that what has come to be must be brought into being by some cause. Now the maker and father of this all it were a hard task to find and having found him, it were impossible to declare him to all men."

Questioning as to whether this maker created the universe upon the model of the eternally existent or upon the transient material thing, he says:

"If now the universe is fair and its artificer good, it is plain that he looked to the eternal, for the universe is

---

[4] Harnack and Mitchell, Encycl. Britannica, (11th ed.), "Neoplatonism."
[5] Citations from "Timaeus" are taken from the English translation by R. D. Archer-Hinds, Macmillan and Co., 1888.

fairest of all things that have come to be, and he is the most excellent of causes.

"If then, Socrates, after so many men have said divers things concerning the gods and the generation of the universe, we should not prove able to render an account everywhere and in all respects consistent and accurate, let no one be surprised, but if we can produce one as probable as any other, we must be content, remembering that I who speak and you my judges are but men, so that on these subjects we should be satisfied with the probable and seek nothing further."

How fundamentally this point of view differs from that of modern science and how accordant it nevertheless is with the greater part of medieval logic in such matters, it is needless to emphasize. Plato places all the emphasis on deductive logic, and his employment of inductive logic is almost subconsciously applied, so little effort is made to control his notions of the causes of things on the basis of observed facts. He is mainly endeavoring to interpret the will of the creative power through his own ideas of harmony, beauty and beneficence.

"Because the Artificer saw that nothing could be fairer than that which has reason, and that without soul reason cannot dwell in anything," Plato deduces "that the universe is a living creature in very truth possessing soul and reason by the providence of God." Because to Plato a sphere is the most perfect figure, the universe is spherical, and because it is made in the image of the eternal, that is of God, it is one and alone. Because rotation on its axis is the most perfect motion, it is so established, and since for this rotation there is no need of feet, he made it "without legs and without feet." "—for its excellence, it was able to be company for itself as acquaintance and friend. For all these things, he created it a happy god."

Confining our attention to those concepts more directly related to subsequent neoplatonic and alchemical views of physical phenomena, it is to be noted that he first formulated notions of the four elements, which, elaborated by his great pupil Aristotle, gave to later times one of the most influential concepts of the nature and changes of matter. The assumption of four elements was at least as early as Empedocles, but his less imaginative ideas were not the ones that directly influenced the neoplatonists of Alexandria.

Plato explains why there should be four and only four elements in a very characteristic logic. After assuming that the universe must be material because it is visible and tangible, he proceeds:

"Apart from fire and light, nothing could ever become visible, nor without something solid, could it be tangible, and solid cannot exist without earth; therefore did God when he set about to frame the body of the universe, frame it of fire and of earth. But it is not possible for two things to be fairly united without a third, for they need a bond between them which shall join them both. The best of bonds is that which makes itself and those which it binds as complete a unity as possible, and the nature of proportion is to accomplish this most perfectly. For when of any three numbers whether expressing three or two dimensions, one is a mean term, so that as the first is to the middle, so is the middle to the last, then since the middle becomes the first and the last, and the last and first both become middle, of necessity, all will come to be the same, and being the same with one another, all will be a unity. Now if the body of the universe were to have been made a plane surface having no thickness, one mean would have sufficed to unify itself and the extremes, but now since it behooved it (the universe) to be solid, and since solids can never be united by one mean, but require two, God accordingly set air and water betwixt fire and earth, and making them as far as possible exactly proportional, so that fire is to air as air is to water, and as air is to water, water is to earth, thus he compacted and constructed a universe visible and tangible. For these reasons and out of elements of this kind, four in number, the body of the universe is created, being brought into concord through proportion; and from these, it derived friendship, so that coming to unity with itself, it became indissoluble by any force save the will of him who joined it. Now the making of the universe took up the whole bulk of each of these four elements. Of all fire and all water and air and earth, its framer fashioned it leaving over no part nor power without.''

Taking these concepts of the nature of the four elements into consideration in connection with the more logical though hardly less imaginative concepts of the elements by Aristotle, it is not difficult to understand that at a period when the ideas of Plato were more determinative of the philosophy of the time than were the ideas of Aristotle, the concept of the nature of the four elements was vague and mystical. Following Pythagoras, Plato conceives of a kind of geometrical basis of the constituting units or particles of the four elements and of the different character of the bounding surfaces of these units as determinative of the four elements. By the breaking down and rearrangement of these bounding surfaces (triangles) he explains why one element may be changed into another, a fact which he accepts as confirmed by experience. The elements are not constant in their properties, but there are different kinds of all the elements.

''Next we must remember that of fire there are many kinds; for instance, flame and that effluence from flame which burns not but gives light to the eyes, and that which remains in the embers when the flame is out. And so with air, the purest is that which is called by the name of ether, and the most turbid is mist and gloom, and there are other kinds which have no names, arising from the inequalities of their triangles. Of water there are two primary divisions, the liquid and the fusible kinds.''

Plato seems to consider that anything that naturally exists as a flowing liquid is a water of the liquid kind, while everything that can be made to flow by the action of heat is a water of the fusible kind; for example:

''Of all the substances which we have ranked as fusible kinds of water, that which is densest and formed of the finest and most uniform particles, a unique kind of brightness of a yellow hue is gold, a most precious treasure which has filtered through precious rocks and there congealed. . . . Another has particles resembling those of gold, but more than one kind; in density it even surpasses gold and has a small admixture of fine earth so that it is harder but lighter because it has large interstices within. This formation is one shining and solid kind of water and is called copper (χαλκός). The earth which is mingled with it when the two through age begin to separate again becomes visible by itself and is named rust" ("ios," that is, verdigris).

Throughout the writings of the alchemists even to the seventeenth century, we find allusions to "waters" and to the congealing of waters in the earth to form rusts or metals, the source of which are plainly to be traced to these curious speculations of Plato. Plato leaves no doubt as to his belief that these four elements are not absolutely distinct substances but that they may be changed from one to another and that they are not to be too definitely characterized.

"For it is hard to say which of all these we ought to call water any more than fire or indeed which we ought to call by any given name rather than all and each severally. . . . In the first place what we now have named water, by condensation as we suppose, we see turning to stones and earth, and by rarifying and expanding this same element becomes wind and air; and air when inflamed becomes fire; and conversely fire contracted and quenched returns again to the form of air; also air concentrating and condensing becomes cloud and mist, and from these yet further compressed comes flowing water, and from water, earth and stones once more."

It will be remembered that Aristotle also conceives of the four elements being transmutable and as substances are made up of these four elements, it is not difficult to understand how the followers of these theories entertained the possibilities of almost any kind of change in the nature of substances if the appropriate agencies or influences might be supplied.

Aristotle in characterizing the properties of the four elements laid great emphasis upon their four constituting qualities—hot, cold, moist and dry. That Plato also associated these properties with the elements is evidenced from the following passage concerning the causes of disease.

"Now the cause whence sicknesses arise is doubtless evident to all. For seeing there are four elements of which the body is composed, earth, fire, water and air, any unnatural excess or deficiency of these or change of position from their own to an alien region, and also, since there are more than one kind of fire and other elements, the reception by each of an unfitting kind, and other causes, all combine to produce discord and disease. For when any of them changes its nature and position, the parts that formerly were cool are heated, and those that were dry become moist and the light become heavy, and all undergo every kind of change."

It may be remembered that Aristotle in his development of the qualities of the elements, discarded the qualities light and heavy as nonessential or as not inherent. The medical theory of disease which during the middle ages and indeed well into the Renaissance was most authoritative, was that of Galen (Claudius Galenus, born ca 121 A.D.) which was largely founded on the conception that conditions of health or disease were determined by normal or abnormal proportions of the four humors, blood, phlegm, yellow and black bile, these being related by metaphysical analogy to the four Platonic-Aristotelian qualities, cold and warm, dry and moist.

The foregoing sketch gives but very incomplete description of the physical basis of the "Timaeus," but will serve to indicate the more important concepts which were particularly influential in determining the fundamental theories of medieval chemistry or alchemy, concepts which were indeed dominant in chemistry at least until the sixteenth century, though gradually supplemented by ideas developed from more practical chemical experiments.

Unreal and fantastic as were the theories of Plato upon

the nature and change of matter, they were nevertheless devoid of any mysticism or unreason such as dominated the natural science of the neoplatonists of the earlier centuries of our era. They were the product of the speculations of a brilliant intellect attempting to fathom the plan of the creator of the universe, under the belief that man has no surer guide for this task than to follow the indications of his own sense of the harmonious, the beautiful, and the desirable, and "that on these subjects we should be satisfied with the probable and seek no further." But the neoplatonists were no longer strict disciples of the Greek philosophers with whom sane reason was characteristic though often imperfect and in error. Egyptian secrecy and mysticism, the superstititous observances and beliefs of Chaldeans, Hebrews and Persians had introduced faith in astrology, in the magic influence of numbers, in exorcisms and invocations, so that the Greek rationalism was well-nigh obscured. The mystical sects which developed in the early centuries of the Christian Church contributed not a little to intensify the factors which tended to diminish the rational development of critical study of causes and effects in nature.

The earliest alchemical writers of whom we have literary remains and of whom we have any items of personal history, as Zosimus, Synesius, Olympiodorus, who lived in about the third to the fifth centuries, belonged to the cult of Gnostics whose traditions and observances rested largely upon a foundation of Jewish, Chaldean and Egyptian mysticism and Alexandrian neoplatonism, and were also influenced by the mysticism of the early Christian Church. This fact has been established by the researches of G. H. Hoffman[6] and confirmed by M. Berthelot[7] and E. von Lippmann.[8]

This sect, which flourished from about the first to the sixth century, is characterized by W. Bousset[9] as composed

[6] Ladenburg, Encyclopedia, art. "Chemie," p. 529.
[7] *Les Origines de l'Alchemie*, Chap. III, p. 57 *ff*.
[7] *Les Origines de l'Alchimie*, Chap. III, p. 57 *ff*.
[9] Encyclopedia Britannica, 11th ed.

of members who "all lived in the conviction that they possessed a secret and mysterious knowledge in no way accessible to those outside, which was not to be proved or propagated but believed in by the initiated and anxiously guarded as a secret. This knowledge of theirs was not based on reflection, on scientific inquiry and proof, but on revelation." Certain it is that a great part of the writings of these earliest Greek-Egyptian alchemists are well described in these terms, as we shall later have occasion to illustrate.

Of the beginnings of development of the cult of Egyptian chemists, doubtless of the priestly caste, to which the original owner of the Theban manuscripts at Leyden and Stockholm probably belonged, we have no definite knowledge. The traditions of the early alchemists name many personages as authorities in the secret and sacred art, many of them doubtless mythical in so far as their connection with chemical arts are concerned. Thus Hermes is commonly referred to as the original founder of the art of alchemy. Hermes was the Egyptian deity called by them Thoth, legendary patron of the arts and sciences. An incredible number of works are said to have been written by him, including works on astrology and magic, and later imposters wrote works which they ascribed to him. The designation of chemistry as the hermetic art is due to this legendary reputation. Also Isis, whose worship had extended from the Egyptians to the Alexandrian Greeks and even to Rome, is associated by legend with alchemy. Another name prominently connected with early alchemy is Ostanes, said to have been a magus-priest and philosopher attached to the court of the Persian king, Xerxes. Another, also named Ostanes, figures as one who practised magic and alchemy at the time of Alexander the Great.

Moses, Miriam the Prophetess, alleged sister of Moses, and Aaron, Cleopatra, Egyptian priestess, not to be confused with the queen of that name—though she also has been asserted to be an adept on the strength of the story of the pearl dissolved in vinegar—these names and others are cited frequently by early alchemists with assertions respecting certain sayings, but nothing definite is known regarding their alleged connection with alchemy or chemistry.

These traditions are chiefly of interest as illustrating how the origin of alchemy is associated by tradition with Egyptian, Persian and Hebrew names, corroborating the evidence that the cult originated at the time when the traditions of these nations were blended with the Greek in the Alexandrian Neoplatonic schools.

The first name which appears to represent a chemical expert whose writings have been preserved fragmentarily in quotations or copies by later writers, is that of Democritus. This person is generally called by alchemical writers Democritus of Abdera, the philosopher who first enunciated an atomic theory. Internal and external evidence, however, make it clear that Democritus, the alchemist, has little in common with the philosopher of Abdera, and that this psuedo-Democritus lived at about the beginning of our era and belonged to the Alexandrian school of neoplatonists. The exact time of his life is unknown. H. Kopp[10] considered that his work, *Physica et Mystica,* was written not earlier than the third century A.D.

Berthelot considers it at least as early as the papyrus of Leyden which was written probably in the third century though evidently copied from earlier writings. Democritus was referred to as a great authority by Zosimus (third century), thus giving the impression that he was earlier than his time.

It may be recalled that Pliny, citing Democritus frequently, refers to a prevalent belief that there were two writers of that name, a belief, however, that he personally was not disposed to credit, attributing all to Democritus of Abdera.[11] Another writer contemporaneous with Pliny,

[10] H. Kopp, *Geschichte der Chemie,* 1843–1844, II, p. 152.
[11] *See* ante p. 25–26.

Columella (died about 65 A.D.), stated that a certain Bolos of Mendes was a writer of the school of Democritus and attributes to him the production of many writings accredited to Democritus of Abdera.[12]

It is, therefore, probable that this earliest alchemical writer of whom we have identifiable writings lived somewhere about the beginning of our era. Whether his name was really Democritus or whether he used that distinguished name to give greater prestige and authority to his writings, as was the practice with many other unknown writers in later periods, we do not know, though the statement of Columella indicates an early belief in the unauthenticity of the writings as ascribed to Democritus. In later periods other unknown writers wrote treatises which they endeavored to pass as works of Hermes, Geber, Lullus, Aristotle, Albertus Magnus, Paracelsus, etc.

There are in existence in manuscripts of dates not earlier than the tenth century and some much later, in Greek and Syrian, quite a number of writings ascribed to this Democritus. In general they are in part practical recipes for alloying or coloring metals to imitate gold or silver, or for dyes, resembling closely the recipes to be found in the papyri of Leyden and Stockholm, and in part mystical, allegorical or symbolic allusions to the art of transmutation, ostensibly intelligible to initiates in the mystic cult, but manifestly intended rather to impress the reader with belief that the writer is the possessor of occult knowledge which he cannot make clearer to the unitiated reader.

The practical recipes of the pseudo-Democritus differ only from the Theban papyri in their less simple and plain directions. They are the same in their intentions of imitating gold and silver by yellow and white alloys of copper, lead, tin, mercury and arsenic; by colored mixtures or varnishes or stains to be superficially applied to give a surface resemblance to gold or silver; and the materials

---

[12] Cf. Berthelot, *op. cit.*, p. 156.

for these recipes when clearly stated are the same as we find in the papyri. There are also recipes for gold "elixirs" and silver "elixirs."

The earliest alchemical work in existence is probably the *Physica et Mystica* of Democritus.[13] Its authenticity rests on the fact that it is cited with great respect by the early Greek alchemists. The earliest manuscript in which it is known is the manuscript of St. Marks of the tenth or eleventh century,[14] though manifestly existing certainly before the fourth century, and probably in some form much earlier. Berthelot has published the text of this work from the manuscript of St. Mark with translations.[15]

This work so well illustrates the twofold basis of the ancient alchemy, the Egyptian practical art, with the mystical obscurity of the secret cult, that it will be worth while to quote it in part.

The treatise begins with a recipe for dying wool that closely resembles some in the Stockholm papyrus.

"Take, to a pound of purple [dyestuff] a weight of two oboles of scoria of iron, macerated in seven drachmas of urine. Place on the fire till it boils. Then removing the decoction from the fire, place the whole in a jar. First withdrawing the purple, pour the decoction upon the purple, let it soak a night and a day. Then taking four pounds of marine lichens [that is orseille] add water until the water is four fingers deep over the lichen and leave it till it thickens; then filter, heat, and pour it on the wool prepared beforehand. Squeeze the loose wool so that the liquid may penetrate thoroughly; then let it stand two nights and two days. Finally let it dry in the shade. The liquid is poured off. Take the same liquid and to two pounds of this liquid add water to reproduce the original proportion. Keep it till it thickens; then, having filtered, put in the wool as at first, and leave it a night and a day. Then take it out, rinse in urine and let it dry in the shade."

---

[13] Cf. Kopp, *Beiträge zur Geschichte der Chemie,* I, p. 128.

[14] Berthelot, *op. cit.,* Chap. VI.

[15] *Collection des Alchimistes Grecs,* I and II, Greek Text, p. 41, translation, p. 43. Kopp published the Latin translation of Pizimenti, *Beiträge zur Geschichte der Chemie,* I, p. 137 *ff.*

Another similar recipe for dyeing in purple follows the treatise, and then comes a brief note on dyestuffs for purple:

"Here is what enters into the composition of purple: the alga which is called false purple, the coccus, the marine dye (orseille), orcanette of Laodicea (anchusa), crimnos (the unidentified dyestuff often mentioned in the Stockholm papyrus), madder of Italy, the phyllantheon of the west (or of the divers ?) the purple worm, Italian pink. These colors have been estimated above all others by our predecessors. Those which do not give fast colors are of no value. Such are the coccus from Galatia; the color from Achaia called lacca, that from Syria called rhizion, the mollusk and double mollusk of Libya, the mollusk called pinna, from the maritime region of Egypt, the plant called isatis, and the dye from upper Syria called murex. These colors are not fast nor valued by us except that from isatis."

These technical notes and recipes are strikingly similar to those we have already quoted and discussed from the two papyri. They might have come from just such laboratory notes of the same period, and if not always clear to us owing to vocabulary difficulties, they are at least free from mysticism.

The next succeeding paragraphs in the manuscript are, however, very different and entirely unrelated to the foregoing.

"Having received these ideas from our master previously mentioned, and recognizing the diversity of matter, we are obliged to harmonize their natures. But our master dying before we were initiated, and at a time when we were still occupied with the knowledge of matter, we were told it would be necessary to attempt to evoke him from Hades, and I forced myself to attain this end by evoking him directly with these words, 'By what gifts dost thou reward that which I have done for thee.' After these words, I remained silent. After invoking him several times and demanding how I could harmonize the natures, he replied that it was difficult to speak without permission of

the daemon (genius), and pronounced these words only—
'The books are in the temple.'

"Returning to the temple, I set about seeking to become possessor of these books, for he had while living
never spoken of the books, dying without testamentary disposition. He had, as had been supposed, taken poison to
part his soul from his body, or, as his son declares, he
swallowed the poison by mistake. But he had intended
before his death to show these books to his son only when
he should have become of age. None of us knew of these
books. Since after seeking we had found nothing, we
would have given much to know how substances and their
natures unite and are blended. When we would have effected the composition of matter, the time having arrived
for a ceremony in the temple, we made a festival together.
Then as we were in the shrine[16] of the temple, suddenly a
certain column opened, but we could see nothing within.
Now neither he (the son) nor we had been told by any one
that his father's books had been so deposited. Advancing,
he led us to the column and we saw with surprise nothing
revealed save this precious formula that we found there—
Nature rejoices in nature, nature triumphs over nature,
nature dominates nature."

The above narrative is so entirely different from what
precedes that it is quite probable as Berthelot has suggested
that they are not parts of the same original writing.

The *Physica et Mystica* then proceeds:[17]

"I also come to bring to Egypt the doctrine of the things
of nature, so that you may be raised above the curiosity of
the vulgar and the confusion of matter.

"Take mercury, fix it with the (metallic) body of magnesia or with the (metallic) body of stimmi from Italy, or
with sulphur apyre (native sulphur), or with aphreselinon
(selenite), or burned limestone, or alum of Melos, or with
arsenicon or what you will. Place the white earth (so
prepared) upon copper (χαλκός, copper or bronze), and
you will have copper without shadow (brilliant). Add yel-

---

[16] ναός, innermost part of a temple; cell.
[17] The Latin text as published by Pizementi and reproduced by Kopp, *loc.
cit.*, begins at this point.

low electron and you will have gold, with gold you will have chropocolla reduced to metallic body. The same result will be obtained if you use yellow arsenicon or sandarach properly treated, and cinnabar wholly transformed. But mercury alone produces the copper without shadow. Nature triumphs over nature.''

The intention of this recipe is very clear. It is a process to give copper or bronze a superficial silver or gold color by the use of mercury alloys or arsenic alloys. In detail, it is ambiguous, largely because the substances used are often named by terms which had no very definite significance with the ancients. Thus ''magnesia'' included white lead, ''cadmia'' (crude oxide of zinc), and the ''body of magnesia'' would then mean metallic zinc or lead, making white alloys with copper. Italian stimmi is the native sulphide of antimony, and the ''metallic body'' of this would be metallic antimony, which as we have seen, the ancients considered a kind of lead. Cinnabar wholly transformed was probably, though not certainly, metallic mercury. The yellow arsenic and sandarach ''properly treated'' probably meant roasted and reduced to metallic arsenic, which also gives a white surface to copper, though there is no evidence that the ancients ever separated the metallic arsenic.

It is also probable that additional obscurity is due to the desire to avoid making the directions clear to the uninitiated public. The use of the substances called gypsum, burned limestone, and alum (which also meant a variety of acid-reacting salts), was probably for the purpose of keeping metallic surfaces free from oxide or other films interfering with perfect contact with amalgams or other alloys.

The following recipe is obscure probably by reason of the use of conventional terms intended to conceal the real substances from general knowledge.[18]

---

[18] Concerning these secret or ambiguous names for inorganic or organic substances, compare E. von Lippmann, *op. cit.*, pp. 15, 28, 225. The fashion among the early alchemists of so concealing the nature of their materials from the public seems to have been inherited from the ancient Egyptian medical practice, as illustrated in the *Papyrus Ebers*, ca 1500 B. C.

"Whiten according to usage cadmia of Cyprus. I refer to that which has been refined. Then make it yellow. You may make it yellow with the bile of the calf, or turpentine, or ricinus oil or radish, or with the yolk of eggs, all substances which turn it yellow. Then apply the mixture to the gold. For gold is obtained by means of gold and the liquor of gold. Nature triumphs over Nature."

The intention of the recipe may have been, as Berthelot suggests, to give a gold color by yellow varnishes to white metals. It would exceed our limits to dwell further upon the technical recipes in the *Physica et Mystica*. The greater number deal with processes for imitating gold and silver by baser alloys, by superficial coloring of white metals or copper, and by superficial varnishes on white metals. They resemble the recipes of the papyri already given, though in general less specific or clear, and interspersed with mystical expressions.

One further extract illustrative of a style of talking which finds many imitators in the later alchemists, even of the seventeenth and eighteenth centuries, will be not without interest. The passage is entirely disconnected from that immediately preceding it in the manuscript, which is a recipe for tincturing a white metal with the color of gold.

"O Nature, producer of Natures, O Nature which charms Natures in marvellous ways. Such are the things which concern great Nature. There are no other natures superior to those in the tinctures; there are none equal nor inferior. All these things are effected in solution. O my colleagues in prophecy, I know that you have not been inclined to unbelief, but to admiration, for you know the powers of matter, whereas the young people are confused and place no faith in what is written because they are dominated by their ignorance of matter, not knowing that the children of medicine when they wish to prepare a medicament proper for a cure do not attempt to make it in thoughtless haste, but first try what substance is warm, what other substance is cold or moist, and in what condition it should be to favor a mean mixture. This is the way that they prepare the medicine destined for the cure. But those who propose to care for the soul and the deliverance from all pains do not perceive that they will be hindered by proceeding with a haste void of discrimination or reason. Indeed, believing that we are employing fabulous and not symbolic language they make no test of different kinds of substances to find out for example if such a kind is useful for cleaning; such another as accessory; such a one for coloring; such a one to effect complete combination; or if such a kind is good to give brilliancy. They do not ascertain if such a substance will resist the action of fire, and if such another by its addition will render a body more resistant to fire; thus, for instance, how salt cleanses the surface of the copper and even its internal parts, and how it corrodes the external parts when scraped, and even its internal parts. And finally, how mercury whitens the surface of brass (aurichalchum) and cleans it, and how it whitens the internal parts (i. e. when alloyed); how it is eliminated from the surface and how it can be eliminated from the internal part. If the young people were trained in these matters they would not go astray in the preparations they undertake. They do not know that one kind of substance alone can be transformed into as many as ten kinds of contrary natures. Indeed one drop of oil may make disappear a great quantity of purple, and a little sulphur can consume many substances.''

The above extract like many passages in later alchemistic writings is obviously intended to impress the reader with the importance of the knowledge possessed by the writer and other adepts, and does indeed convey the impression that these people were more or less familiar with chemical operations while conveying no definite information as to methods or applications that could be of practical utility to the reader.

The various writings—Greek, Syrian or Arabic—which are attributed to Democritus may have been much added to or modified by their translators or copyists in the course of centuries. Accepting them on their face, however, as

THE STORY OF EARLY CHEMISTRY

the chemical knowledge of this pseudo-Democritus, they show him as a person with a wide experience of chemical operations, and of the substances used in ancient days in the arts of chemistry.

A great number of recipes attributed to Democritus are given, of which many are clear, and the purpose evident, many are apparently matters of fact but the descriptions are not clear. In many the purpose is not plain, and others are intentionally mystifying, many processes essentially simple being made complicated by reason of operations which probably find their basis in superstitions. An illustration of the last-mentioned kind is, for example, in the preparation of "our cinnabar."

"Take mercury and put it in a marmite of clay, with native sulphur above and beneath the mercury. Cover with a clay cover and seal with a lute resistent to fire. When the lute is dry, heat it in a glass furnace three days and nights. After this, take the marmite and you will find a red substance. Take this, work it, grind it in sea water, expose it to the sun for three days and let it dry. When finally dried, expose it to the sun with urine from an infant at breast, during sixteen days and as many nights. Dry it and put in a glass vessel. Preserve it for use. This is our cinnabar." [19]

The use of a name for a reagent which is intended to mislead or to conceal the truth from those not adepts, may be illustrated in the following directions for imitating the emerald.

"Take white lead (cerusa) one part, and of any glass you choose two parts, fuse together in a crucible, then pour the mixture. To this crystal add the urine of an ass and after forty days you will find emeralds."

Assuming that the desired green color of this brilliant lead glass was derived from copper, as is probable, the copper derivative used is masked under the designation of asses' urine. [20]

[19] Berthelot, *La Chimie au Moyen Age*, II, p. 31.
[20] Berthelot, *op. cit.*, p. 29.

Among the multitude of recipes of a more matter of fact character, the following may illustrate.

How copper becomes white like silver.

Clean the copper properly and take mercury and white lead (cerussa), rub it strongly, and the color will become like that of silver.[21]

Diplosis of gold.

Take a mithgal of soft copper of Cyprus, ten mithgals of gold, ten mithgals of silver and fifteen of salammoniac. Scrape the metals and put them in a crucible. Fuse them and put them into water of couperose, it will come out good gold.[22]

This is a recipe for gold alloy retaining the color of a purer gold. The ammonium chloride evidently was for the purpose of cleansing the metals to facilitate alloying.

Fabrication of asem.[23]

Fix according to custom the mercury obtained from arsenikon (orpiment) or sandarach (realgar), or prepared as you know how; project it upon copper and iron treated with sulphur and the metal will become white.

The same effect is produced by magnesia whitened, arsenikon transformed, cadmia calcined, sandarach unburned, pyrites whitened, and cerussa digested with sulphur. You can soften iron by mixing with it magnesia or a small portion of sulphur, or a little magnetic stone, for the magnetic stone (lodestone) has an attraction for iron. Nature charms Nature.[24]

Here again we have the superficial whitening of copper by the action of reduced arsenic, and by various other substances which by reduction give white metals, as zinc and lead. The ambiguity attending the nomenclature of minerals renders the interpretation sometimes uncertain.

The earliest alchemical writer whose personal identity is

---

[21] Berthelot, *op. cit.*, p. 28.
[22] Berthelot, *op. cit.*, II, p. 67. Berthelot's translation from the Syriac manuscript gives sal ammoniac. This would imply an interpolation of about the period of these manuscripts as no such salt was known to the time of Democritus himself.
[23] "Asem" used as in the Stockholm papyrus to designate silver.
[24] Berthelot, *Collection des Alchimistes Grecs*, II, p. 53.

known is Zosimos, called the Panopolitan, or Zosimos the
Theban. He appears to have written and taught in Alexandria and lived about 300 A.D.. There exist quite a
number of writings attributed to him. He is credited by
later writers with having been the author of an encyclopedic work on alchemy, and writings now extant may be
disconnected fragments of this work.

Zosimos is in his philosophy and chemical knowledge and
points of view very similar to pseudo-Democritus whom he
often cites with evident respect. Like the latter, he seems
to be familiar with the practical chemistry of the Alex-
andrian-Egyptian school, and his writings are a similar
mixture of laboratory directions, chemical apparatus and
methods and mystical symbolism. It has been previously
noted that he belonged to the cult of Gnostics.

An illustration of this mystical and mystifying symbolism
manifestly referring to the transmutation of baser metals
into gold or silver, though utterly unintelligible as to materials or methods, is found in a treatise of Zosimos ''on the
virtues and composition of the waters.'' By the waters, it
must be understood that Zosimos means with Plato all
liquid or fused or fusible substances, as fused metals.

The text of this passage is translated by Berthelot from
the manuscript of St. Mark's (tenth century) previously
alluded to.[25]

''The composition of the waters, the movement, growth,
removal, restoration of the bodily nature, the separation
of the spirit from the body and the fixation of the spirit
upon the body, operations which do not result from the
addition of foreign natures drawn from without, but which
are due to its own nature acting upon itself derived from
a single kind only, as with hard and solidified minerals
and with liquid extracts of the tissues of plants, all this
uniform and many colored system comprises the manifold
and infinitely varied investigation of all things, the investigation of Nature, subordinated to the lunar influence and

---

[25] Berthelot, *op. cit.*, II, Greek text, p. 107 *ff*. French translation, p. 117 *ff*.

to the measure of time, which govern the term and the growth according to which nature is transformed.

"While saying these things, I fell asleep and I saw standing before me at an altar shaped like a dome (φίαλη),[26] a priest sacrificing. There were fifteen steps to mount to this altar. The priest stood there, and I heard a voice from above saying—'I have accomplished the act of descending the fifteen steps walking toward the darkness and the act of mounting the steps going toward the light.[27] It is the sacrifice that renews me eliminating the dense nature of the body. Thus by necessity consecrated, I become a spirit.' Having heard the voice of him who stood at the dome-shaped altar, I asked him who he was. In a shrill voice he answered in these words, 'I am Ion, priest of the sanctuaries, and I undergo intolerable violence. Some one has come hastily in the morning and has done violence upon me, cleaving me asunder with a sword and dismembering me according to the rules of combination. He has removed the skin from my head with the sword which he held; he has mixed my bones with my flesh and has burned them with the fire of the treatment. It is thus I have learned of the transformation of the body to become a spirit. Such is this intolerable violence.'

"While he yet conversed with me, and I forced him to speak, his eyes became like blood and he vomited all his flesh and I saw him (changed to) a little imitation man, rend himself with his teeth and sink down.

"Filled with fear, I awoke and reflected—'Is not this the composition of the waters?' I was persuaded that I had rightly understood and I fell asleep again. I saw the same dome-shaped altar and at the upper part a water boiling and many people circulating continuously. And there was no one outside of the altar whom I could question. I then moved toward the altar to see this spectacle, and I perceived a little man, a barber, whitened with years,

---

[26] The Greek word φίαλη was used also for the dome-shaped receiver of glass placed over distilling apparatus to act as a condenser of vapors. *See* Berthelot, *Introduction*, pp. 132–134. The word thus conveys a double sense, a popular and a technical concept.

[27] Very probably these fifteen steps indicate thus obscurely the various operations involved in laboratory operations, fusion, fixation, distillation, sublimation, projection, crystallization, etc.

who asks me, 'What dost thou look upon?' I answered that I was surprized to see the agitation of the water and of the men burned yet living. He answered in these words, 'This spectacle that thou seest is the entrance, the departure and the mutation.' I asked him, 'What mutation?' and he replied, 'This is the place of the operation called maceration, for the men who wish to obtain virtue enter here and become spirits after having escaped from the body.' Then said I, 'Art thou a spirit?' and he answered, 'Yes, a spirit and a guardian of spirits.'

"During our conversation, the boiling continuing to increase and the people uttering cries of lamentation, I saw a man of copper holding in his hand a tablet of lead. Looking at the tablet, he spoke the following words, 'I command all those who have submitted to the punishment to be calm, to take each one a tablet of lead, to write with their own hands, to keep their eyes lifted, and their mouths open until their vintage be developed.'

"The act followed the word, and the master of the house said to me, 'Thou hast contemplated, thou hast stretched thy neck upward and seen what has been done.' I replied that I had seen, and he explained to me, 'He whom thou seest is the man of copper, he is the master of the sacrifices and is the sacrificed. It is he who vomits his own flesh. Authority has been given him over this water and over the people here punished.'

"After this vision, I awoke again and said, 'What is the meaning of this vision? Is not this water, white, yellow and boiling, the water divine?' And I found that I had well comprehended. . . . In the dome-shaped altar all things are blended, all are dissociated, all things unite, all things combine, all things are mixed and all are separated, all things are moistened and all are dried, all things flourish and all things wither. Indeed for each it is by method, by measure, by exact weight of the four elements that the mixing and the separation of all things take place. . . .

"In short, my friend, build a monolith temple as of white lead (cerussa), as of alabaster (usually quicklime), having neither commencement nor end in its construction. Let it have in its interior a spring of pure water, sparkling like the sun. Observe carefully on which side is the entrance to the temple, and taking in your hand a sword, seek then the entrance for the place is narrow where the opening is to be found. A serpent is lying at the entrance guarding the temple. Seize him, immolate him, flay him, and taking his flesh and his bones, separate his members. Then joining the members with the bones, make of them a step to the entrance of the temple, mount upon it, and enter. Thou wilt find what thou seekest. The priest, this man of copper, whom thou seest seated in the spring gathering to himself the color—do not consider him as a man of copper, for he has changed the color of his nature and has becôme a man of silver. If thou wishest, thou wilt soon have him a man of gold.

" . . . Relying upon the clearness of these concepts of intelligence, transform the nature and consider manifold matter as being one. Never reveal clearly to any one any such property, but be sufficient unto thyself for fear that in speaking thou bringest destruction on thyself."

Certain things are clear from this obscure description. Transmutation of base metals to silver and gold is the general theme, and the suggestion of manifold matter being one is evidently the fundamental notion of the essential unity of matter which underlay the philosophy of Plato and Aristotle, and was perpetuated by chemical philosophers of later schools. It has never been entirely absent from chemical speculation, and in a different sense is still existent in theories of matter. The "temple" may be interpreted as the laboratory of the metal worker, once secret and sacred in Egypt. The altar, dome-shaped, is probably the apparatus in which the experiments were performed—furnace and crucibles with the balloon-shaped receiver or condensor of substances given off by the heating. The "men" are the metals or other constituents which enter into the process and which are freed from their bodies and become spirits or the reverse. This change meant with the ancients, the giving off of gaseous or volatile matter, leaving the nonvolatile, or the contrary process, the fixing of the spirits by the body. Sulphides or oxides of the metals reduced by any process yielding the metal was a separation of the spirit, the metal being the body. The "body of magnesia" or the "body of stimmi" (sulphide antimony), were metals obtained from "magnesia" which was a term covering many substances—white lead, pyrites, magnetic oxide of iron and even sulphide of antimony which is the "stimmi" of the ancients. It will be recalled also that the ancients did not know how to dicriminate distinctly between lead and antimony or zinc, all being generally called lead. The curious figure of the Ouroboros, or serpent, which appears so often in text or illustration, here seems to symbolize difficulties of some kind which are to be conquered by the successful adept.

The following is a specimen of alchemistic philosophy from Zosimos:[28]

"Democritus has named the four metallic bodies, sub-stances, meaning by that copper, iron, tin and lead. Everybody employs them in the two tinctures of gold and silver, and all substances undergo the two tinctures. All the substances have been recognized by the Egyptians as produced by lead alone, for it is from lead that the other bodies are derived. He (Democritus) has then called substances matters resistent to fire, and nonsubstances matters which do not resist it. Indeed nonsubstances act in a suitable manner independently of fire. He said that they are engendered by the action of apparatus, and of combustion, whilst the true residue of the preparation prepared without the action of fire produces a stable tincture in white and yellow. The use of the volatile preparation obtained by the flame destroys the yellowing of defective molyb-dochalc (a lead and copper alloy) in that it makes it disappear. Upon this point, it is necessary not to deceive oneself. See how he expresses himself in this respect.

" 'Bring it to a waxy consistency, spread with half the preparation destined for the heating, and stain with the remainder, so that the color may be fixed without the help of fire. Sulphurous matters not resistent to fire are called

28 Berthelot, *Collection des Alchimistes Grecs*, II, p. 167 (translation).

nonsubstances. But the use of suitable liquids communicates to them the property of resisting fire and remaining stable, for water opposes the action of fire. It is for that reason that he says, "Nature acquiring in itself the contrary property becomes solid and fixed, dominating and dominated." Thus it acquires in itself the sulphurous quality, that which gives its name to the water of native sulphur.' Why does he speak also of the opposite? It is because water is the opposite of fire. Its liquid quality prevents matters submitted to fire from evaporating or volatilizing. They are as if enveloped in the humidity and retained until they are tinctured. Water retains because it is liquid. This is why he says, 'Nature acquiring in itself the opposite quality' etc. It has been explained how, by means of liquids, products are obtained which resist fire, but the liquids, these are the water divine."

A Syrian manuscript of the fifteenth century, in possession of the University of Cambridge,[29] contains a treatise attributed to Zosimos. It is difficult to say to what extent this work is authentic and to what extent it has been extended or interpolated. It is, however, a much more extensive work than any among known Greek manuscripts. It contains a great many recipes similar in objects and style to those of Democritus and of the Theban papyri, and these are interspersed with much of the mystical and obscure material which characterizes the Greek fragments of Zosimos and of the pseudo-Democritus. Compared with the similar writings of Democritus, Zosimos appears to be addicted to even less clear and more obscure and mystical descriptions. Nevertheless, it appears evident that he is experienced in the operations of Egyptian metallurgists and not like most of the other Greek alchemists merely mystical commentators.

A passage in this Syrian work of Zosimos is illustrative of his style and includes an interesting fable which if allegorical or symbolic is not simple of interpretation.

"Those who have written upon the work of the stones

---

[29] Ms. M. M. 6. 29.

have also defined mercury; they not only do not call it simply zioug, but they say further that it is formed of silver and ferruginous stone. Those who have written upon preparations have also defined it in saying, 'The zioug vivus (quicksilver) which is formed by cinnabar they have called tinctorial mercury. That formed by copper they have called water of copper and water of Aphroud; so also they have called the mercury drawn from silver, water of silver, foam of aphroseline (selenite ?) and dew. That which is obtained from tin some have called water of the river, others bile of the dragon.'

"We will now speak of this subject. In a place in the far west, where tin is found, there is a spring which rises from the earth and gives rise to it (tin) like water. When the inhabitants of this region see that it is about to spread beyond its source, they select a young girl remarkable for her beauty and place her entirely nude below it, in a hollow of the ground, in order that it shall be enamoured by the beauty of the young girl. It springs at her with a bound seeking to seize her; but she escapes by running rapidly while the young people keep near her holding axes in their hands. As soon as they see it approach the young girl, they strike and cut it, and it comes of itself into the hollow and of itself solidifies and hardens. They cut it into bars and use it. This is why they call "water of the river" the mercury drawn from tin; they call it thus, because it runs like water which throws itself into lakes and which has the appearance of a dragon furious and venomous."[30]

There is room for doubt as to whether these fanciful appellations really arose from this fable or whether the appellations are of earlier origin and this explanation of them is an attempt to account for them by later invention.

There are a number of alchemists or commentators upon alchemy who have left fragments of their writings in the manuscripts in Greek, existing in the libraries of Europe. Some of these may have been contemporaneous with Zosimos, but others are later. The principal writers whose

---

[30] Berthelot, *La Chimie au Moyen Age*, II, pp. 244, 245.

names are known to us are, presumably in the fourth century, Pelagios, Pebechios, Heliodorus, Synesius, in the fifth century, Olympiodorus, in the seventh century, Stephanus of Alexandria, while many others of less celebrity wrote in the sixth and seventh centuries. In general, it may be said that the later Greek alchemists added nothing of importance to the knowledge to be gleaned from the pseudo-Democritus or Zosimos. With the lapse of time, these writings give the impression that their writers lack familiarity with the operations of chemistry and metal working, and are more and more lost in a mystical philosophy. Their philosophy caused them to believe that the original four elements of which all bodies were constituted might be transmuted into one another by depriving them of certain properties or qualities, and by analogy any substances might be changed to other substances. Naturally, they considered that substances most readily changed into gold or silver were those substances which were most like these in their properties and these were the four base metals known to the ancients, lead, copper, tin and iron. If lead, for instance, might be deprived of its softness, its ready fusibility, and be colored or tinctured, it might not only resemble gold, it might be gold. And so with others. As tradition told them that such transmutations had taken place by the skill and mystic knowledge of the masters, they might succeed could they but interpret aright the oraculor indications or secret formulas of the authorities. It must be remembered that the attitude of the middle ages generally was to have great faith and reverence for authority. The whole spirit of the time was to look to the past for all wisdom and knowledge. This was true in the domain of religion, medicine, philosophy and so also in the philosophy of chemistry which was then alchemy. It is true that certain extensions of Plato's theories of matter had been developed to explain facts observed in chemical operations. For instance, the concept that all metals were composed of mercury and sulphur, not common mercury or sulphur, but the "mercury of the philosophers" and the "sulphur of the philosophers" that is hypothetical substances which carried the essential qualities of the common substance, but were a kind of quintessence, one might say the spirits or souls of mercury and sulphur. This notion the beginnings of which it is difficult to trace[31] became one of the corner stones of alchemical belief in later centuries among the Arabian and later European alchemists.

The concept of the "philosopher's stone" which appears under many names, was that of the existence of some substance which should act as a ferment just as yeast acts upon dough, some mystic substance which added to baser metals should induce the transmutation of larger quantities of these to gold or to silver. An idea of this character is of very early origin, but any definite ideas as to the nature of this substance are lacking, and in the later alchemists, they take an infinity of forms. The philosopher's stone first appears about the seventh century in literature, but it may be earlier. In the early centuries of alchemy, there was also developed a mass of symbolism which lost nothing of complexity and obscurity with the development of alchemy. Thus, the egg, symbol of the round universe, or of eternity; the "egg of the philosophers" consisted, like the physical universe, of four components, white and yolk a skin and shell. These four constituents again are sometimes said to typify the four metals which form the basis for transmutation, copper, tin, lead and iron.

The Greek alchemists have given us several treatises on the nomenclature of the egg; they do not agree entirely, but are nevertheless similar enough to show their common origin. One of these is in the earliest manuscript, that of St. Mark's, in the tenth or eleventh century. The following is from a different manuscript copied in 1478.[32]
"Nomenclature of the Egg. This is the mystery of the art.

---

[31] This theory is probably also of Alexandrian-Greek origin. Cf. Von Lippmann, *op. cit.*, pp. 380, 381.
[32] Berthelot, *Collection des Alchimistes Grecs*, I, p. 21.

"1. It has been said that the egg is composed of the four elements, because it is the image of the world and contains in itself the four elements. It is called also the 'stone which causes the moon to turn,' 'stone which is not a stone,' 'stone of the eagle' and 'brain of alabaster.' [33]

"2. 'The shell of the egg is an element like earth, cold and dry; it has been called copper, iron, tin, lead. The white of the egg is the water divine, the yellow of the egg is couperose, the oily portion is fire.

"3. The egg has been called the seed and its shell the skin; its white and its yellow the flesh, its oily part, the soul, its aqueous, the breath or the air.

"4. . . . (Seems interpolated and disconnected from the rest, part of a practical recipe but not intelligible.)

"5. The yellow of the egg has been called at first, attic ochre, vermillion of Pontus, soda (nitron) of Egypt, blue of Armenia, safran of Cilicia, Cheledony. The white of the egg mixed with water of sulphur is vinegar, water of alum, water of lime, water of ashes of cabbage, etc."

The treatise in the earlier manuscript is more extensive, but no more illuminating as to the reasons for such strangely grouped synonyms for the white or yellow of the egg as the above.

Another symbol which enters throughout all alchemical literature and graphic representation is the serpent Ouroboros in the attitude of biting his tail—symbol of the eternal cycle of world changes, as also of the cycle of chemical transformation, distillation, and condensation. This symbol is thus described in the same manuscript as the foregoing upon the egg.[34]

"1. Here is the mystery: the serpent Ouroboros this composition which in its ensemble is devoured and melted, dissolved and transformed by the fermentation or putrefaction. It becomes a deep green and the color of gold is derived from it. It is from it that is derived the red called

---

[33] Alabaster is often quicklime. Here perhaps as Berthelot suggests, meaning the lime from eggshells as this definition of alabaster appears in the early alchemical lexicon of the manuscript of St. Mark.

[34] Berthelot, *loc. cit.*, p. 171.

the color of cinnabar. This is the cinnabar of the philoso-phers.

"2. Its stomach and back are the color of saffron, its head is a deep green, its four feet constitute the tetrasomie (term extensively used to signify the four base metals). Its three ears are the three sublimed vapors. (Probably here sulphur, mercury and orpiment).

"The One furnishes the Other its blood; and the One gives birth to the Other. Nature rejoices in nature; nature triumphs over nature; nature masters nature; and that not for a nature opposed to such another nature, but for one and the same nature proceeding of itself by the process, with trouble and great effort.

"4. But thou, my dear friend, apply thy intelligence to these matters and thou wilt not fall into error; but work seriously and without negligence, until thou hast seen the end (of the process).

"5. A serpent is stretched, guarding this temple, and he who has subdued it commences by sacrificing it, then roasts it, and after removing its flesh up to the bones, make of it a step to the entrance of the temple. Mount upon it and thou shalt find the object sought. For the priest at first a man of copper has changed color and nature and has become a man of silver; a few days later, if thou wishest, thou wilt find him changed to a man of gold."

This is a typical description with alchemists early and late, and is probably about as intelligible as it was intended to be. It is evident enough that the whole passage refers to the transmutation of the base metals and that the sym-bolism of the serpent may be interpreted in vaguely ex-pressed references to the recognized neoplatonic theories of matter, while chemical operations and apparatus are still more vaguely indicated in different passages. In ad-dition to all that, fanciful designations or secret names to conceal operations from the general public were so exten-sively employed by the early chemists and by later imi-tators and impostors that the definite understanding of the alchemical vocabulary is at the present time almost hopeless, though often inferences may be drawn with a fair degree of probability. But it is quite evident that the early alchemists added little, if anything, to the knowledge of chemistry at the time of Pliny, Dioscorides and the time when the papyri of Leyden and Stockholm were originated. Zosimos perhaps more than any other of the Greek alchemists has given descriptions of apparatus and of their nomenclature, a subject almost ignored by the ancient chroniclers from Theophrastus to Pliny and Dioscorides.

The manuscript of St. Mark is perhaps the earliest manuscript which gives in connection with descriptions the sketches of apparatus and tables of alchemical symbols. Berthelot has reproduced these and others from later sources in the *Introduction a l'étude de la Chimie*.[35] These figures of apparatus are all extremely crude—rather diagrammatic than realistic.

The Syrian manuscripts also give a long list of signs for chemical substances, which generally speaking are similar to those given in the manuscript of St. Mark, though they are not in all cases identical, and many are written differently although essentially the same and evidently of common origin. It appears to be demonstrated beyond doubt that the Syrian alchemy is merely the alchemy of the Alexandrian schools transplanted and preserved without notable change by the writers of the Syrian schools which flourished from the fifth and sixth centuries until they were abolished by Moslem fanaticism about the eleventh century.

From the descriptions of Zosimos and others, we learn that such apparatus may consist of pottery, metal or glass, the latter having the advantage of transparency as well as being impervious to certain vapors as quicksilver. Parts of the apparatus are joined together by clay, gypsum, wax or fats and oils, according to conditions. Heating processes are conducted by the sun's heat, by the warmth of manures of various kinds, by baths of hot ashes (sand baths)

---

[35] With respect to symbols and signs, *see* also Von Lippmann *op. cit.*, p. 347 *ff.*, where many of these signs with interesting notices are brought together.

or by water baths. For higher temperature, charcoal was the common fuel in the laboratory. The nonvolatile residue after heating, having lost its spirit or soul (pneuma) was dead. (The caput mortuum of the Latin chemists of later days.) That which was dead might by other processes have its spirit restored and be resuscitated or resurrected.

The Greek alchemists exerted no considerable influence directly on western science though their works were as we have seen kept alive to a certain extent by copyists through the Middle Ages. In Constantinople there seems to have been among the Byzantine alchemists somewhat greater activity than in the west, but as for the direct influence of Greek writers on the later Middle Ages or early Renaissance, it may be considered as almost lost to any but a few scattered disciples whose activities were insignificant and without any distinct impression on their times.

Under Mohammedan patronage, however, as has been stated, Syrian alchemy, transplanted to Asia Minor and Persia, after the fall of Alexandrian schools, was assimilated by the Arabians, and in the westward sweep of Arabian conquest was cultivated, finding in the Arabian universities of Spain a fertile soil for its cultivation. It does not appear that Arabian culture had developed any notable chemical or alchemical philosophy until it came into contact with Syrian culture.

It is Arabian alchemy that preserved the traditions and literature of the Alexandrian-Greek alchemists, derived from the Syrians during the long period when the culture of Christian Europe was inhospitable to its development. From such Syrian and Arabian manuscripts as have been preserved and examined, it does not appear that during the centuries of their alchemical activity any very notable additions were made to the practical chemistry known to the ancients of the times of Pliny, Dioscorides or the writers of the Theban papyri. Nor was the development of the theories of matter and its changes in the direction of a distinct advance over the Neoplatonism of their Alexandrian masters.

The Arabian writers seemed to have no thought of challenging the authority of the traditional masters of the art.

The first Moslem writer on alchemy cited by later Arabian authors was Khaled ben Yezid ibn Moaonia, Prince Oneeyade, who died in 708 A. D., reputed to be a pupil of the Syrian monk, Marianas.[36] No remnant of his writings of any significance has been preserved.

The earliest Arabic manuscript on alchemy now known is the *Book of Crates,* which is manifestly a translation of a Greek original, probably also by way of a Syrian translation, though the original in Greek is not now extant. The *Book of Crates* is referred to in a Syrian manuscript of writings attributed to pseudo-Democritus. The copyists of those days took so many liberties that it is not impossible that the name Crates may itself have been a corruption of Democritus, as suggested by Berthelot,[37] for though the *Book of Crates* is in the Syrian manuscript quoted in a writing accredited to Democritus, so also the work itself contains references to Democritus.

The Arabic manuscript containing the *Book of Crates* is, according to the translator, M. Houdas, a copy not earlier than the sixth or seventh century after the Hejira, the thirteenth or fourteenth century A. D. Based on internal evidence, the work from which it was copied was, in the opinion of M. Berthelot, written about the ninth century of our era. The contents of the book show that it is mainly a translation from the Greek, and it is of much the same character as the Greek alchemical manuscripts, lacking, to be sure, the specific recipes which are common to pseudo-Democritus and Zosimos, but otherwise very similar. The same Egyptian and Hebrew and other authorities are cited, and the same allegorical and obscure lucubrations are in-

---

[36] Berthelot, *La Chimie au Moyen Age,* III, p. 2. The name is given by Von Lippmann as Khalid ibn Jazid ibn Muawijah (635–704 A. D.), *op. cit.,* p. 357.

[37] Berthelot, *op. cit.,* III, p. 9.

dulged in. References are also made to revelations they
are careful not to reveal. Among the Arabian manuscripts,
included in this same collection at Leyden, are several
others which, like the *Book of Crates,* are manifestly
founded on the Alexandrian alchemy without any evidence
of original extension or development. Besides these, however, are several works attributed to Djaber ben Hayyan
Eç-Confy, who enjoys the reputation among later Arabian
writers as the grand master of the art. It is this Djaber
who among European alchemists and chemists of the late
middle ages and the Renaissance, under the name of Geber
or Gheber, was credited with many chemical writings which
modern criticism has conclusively shown to have been in
no way related to the real Gheber or Djaber.

Djaber was a writer of the eighth or ninth century,
looked up to with reverence for his learning by the Arabian
writers. His contributions to alchemy and chemistry are,
however, not improtant. The false Gheber was a writer,
of the thirteenth or fourteenth centuries, whose personality
is unknown, who possessed a much more advanced knowledge of chemistry and who, for his greater security or in
order to obtain greater prestige for his writings, chose
to have them accepted as translations of Arabian works of
Gheber (Djaber). As a matter of fact, they were probably
written in Latin, following no Arabian original. This
judgment, long suspected by historians, has been finally
confirmed by Berthelot, through his publication with translation of Arabian manuscripts of the real Djaber, thus
enabling a critical comparison of the two writers. Not
much is known with certainty as to the personal history of
Djaber. Arabian writers differ as to the place of his birth
and the time of his activity, though it is generally accepted
that he was the author of a great number of works on
many subjects, some of them on magic and on alchemy. It
is thought that he lived about the eighth or the beginning
of the ninth century. Little of his work now remains, nor
do later historians state what discoveries or advances in

any science Djaber made to justify the high reputation in which he stood with his successors.

From the half dozen treatises which are published by Berthelot, one can obtain a fair idea of the kind of writing which characterizes the real Djaber. His style is diffuse and verbose. He is interested in the philosophy of matter, its constitution and change rather than in experimental manipulation or phenomena. His allusions to the great work of transmutation are like his Greek predecessors, vague, mystical and obscure. His citations of authorities are to Aristotle, Pythagoras and Plato, rarely Democritus, Hermes or Stephanus. He evidently is extremely egotistic and continually boasts of the superiority of his knowledge and his writings. In this he may have exerted an influence upon later alchemists, for this is a common characteristic of the later Latin-writing alchemists.

From an examination of these works of Djaber, there is not found anything that suggests a real advance over the Greek alchemists, either in knowledge of chemical facts or in theories, though it is easy to recognize an individuality in style and in emphasis and development of notions of matter. Thus while recognizing the four elements and their Aristotelian qualities, he lays particular stress upon the neoplatonic idea of body and spirit which often occurs in the Greek alchemists. He also lays great emphasis on the equilibrium of "natures." He says:[38]

"God, after having created all things from the four elements: fire, water, air and earth, causes the four qualities to depart from the ancient worlds: heat, cold, moisture and dryness. The combinations of these elements have produced fire, which contains heat and dryness; water, which contains cold and moistness; air, which contains heat and moisture; earth, which has cold and dryness. It is with the aid of these elements that God has created the superior and the inferior world. When he has established equilibrium between their natures, things persist in spite of time, without being consumed by the two luminaries, nor rusted

---

[38] Berthelot, *op. cit.*, III, p. 147.

by the waters of ponds: such is pure gold that nature has digested and purified in all its parts without having need of drugs, analyses or refining. I am telling you, if you are clear seeing, the theory and practice in two great chapters. I have shown you the necessity of the equilibrium of natures in that which concerns the work. The thing is rarely necessary outside of that. Know then that the equilibrium of natures is indispensible in the science of balances and in the practices of the work.''

This importance of the balance or equilibrium of natures is a subject he often alludes to, but nowhere makes definite.

The following extracts are from a treatise entitled the *Book of Mercy*. This work appears to have been edited by a follower of Djaber, though credited by this disciple to Djaber. It shows a rather more orderly arrangement, and its style seems more influenced by Aristotle's logical form than the other works of Djaber.[39]

''The mass of corporeal things is only the place of sojourn and refuge of spiritual things, in itself it has neither force nor utility, when the acting force has ceased to be in it. The body which remains as substratum is only the place of sojourn and refuge of the spirit which has left it, and it has force only from the spirit which can leave it. If returned to it, it will certainly combine with it. . . . Things the most stable are those which contain most of body and less of spirit; such are gold, silver and analagous substances. Things the most fugacious among bodies are those which contain the most spirit; such are mercury, sulphur and arsenicon. All bodies contain spirits and all spirits contain bodies, but the name that one gives to them is taken from the preponderating components—mercury, sulphur, arsenic, gold, silver, the two leads (black and white, that is lead and tin; cf. Pliny); copper and iron, are considered as the mineral elements of the world, and all stones and earths are produced from these.

''In the whole world things are mixed with one another. You will not find fire which does not contain some cold,

---

[39] Berthelot, *op. cit.*, III, p. 176.

nor cold which does not contain some heat; no dryness without a little humidity, no humidity without dryness. No more will you find spirit which does not contain a little body, nor body which does not contain a little spirit. Sometimes these two elements cannot be separated when one of them is too abundant, and the other too much lacking, so that there is a transformation and absorption of the part which is in less quantity by the part that predominates. It is as if we let fall some drops of honey into the sea, no created being will ever be able to separate this sugary part. God alone could do that. Nevertheless, nobody would be justified in saying the sea possessed a sugary taste. This is why some one has said that the work is produced by every kind of thing. If he says a thing which is possible; or if further he says that the natures are found in everything, that is possible in two ways, everything coming from another in potentiality and not in accomplishment. When things meet a force more intense than their large mass, the whole mass takes the nature of this force: for example, a small quantity of ferment transforms a considerable mass of dough."

The body of the writings of Djaber that have been translated at the instance of M. Berthelot and published by him, are fine-spun metaphysical discussions upon the nature of matter and its changes and the application of these. There is very little allusion indeed to anything conveying any comprehensible idea of actual substances or methods. There are passages which refer to transmutation of base metals into silver and gold, but the emphasis upon these is not so great as with the earlier Greek alchemists. The work entitled the *Book of Mercy* is, as above stated, not by Djaber, but by a disciple of his, mentioning Djaber in the third person. This manuscript is notable, however, in that it begins with a paragraph denouncing the vanity of the attempts to make gold and silver.

"The Book of Mercy by Abou Musa Djäber ben Hayyan El Dumaouï El Azdi Eç Confi. May God be merciful to him.

"In the name of God gracious and merciful!

"Abou Abdallah Mohammed ben Yahia reports that Abou Mousa Djaber (may God be merciful to him) has said, 'I have seen that the people devoted to the search for the making of gold and silver were in ignorance and on a false road. I have also perceived that they may be separated into two categories—deceivers and their dupes. I have had pity for both classes who squander uselessly the goods which the Most High has granted them, who fatigue their bodies in vain, who let themselves be turned away from the care of acquiring those good and beautiful things necessary to daily life, and who neglect amassing a store of good works useful at the day of meeting to which all men ought to help. I have pitied these victims who consume their bodies and wealth through long days and who fatigue themselves to the detriment of their religion and faith, to obtain a slight portion of goods of this world. Their sad situation has moved me to compassion. I have tried to replace them upon the right road; by turning them from this occupation I should have done a pious work for which God will recompense me in the other world. God is the dispensator of all favors and all wisdom.'" [40]

In the same work, however, there are vague allusions to the red elixir and the white elixir, terms conventionally used by alchemists to indicate preparations supposed to convert base metal into gold and silver.

"Make so that your combination of natures may be obtained by the aid of the spirits and their special bodies, and then commence the true and sure operation to make a homogeneous whole, so that the spiritual element of the preparations does not become separated from the corporeal element and vice versa. The elixir should become red, for the nature of gold, and white, for the nature of silver. This is what the philosophers mean by the words, 'Gold can only come from gold, silver from silver, and a child from the father.' The red elixir is warm and dry and of the same nature as gold; this is why they consider it as of gold. The white elixir is cold and dry of the same nature as silver, and for them it is of silver. This is why they say, 'our gold is not common gold, nor our silver common

---
[40] Berthelot, *La Chimie au Moyen Age,* III, p. 163 *ff.*

silver.' Their gold and their silver are tinctured by the elixir and superior to common gold and silver."

It seems not unreasonable to presume that this work of Djaber was edited at a later period when alchemical works were under the suspicion of the Mohammedan church.

In the *Little Book of Pity*,[41] allusion is made to an analogous agent to the elixir, the "imam."

"Establish the equilibrium, the parallel, with the aid of fire of three degrees, namely, the incipient fire, the medium fire, the extreme fire, which melts the elixir; the solid will melt like wax and afterwards harden in the air. It will penetrate and be introduced like a poison. The result will conform to the operation, if the substance is excellent as I have already told you. The operation will be only rapid with the preceding substance, it will be very solid, excellent and very pure. Only one part will suffice for a million. If, with an excellent substance you commit some negligence in the operation, the result will be in proportion to this negligence. Preserve the elixir in a vessel of rock crystal or gold or silver, glass being subject to breaking. Implore the help of God in all things and you will be happy and on the good road."

The high reputation in which the name of Djaber was held by later alchemists seems to be due to the appeal of his metaphysical philosophy of nature and perhaps to its mystical obscurity as well, for there is no evidence of any important achievement of his, either in the direction of theory or in practical advances in chemical knowledge.

In the twelfth or early thirteenth century, unknown Arabian writers on alchemy issued treatises under the names of Aristotle, Rhazes and Avicenna, which were accepted by the encyclopedists of the thirteenth century and by their successors as genuine. These works such as the *De Perfecto Magisterio,* by a pseudo-Aristotle, the *De Aluminibus et Salibus,* attributed to Rhazes (Alrazi) and the *De Anima* falsely credited to Avicenna, were often cited by Vincent of Beauvais, Albertus Magnus, and Roger

---

[41] Berthelot, *op. cit.,* III, p. 137.

Bacon. They are just such mixtures of general information about chemical substances and chemical philosophy as are found in the writings of the Faithful Brothers.[42] References to some of their works will be met in the consideration of the thirteenth century encyclopedists.

Arabian alchemists were numerous from the ninth to the fourteenth century. Von Lippmann enumerates about sixty Arabian authors who wrote or were reputed to have written on alchemy[43] during that period. Of the contents of many of these writings very little is known. From such writings as have been accessible, Von Lippmann expresses the judgment that neither the Syrians nor the Arabians enriched the knowledge of chemistry with a single new and original thought, being dependent on the authority of the Greek alchemists and producing only increased confusion by their efforts to explain what was to themselves incomprehensible.[44]

M. Berthelot in his researches has shown clearly the Greek origin of the Arabian alchemy, the connection of their practical chemical knowledge with that of Greek-Egyptian sources, and that much of the later chemical advances previously attributed to them were of later origin, and perhaps due to European chemists of the thirteenth and fourteenth centuries. Thus there is no known reference in Arab texts to alcohol (meaning the liquid which we call by that name), nor to nitric acid, aqua regia or sulphuric acid, inventions attributed to them by Berthelot himself in earlier writings.[45]

Kopp also, referring to the Arabian alchemists of the eleventh to thirteenth centuries, says that from such writings as were accessible at his time one learns no new facts, and though by preserving and transmitting chemical knowl-

---

[42] *See* post., pp. 210 *ff.*
[43] Von Lippmann, *op. cit.*, p. 396 *ff.*
[44] Von Lippmann, *op. cit.*, p. 424.
[45] *Origines de l'Alchimie*, p. 209. Cf., for example, Berthelot on alcohol, *La Chimie au Moyen Age*, I, p. 136 *ff.* and p. 165, where he says that the first indications of the mineral acids, clearly expressed, are in the fourteenth and fifteenth centuries.

edge they contributed to the advance of the science, yet their writings are without interest in the history of the development of chemistry.[46]

Nevertheless, though the Arabians seem to have exhibited little originality either in chemical thought or in chemical invention, it is none the less true that their activities furnished the foundation for the chemistry of Europe. Their theories and their practices as elaborated from the Alexandrian and Byzantine alchemists were adopted and assimilated by Christian Europe without great changes during the twelfth to the fourteenth centuries.

That curious occult philosophy which constitutes the basis of alchemy in the modern sense of the term, derived from the Greek neoplatonists and transmitted mainly through Arabian disciples, was to find a recrudescence with, if possible, more extravagant manifestations of credulity, mysticism and charlatanism in the western alchemists of the fourteenth to the eighteenth centuries, a development greatly fostered also by the revolt from authority which culminated in the Protestant Reformation and was facilitated by the printing press in the latter part of the fifteenth century.

---

[46] Kopp, *Geschichte der Chemie*, I, p. 58.

CHAPTER V

THE CHEMICAL KNOWLEDGE OF THE MIDDLE AGES

When in 1530 Henry Cornelius Agrippa in his work on *The Vanity of the Arts and Sciences* quoted the proverb, "Every alchemist is a physician or a soapboiler," he expressed in epigrammatic form a not unimportant classification for his time, as also for centuries before. By alchemists he meant all chemists, and there were indeed two classes of chemists, those who were scholars learned in the natural philosophy of the time and versed in the doctrines of Plato, Aristotle, Galen or of the Alexandrian neoplatonists, and those on the other hand who with no pretensions to be philosophers, were engaged in the practical arts of chemistry in its various applications.

It was by the scholars or "philosophers" that were principally written the manuscripts which constitute the literature of natural science including chemistry, and by these that the fantastic and largely metaphysical chemical philosophy of the period was transmitted and elaborated. The artisans in chemistry of the middle ages, on the contrary, were not writers of books. They were busied with perfecting their chemical arts, perhaps at times also seeking in secret to attain the vain aims which the philosophers had led them to believe might be attained, such as the elixir of life or the real transmutation of the metals. When these artisans recorded their knowledge it was not for public information, but for the use of themselves or their associates, brief laboratory notes or recipes which should be clear enough for the purpose but with no intention to instruct the general public. On the contrary, they often took special precautions against being too easily understood by the uninitiated. Thus, as with the early Greek alchemists, terms were sometimes used to mask the real constituents. Sometimes even anagrams were employed so that the casual reader should not comprehend. In this respect such collections were of similar character to the Theban papyri previously noted. Naturally enough, and especially before printing was invented, this class of manuscripts was not widely distributed and not often preserved in the libraries.

Nevertheless such collections of recipes have been from time to time discovered. Naturally also the sources of them are obscure. They generally bear evidence of having been a growth by accessions and interpolations, often more or less confused by careless or ignorant translators or copyists.

But such as they are they often give a definiteness and significance to the very often vague descriptions of the learned but nontechnical philosophers and encyclopedists who were nevertheless the principal distributors of information as to the progress of science in the middle ages.

The earliest collection at present known of these technical recipes, after the papyri of Leyden and Stockholm, is a Latin manuscript dating from about the eighth century. It was first printed by Muratori in his *Antiquitates Italicae Medii Aevi* (Milan, 1738), and is described by Berthelot.[1] The entire title fairly summarizes its contents. Translated it reads—"Compositions for coloring mosaics, skins and other things, for gilding iron, concerning minerals, for writing in letters of gold, for making certain cements, and other documents relating to the arts." It is usually referred to as *Compositiones ad Tingenda*. Parts of this manuscript were manifestly copied from the Greek, and Berthelot calls attention to a case where a certain recipe is transcribed from Greek into Latin letters without translation—the evident work of a copyist who did not understand the meaning of the Greek and apparently knew only

---

[1] Berthelot, *La Chimie au Moyen Age*, I, p. 7 *ff.*

the Greek alphabet.[2] Byzantine Greek seems to have been the first source which furnished the basis of the contents, though probably added to from later sources.

The recipes deal with the coloring of glass used in mosaics, sometimes of the entire body of the glass, sometimes superficially. The essential coloring matters employed are often metallic compounds, as tin (oxide) for milky white, cinnabar, or burned copper, litharge for red, and mixtures of oils and resins evidently applied as varnishes for superficial coloring. Recipes for making glass and a description of the glass furnace, the gilding or silvering of glass for mosaics are also given. Processes are given for dyeing leather in purple, green, orange, red and yellow. Various minerals and chemicals are mentioned. In general these are the same that we find in Pliny, Dioscorides and the Theban papyri, alums, sulphur, soda, vinegar, afronitron, cadmia, flowers of copper, white lead, ochre, cinnabar, etc. and are written in a nomenclature that makes clear that the recipes are derived from Greek or Latin sources, and not from Arabian. Berthelot has called attention to one recipe almost literally identical with one in the papyrus of Leyden:[3]

> Chelidony 3 drachmas, fresh and clear resin 3 drachmas, gum of gold color 3 drachmas, brilliant orpiment 3 drachmas, bile of the tortoise 3 drachmas, white of egg 5 drachmas. The whole makes 20 drachmas. Add 7 drachmas of safran of Cilicia. You can write with it not only upon parchment or paper but also on a glass vessel or on marble.

It is of interest to note the use of the word vitriol (vitriolum) as applied to the impure sulphate of iron produced by the weathering of pyrites. This substance was known, it will be recalled, to Pliny and Dioscorides, but the name for it was chalcanthum, green or blue. It is worthy of note also that the preparation of cinnabar by uniting mercury and sulphur occurs in this manuscript seemingly

---

[2] Berthelot, *op. cit.*, I, p. 9.
[3] Berthelot, *op. cit.*, I, p. 10,

the first notice of this synthetic preparation. Though it is not mentioned by Pliny or Dioscorides it may be of as ancient origin.[4]

In this eighth century manuscript appear two recipes for brandisium, alloys of copper, tin and lead. Berthelot considers this the first known mention of the word whence our "bronze" is derived. The question of the origin of bronze has been a subject of much speculation and debate. The ancients used for bronze, copper or common alloys of copper, the Greek χαλκός or Latin aes. The term "orichalcum" or "aurichalchum" applied to golden colored bronze and later to brass was also in use. The word brandisium may be derived from the city in Italy, Brindisium (modern Brindisi) or possibly from the Greek βροντήσιος and ultimately from βροντή, thunder, and the legendary thunderstone with magic powers, Brontea or Brontia.[5]

A work entitled *Mappae Clavicula,* or "little key to painting," exists in two manuscripts. The earliest is of the tenth century in the library of Schlettstedt. This manuscript has not been published, though Berthelot had the advantage of the studies of M. Giry who first (1878) gave an account of it. This manuscript also, it is of interest to observe, shows no trace of Arabian sources, but like the *Compositiones ad Tingenda* is based upon Greek and Latin sources only. The later manuscript was written in the twelfth century and was published by Albert Way in 1847 in the *London Archaeologia* (Vol. 32). On account of the presence of two old English words in the text it is probable that it is edited by an English writer. Berthelot gives reasons for believing that this latest manuscript may have been edited by Adelard of Bath, an English scholar who had studied in Caen, Salerno, and in Egypt and who wrote many works interpreting Arab science. Among titles attributed to his authorship is one entitled *Mappae Clavicula.* Adelard lived in the first third of the twelfth century and

---

[4] Cf. Berthelot, *op. cit.,* I, p. 17.
[5] *See* Von Lippmann, *Entstehung und Ausbreitung der Alchemie,* article "Bronze," pp. 559–569, for extended discussion of the origin of the word.

this would place the date of this writing at about 1130 A. D.[6]

That this twelfth century manuscript is copied from an earlier version which has been much amplified, is made evident by the fact that this treatise is preceded by a table of contents in which the recipes are listed in consecutive numbers. This table contains 209 recipes with their titles. The work itself however contains 293 numbered recipes. The table agrees with the work itself as far as number 51, but thereafter the numberings bear no relation to those in the table of contents.

The recipes in the *Compositiones ad Tingenda* are largely included in the *Mappae Clavicula,* while the twelfth century manuscript contains later additions including Arabic names. It is during the twelfth century that Christian Europe first seems to have assimilated the results of Arabian chemistry and it is probable that these manuscripts had their origin either in Italy or the south of France.[7]

While the original work may have been confined to the art of painting or of coloring metals or other substances, in its ultimate form the *Mappae Clavicula* includes a great variety of recipes on all kinds of subjects without system or order of arrangement, some of them being even merely mystical and magical formulas. The great majority are however practical laboratory notes, not citing authorities, nor attempting any philosophical explanations such as are found in the Alexandrian-Greek alchemists, the Arabian alchemists, or in the thirteenth century encyclopedists.

Many of these recipes are similar to those in the Stockholm and Leyden Papyri, some indeed are practically identical, dealing with the same variety of subjects, imitation gold and silver, writing in gold and silver letters, dyeing skins, and in general all kinds of recipes pertaining to the arts practised by the Greek-Egyptian chemists. Many of

---

[6] Cf. Berthelot, *Archéologie et Histoire des Sciences,* 1906, p. 172 *ff*.
[7] Cf. Berthelot, *op. cit.,* I, p. 65; Von Lippmann, *op. cit.,* p. 470.

these recipes are complex mixtures containing metallic compounds, sometimes filings of gold and silver, often mixed with a great variety of oils and resins, vegetable colors, albumin of eggs, evidently used as paints and varnishes of all colors for application to articles of metal, glass, wood, etc. There are recipes for mixtures to set fire to the ships or houses of an enemy, such as found in the *Book of Fires* of Marcus Graecus, a work probably of about the same period. Into the composition of these mixtures enter sulphur, naphtha, resin, and oils. There is no reference to saltpeter as a constituent in either of the manuscripts, nor any description of such a mixture as black powder—which is found in the *Book of Fires* of Marcus Graecus. An item of particular interest is one of the earliest references to alcohol, though not described under that name. It is found only in the twelfth century manuscript and is recipe No. 212 of the *Archaeologia* text,[8] under an entirely irrelevant title:

Ad bonum argentum solidandum medium oboli. De commixtione puri et fortissimi xknk cum iij qbsuf tbmkt, cocta in ejus negocii vasis fit aqua quae accensa flammans incombustam servat materiam.

The solution of the anagram as first shown by Berthelot is simple, as each letter is to be substituted by the letter next preceding in the alphabet. The reading then of the anagram is "vini cum 3 partibus salis," and the translation is:

By mixing pure and strongest wine with three parts of salt and heating in a vessel customary for that purpose, a water is produced which when kindled inflames, (yet) leaves the material unburned.

Aqua (water) it may be recalled, was a generic name for liquids with the ancients. This according to Berthelot confirmed by Von Lippmann is the first definite reference to the separation of a combustible liquid by the distillation

---

[8] *Archaeologia*, London, Vol. 32, p. 227.

of wine. The statement that this "water" protects the substance upon which it burns from taking fire is easily explained on the supposition that the alcohol was moderately dilute.

Another early description of the distillation of alcohol is among recipes by Magister Salernus (who died soon after 1167, A.D.) contained in a compendium of Salernitan Medicine. The "aqua ardens" (burning water), is there said to be made "after the fashion of rose water" as follows:

Place in the *cucurbita* one pound (white, or) red wine, one pound powdered salt, four ounces native sulphur, four ounces of tartar (from wine). The liquid distilling is collected. A cloth saturated with this liquid will maintain a flame without suffering injury. Cotton does the same without loss of substance.[9]

It will be recalled that Pliny[10] records that the Falernian wine was capable of being kindled into a flame—though he gives no indications as to the circumstances under which this took place. Still earlier remarks of Aristotle and of his pupil Theophrastus indicate that under circumstances wine could yield a flash or flame when poured on the fire as in libations. Though the ancients and Arabian chemical writers possessed knowledge of distillation processes, they give no evidence of the accomplishment of the separation of alcohol. This was doubtless because their condensation methods while adapted to distilling water, vinegar and to the liquids of relatively high boiling points, were not adapted for the condensation of the more volatile alcohol vapor.[11]

Following these first known descriptions there appears in the Latin manuscript of the *Book of Fires* of Marcus Graecus a further description written in about the twelfth or thirteenth century. The copy in which it appears is

[9] *See* V. Lippmann, *Chemiker Zeitung*, 1917, p. 884, and 1920, p. 625.
[10] *See ante*, p. 74.
[11] The history of alcohol is given by Berthelot in *op. cit.*, 1893, I, p. 136 *ff.*, and by Von Lippmann in *Abhandlung und Vorträge zur Geschichte der Naturwissenschaften*, 1913, II, pp. 203–225.

apparently of date between 1250 and 1300. Translated it reads as follows:[12]

You may make burning water (*aquam ardentem*) thus:
Take black wine thick and old, and in one quart of it mix two scruples of native sulphur very finely powdered, one or two parts of tartar extracted from good white wine and two scruples of coarse common salt, and put the above into a *cucurbita* well leaded (that is luted), with an alembic superimposed and distil the *aqua ardens,* which you should keep in a closed glass vessel.

It is of interest to note that though the Latin manuscript in which appear these notices of the separation of alcohol both contain evidence of Arabic influences, yet thus far no such definite knowledge of the process has been found in any Arabian manuscripts of earlier or even contemporary dates. It is probable that its separation was effected by Italian or Spanish chemists who, while they served as mediators between Arabia and Latin scholars, were themselves originators of much that was later attributed to Arabian chemists.

A later manuscript of the *Book of Fires* at Munich, written in 1438, is still more explicit in some respects:

Aqua ardens is made thus:
Take best old wine of any color whatsoever in a cucurbita and alembic with joints well luted and distil with gentle fire. That which distils is called aqua ardens. Its virtue and property is such that if a linen cloth is dipped in it and kindled it will give a great flame. When consumed the cloth will remain entire as it was at first. If you introduce your finger in it and light it it will burn like a candle without injury. If you dip a lighted candle in this water it will not be extinguished. And note that that part which is first yielded is good and inflammable, but that which comes after is useful in medicine. From the first also is made a wonderful collirium for macula (spots) or pannum (film) of the eyes.

Von Lippman records that in about 1250 alcohol was first used as a medicine, two Italian physicians, Vitalis de

---

[12] Berthelot, *op. cit.,* I, p. 117.

Furno and Thaddaeus of Florence, being the first who are known to have so employed it.[13]   Albertus Magnus (ca 1260) refers to the fact that by "sublimation" of wine there is produced a light inflammable, supernatant liquid. Arnoldus Villanova, physician and chemist, also describes it and its uses in medicine in about 1300.   He calls it aqua ardens or aqua vini and says that some call it aqua vitae. The latter title had been in use by early alchemists as applying to the supposed elixirs of long life.   The name "alcohol" however was not used for this substance until introduced by Paracelsus in the sixteenth century.   The word "alcool," or "alkohol" or "kohol," with other spellings, was an Arab term designating various very fine powders as of antimony sulphide, and was used by them only in that sense.   The terms "alcohol" and "alcool" are also used by Paracelsus in that sense.   He indeed defines the term "alcohol" as "the most subtle part of anything."   It is doubtless in that sense that he applies it in his *alcool vini*, that is, the most subtle part of wine, and it is always as "alcool vini" or "alcohol vini" that he uses this term, never "alcohol" alone.   Later chemists dropped the "vini" and let the alcohol stand alone for the name.   Paracelsus leaves no doubt as to what he means, for in his *Von Offenen Schaden*,[14] in a prescription for excessive perspiration, the directions are:

"Rec. Theriacae drach. II, alcool vini (id est vini ardentis) unc. II," etc., and elsewhere[15] he speaks of "alcool vini (id est vino ardenti)."

From about 1250, under the names of "aqua ardens," "aqua vini," aqua vitae," and in the sixteenth century as "alcohol vini" or finally simply as "alcohol" the application of alcohol to medicine and to other arts extended rapidly.

In the twelfth century *Mappae Clavicula* occur three rec-

---

[13] Von Lippmann, *Abhandlungen und Vorträge zur Geschichte der Naturwissenschaften*, Leipsic, 1913, II, p. 212.

[14] Paracelsus, *Chirurgische Bücher*, Strassburg, 1618, p. 618b.

[15] Paracelsus, *Opera*, Strassburg, 1616, Bd. I, p. 178a.

ipes concerning sugar and confections made from that and from honey. Although the use of sugar in the orient was of much earlier date, any technical description of its refining, or of products made from it are apparently lacking in Arabian literature. The first of such descriptions are found in the work issued about 1150 by Matthaeus Platearius, an Italian writer on medicinal simples, a work which achieved a wide recognition and is an important source for writers of following centuries. There is a description in Platearius so similar to the recipes of this twelfth century manuscript as to suggest either a common source or that the latter are derived from Platearius.[16]

"Sugar is obtained in the following manner: When the canes in which it is formed are ripe, the tips are cut off for about two handbreadths, and planted like grass stalks in the earth. The rest is cut up, the pieces expressed in a mill and the juice conducted through wooden pipes into small vessels. It is then cooked down in a kettle, whereby a great mass of scum rises, and is then ladled out into round dishes. These are set aside in special houses, covered immediately with straw and then sprinkled with cold water. If moistened with but little water the sugar remains yellow and is called honey sugar (Zuccara Mellita) which, because it is of warmer nature cannot be given in violent fevers. In the same vessels (sprinkled with more water), however, in which the sugar at the bottom has this character, further above it is white and good, and boiled to dryness with vinegar and formed into cones furnishes an unexcelled remedy for fever and stomach complaints. It may be again boiled [after again dissolving]; the oftener it is boiled and purified the finer and whiter it becomes but the less of it remains."

Penidium (from the Persian fanid) is thus described:

"Sugar and water are boiled down strongly so that a drop brought on to a stone solidifies and the mass can be broken by the fingers. The whole is then poured upon a polished stone plate, allowed to cool somewhat, rolled to-

---

[16] From E. von Lippmann, *Geschichte des Zuckers*, Leipzig, 1890, pp. 174, 175.

gether, hung on a well fastened hook and twisted and pulled continually until it is quite white. As soon as it no longer sticks to the hands it is cut up into pieces with shears. Powder cannot be mixed with it, because it does not hold them, yet such may be sprinkled upon the finished product—for instance flour—so that it looks a fine white. Penidium is an excellent remedy for fevers, dry cough and chest diseases, and also when moistened with Tragantha water heals cracked lips.''

Various mixtures of sugar with other substances are described as medicines. The similar recipes in the *Mappae Clavicula* are as follows:[17]

''Compositio sisami. Honey, white and pure, is placed in a tinned (stannato) vessel on a moderate fire constantly stirred with a spatula, and alternately removed from and to the fire and stirred a long time, and again placed on and taken away from the fire, stirring without intermission until it becomes thick and viscous (conglutinosum). When it shall have become sufficiently thickened let it cool gradually. It is then poured upon marble: then suspended to an iron hook, and pulled frequently and gradually and folded until it becomes white as it should be, then twisted and shaped and placed upon marble. Then keep it for use.''

This preparation from honey instead of sugar is very similar in description to the directions of Platearius for penidium.

The second recipe in the *Mappae* is entitled, *De Zuchara.* It is a clarification of raw sugar.

''By a like action and boiling of sugar in a tinned vessel, a little water added to it when boiled, skimmed and well strained in a strainer; and with addition of such kinds of things as you know (adhibitis quibus scio speciebus), with incessant agitation, brought to thickness. Pour it out thinly on a marble smeared with a little oil, and when carefully cooled on the marble, separate it by hand from the marble and keep it for use.''

---

[17] *Archaeologia, London,* Vol. 32, p. 241.

This would give a clarified sugar in the form of barley-sugar.

The third recipe is *De Penidiade.*

"Penidias is made like sisamum after skimming and straining sugar, but without stirring—well boiled and placed in a hook as has been described and softened (malaxando)—fashion it and cut in pieces with scissors." [18]

It may be inferred from these recipes that such preparations of sugar and sugar candy were very popular even in those times.

Technical recipes of a different character are found in a work attributed to Marcus Graecus, though nothing is known as to the identity of the supposed author. That it was attributed to Marcus the Greek is of interest as lending additional probability to the assumption that the original compilation is due to the Byzantine Greek chemists. The work is entitled *Liber Ignium ad Comburendos Hostes,* or *Book of Fires for Burning Enemies.* It is indeed largely a description of mixtures ordinarily included under the designation of Greek fires. That some mixtures of this character were known in ancient times is manifest from early writers. Livy speaks for instance of Bacchantes carrying torches which took fire by dipping in water, and that writer says this was because they contained sulphur and quicklime. Julius Africanus (third century A. D.) gives a more specific account of a mixture kindling spontaneously when exposed to sunshine.[19]

"It is prepared as follows: native sulphur salt of the mountains, ashes, brontesinos (thunder-stone) pyrites, equal parts. Mix in a black mortar at noon with the juice of the black mulberry and bitumen of Zacynthus, a natural liquid, in equal parts, to a pasty consistency. Add with care a little quicklime, grind carefully at noon. Guard your face for the material may take fire suddenly. Enclose it in a copper box with a cover, and keep it and do not expose it to the sun. If you wish to set fire to the arms of

---

[18] Cf. also extracts on sugar in Bartholomaeus Anglicus, *see* post, p. 236.
[19] Berthelot, *op. cit.,* I, p. 95.

the enemy, secretly spread over them this preparation at night. When the sun rises all will be burned.''

Von Lippmann considers this passage as an interpolation of perhaps the seventh century. It is doubtless hardly worth while to take this formula as accurate or reliable, but it is evidence of the existence of some such mixtures for use in warfare. The *Book of Fires* is supposed to be based upon the experience of the Greeks and the work was supposed by Kopp[20] and Hoefer[21] to have been written in the eighth century. The reasons given for this assumption have not stood the light of later researches, and there is no identifiable reference to the work nor to this Marcus until the thirteenth century. The *Mappae Clavicula* text of about 1130 A. D. contains some recipes of very similar character, suggesting the existence of some such source as this at that time. The earliest manuscript of the *Book of Fires* thus far known is apparently of the latter part of the thirteenth century.[22] The existing texts also give evidence by the presence of Arabic names of some Arabian mediation which would suggest that the work in its present form is certainly not earlier than the eleventh or twelfth century when Arabian influence makes itself felt upon Latin writers. These recipes may then be taken as an accumulation of early Greek origin, with gradual alteration and additions possibly as late as the thirteenth century. It will be of interest to illustrate the character of the compositions described in this work. The opening recipe of the early Paris manuscript[23] is the following:

''Take pure sandarac (the resin) 1 lb., liquid (gum) armoniac, 1 lb., rub them together and put in a glazed earthen vessel carefully closed and luted with sapia (the lute of the philosophers), then let it be placed over the fire and liquefied. These are the signs of (completion) of this liquid, that placed upon wood it seems of the consistency of butter. Then add four lbs. of Greek pitch (''Alkitram''

[20] *Geschichte der Chemie*, III, p. 220.
[21] *Histoire de la Chimie*, 2d ed., p. 304.
[22] Berthelot, *op. cit.*, I, p. 89 *ff.*
[23] Berthelot, *op. cit.*, I, p. 100.

—Arab word meaning bitumen or a liquid pitch). It is forbidden to do this under a roof since danger would threaten.

"If you wish to use this on the sea, place about 2 lbs. of this oil in a goat-skin bottle if the enemy is near, more if he is distant. Attach the bottle to an iron dart (veru). Provide a piece of wood of size proportionate to the dart, and this (wood) should be rubbed with grease on the lower side. Set fire to this wood at the shore and place upon it the bottle. The oily matter burning upon the dart and the wood will run over the water and burn whatever it meets.

"Another kind of 'fire' which sets fire to the houses of the enemy whether situated in the mountains or in other similar places:

"Take balsam or petroleum, 2 lbs., the pith of Canna ferula, ½ lb. [described by Pliny, Liber XIII, Chapter 42, as a tall jointed reed with a fungous kind of pith], sulphur 1 lb., melted mutton fat, 1 lb., either the oil of teribenthine or the oil of bricks,[24] or the oil of anise. All being mixed prepare an arrow (sagitta) with four openings (or cavities) and fill with the above composition. Set fire to it and shoot it with the bow. Then the grease being melted and the composition kindled it will set fire wherever it falls and if water is thrown upon it it only augments the flames."

Another mixture suggests the torches of the Bacchantes of Livy which were inflamed by wetting.

"Here follows another kind of fire with which Aristotle destroyed the houses situated in the mountains and so that the mountain itself settled down.

"Take of balsam 1 lb., pitch 5 lbs., oil of eggs and quicklime equal parts, (in all) 10 parts. Grind the lime with the oil so as to make one mass. Smear with this mixture the stones, herbs and any growing things, during the dog days. Bury them in manure under ditches in that place. At the first autumn rain falls, the earth will take fire and its fire will burn the inhabitants, for Aristotle asserts that the fire of this lasts nine years."

Though this tradition is falsely attributed to Aristotle

---

[24] The oil of bricks is described in a Munich manuscript of the *Book of Fires* (written 1438 A. D.) as made by pounding up bricks into small pieces, soaking in olive oil and distilling, thus producing an oily product modified by "cracking." Berthelot, *op. cit.*, I, p. 102.

and the whole affair is legendary yet it may be assumed
that it has a basis of fact since it is known that the
action of water upon quicklime and confined masses of
combustible materials can produce inflammation. The
Book of Fires also contains the first unmistakable reference to saltpeter and to its use in explosive mixtures,
and also to black powder. It is here called sal pet-
rosus, or salt from stones. It is not improbable that
its use in inflammable mixtures was known to Byzantine Greeks at a much earlier period. If so they probably kept this knowledge to themselves. Whether this
substance was known to the Arabian chemists previous
to the date of the manuscript of the *Book of Fires* is a
matter of considerable doubt. Berthelot[25] thinks it probable, though no direct evidence is as yet available. Von
Lippmann[26] considers that the knowledge of it is due
not to Arabian but to Italian chemists. Though some
of the mixtures that are called Greek fires are mentioned in the *Mappae Clavicula,* the explosive mixtures
resembling black powder are not mentioned by any
known European or Arab writers previous to about
1250 A. D.

The first reference to saltpeter is found in No. 12 of the
recipes of Marcus Graecus:[27]

"Note that the composition of a fire for flying in the air
is twofold, of which the first is:

"Take one part of colophony and as much native sulphur, (?) parts of sal petrosum. These well pulverized and
saturated with oil are dissolved in linseed oil or, which is
better, in laurel oil. It is then put into a reed or hollow
stick and kindled. It rushes out suddenly to whatever place
you will and burns everything.

"No. 13. The second method for 'flying fire' is thus effected:

"Take 1 libra of native sulphur, 2 libra of charcoal of
linden or willow, 6 libra of sal petrosum which are all three
well mixed on a marble stone. Afterwards place the pow-

---

[25] Berthelot, *op. cit.*, I, p. 98.
[26] *Entstehung und Ausbreitung der Alchemie*, p. 487.
[27] Berthelot, *op. cit.*, I, pp. 108, 109.

der at pleasure in an envelope for flying (rocket) or for making thunder (tonitrum).''

Note that the envelope for flying should be long and slender and filled with the powder well packed. But the envelop for making thunder should be short and thick and half-filled with the said powder and strongly tied at both end with iron wire. (filo ferreo).

"No. 14. Note that sal petrosum is a mineral of the earth and is found in efflorescences upon the stones. This earth is dissolved in boiling water, afterwards purified and filtered ("distilled per filtrum") and is permitted to heat for a whole day and night and you will find at the bottom scales of the salt, solid and clear."

Another recipe in the same manuscript for black powder is No. 33.[28] It is here called also flying fire, and is made from sal petrosum, native sulphur, and from charcoal of grapevine or of willow. "And note that with respect to sulphur you should take 3 parts of charcoal and with respect to charcoal 3 parts of sal petrosum."

It is interesting to note that a Syrian-Arabian manuscript based probably upon much earlier writings of the tenth or eleventh century but written in the sixteenth century contains several mixtures for black powder[29] for various applications, for example, these formulae:

"For priming of firearms, 10 of salpeter, 1 of sulphur, 1 of charcoal—grind them together.

"For rockets and war machines, 10 drachmas of salpeter, 2 of charcoal, 2 of sulphur—reduce to powder.

"For petards or crackers, 10 drachmas salpeter, 3 of charcoal and 1½ of sulphur."

These items evidently are late interpolations, for the use of black powders in firearms was not earlier than the fourteenth century.

The English scholar Roger Bacon has often been popularly credited as being the discoverer of black powder. That he knew of black powder and that it was composed of

28 Berthelot, op. cit., I, p. 119.
29 Berthelot, op. cit., II, p. 198.

sulphur, charcoal and "sal petrae" is certain, as appears from his writings of unchallenged authenticity. But in these writings he refers to this knowledge as common property and his references to it indicate that he has derived it from other sources than his own experience.

Thus in his *Opus Majus* (1267–1268 A. D.) speaking of various important results of experimental science he says:[30]

"There are certain things which undergo change by contact alone and so destroy life. Thus Malta, which is a kind of bitumen and is in great abundance in the world, when projected upon an armed man sets him on fire. This fact the Romans experienced with heavy slaughter in taking places by storm as Pliny testifies in the second book of his *Natural History,* and as histories confirm. Similarly yellow petroleum oil, that is an oil originating in rock, sets fire to whatever it meets if rightly prepared, for a fire made from it can with difficulty be extinguished for water does not extinguish it. Certain things disturb hearing so much that if suddenly operated at night and with sufficient skill neither city nor army could endure it. No thunder clap could be compared with such. Certain things inspire such terror at sight that the flashes from stormclouds disturb far less—beyond comparison; by works such as these Gideon is believed to have operated in the camp of the Midianites. And an experiment of that character we take from that boyish trick (ludicro puerile) which is performed in many parts of the world, namely that by a device made of a size as small as the human thumb, by the force of that salt called sal petrae, such a horrible noise is produced in the rupture of such a small thing as a little parchment that it is felt to surpass the noise of violent thunder and its light surpasses the greatest flashes of lightning."

In a fragment of the *Opus Tertium* (ca. 1268) discovered by Prof. P. Duhem, Roger Bacon refers again to these explosive toys and states that their contents was a mixture of salpeter, charcoal and sulphur.[31] These references of Bacon's to such mixtures of inflammable and explosive

---

[30] *Opus Majus,* Bridges ed., II, pp. 217, 218.
[31] A. G. Little, *Part of Opus Tertium,* Aberdeen, 1912, p. 51.

mixtures seem to indicate that his knowledge of them was general and derived from his readings of other works, among which were probably some such as the *Book of Fires*.

The claim for Roger Bacon as the inventor of gunpowder rests mainly however on a portion of a work entitled *De secretis operibus naturae et de multitate magiae*. This is a short treatise on remarkable inventions. The first few chapters contain very interesting examples of Roger Bacon's scientific imagination. The last few chapters are so very different in character that modern scholarship suspects the genuineness of their authorship. Certainly the chapters in question differ in style and content so greatly from his well authenticated writings as to strongly confirm this suspicion. The passage in question has to do with the composition of gunpowder but the language is unclear and is made more so by the use of a secret cipher which has long puzzled chemists. Lieutenant Colonel Hime[32] has given much study to the subject and has presented an attempted solution. The passage runs thus:

"Item pondus totum sit 30. Sed tamen sal petrae LURU VOPO VIR CAN UTRIET Sulphuris: et sic facies tonitrum et coruscutionem si scias artificium." Hime transposes the letters thus R. VII PART V NOV CORUL V ET, and makes the sentence read:

"Sed tamen sal petrae R (cepie) VII part (es), V Nov (elle) corul(i) V et sulphuris."

The whole paragraph would then read translated:

"Let the whole weight be 30. But take of salt peter VII parts, V of young hazelwood and V of sulphur, and thus you can make thunder and lightning if you know the trick."

These proportions would give when calculated to percentage composition a less efficient mixture than those quoted from the *Book of Fires*, as shown in the following comparison:

[32] *Roger Bacon Commemoration Essays.* Oxford, 1914, p. 321 *ff.*

ANALYSIS OF EXPLOSIVE POWDERS

	Salpeter	Charcoal	Sulphur
Marcus Graecus 1st recipe.....	66.7	22.2	11.1
Marcus Graecus 2nd recipe.....	69.2	23.1	7.7
Supposed R. Bacon's recipe.....	40.2	29.4	29.4
Modern military blck powder...	75.00	10 to 15	10 to 12.5

Even should we grant that Bacon wrote this cipher and that it is here correctly interpreted there is little basis to assume for Bacon the original invention. On the other hand a student of Roger Bacon's works of recognized ability, M. Charles,[33] has expressed his conviction that the last six chapters of the above work are apocryphal and Prof. M. M. P. Muir[34] also has recently voiced his doubts of their authenticity on the basis of internal evidence. In the present state of our knowledge therefore there seems no adequate reason to ascribe to Roger Bacon any other than an early knowledge and appreciation of the advance in some chemical arts, of which his great contemporary scholars, Vincent of Beauvais, and Albertus Magnus were not yet cognizant.[35]

The *Liber Sacerdotum* or *Book of the Priests* the text of which is published by Berthelot[36] is a work translated into Latin from the Arabic. It is evidently based largely on

---

[33] *Roger Bacon, Sa Vie, Ses Ouvrages, Ses Doctrines d'apres des Textes Inedits*, Paris, 1861.

[34] Muir, *Roger Bacon Commemoration Essays*, p. 301, Oxford, 1914.

[35] Since the above was written the work by Lynn Thorndike on the *History of Magic and Experimental Science* has appeared, containing an elaborate discussion of Roger Bacon, and as an appendix a chapter on ''Roger Bacon and Gunpowder.'' In this the author expresses his doubt of the authenticity of the *Epistola de secretis operibus*. Much of it sounds like a brief compilation from Bacon's three works of 1266–1267, concocted by some one else later. And Hime's interpretation of the cipher is subjected to searching criticism, concluding as follows: ''And now what becomes of Colonel Hime's assertion 'Since therefore charcoal is one of the subjects of these two chapters, it becomes all the more probable that saltpeter forms another'? We may alter it to read thus; 'since charcoal is not a subject of either of these chapters, it becomes all the more improbable that a method of refining saltpeter is disclosed in them in cipher.'

[36] Berthelot, *op. cit.*, I, pp. 179–228.

Greek-Egyptian sources. The work was probably edited about the tenth or eleventh century, and attributed to a Joannes, a person of unknown identity. The manuscript as published by Berthelot appears to be an elaboration of about the twelfth century by a Spanish editor. It is a collection of about 200 recipes, without system or order, carelessly copied, and in rather bad Latin. In content it resembles somewhat the Theban Papyri and the *Mappae Clavicula,* in being largely devoted to imitating gold and silver in cheaper alloys, to methods of superficial coloring of metals or other substances, to inks, to mixtures for decorating glass or pottery, for imitating precious stones, or semiprecious stones, for purifying various chemical substances. The meaning and purpose of many of these recipes are obscure—owing to unclear descriptions—and sometimes evidently to a desire to conceal definite information from the general reader. The directions in this work call for the use of a great many constituents of mineral, vegetable and animal origin. While on the whole not very instructive as to accurate information or processes or purposes, yet they evidence the fact that the activities of the Arabian chemists were very considerable, though their originality is not manifest in notable discoveries. A few illustrations will perhaps serve to a better understanding of the character, though they represent the least obscure class of recipes.

No. 40 is apparently a recipe for making a sort of mixture for coating silver articles to resemble gold.

No. 40.   How silver is turned into gold.

Almagra,[37] (defined sometimes as a red earth, sometimes as brass, or as copper bole), acimar (that is, flos aeris, copper oxide), Atramentum ustum, roasted vitriol, roasted brass (mixture of copper and zinc oxides), rock salt, almisadir (sal ammoniac), saffron root or saffron itself—equal parts. All these are mixed with urine and dried in the sun. With this powder mix filings

---

[37] A note in the manuscript defines *Almagar* (sic) as *"berillus,* namely, a red with which walls are painted."

or very thin flakes of silver leaf and heat in the manner of gold, that is in a crucible well covered. Then heat again with filings or flakes. Do this seven times and it will be what you have wished. With this unite just as much gold and it will be the best gold after you have decorated.[38]

No. 163. To make best gold.[39]

Of bronze (aeris) 3 parts, of silver 1 part, melt together and add orpiment, not roasted, 3 parts. When strongly heated let it cool and put in a pan and cover with clay and roast until cerusa is made. Take it out and melt and you will find silver. If too much roasted, electrum will be made, to which if 1 part of gold is added it will make the best gold.

In this recipe the reference to cerusa (white lead) is puzzling. Unless lead was frequently present in bronze its formation is difficult to explain. It is possible that white fumes of arsenious oxide from the orpiment may have here been mistaken for it or it may be a blunder in translation into or out of Arabian.

166. To make a gold-colored water.

Kibrit (sulphur) 1 (pt), sulfur (manuscript gloss says "id est auripigmentum") Asphar (?) 1, quicklime, 1 part. Place in a pot (cacabo) with ox urine and heat 1 hour and you will see a golden color. Put in a glazed jar and put this water into your operations.

Essentially this seems to be a solution of persulphide of lime and perhaps other constituents. The directions are not very satisfactory as to ingredients.

183. For making oil of eggs.[40]

Take eggs and cook in water. Place the yolks in a pan, roast gently and squeeze them out. This is the oil of eggs.

154. Take two parts quicksilver and one of sulphur and put in a new dish and place in the furnace and heat with a moderate fire as much as suffices and then collect it. You will find what is pure.[41]

This is manifestly the preparation of cinnabar.

---

[38] Berthelot, op. cit., I, p. 195.
[39] Berthelot, op. cit., p. 218.
[40] Berthelot, op. cit., I, p. 222.
[41] Berthelot, op. cit., p. 216.

95. "Sulphur turns quicksilver red. The severity of the fire generates black at first, then a yellow (or red) color."

159. "Alchool—that is a most subtle powder." [42]

The Book of Stones falsely attributed to Aristotle—*Aristoteles de Lapidibus,* is a work which occupied a prominent place in the middle ages as a source of information on mineralogy. The work according to Ruska[43] was originally compiled by a Syrian acquainted with Persian and Greek traditions sometime before the middle of the ninth century. Written in Arabian or translated into Arabian at an early date, it is the oldest known Arabian authority upon mineralogy. It has been rewritten and expanded by various scholars at various times, so that the existing manuscripts in Hebrew, Arabic and Latin languages differ widely in content. In these various versions it served as a basis for later writers and especially either directly or indirectly for the encyclopedists of the thirteenth century—as Bartholomaeus Anglicus, Vincent of Beauvais, Albertus Magnus and writers of less importance.

A Latin manuscript preserved in Lüttich has been published by Valentin Rose.[44] This is an early fourteenth century copy of a version edited with elaboration and additions by a Spanish-Arabian writer probably not earlier than the twelfth century.[45] The content of this work is naturally quite different from that of the laboratory manuals above described. It is a catalogue of minerals and precious stones, with a summary of their more obvious physical properties, their virtues—medicinal, or occult—for the ancient habit of assigning mystical and supernatural properties to all kinds of materials in nature—so well illustrated in Pliny's records—was well maintained in Arabian natural science, as it was by the early Greek alchemists. Though

---

[42] *See* ante, p. 189.

[43] Ruska, Dr. Julius, *Untersuchungen über das Steinbuch des Aristoteles,* Heidelberg, 1911.

[44] "Aristoteles de lapidibus und Arnoldus Saxo" in *Zeitschrift für deutsches Alterthum,* 1875, XVIII, pp. 321–455.

[45] Cf. Ruska, *op. cit.* p. 320 *ff.*

written in Latin, nearly all the names of the minerals are in Arabic, relatively few being also translated into Latin equivalents. Illustration of these descriptions are of interest as showing what kind of mineralogy served as a basis of much of the conceptions of minerals in the great thirteenth century encyclopedists.

"Description of the stone azurium. This stone is cold and dry and soft and of beautiful color. When this stone is mixed with gold the beauty of the gold and of the stone is increased and made durable and one color brightens and illuminates the other. And this stone contains gold mixed with it.

"The nature of this stone benefits the eyes when mixed with other powders. And when some of this stone is placed upon a fire without smoke, the flame is tinged by its color. And when calcined, fire becomes concealed in it.[46]

"There is a stone, called by the Greeks elsbacher, and named elbasifer kaker, and the description of it is that it is poison (stone). This stone is of great dignity and nobility. It is soft to the touch as found. The nature of it is warm and not very moist. It is subtle and smooth and a valuable property of it is that it cures from all poison of whatsoever kind whether deadly or not, both from poisons that come from the earth or from those produced by the bites of worms or reptiles. It also cures wounds and snake bites. Since we are speaking of poison, it is proper that we speak of its name and give its description because poison does not kill man by its coldness or its heat but by its property of evil for it penetrates even to the blood of the heart and liver and when it reaches the blood it makes it liquefy, resembling the water running out from flesh that is salted and this blood runs in the veins obstructing the passage of the living body and spreads through the whole body like grease (sagimen) upon water."[47]

"Description of the Stone called Elzarmeth." (Interpreted in a gloss to the manuscript as "auripigmentum.")

"This stone is found of many shades of red and yellow. Mixed with lime it removes hairs, skin, and flesh, and when

[46] V. Rose, *op. cit.*, p. 366.
[47] V. Rose, *op. cit.*, p. 362.

combined in this way they are a deadly poison. If the red
and the yellow be each calcined by itself until it becomes
white (that is roasted to arsenious oxide) and placed with
a little borax upon red copper and heated a while at the
fire, it (the copper) will be whitened and purged from its
corruption and made more beautiful. These stones have
many ores (mineras). If elzarivech (sic) is burned and
then made into powder it is able to cure cancer and fistula.
Also it enters much into furnace operations (opera ig-
nea).'' [48]

The above indicates use of arsenious oxide for cure of
cancers, etc.

"Description of glass (vitrum) which is called zegeg.

"Glass is of many colors. It is produced from many
stony and sandy minerals. When it is placed in the fire
with magnesia (in another manuscript *magnes*) they melt
and form one body by virtue of the lead and magnesia (or
magnes), and when drawn from the fire and exposed to the
wind (vento), before receiving a second temperate heating, the body of it is easily broken, and as there are many
colors of glass, there is found a certain kind that is so
white as scarcely to be distinguished from crystal (that is,
quartz), and this is the best. From this is derived red, yellow, green and violet. For it is a soft and fragile stone;
and just as a foolish man is bent by the sayings of anybody; so glass is of all colors for it receives all colors by
the heat of the fire and again is made stone when exposed
to the air. Its nature is warm in the first degree and dry
in the fourth degree. It is convertible into the nature of
any other stone, as glass becomes stone when brought into
cold air. (Referring to its use in artificial or imitation precious stones.) When tinctured in a temperate fire it is well
colored but if the fire is excessive or too feeble it is not
well colored. And just as flesh is pulled by beasts so glass
attracts iron to itself by virtue of its heat and dryness.'' [49]

The magnetic oxide of iron or the lodestone by its attraction at a distance made a great impression on the minds
of the ancients and there was a natural tendency to exag-

---

[48] V. Rose, *op. cit.*, p. 373.
[49] V. Rose, *op. cit.*, p. 381.

gerate its powers which lost nothing in the middle ages. The description of it here is long but nevertheless so well exemplifies the character of much of the natural science of the period that it is here given in extenso:

"Description of the stone called elbeneg, that is *magnes* or *calamita,* attracting iron. Its nature is warm and dry, and it is a stone which iron obeys. For no one who has sense and memory could believe but that iron is stronger than other things; also that it is stronger than other stones in so far as sustaining the action of fire and of sulphur and of strong hammering between two irons and for manufacturing. Also they make from it weapons against all men and beasts and man avails himself of it in all his operations except concerning plants. So also those are safe who work with it upon other metallic (?) bodies (corpora).

"Whenever that stone approaches iron it draws it to itself so that it is seen that iron has in it a spirit, for *magnes* causes it to move as if it had a living spirit in it. And it comes to this stone and attaches itself to it. And such is the obedience of iron to this stone that if many needles should be fixed in the earth and this stone should approach them, all the needles would attach themselves to the stone, or if one were attached to the stone the others would attach themselves to this one so that one would hang from another.

"The best of this kind of stones is black mixed with reddish. There is a great force hidden in this stone, for if placed in some large vessel full of quicklime untouched by water, and the vessel is so great that the force of fire may be concealed in it, and the vessel be placed in a brick-maker's furnace when it is first fired, then taken out and permitted to cool, and then this stone taken out and placed similarly in another vessel as before and in the furnace just as before, and so it be done three or four times, and the stone removed and put in a clean place where neither wind nor water nor dampness can touch it; and pieces of this are taken which weigh 10 drachmas, and if now one such piece is taken and the same weight of alkibric (sulphur) added and strongly stirred and mixed; then thrown into water

very great heat will be generated burning whatever combustible may be near.

"If this stone before being calcined is placed in the water
of onions or of garlic and should remain there three days
it entirely loses its power; but it will recover it if it is
placed in goats' blood for three days so the blood be renewed each day. And he who wishes to deprive it of its
power which it possesses from heating, let him put on it a
little goats' blood and he will thus deprive it of its power.

"The mine (or source) of this stone is on the shore of
the sea near the country of India. When ships pass near
the mountain where this stone exists, it is not possible for
iron to stay in them, but that it leaps out; flying out, now
above, now below, it does not stop until it reaches the *magnes*. Similarly the nails of ships are pulled out whence it
is commonly accepted that ships passing through that sea
should not be endangered by being joined with iron nails
but by bolts of wood, or in some other way, for either they
are broken up by the removal of the nails or they are even
drawn to the mountain from which it is impossible to separate a ship when once adhered.

"If any poison containing iron filings should be given
to anybody in drink or if any one is wounded with a poisoned iron, powdered magnes finely rubbed with milk may
be given and he who has drunk iron filings, or poison mixed
with iron will be purged of that poison. But a wound made
by a poisoned iron may be sprinkled with the powder of
this stone and will be cured through the power of God.

"Iron therefore obeys this stone by the virtue that is contained in it. The good properties that God has given it
must be made manifest to those who believe in Him just
as He, by His might (ex se), overcomes bodies that seem
to men perfect, strong and lasting. May He be blessed
through the ages."

This discussion of the lodestone affords an excellent illustration of the curious medley of facts, ancient fables,
exaggerations, and distorted description that characterize
so much of the natural science of the middle ages and early
renaissance. It is probable that the above description of
chemical operations is a distorted and garbled version of

some actual operations. Repeated copying and interpreting of early manuscripts by scribes or editors themselves ignorant of the actual processes often obscure the meaning of passages of technical character, even if they were at first clear and intelligible, which as we have seen was not always the case.

Chemistry in the tenth century writings of the Ichwan el safâ—the "Lautern Brueder" or "Faithful Brothers."

This was a society of Arabian scholars founded in Basra about 950. Their writings were issued about 975 to 1000 A. D., and were strongly influenced by Aristotelianism and Neoplatonism. In 1160 their works were publicly burned in Bagdad, as the Mohammedan church of that period was very suspicious of any writings which might threaten the orthodox doctrines of the church, as was also the Christian church of this and later centuries. The chemical philosophy of these writings has been summarized by Dieterici, editor of the Arabian text of these writings, in his treatises on the philosophy of the Arabs of the tenth century,[50] from which the following abstract is derived: The writings of the Faithful Brothers are of especial interest in the history of chemistry because they summarize the Arabian chemical philosophy at the period previous to the mingling of Arabian and western ideas which occurred during the eleventh and twelfth centuries. The fundamental concepts of the four elements of Plato and Aristotle, as constituting matter, are at the basis of their theories as with the ancient Greeks and Romans. These concepts, however, were developed in certain directions and systematized to some degree on the basis of observation and experiment.

With the Arabs, water, air, and earth were components of all minerals, while fire was not so much a constituent as a regulator of the union of the other three. Plants and animals also, in so far as their material composition is concerned, contain the same constituents. All minerals con-

---

[50] Fr. Dieterici, *Die Philosophie der Araber im IX und X, Jahrhundert n. Chr.*, 2te Theil, Leipzig, 1879.

tain earth as their body, water as spirit, air as soul, all combined and regulated by fire or heat. Minerals were subdivided into seven classes:

1. Stoney, but fusible and solidifying on cooling. Such are gold, silver, copper, iron, tin, lead, glass, etc.
2. Stoney, but not fusible, as the diamond, hyacinth, cornelian, etc.
3. Earthy, soft, not fusible, but easily separated, friable. Such are salts, talc, vitriols, etc.
4. Watery and escaping from fire (volatile) as quicksilver.
5. Aerial—or oily—consumed by fire, as sulphur, arsenic, (sulphides of arsenic probably referred to).
6. Vegetablelike, as coral, which grows like a plant.
7. Animal-like, as pearls.

Pearls are also, like amber and manna, considered as originating from dews formed in the air and condensing under conditions in various places. It was fancied even that the oysters came up at times and opened their shells to the air to receive the dew which formed the pearls.

All the metals are composed of the same constituent materials, mercury and sulphur, and only indirectly of the four elements. Thus differences result from the proportions and the grades of purity of the mercuries and sulphurs and the degrees of perfection in their combination as the result of their heating or digestion in the earth.

The various waters which mingle in the interior of the earth, are by heat volatilized to the upper strata in crevices and cavities, there becoming condensed and thickened by cooling, and, again percolating downward, mix with earthy particles and by the heat of the earth are changed to quicksilver.

The aerial and oily parts mixing with earthy particles, become viscous and heavier and by heat eventually produce sulphur. The quicksilver and the sulphur, mixing in the earth under the influence of heat and time, form the metals and minerals. If the quicksilver is clear and the sulphur pure and they are perfectly united in the proper proportion and the heat is favorable and time is adequate for completion of the process, gold is produced. If this process is interrupted so that time is not allowed for perfection and too early cooling take place, silver is formed. If the heat is excessive and the sulphur or mercury contain an over proportion of earth, copper is the result. If cooled before well combined, tin is formed. If cooled before properly combined and there is too much earth present we have iron. If too much quicksilver and too little sulphur and the combination is imperfect from inadequate heating, lead is the product. If the heating is too great so that both the constituents are injured by burning, antimony (stimmi) results.

It is very difficult to know positively whether the Arabians refer here to metallic antimony or not. Stimmi with ancient and medieval writers generally means the native sulphide, yet that they used metallic antimony, but generally confused it with lead, is also certain. Yet classifying the sulphide of antimony here among the metals seems to be hardly reasonable. In this connection, however, it should be remembered that the word for metals originally meant the mines themselves, and later was used to represent the products of the mines and that at no time with the ancient or medieval writers was there any recognition of the existence of metals as elementary substances, nor were they fundamentally distinguished from other minerals.

The common metals gold, silver, lead, copper, tin, iron, being similar from so many points of view, were from an early period considered as minerals especially closely related. Their fusibility; their cooling again to the same solid condition; the fact that they could be melted together to form other kinds of metal (alloys); their malleability either in the cold or at furnace heat; their adaptability to so many common uses, coins, statues, jewelry, tools, etc. easily gave rise to the idea that they possessed a constitution more alike than was the case with minerals in general. There existed, however, no philosophic distinction between the metals and the minerals. Thus glass melts and solidifies again to a hard mass. It is also malleable and ductile in the heat. This perhaps is why we find it occasionally listed with metals in medieval writers. So also quite frequently electrum and bronze or brass are described as separate metals, even by writers who know they can be obtained by processes involving melting together silver or gold or copper and tin, etc. Mercury, on the other hand, is generally excluded from the metals though it is known to alloy with them or to dissolve them.

An English chemical writer of the thirteenth century, Richardus Anglicus in his *Correctorium Alchemiae*,[51] attempts to distinguish formally between metals and other minerals. In general his chemical philosophy is the current Arabian. Minerals are divided, he explains, into two classes: the metals, which owe their origin to mercury, as gold, silver, copper, tin, lead, iron, or "major minerals;" and those which do not owe their origin to mercury, such as salts, atramenta, alums, vitriol, arsenic, orpiment, sulphur and the like, called "minor minerals," but are not metallic bodies. The metals owe their origin to mercury and sulphur of different degrees of purity.

Those minerals which are not from mercury, and those salts which are soluble in water, as alums, chalcanthum (sulphates of iron and copper), common salt, sal petrae, and some substances insoluble in water alone, as orpiment, arsenicum, sulphur and other sulphurous minerals, result from the "aqueosity of sulphurs mixed with viscous earths firmly united by a fervent heat, whence they are rendered unctuous and afterward solidified by cold." The medieval chemical philosophers generally do not devote so much attention to the fundamental composition of nonmetallic minerals, and the classification here given by Richardus as to their origin is by no means in accord with others, especi-

---

[51] *Alchemiae Gebri*, etc. Bern, 1545, pp. 223–227. There has been much doubt as to the identity and date of this Richard, but the strong probability is that he died in 1252. Cf. Ferguson, *Bibliotheca Chemica*, II.

ally as to the absence of mercury as a constituting substance in this class of substances, the philosophy of the Arabs as shown in the Faithful Brothers being directly opposed in that particular.

This concept of the nature of the metals which assumes that they are all essentially of the same constitution differing only in the relative proportion and purity and degree of "ripening" by heat, gave encouragement to the alchemical experimentors of the time to hope that by the use of artificial admixtures and by varied conditions of heating of the less perfect metals, it might be possible to complete the perfection of them and thus to actually transmute the baser metals into the noble or more perfect metals, gold and silver.

In the noble metals and in many minerals the elements were believed to be so well combined that heat could not separate them. Other minerals, as sulphur, orpiment, asphalt, etc., when heated in the air are partly broken down, the aerial element, not being so firmly united to the earth, being driven off as vapor and mingling with the particles of the atmosphere. This process was interpreted by the Greek alchemists and their Arabian successors as the separation of the spirit from the body, and such substances as were volatilized or burned with formation of gaseous products—as sulphur, arsenic (sulphides), sal ammoniac, quicksilver—were called spirits, while the metals and minerals which, when heated in the air did not volatilize nor disappear in gaseous products, were called bodies (corpora).

The influence of the planets and other heavenly bodies upon the generation of metals and minerals was considered of importance, as also their influence upon the growth and development of organic life, including man.

This theory of the composition of metals from quicksilver and sulphur in this Arabian chemical philosophy is so authoritatively asserted that, as von Lippman suggests, it is very probable that it was not new with them, but was derived from Greek-Alexandrian sources through the Syrian mediation.

The books of the Faithful Brothers contain much in the way of chemical facts that shows a knowledge based upon practical experience in chemical operations. Thus operations of distillation are spoken of in the preparation of waters of roses and violets, and of sharp vinegar, though there is no indication that the methods or apparatus were other than those given for instance by "Zosimos" in the Greek manuscript of St. Mark. In discussing the metals, many properties are described:

"Gold is a substance of well proportioned native and perfect mixture. Soul, spirit, and body have become one in it, therefore it does not change by any happening, nor does it decay. Its yellow color comes from its fire, its purity and luster from its aerial element, its softness from unctuous moisture and its weight from its earthy constituents. Gold alloys with copper and with silver. From these it can be separated by strong heating with 'Markasite' (pyrites) which is a kind of sulphur which is not consumed by fire like other sulphurs. The gold is by this unchanged while the copper is burned away. (This is purification of gold by the ancient process of cementation.) Gold is dissolved by quicksilver, but by heating the quicksilver can be driven off. By malachite and borax (tinkar) gold can be soldered.

"Silver alloys with copper and lead from which it can be separated by heating with soda (nitrum) and other substances. It is burned by long continued strong fire, 'decays' if buried in the earth. Sulphur blackens it. It is softened and dissolved by quicksilver.

"Tin, white like silver, but soft and of bad odor, creaks when bent. It can be burned and the product is not poisonous but useful in medicine. By heating with salt, arsenic (sulphide), markasite and twigs of myrtle it may be changed to silver.

"The Emerald and Topaz are stones of the same class, dry and cold. They are found in gold mines. The best are the clearest, greenest and most transparent. By long gazing upon the emerald, weakness of the eyes is cured.

Worn in seal ring or as a button on the girdle it protects from epilepsy.

"Malachite is a stone originating in copper mines. Its nature is cold and tender; it rises as a vapor with the sulphur originating in copper mines; is green like the rust of copper (verdigris); when it arrives at some place in the mine its particles bake together one upon another and it becomes a body. This stone is of green but cloudy color. It is of poisonous character, the dust of it produces sores in the bowels and inflames the eyes. Malachite is an enemy to topaz although similar. When lying near it, it clouds its color and spoils its luster.

"Quicksilver is a moist liquid unquiet in the heat. Mixed with mineral bodies it softens, weakens and makes them brittle. Heated it leaves them again hard, just as water mixed with clay leaves it when heated.

"Salts, alums, soda, glass, etc., some of which have agreeable tastes, others bitter, others hot or astringent, are minerals derived from moistnesses mixed with earths and baked and hardened by fire, by the sun's heat or the interior heat.

"The diamond is cold and dry to the fourth and highest degree. Seldom are these two qualities (natures) so united in one mineral, therefore when rubbed upon other minerals it breaks and cuts them. Only a kind of lead is an exception which in spite of its softness and ugly form acts upon the diamond, breaking and wearing it off, just as the small gnat has power over the elephant.[52]

"The lodestone (magnes) is an example for the intelligent. Through iron is extremely hard and dry, more so than minerals, plants or animals, it moves to this stone and clings to it like the lover to the beloved. The creator moves these two together as the body of itself has no such power."

As sources of acids (acetum) used by the Arabs are mentioned not merely vinegar, but the juices of unripe

---

[52] The power of lead to break the diamond often repeated by Arabian writers depends according to v. Lippmann on a misunderstanding of the custom of melting the diamond into lead for the purpose of holding it when cutting or splitting the stone. V. Lippmann, *op. cit.*, p. 385.

citrons, lemons, oranges, tamarinds, and from the ripe fruits of the oak and cypress, and nutgalls. Tannin solutions were thus not sharply differentiated from the vegetable acids, or vinegar.

The theories of chemical composition as formulated in this tenth century work are with slight variations the theories which were maintained through the fifteenth century in Europe, as we shall have many illustrations in the future. The practical knowledge as here illustrated is not a great advance over the chemistry as known to Pliny or as shown in the Theban papyri, though more specific in many details.

An important name in chemistry to the writers of the thirteenth and later centuries is Avicenna. The importance of Avicenna in the history of medicine is beyond question. He was largely determinative of the theory and practice of medicine in the middle ages. His significance in chemistry is however not great, and such as it is, it is due not so much to his own contributions as to works published under his name by unknown writers of much later date. Avicenna lived from 980–1036. The most influential chemical work attributed to him was a work entitled *De Anima in Arte Alchemiae*. It is possible though unproven that it is based upon some original writing of Avicenna. Berthelot, who has published an extended analysis of the work,[53] considers this possible, though other students of the history of those times consider it as composed not earlier than the twelfth century. Certainly in the form in which it is known in manuscript or print it was written in Latin by a Spanish writer, as it contains Spanish words as for instance *plata* for silver, and it contains mention of names of several writers not earlier than the twelfth century or the beginning of the thirteenth, some of them being Christian churchmen. That it is the work of an Arabian scholar in Spain is evident from the many references to Arabian authorities some of whom are not known

---

[53] Berthelot, *op. cit.*, I, Chap. VI, p. 293 *ff.*

otherwise. That it was a well known and highly estimated authority in the thirteenth century is evidenced by the frequent quotations from it by Vincent of Beauvais (about 1250), and by references in the works of Roger Bacon and other writers of the thirteenth century.

The chemical philosophy of the pseudo-Avicenna is practically the same as that of the writings of the Faithful Brothers previously discussed. The chemical facts contained in it appear to present no important advances over preceding writers. Like other Arabian chemists the chemical philosophy is based upon the theories of matter of Plato and Aristotle, and upon mercury and sulphur as the constitutents of metals. It contains much mysticism, astrology and much is incomprehensible. The reality of transmutation of metals is recognized, as when he says that the best gold is made by the philosopher's stone, but it is also stated that "certain ones make false gold and silver. They stamp out (stringunt) and harden tin, whiten it and call it silver. So also they take sublimed orpiment (arsenious oxide), digest it in manure, and mix it with salammoniac and incorporate it with copper by treating it in the furnace *per decensum* with addition of red mercury (oxide) and they say this is gold. But there are seven signs by which gold is recognized: by its fusion, the touchstone, its density, its taste, the action of fire, etc." [54]

The *De Anima* describes briefly many common minerals and salts, alums, vitriols and fluxes (borax) and processes and apparatus for washing, calcining, hardening and softening, sublimation, fusion and solution much in the same way as is done in some of the books of recipes already noted, though there is little that gives evidence of any thing other than a resumé of previous writers, whose chemical knowledge has been previously noted.

When we consider how important were the contributions of Arabian scholars in other domains of science as as-

[54] Berthelot, *op. cit.*, p. 304. Vincent de Beauvais, Lib. VIII, chap. XIII, gives as the seven tests for gold: solution, the touchstone, density, taste, the action of fire, fusion, sublimation.

tronomy and mathematics it seems strange that their contributions to chemical science and practice were so unimportant. The inference seems clear that the domain of
chemical science of the time, founded on the mystical al-
chemistry of the Alexandrian schools did not attract the
ablest scholars, so that except for the work of artisans in
the various trades the field of chemistry occupied the attention of students of inferior acumen and initiative.

A work on the arts of the Romans, *De Artibus Romanorum,* by Heraclius, a person of unknown identity supposed to have been a monk and to have lived in Rome about
the tenth century, is another treatise dealing to some extent with chemical arts. The work as it is known to us is
in three parts, the first two seeming to be of about the
tenth century while the third is obviously of much later date,
probably of the twelfth or the early part of the thirteenth
century. Lessing first called attention to it in 1774 in his
treatise on *The Age of Oil Painting.* Mrs. Merifield, in
her noted work *Original Treatises on the Art of Painting,*
(London 1849), first published the original text, and Ilg,
in 1873, published the text with German translation.[55]

The earlier portion of the work deals with gold and other
colors for manuscripts and miniatures, with the making of
artificial precious stones from glass, with various colored
glass enamels on pottery or glassware. Of the colors used
some are of plant origin, some mineral. Thus a green color
is produced from the leaves of a nightshade, *solanum nigrum,* ground with gypsum and water and afterwards dried
for use, or by copper and honey and vinegar (verdigris).
Gold color is produced by rubbing gold leaf with wine and
afterward mixing with a glue or gum for application;
or fish glue is applied and gold laid on in leaf form. Artificial gems are made from Roman glass, which is introduced
melted into forms or molded in earthenware and pressed

---

[55] The writer is indebted to E. v. Lippmann, *Chemiker Zeitung,* 1916, pp. 3
ff, 26 *ff,* 48 *ff,* and the same author's work on *Entstehung und Ausbreitung
der Alchemie,* pp. 472, 473, for his analysis of the chemical content of Heraclius.

while soft with a spatula to fill the mold perfectly and to free from air bubbles. Enamels on pottery are prepared from a clear powdered glass, mixed with various coloring substances and gum and burned on. Also various colored glasses are powdered and fixed with gelatin or gum and then burned on. It is of interest that no reference to lead glazes is made in these earlier sections.

The third section of the work is of later date, twelfth or thirteenth century, and much more extensive. The discussion of glass begins by quoting Isidorus as to the origin of the word vitrum from visui because it is transparent to vision, and quotes Pliny's fable of its discovery by Phoenician sailors by their using lumps of soda from their cargo when cooking their meals upon the sands. The description of glass making is quite detailed and suggests that the work of Theophilus the Monk (twelfth century) might have been available to the writer.

Glazes for pottery are described much as in the earlier divisions of the work, but here lead glazes are described.

Methods of gilding silver, copper or brass are given much as in the more ancient writings. So also solders for gold and silver, though borax is included indicating Arabian influence at the time of this writing.

The preparation of painter's pigments is also treated, with many vegetable as well as mineral colors in use.[56]

In 1781 Gotthold Ephraim Lessing published the Latin text of a work entitled *Diversarum Artium Schedula* from a manuscript in the Ducal Library at Wolfenbuettel. The author in the prologue calls himself "Theophilus, humble priest, servant of the servants of God, unworthy of the name and profession of monk." Nothing is known of his personality, though it is generally supposed by modern writers that he wrote in the twelfth century, though whether at the beginning or the end of that century, is variously estimated. That he was German has been inferred prob-

---

[56] Among these "*vitrum*" is noted, meaning woad. Ilg, as noted by v. Lippmann, has translated this word by "glass" which is the commoner meaning of this word.

ably from the fact that in this work a large proportion of the processes is devoted to the subject of metallurgy, especially developed in Germany, and that in his prologue he credits Germany with especial skill in the metals. He professes, however, to give an account of the various arts which have to do with decoration of churches, and with utensils or ornaments which are used in ecleciastical observances; of the kinds and mixtures of pigments and colors known to Greece; of whatever of ductile or liquid or polished substance distinguish the works of the Arabians; whatever of diversity in vessels, of gems or bones or works in gold distinguishes Italy; whatever France accomplishes in a variety of costly windows.

The work of Theophilus the Monk, as he is called to distinguish him from the early Greek alchemist Theophilus, is very notable among medieval writings for the clear and exact descriptions of the many processes which he describes. As a source of specific information on many technical chemical operations, it has no parallel until the pseudo-Geber at the beginning of the fourteenth century, and in clearness and definiteness it is not excelled by him.

Strangely enough the work of Theophilus seems not to have been known to the encyclopedists of the thirteenth century, Bartholomaeus Anglicus, Vincent of Beauvais or Albertus Magnus, nor indeed even to George Agricola in the sixteenth century. Though the text of his work was published by Lessing in 1781, the eighteenth and nineteenth century historians Gmelin, Hoefer, and Kopp make no mention of him. His work is referred to by Cornelius Agrippa in 1530, who credits him with an excellent treatise on glass making,[57] and by Morhof in 1688. In 1847 a good English translation was published by Robert Hendrie, (London, 1847) and a German translation by Ilg (Wien, 1874).[58]

---

[57] *Vanity of the Arts and Sciences,* Chap. 90.
[58] The Latin text available to the present writer is from the edition of Lessing's collected works edited by Karl Lachmann, 3d ed., Leipzig, 1898, Vol. 14, pp. 47–124.

The first of the three books comprising the work is devoted to the pigments used in painting, to their mixtures for special purposes, to the vehicles used in the application, oils, white of egg, glue, etc., to the preparation of gold leaf or tinfoil, or fine powders of gold or silver and their application in painting or illuminating and to the methods of making certain colors as cerussa (white lead), cinnabar, copper-greens, etc. While the materials used and the processes described were in the main used by the ancients, yet the descriptions are generally so much more specific than previous data generally that it is said that Arnold Böcklin, the eminent painter, made use of these recipes "in his partly successful attempts at producing beautiful and at the same time permanent pigments."[59] The following description of the preparation of cinnabar will illustrate his style.[60]

"If you wish to prepare cinnabar, take sulphur of which there are three kinds, white, black and yellow, to which, after crushing upon a dry stone, add two parts of quicksilver, correctly weighed on the balance; mix with care, put into a glass flask, covering that on all sides with clay and close the mouth so that no vapor may escape, and place it by fire that it may dry. Then place it among burning coals and directly that it begins to be heated you will hear a crackling ("fragorem") within, in which way the quicksilver mixes with the burning sulphur. When the sound has ceased, immediately remove the flask and opening it take out the color."

Book II is devoted to the manufacture of glass, to the making of glass articles, glass blowing, colored glass, decorating glass articles with painted patterns or with gold or silver—whether burned into the glass or merely laid on—imitation precious stones, etc. The description of the mode of constructing the glass furnaces is quite detailed as to plan, materials, dimensions, openings for working and for draft, etc. The utensils are also similarly

---

[59] E. von Meyer, *History of Chemistry*, London, 1906, p. 49.
[60] Book I, Chap. XXXII.

described—from the pipes for blowing glass to the tongs, bellows, etc.

His directions for making the glass are to take beech-wood, dried, and burn it in the furnace. Then take two parts of these ashes and a third part of flint, carefully cleaned from earth and stones, and mix them in a clean place. Then put them in the furnace and when they become heated stir at once so that they shall not conglomerate in melting, and do this for the space of one night and day.

Detailed directions are given for making glass articles of various kinds, glass plates, flasks, colored glasses, which he says the French make with great skill. There are also directions for painting, gilding glass articles, and for burning on the colors or gold in the special furnace for burning on colors. Artificially colored gems and their polishing are described. The execution of stained-glass windows for patterns in color is carefully described. It consists in cutting out pieces of colored glass plates to pattern, setting them in lead frames and then soldering these together by the use of a solder of four parts tin to one of lead. The details of this process are given with minuteness, and he well describes the style of stained-glass windows which characterized the twelfth century cathedrals. The earliest date known of this use seems to be about 1140. It is also to France that the earliest development of this art is credited.[61] As Theophilus also mentions France as being most expert in making beautiful windows, it is evident that the art was well established at the time this work was written, which is variously estimated from 1100 to 1175 A.D.

The third book deals with metals and metal working and constitutes more than half of the work. This section is of particular interest by reason of the fact that many processes which ancient or earlier medieval writers refer to or describe vaguely are described with such detail as to be

---

[61] Cf. A. Kingsley Porter, *Medieval Architecture*, N. Y., 1909, II, pp. 106–109.

clearly understood. Indeed many operations are much
more clearly described than by any subsequent writer before the pseudo-Geber (about 1300 A.D.), and some processes are described better than any writers before the time
of Biringuccio and Agricola in the sixteenth century. The
furnaces and tools are described at some length and also the
making of various articles from cups to organ pipes. Of
special interest from the chemical point of view are the
methods of smelting, purification and separation or parting
of the metals. The recovery of gold from the sands of the
Rhine is thus described:[62]

"There is gold sand which is obtained on the shores of
the Rhine in this way. They dig up the sand in those places
where there is hope of finding it and place it upon wooden
tables. Then water is poured over it frequently and carefully, and, the sands floated off, there remains the finest
gold which is placed by itself in a small vessel. When the
vessel is half full quicksilver is introduced and strongly
rubbed down by hand until thoroughly mixed, and the fine
quicksilver thus added is wrung out. That which remains
is placed in a melting pot and melted."

It will be recalled that ancient writers were familiar
with the use of mercury for recovering gold from mixtures
by wringing through skin, though they do not generally refer
to the necessary further operation of heating the remaining amalgam to expel the mercury.

How to separate gold from copper.[63]

"If you should break any kind of gilded copper or silver
vessel, you may recover the gold in this method: take
bones of any kind of animal, such as you may find in the
streets and burn them. When cooled grind them very fine
and mix with a third part of beechwood ashes and make
testas (cupels) such as we have above described in the
purification of silver, which you will dry either with fire or
in the sun. Then carefully scrape the gold from the copper
and wrap these scrapings in lead hammered thin, and, one
of these cupels being placed in front of the furnace in the

<hr>

[62] Liber III, Cap. XLVIII.
[63] Liber III, Cap. LXVIII.

coals, put in it, white hot, the folded lead with the scrapings and, covering it with coals, melt it. When it is liquified, in the manner in which it is customary for purifying
silver, occasionally renewing the coals and adding lead, occasionally uncovering and carefully blowing, heat until, the
copper being entirely consumed, the gold appears pure.''

The following passage "On Heating Gold" [64] is characterized in the edition of *Georgius Agricola's De Re Metallica,* by H. C. and L. H. Hoover, as "the first entirely satisfactory evidence on parting." [65]

"Take gold, of whatsoever sort it may be, and beat it
until thin leaves are made, in breadth three fingers, and as
long as you can. Then cut out pieces that are equally long
and wide and join them together equally, and perforate
through all with a fine cutting iron. Afterwards take two
earthen pots proved in the fire, of such size that the gold
can lie flat in them, and break a tile very small, or clay of
the furnace burned and red, weigh it, powdered, into two
equal parts, and add to it a third part salt for the same
weight; which things being slightly sprinkled with urine,
are mixed together so that they may not adhere together,
but are scarcely wetted, and put a little of it upon a pot
about the breadth of the gold, then a piece of the gold itself, and again the composition, and again the gold, which
in the digestion is thus always covered, that gold may not
be in contact with gold; and thus fill the pot to the top and
cover it above with another pot, which you carefully lute
round with clay, mixed and beaten, and you place it over
the fire, that it may be dried. In the meantime compose a
furnace from stones and clay, two feet in height, and a
foot and a half in breadth, wide at the bottom, but narrow
at the top, where there is an opening in the middle, in
which project three long and hard stones, which may be
able to sustain the flame for a long time, upon which you
place the pots with the gold, and cover them with other
tiles in abundance. Then supply fire and wood, and take
care that a copious fire is not wanting for the space of a day

---

[64] Liber III, Cap. XXXII.
[65] Georgius Agricola *De Re Metallica,* H. C. and L. H. Hoover, p. 459,
footnote. The translation here given is by Hendrie as quoted by Hoover.

and night. In the morning taking out the gold, again melt, beat and place it in the furnace as before. Again also, after a day and night, take it away and mixing a little copper with it, melt it as before, and replace it upon the furnace. And when you have taken it away a third time, wash and dry it carefully, and so weighing it, see how much is wanting, then fold it up and keep it.''

Also the description by Theophilus of the refining of copper is characterized by the same authority as the first notice of the process of ''poling,'' essential in the production of malleable copper.

''Of the Purification of Copper. Take an iron dish of the size you wish, and line it inside and out with clay strongly beaten and mixed, and it is carefully dried. Then place it before a forge upon the coals, so that when the bellows act upon it the wind may issue partly within and partly above it, and not below it. And very small coals being placed round it, place the copper in it equally, and add over it a heap of coals. When by blowing a long time this has become melted, uncover it and cast immediately fine ashes of coals over it, and stir it with a thin and dry piece of wood as if mixing it, and you will directly see the burnt lead adhere to these ashes like a glue, which being cast out again superpose coals, and blowing for a long time, as at first, again uncover it, and then do as you did before. You do this until at length by cooking it you can withdraw the lead entirely. Then pour it over the mould which you have prepared for this, and you will thus prove if it be pure. Hold it with the pincers, glowing as it is, before it has become cold, and strike it with a large hammer strongly over the anvil, and if it be broken or split you must liquefy it anew as before. If, however, it should remain sound, you will cool it in water, and you cook other (copper) in the same manner.'' [66]

The parting of gold and silver by means of sulphur is first clearly described by Theophilus.[67]

---

[66] This also is Hendrie's translation as quoted in Hoover's *Agricola*, p. 536, footnote.

[67] Liber III, Cap. LXIX. Hendrie's translation as quoted in Hoover's *Agricola*, footnote, p. 461.

"How gold is separated from silver. When you have scraped the gold from silver, place this scraping in a small cup in which gold or silver is accustomed to be melted, and press a small linen cloth upon it, that nothing may by chance be abstracted from it by the wind of the bellows, and placing it before the furnace, melt it; and directly lay fragments of sulphur in it, according to the quantity of the scraping, and carefully stir it with a thin piece of charcoal until its fumes cease; and immediately pour it into an iron mould.   Then gently beat it upon the anvil lest by chance some of that black may fly from it which the sulphur has burnt, because it is itself silver.   For the sulphur consumes nothing of the gold, but the silver only, which it thus separates from the gold, and which you will carefully keep. Again melt this gold in the same small cup as before, and add sulphur.   This being stirred and poured out, break what has become black and keep it, and do thus until the gold appears pure.   Then gather together all that black, which you have carefully kept, upon the cup made from the bone and ash, and add lead, and so burn it that you may recover the silver.   But if you wish to keep it for the service of niello, before you burn it add to it copper and lead, according to the measure mentioned above, and mix with sulphur."

The niello (or nigello) above alluded to is similar to the material as described by Pliny for blackening the surface of silver vessels[68] a fused mass of silver, copper and sulphur.   Theophilus[69] directs to take two parts pure silver, one part copper, and a weight of lead equal to that of the copper, and cover with sulphur and melt together with constant stirring.   When thoroughly melted the mixture is poured into an iron vessel and as soon as cool it is hammered a little, warmed a little and again hammered and so on until it is entirely hammered thin.

"For the nature of nigello is such that if hammered cold it liquifies, breaks and springs back (resilit).   It must not be heated to redness because it then melts and runs into

---

[68] *See* ante, p. 63.
[69] Liber III, Cap. XXVII.

the ashes." The nigello thus made then is placed in a deep and strong vessel, covered with water and powdered with a rounded hammer, taken out and dried and the finest is put into a goose quill and closed up, and the coarser material is again similarly crushed, dried and put into other quills.

The use of this nigello emphasized by Theophilus is for inlaying metal articles in decorative patterns and the method of application is also described in detail.

Very many processes connected with the chemistry of the metals and their compounds and alloys known to the ancients are described with similar detail by Theophilus, the preparation of cinnabar, white lead, verdigris, brass, gold leaf and tin foil or stanniol, cements, varnishes, etc. It is noticeable that no mention is made of any processes of metal working in which the mineral acids—aqua fortis, or aqua regia— are employed. Saltpeter, so much used in fusions in later times is not mentioned, nor does alcohol under any name enter into any operations. These omissions go far to confirm the assumption that this work is not of later origin than the twelfth century. The term "calamina" instead of the ancient "cadmia" for the ores of zinc used in the making of brass first appears in Theophilus.[70] With thirteenth century writers calamina is the more commonly used term. We may realize that this work is of the middle ages in spite of its almost modern style of description, from the account by Theophilus of the preparation of Spanish gold— "aurum hispanicum:" [71]

"There is also gold which is called Spanish which is made from red copper, ashes of basilisks, human blood and vinegar. For the pagans (Gentiles), whose skill in this art is probable, produce basilisks for themselves in this manner. They have a subterranean house, above, below, and on all sides of stone with two openings so small that scarcely any is visible. Through these it is said they put two cocks (galli) of twelve to fifteen years old and give them

---

[70] Hoover, *Agricola*, p. 112, footnote.
[71] Liber III, Cap. XLVII.

sufficient food. These when they have become fatted, from the heat of their fatness, mate and lay eggs. Which being laid, the cocks are ejected and toads introduced which keep the eggs warm, and to which is given bread for food. From the hatched eggs there come forth male chickens like hen's chickens, which after seven days grow serpents' tails, and if the house were not paved with stone they would enter the earth. Their masters guarding against this have round brass vessels of great size perforated on all sides, the mouths of which are narrow, in which they place these chickens, close the openings with copper covers and bury them in the earth and a fine earth entering through the openings they are nourished for six months. After this they uncover and apply an ample fire until the creatures within are completely consumed. This done and when cool they throw out and pulverize them, adding a third part of the blood of a red haired man (hominis rufi) which blood is dried and powdered. These two put together are mixed with sharp vinegar in a clean vessel. Then they take thinnest plates of purest red copper and spread entirely over them this preparation and place them in the fire. When they become red hot they take them out, quench in the same preparation and wash, and this they repeat so long until the preparation penetrates through the copper and then it takes on the weight and color of gold. This gold is fit for all works." [72]

As the glance of the fabled basilisk was believed to be fatal, the elaborate precautions taken in maturing and burning them are easily understood. We may perhaps infer that this curious example of superstitious alchemy is of Arabian origin from the designation of this gold as Spanish gold, Spain being then the meeting place of Arabian and western chemistry and alchemy. Of about one hundred and forty recipes in the *Schedula Diversarum Artium* the foregoing is the only one which is of that legendary character.

---

[72] This translation is from the Latin text of Lessing. Since the above was written, a translation of this passage following Ilg is published by Thorndike in his *History of Magic and Experimental Science*, I, p. 770. It differs only verbally from the above.

CHAPTER VI

CHEMISTRY IN THE THIRTEENTH CENTURY

The thirteenth century is distinguished by a remarkable development of culture in Europe.[1] The crusades covering a period from the end of the eleventh century to the middle of the thirteenth, exerted a great influence to that end. They brought western civilization into contact with Arabian culture, and opened to western scholars freer access to Constantinople and its treasures in manuscripts of Grecian classical literature as well as to later Byzantine developments. The crusades therefore functioned in that respect as a great international world fair. As we have seen the twelfth century was especially notable in the history of chemistry for the introduction of Arabian texts to European scholars and for the circulation of many such works in Latin translations.

The thirteenth century witnessed the founding of a large number of universities and the intellectual impulse brought forward men of eminence in many fields of thought, as for example Dante, Francis of Assisi, Roger Bacon, Albertus Magnus, Vincent of Beauvais, Marco Polo. Universities founded in the latter part of the twelfth and in the thirteenth centuries were, among others, those at Naples, Montpelier, Paris, Salamanca, Padua, Oxford, Cambridge, Toulouse, Sevilla, Orleans, Piacenza, Arezzo, Siena, Valladolid. Schools of earlier date which had existed as schools of law or medicine as Bologna and Salerno, were constituted as universities in the same period. Says J. R. Green:[2]

---

[1] *See* the interesting work of J. J. Walsh, *The Thirteenth, Greatest of Centuries*, New York, 1911.

[2] *History of the English People*, I, pp. 198, 205.

"The establishment of the great schools which bore that name (university) was everywhere throughout (Europe) a special mark of the impulse which Christendom gained from the crusades. A new fervour of study sprang up in the west from its contact with the more cultured east. Travellers like Adelard of Bath brought back the first rudiments of physical and mathematical science from the schools of Cordova and Bagdad. . . . To all outer seeming they were purely ecclesiastical bodies. The wide extension which medieval usage gave to the word 'orders' gathered the whole educated world within the pale of the clergy. . . . The revival of classic literature, the rediscovery as it were of an older and a greater world, the contact with a larger freer life whether in mind, society, or in politics, introduced a spirit of skepticism, of doubt, of denial into the realms of unquestioning beliefs."

One result of the new impulse was a renewed interest in natural sciences, particularly manifested in the translation and circulation of the works of Aristotle. Several influential scholars fostered the spread of the doctrines of Aristotle, notably Robert Greathead, Bishop of Lincoln, who influenced Greeks in Italy to translate Aristotle's works into Latin, Thomas Aquinas, who encouraged a translation by William of Moerbecke (archbishop of Corinth). Albertus Magnus and Roger Bacon, both appreciative students of Aristotle, exerted much influence to spread the knowledge of Aristotle and also to encourage the interest in natural sciences.

It was not without difficulties that the reëstablishment of the authority of Aristotle was effected. Some of his doctrines such as his concept of the eternity of the physical universe, and other ideas which seemed in conflict with the doctrines accepted by the church, excited some opposition. In 1209 the works of Aristotle were condemned and forbidden. In 1210 at the Provincial Synod at Paris the teaching of Aristotelian doctrines of natural philosophy was forbidden—*nec libri Aristotelis de naturali philosophia nec commenta legantur Parisiis publice vel secreto.*

These objections were in part due to the existence of works believed to be by Aristotle, but which were really not his, being productions of the Neoplatonic philosophers Plotinus and Proclus. With the appearance of translations from better Greek original works instead of from Arabic translations, this opposition gradually disappeared, so that in 1231 Pope Gregory IX ordered that books of Aristotle should only be used after being inspected and thus cleared of suspicion. By 1254 the study of Aristotle was again established in the University of Paris.[3]

In so far as the influence of this great revival of interest in the sciences of nature concerns the development of chemistry in the thirteenth century, we must note the appearance of a number of works, encyclopedic in character, which brought together and made accessible to a wide public the knowledge and speculations of ancient writers, Greek and Latin, as well as of their later Arabian interpreters and followers. Especially important as recorders and distributors of such chemical facts and ideas are Vincent of Beauvais, Albertus Magnus and Roger Bacon, while not so important in so far as content is concerned but influential on account of the wide use made of his works was Bartholomaeus Anglicus. These writers brought together the chemical science of the period from all authorities then recognized, from the early Greek philosophers to Dioscorides, Pliny and many other ancient writers, and many Arabic writers and other medieval authorities as Isidorus Hispalensis, Rhazes and some works of later origin attributed, though falsely, to those writers.

From these works can best be seen in what, to the most prominent scholars of the thirteenth century, chemistry consisted. It must be remembered however that not yet were the phenomena of matter classified as chemistry in the sense in which we use the term. They speak of alchemy

---

[3] Cf. Deussen, *Allgemeine Geschichte der Philosophie* "*Wachsende Autorität des Aristoteles,*" II, 2, p. 425 *ff.*

and the alchemist meaning the workers in metals chiefly, and always with the subject more or less clearly in mind of the transmutation of the metals as one of their principal aims. Because Bartholomaeus Anglicus is apparently the earliest of these writers, his work is deserving of attention. Bartholomaeus Anglicus composed an encyclopedia—*Liber de proprietatibus rerum*—much less bulky than the great works of Vincent of Beauvais or Albertus Magnus but perhaps of equal influence at the time. It was written probably about 1240.[4]

This encyclopedia, appearing apparently a little before the more comprehensive works of Vincent of Beauvais and Albertus Magnus, evidently had much influence in its time.

Says Langlois:[5] "Its success was prodigious during the latter centuries of the middle ages. It was in great favor and use in the universities and manuscript copies of the thirteenth to fifteenth centuries are still numerous in many

---

[4] Bartholomaeus Anglicus and his work has been discussed among others by Leopold Delisle in the *Histoire Litteraire de France*, XXX, 1888; by Robert Steele, *Medieval Lore from Bartholomaeus Anglicus*, 1905, reissued in the King's classics, London, 1907; and by Ch. V. Langlois in his *La Connaissance de la Nature et du Monde au Moyen Age*, Paris, 1911. Mr. Steele considers 1260 the probable date while M. Langlois gives good reasons for considering 1240 as more probable. Steele credits Bartholomaeus with citing Albertus Magnus among his authorities on plants and herbs, and a text Steele has used includes Albertus among the 94 authorities listed. On the other hand, among the list of 106 authorities quoted by Delisle, and a similar list printed in the Strassburg (Latin) text of 1505, Albertus is not named. Valentine Rose, who in 1875 (*Zeitschrift für deutsches Alterthum*, XXIII, pp. 321–455) published a discussion of the *De Lapidibus* attributed to Aristotle and the *De Lapidibus* of Arnoldus de Saxonia with special reference to these treatises as sources for Vincent of Beauvais, Bartholomaeus Anglicus and Albertus Magnus, states that Bartholomaeus makes no use of Albertus either in the books upon animals or plants or otherwise, although his name appears by false reading for Alfredus in some printed texts (*op. cit.*, footnote, p. 340).

Delisle (*op. cit.*, p. 357) also says that none of the hundred authors cited by Bartholomaeus is later than the commencement of the thirteenth century. In this book on herbs and plants in the 1506 Latin text the abbreviation *Al.* is frequently used, and the name Alfredus is frequently used also, but the name Albertus does not appear. This would seem to bear out the above statement of Rose and to explain the possibility of a misinterpretation of certain references to Alfredus as by Albertus, and this admitted, the reasons for placing Bartholomaeus as later than Albertus disappear.

Professor Thorndike in his *History of Magic and Experimental Science* in the very excellent chapter on Bartholomew of England says: "On the whole it seems possible that Bartholomew wrote his work as early as 1230."

[5] Langlois, *op. cit.*

libraries of Europe.'' Delisle enumerates eighteen manuscripts in the Bibliotheque Nationale and Steele lists twenty-one printed editions of the Latin text between 1480 and
1609, and of translations, two impressions in Dutch, twenty-one in French, three in English and two in Spanish.[6]
Little is known of the personality of Bartholomaeus. That
he was a Minorite friar, that he is said to have been, like
Roger Bacon, a pupil of Robert Grosseteste, and that he
was for some time in Paris and lectured there on the
Bible, and that in 1230 the general minister of the Franciscan order in Saxony requested the Provincial of France
to send him as a teacher of the Minorites in that new province,[7] comprises about all that is known of him, except his
works. Steele,[8] states that there is in Roger Bacon's *Opus
Tertium* (1267) a passage that may be a quotation from
the *De Proprietatibus*. Upon subjects relating to the properties of matter, the elements, minerals, metals, colors,
gems, etc., the sources utilized by Bartholomaeus are,
though far fewer, those utilized by Vincent of Beauvais—
Theophrastus, Plato, Aristotle, pseudo-Aristotle, Pliny,
Dioscorides, Isidorus,[9] Avicenna, Rhases, pseudo-Avicenna,
pseudo-Rhases, Averrois, and a work entitled *De Natura
Rerum* (supposed to be that of Thomas de Cantempre).
The Book upon Stones and Metals cites mainly from Isidorus (seventh century), Dioscorides (first century, B.C.),
the ''Lapidarium'' Platearius (ca. 1150)—especially with
respect to their medicinal properties and uses—Avicenna,
and others rarely.

In so far as concerns the information contained in this
work upon subjects related to chemistry there is nothing

[6] Steele, *op. cit.*, 1907, pp. 181, 182.

[7] Langlois, *op. cit.*, p. 114.

[8] Steele, *op. cit.*, 1907.

[9] Isidorus Hispalensis or Isidorus of Seville was a writer of the seventh
century who wrote a work on the origin and signification of words. For such
definition and descriptions as pertain to natural science he was dependent
upon Greek and Latin authors with no infusion of Arabian science. Throughout the middle ages Isidorus was a much respected authority. A modern
Latin text, edited by W. M. Lindsay, was published in Oxford in two octavo
volumes in 1911.

of importance that is not contained in the Arabian encyclopedias of the tenth century. Its importance for the history of chemistry depends solely on the fact that with its wide circulation as a handy text in the universities, it helped greatly in familiarizing a large public of the western world with current chemical theories and many chemical facts as then understood.

It may be of interest to quote a few illustrations of the style and method of his treatment of such topics.[10]

"Aurichalcum,[11] as says Isidorus, is called thus because, although it is bronze (aes) or copper (cuprum), it has superficially the luster of gold. For aes is called in Greek calchum. Aurichalcum thus has the hardness of bronze or copper. From a mixture of bronze and tin and orpiment and some other medicines in the fire it is brought to the color of gold, as says Isidorus. It has the color and likeness of gold but not the value. Vessels and works of art of various kinds, beautiful when new and presenting the appearance of gold, gradually lose their first brilliancy and become red and thus show by their coppery color and odor the material of their origin. In such vessels food and wines when long preserved acquire a horrible taste from the corruption and odor of the brass. Yet salves for the eyes are medicines which are profitably kept in them and are improved by the strength of the bronze, as says Platearius."

The idea that brass or bronze vessels are especially adapted for keeping ointments for the eyes is of ancient origin, for a Syrian work on medicine originating from the early centuries of our era prescribes that certain eye ointments be kept in brass vessels.[12]

"Glass,[13] as says Avicenna, is among stones as is a fool among men for it takes on any color.[14] It is called vitrum, as says Isidorus, because it is clear and transparent to

---

[10] Text used is the edition of Strassburg, 1505 A. D.
[11] Lib. XVI, Cap. 5.
[12] E. W. Budge, *The Syriac Book of Medicines*, London, 1913, II, p. 95.
[13] Bartholomaeus, XVI, p. 100.
[14] Cf. Vincent de Beauvais who attributes the statement to "Alchemista." Cf. also the De Lapidibus attributed to Aristotle, pp. 286, 287.

vision (visui). On account of the transparency of its surface glass is pervious to light. In other metals and minerals that which is contained inside them is hidden, but
in a glass the nature of whatever liquid is contained in it
is made manifest as if made visible to closed eyes, as says
Isidorus. Glass was first found near Ptolomaida on the
shore near the river Belus whose source is in the roots of
Mt. Carmel, where sailors landed their ship. For when
the sailors made a fire with lumps of soda (nitrum) on
the sands of that river, from the soda and the clear sand
there flowed out rivulets of a new liquid and (thus) they
explain the origin of glass, as says Isidorus.[15]

"In the actual method glass is made from the ashes of
trees and herbs burned by the greatest strength of fire with
which ashes, sometimes nitrum (soda), sometimes brass,
sometimes both are mixed and thus are changed into a
vitreous mass."

This statement of the use of brass (or copper) in the
making of glass may be derived from the statement of
Theophrastus, who mentions the beautiful color of some
glass made by the use of bronze or copper ($\chi\alpha\lambda\kappa\acute{o}s$).[16] "Its
powder cleans the teeth, is good for stone of the bladder
and kidneys when drunk with wine, as says Avicenna."
We may safely infer that this surprising statement results
from an error in copying manuscripts by scribes ignorant
of the subject matter—an error which has been noted not
infrequently in medieval manuscripts, namely of writing
vitrum (glass) instead of nitrum (soda). Such an error
would explain this statement of Bartholomaeus.

"Zucarum or zucara (sugar) is made from certain canes
and reeds which grow in swamps near the Nile, and it is
the juice of these canes called sweet cane (cana mellis)
from which is made zucarum by boiling, just as salt is
made from water. For the ground canes are first placed
in a cauldron and cooked with slow fire until it (the juice)
is thickened, and first there is seen to pass off from the
whole mass [a portion] in foam, and afterwards the thicker

---

[15] This is Pliny's narrative alluded to by him as a fable. *See* ante, p. 72.
[16] *See* ante, p. 31.

and better residue sinks to the bottom, and what is light and foamy remains above and is porous and less sweet and does not crackle between the teeth when masticated but disappears quickly. But the good is the opposite (econverso), for the good, placed in round vessels in the sun, is made hard and white. The other is yellow and is warmer and therefore not to be given in acute fevers, while the good sugar is temperate in its qualities and therefore, as says Isaac[17] in 'Dieta,' has a cleansing, solvent and diluent virtue and removes wateriness of the stomach without corrosion, cleanses the stomach, soothes the lungs, clears the voice, removes cough and hoarseness, restores lost humidity, and tempers the sharpness and bitterness of certain kinds of aromatics and therefore is of the greatest service in medicine as in electuaries, powders and syrups, as says Isaac.'' [18]

These extracts will illustrate the style of treatment of such subjects and the care with which he quotes the authorities for his statements. The scope of the work is well described by its title *On The Properties of Things*.

The constitution of matter by the four elements, and the generally prevalent notions of the constitution of the physical universe, metals, stones, gems, medicines, man and his manners as well as his anatomy, geography, plants, trees, birds, fishes and other animals are treated in the form of a condensed encyclopedia of what was then understood by natural science. It is not difficult to understand the favor in which this work was received in the many new universities of the thirteenth century.

Vincent of Beauvais (Vincentius Bellovacensis) is noted for his stupendous encyclopedia of human knowledge entitled *Speculum Majus* or *Greater Mirror* a vast collection of citations from recognized authorities upon the whole range of learning of his time. Of his life or personality little is known. A native of Burgundy, reputed to have been a tutor to the princes at the court of Saint Louis, he was

---

[17] Isaac Judaeus, an Arabian writer on medicine of the tenth century, who wrote a treatise on diet. Cf. Haeser, *Geschichte der Medicin,* I, p. 573.
[18] Lib. XVII, Cap. 197. Cf. *Platearius,* p. 260.

a member of the Dominican order, and died in 1256 or 1264 at the cloister of that order in Beauvais. The great importance of this work for his time lay in the conscientious care with which a great number of authors ancient and medieval were drawn upon. His own commentaries are of less importance. As says M. Daunou,[19] "of all the works of the thirteenth century his is the one which can throw most light upon the general character and many details of the literary history of that epoch." M. Daunou also cites Cuvier as testifying that in those parts relating to animals the notices of Vincent of Beauvais are more precise and more accurate than those of Albertus Magnus, "he had better copies of Pliny and he also knew better how to draw upon the *Origines* of Isidorus of Seville."

The importance of the *Speculum Majus* in the later centuries was shown by the early date at which his entire works comprising several bulky folios, were printed in at least four editions between 1472 and 1485. That portion of his work which contains the greater part of material relating to chemistry is the *Speculum Naturale* (Mirror of Nature) written about 1250 A. D. In this work he assumes to follow in arrangement the chronological order of creation—beginning, therefore, with angels, and including all created things of the physical universe. Over 300 authors are quoted in the *Speculum Naturale,* many of these being known only through his citations, the manuscripts from which he drew being no longer extant. Authors drawn upon in those portions dealing with matters of chemical interest include Aristotle, Plato, Theophrastus, Vitruvius, Democritus, Columella, Galen, Pliny, Dioscorides, Seneca, Isidorus, Platearius, Avicenna, Rhazes, (and works attributed mistakenly to these writers,) Albumasar, Arrenois, and many others less prominent. Several works quoted by him are not otherwise accessible as for instance the work *De Aluminibus et Salibus* attributed to Rhazes, though

---

[19] *Histoire Littéraire de France,* XVIII, pp. 449–519, in a comprehensive article upon Vincent and his works.

written at a much later period. Various works are referred to by title, the authors being presumably unknown by name. Such are for example *De Natura Rerum* (Isidorus?), *Philosophus, Doctrina Alchemiae.*

In the *Speculum Naturale* there is thus brought together a very compendious collection of the ancient and medieval authorities upon subjects relating to chemical themes and data. There is no attempt at digestion of these citations by the author, who evidently had no experience of his own in such matters, as his own commentaries on these subjects are of little value and give no evidence of personal experience.[20]

While it would be vain to attempt here to convey an adequate idea of the entire scope and character of this encyclopedic work, it may serve to assist in some understanding of its nature if we quote some illustrations, choosing for this purpose extracts from later and less known authors, rather than from the works of Aristotle, Pliny or other well known authors.[21]

From the work probably of the twelfth or early thirteenth century entitled *De Aluminibus et Salibus,* incorrectly attributed to Rhazes (Alrazi) who lived about 850–927 (?), Vincent quotes with respect to salt.[22]

"Salt ('sal') is a water which the dryness of fire has solidified and the nature of which is dry and warm. It has the property of liquefying gold and silver in the vehemency of the fire and augmenting in them their natural colors, namely in gold, red, and in silver, white. It converts them from their bodily nature to a foamy nature (spumalitas), and frees them from their impurities and consumes their foulness of a sulphurous nature, when the bodies (that is, metallic) are roasted with it. This does not take place with anything else.

[20] Cf. Berthelot, *op. cit.,* I, pp. 280–286, where is given a brief analysis of the *Speculum Naturale.*
   See also the chapter on Vincent of Beauvais in Professor Thorndike's *History of Magic and Experimental Science.*
[21] References are to the Nuremberg edition of the *Speculum Naturale* printed 1485 in two folio vols.
[22] Liber VI, Cap. 86.

"It (sal) is found in the ashes of all plants, in the calces of stones, in the bones of animals and in all things. There-fore the wise have called it the silver of the common people, (silver) on account of its whiteness, and of the common people because all men have need of it. . . . He who knows salt and its solutions and its solidifications knows the hidden secrets of the wise in alchemy. It whitens and cleanses and resolves bodies, and the spumus melts and also solidifies and preserves them and protects from burn-ing by fire.

"There are many salts and all when completely purified turn to Sal Harmonicum which is of all salts the best and most splendid, unchangeable and not fleeing from fire. It is indeed an oil (oleum) which the dryness of fire has solidified and the nature of which is warm and dry, subtle and penetrating, pouring forth (?) (profunduus) and it is a flying foam ("spumus volans") useful for the elixir, for without it the elixir cannot be completed nor matured nor come forth."

In the above description it is evident that the various salts referred to are the different kinds of common salt such as described by Pliny and Dioscorides. "Sal Har-monicum" is manifestly the superior commercial grade of salt from the region of Egypt near the temple Ammon, called by Pliny the oracle of Hammon.[23] It is specially characterized by the unchangeability and nonvolatility by the fire, thus making certain that it is not ammonium salts that are here referred to. The terms "spumus," "spuma," and "spumalitas," were used as applying to a condition not only of froth or foam, but to light powders—efflor-escences—and to such powdery sublimates as collected in the flues or walls of furnaces. Litharge thus formed in the reduction of silver was sometimes called "Argenti spuma."

Isidorus is quoted[24] concerning nitrum (sodium carbon-ate) that "it differs little from salt but has specific virtues in medicine, that afronitrum is the foam of nitrum (spuma nitri). It is collected in Asia distilling in caves, then dried

23 *See* ante, p. 48.
24 *Speculum Naturale*, VI, p. 91.

in the sun. The best is considered that which is lightest and most friable.''

Isidorus it will be remembered derives his information from Greek and Latin writers earlier than the seventh century. The work *Liber de Naturis Rerum* is quoted by Vincent for a definition of flame (flamma). ''Flame is a burning smoke (fumus ardens) as flame it is visible, though created from heat and vapor each of which is invisible.'' [25]

The subjects of the metals, their properties, and the operation for their preparation so important in all early chemistry are naturally extensively treated. Isidorus is cited:[26] ''There are seven kinds of metals, namely, gold, silver, copper (aes), electrum, tin, lead and iron which subdues all things. The *Alchemia de Anima* of (pseudo-) Avicenna is quoted:[27]

''There are seven things that can be elongated by hammering at the furnace, namely Sol, that is gold, luna (silver), tin, copper (aes), iron, lead.[28] These are formed in nature under the earth. Gold is generated in the earth by the great heat of the sun from excellent quicksilver and red and pure sulphur by digestion in the rocks for a hundred years or more; silver from pure quicksilver and pure sulphur digested for a hundred years. But copper (here *cuprum* instead of *aes*) from impure quicksilver and impure sulphur digested for a hundred years. But gold indeed is excessively digested and hardened, therefore neither fire nor water nor earth destroys it. But silver is crude and not well digested, therefore the earth speedily destroys it. Copper indeed can be burned up, therefore earth does not destroy that for many years but fire consumes it quickly. Lead, the philosophers say, is made under the earth from impure and thick quicksilver and from the worst sulphur and is a crude mixture and not well digested. And lead is indeed of such very bad nature that with its odor (or tincture? odorem) it renders gold breakable, and hardens

[25] Vincent de Beauvais, VII, 73.
[26] Vincent de Beauvais, VIII, 3.
[27] Vincent de Beauvais, VIII, 4.
[28] While the text says seven, only six are named—electrum being perhaps an accidental omission.

quicksilver and indeed dissolves gold, (si dent odorem de plumbo vertitur in calcem?). Tin, however, is made from excellent and pure quicksilver, but from the poorest sulphur impure and not well digested. Iron is from thick quicksilver and thick red sulphur, and is not sufficiently digested.''

This description of the nature of the metals gives evidence of familiarity with some properties of the metals, and their behavior. In general it differs but little except in minor details from the account of the origin of metals and their properties in the Arabian writings of the Faithful Brothers of the tenth century.[29] Particular facts as to metals recorded in this treatise were, however, known to the ancients.

The notion of the origin of the metals from quicksilver and sulphur was also in the writings of the Faithful Brothers supplemented by the theory of the origin of quicksilver from water and earth, and of sulphur from aerial or oily elements with earth. Vincent quotes from the *De Aluminibus et Salibus* attributed to Rhazes with respect to this theory.[30]

''Mineral bodies are vapors which have coagulated in nature in the course of long lapses of time, and the first things which coagulate are quicksilver and sulphur, for these and not water or oil (oleum) are the elements of minerals, for the first of these (quicksilver) is generated from a water and the other (sulphur) from an oil. Upon these things there operates a gentle digestion constantly with heat and moisture until they are solidified and from them (metallic) bodies are generated by gradual mutation in thousands of years. For if they remain in their minerals, nature purifies them until they arrive at a kind of gold or silver. But by the subtlety of the artist, transmutation of this kind is made in one day or in a brief space of time.''

From another work *Doctrina Alchemiae* Vincent also quotes:

---

[29] *See ante*, p. 210 *ff.*
[30] Vincent de Beauvais, VIII, 6.

"Furthermore by the art of Alchemy mineral bodies and especially metals are transmuted from their own kinds to others, for this science arises from that part of natural philosophy which deals with minerals just as agriculture has to do with that part which deals with plants."[31]

Of the properties of iron "in alchemy" the *De Alumini-bus et Salibus* is quoted.[32]

"Iron belongs to the domain of Mars, its nature is warm and dry, of sour taste and of vehement strength expelling and resisting fire. It is liquefied by four things, namely arsenic, lead, magnesium and markasite."

If we remember that arsenic meant usually the sulphide, orpiment, and sometimes realgar also, and that by magnesia very frequently was meant native sulphides of lead, zinc and other metals, and marcasite usually meant sulphides of the character of various colored pyrites, the above statement records the production of fusible ferrous sulphide when iron and these sulphides are heated together. The inclusion of lead in the list may also perhaps be explained by an ancient habit of occasionally using the same term for a metal and its principal ores in metallurgy as is sometimes seen in Pliny. Thus galena, the native sulphide of lead, heated with iron would also "liquefy" it as do the other sulphides.

"Glass, says Razi (Rhazes)[33] in his *Liber de Animalibus*, is from parts of quicksilver. Coldness and dryness dominate its nature. It liquefies iron and all bodies (corpora), and causes them to run in fusion," and from *Alchemista* Vincent quotes in the same chapter, "Glass is among stones as are the foolish among men for it receives all colors.[34] It is liquefied easily by fire and quickly returns again to its stony condition. It softens and cleanses and liquefies all bodies, and is removed from them by fusion just as salt is by washing. Hence salt and glass are things in which lies the whole secret of the art nor is it possible to

---

[31] Vincent de Beauvais,
[32] Vincent de Beauvais, VIII, 54.
[33] Vincent de Beauvais, VII, 79.
[34] Cf. the *De Lapidibus* of pseudo-Aristotle, p. 207 and Bartholomaeus Anglicus, ante, p. 235.

produce the 'stone (philosophers' stone) without them, particularly without salt.''

When Pliny describes the manufacture of glass he says it is made from soda (nitrum) and sand with addition of magnes lapis, which has an attraction for glass as well as for iron.[35] He probably confused the magnetic oxide of iron with pyrolusite, manganese dioxide, which found early use in glass making both for decolorizing as well as for coloring. It is a notable fact that later medieval writers also make no allusion to the use of any lime-containing mineral in the manufacture of glass, but usually speak of glass as made from sand and soda or the ashes of plants. Analyses of ancient glass have shown however that they are generally soda-lime glasses. Lead also was used in glass by the ancients. Berthelot analyzed a glass vase of the fourth dynasty of Egypt and found it to contain about 25 per cent lead;[36] and both Kopp and Schliemann note its frequent occurrence in ancient glass. In Vincent de Beauvais[37] it is alluded to very incidentally and with no apparent understanding of the reasons for its use: *"Ex causa supradicta factum semper est decoloratum quod autem fit ex plumbo et terra arenosa subtile aut ex cinere filicis coloratum est."*

In the *Aristoteles de Lapidibus,* lead is also alluded to very casually as a constituent of glass.[38] One cannot fail to be impressed by this failure of later medieval writers to note adequately the real composition of glass. It shows how these writers are prone to depend on the writings of earlier authorities without attempting to improve upon them on the basis of actual technical experience which must have been not difficult of access. The notion of body and spirit in the sense of nonvolatile and noncombustible as compared with the volatile and combustible (in the limited sense of conversion into gaseous products of combustion)

---

[35] *See* ante, p. 72.
[36] Berthelot, *Archéologie et Histoire des Sciences,* pp. 17, 18.
[37] Vincent de Beauvais, VII, 77.
[38] *See* ante, p. 207.

derived from the neoplatonic Greek alchemists, and emphasized by Arabian followers[39] is emphasized also by the later writers. Thus the pseudo-Avicenna's *De Anima* is quoted as stating that the mineral spirits are sulphur, orpiment, sal hammoniacum, and mercury. These can be volatilized and bodies, as gold, silver, copper, etc. cannot.[40] In the same chapter is also cited the work *Doctrina Alchemiae*: "Spirits are four in number, namely, sal hammoniacum, sulphur, quicksilver and arsenicum, but bodies are six, that is, gold, silver, copper," etc. In a Latin manuscript of the *Book of Seventy* published by Berthelot,[41] accredited to Djaber and believed by Berthelot to be a translation, not without later corruptions and additions, of a work of Djaber (eighth to ninth century), there is a very similar passage.

"I say therefore first, that spirits (spiritus) are four and bodies (corpora) are seven. The four spirits are quicksilver, sulphur, orpiment, and sal armoniacum. The volatilization of these has differences, for all are volatile, but on account of their conditions they are themselves different (sed propter causas eorum sunt ipsa diversa). The seven bodies (corpora), are lead, tin, gold, iron, silver, copper (or bronze, 'aes'), glass (vitrum). Quicksilver is not among these for I have placed that among the spirits." [42]

With respect to the inclusion of "sal hammoniacum" or "sal armoniacum" among the spirits, it is evidently ammonium salts, chloride or carbonate or both, that are here alluded to. As has been previously stated there is no evidence that the ancients knew of ammonium salts. Somewhere about the time of Djaber however the knowledge of these came to the Arabians, and was originally in Latin described as a salt from Armenia. Von Lippmann[43] states that the Arabians obtained their knowledge of our sal ammoniac from later Alexandrian chemists, and that its

[39] Cf. extracts from Djaber's *Book of Mercy*, ante, pp. 178–180.
[40] Vincent de Beauvais, VIII, 60.
[41] *Archéologie et Histoire des Sciences*, p. 310 *ff.*
[42] *Op. cit.*, p. 357.
[43] *Entstehung und Ausbreitung der Alchemie*, p. 392.

occurrence in the volcanic regions of nearer Asia and of China was known to early Arabian geographers. Its volatility and its purification by sublimation were understood by the Arabians. It is often difficult however to tell in the case of medieval Latin writers whether in writing sal ammoniacum, sal hammoniacum, sal armeniacum, etc., they mean with Pliny the superior grade of common salt, or sal ammoniac, as the confusion of spellings and signification is great and often no clue is given as to properties of the salt alluded to.[44]

The notions of the thirteenth century with regard to the process of combustion, were comprised in the idea that the sulphurous constituent of bodies is what disappears in combustion. Vincent quotes[45] *Alchemista*:

"Fire which calcines bodies without melting them has the property of burning the less strong part of them, namely the sulphureity (sulphureitatem) leaving the stronger part unchanged, until it builds up (erigit) the body (that is, the metal) and cleanses it from blackness."

The opinion of Vincent himself regarding the possibility of transmutation of the metals, based not upon any work of his own, to be sure, but upon his extensive reading of the works he has studied is expressed rather positively. He has quoted several authorities upon the question as for example from the *Liber Metheorum* which says:

"Let the artisans of alchemy know that it is not possible for species to be transmuted, but they can make things similar to these, as by tincturing white [metal] to a yellow color so that it may seem to be gold, also by removing the impurities of lead so that it may seem to be silver; but it will always be really lead: but they may produce in it such qualities that they may deceive men in it. For the rest, I do not believe it is possible that a specific difference in any innate quality can be removed. But there is effected a removal (or change, "expoliatio") of its accidental qualities as color, flavor or weight. The works of art also are

---

[44] *See* ante, p. 48.
[45] Vincent de Beauvais, VIII, 90.

not of the same kind as the works of nature, nor so certain although they may be kindred and similar. For art is more feeble than nature, nor can it overtake it without great labor. And further the proportions of the composition of these substances will not be the same for all. Therefore it will not be possible to transmute these [compositions] into others unless by chance they are first reduced to their primal matter.''

The idea here suggested that different kinds of metals or minerals might be changed into others if they could first be reduced to the primal matter is met with in various writings of the time. Vincent says himself :[46]

''From the foregoing statement it may be seen that alchemy may be to a certain degree false (or fraudulent, 'falsa') nevertheless it is true that by the ancient philosophers and by artizans in our time it has been proved to be true.''

Vincent is quite typical, in this statement, of the best thinkers of his period—in admitting the fact of transmutation of the metals as possible, although they know that there is very much imposture in the art, and they often express their doubt as to the reality of the claims of the alchemistical workers to be able actually to perform this transmutation.

With respect to the medieval authorities which Vincent has brought together dealing with the theories and art of chemistry, the statement of Berthelot seems justified:

''The 'Doctrine of Alchemy' and all authors cited by Vincent of Beauvais revolve in the same circle of doctrines and facts nearly as do modern scientific writers of any particular epoch.'' [47]

It is impossible here to convey an adequate concept of the mass of material brought together in the *Speculum Naturale*. It comprises a very complete compendium of the chemical knowledge and concepts of the alchemical writers and natural philosophers up to its period. It does not,

---

[46] Vincent de Beauvais, VIII, 85.
[47] *La Chimie au Moyen Age*, I, p. 282.

however, include the kind of information that is contained in collections of recipes such as the *Book of Fire* by Marcus Graecus, the *Compositiones ad Tingenda,* the *Diversarum Artium Schedula* of Theophilus the Monk, etc. Whether such works were not accessible to him or whether they were considered as not pertaining to the liberal arts, and beneath the consideration of scholars, is a matter of conjecture, though the latter alternative is not improbable.

Contemporaneous with Vincent of Beauvais was a scholar of greater influence and renown—Albert von Bollstedt, better known in later times as Albertus Magnus, on account of the great reputation he held for learning and wisdom in many fields. "Great in the magic of nature, greater in philosophy, greatest in theology" was said of him by Johann Trithemius, abbot at Spanheim and at Würzburg (1462-1516), and mentioned by Paracelsus as one of his early teachers. Albertus was born in Bavarian Swabia in 1193, is known to have studied in Pavia, and to have taught theology in Cologne and in Paris. As Provincial of the Order of Dominicans, to which he belonged, he traveled throughout Germany; in 1260 was made Bishop of Regensburg and died in the cloister of that order in Cologne in 1280. Like Vincent of Beauvais he was a very prolific writer. His collected works were printed at Lyons in 1651 in 21 folio volumes and have been published with modern revisions in Paris, 1890–1899 in 38 volumes.[48]

Though there is no evidence that Albertus had any practical experience in subjects relating to chemistry, other than was acquired by a scholar who had traveled and talked with men who had some technical experience, yet he was a student of literary records and his writings in so far as they include related topics are valuable in the same way, if not to the same extent, as the encyclopedia of Vincent of Beauvais. As an earnest and sympathetic student of Aristotle, in his general views of the nature and changes

[48] Albertus Magnus, *Opera Omnia*, 38 vols., Paris, 1890–1899. It is this edition that is referred to in the following paragraphs.

of matter, the four elements, etc., he is a follower of Aristotle. He depended also largely upon the current latinized versions of Arabian chemistry and mineralogy.

His method of treatment of this material was different from that of Vincent. The latter, as we have seen, quotes quite literally, or in a form more or less condensed, from his authorities. Albertus, however, speaks in his own words from a more scholarly digestion of his authorities. Rarely does he refer to experiences of his own, and when he does he indicates that he speaks rather from casual observations than intimate knowledge. His writings on subjects of chemical interest are scattered through his works, notably, in so far as concerns his general theories of matter, in his treatises in meteorology and physics. Of more especial chemical interest is his work in five books, *De Rebus Metallicis et Mineralibus*. The brief work entitled *Libellus de Alchemia,* included in his printed collected works, is now recognized as falsely attributed to Albertus. Not only are the contents and style at variance with his other work in its assumptions of a wide experience in alchemical operations, but it cites in its text authors of later date than Albertus, as Arnald of Villanova (died 1312 or 1314), Jean de Meun (1280–1365) and Philip Ulsted, an alchemist who lived about 1500.[49] Several other works of less importance were issued as written by Albertus but are obviously not genuine though accepted as such during the uncritical later centuries. As, however, these works were popularly assumed to have been written by him, Albertus acquired a reputation as alchemist which was wholly undeserved. Other works thus attributed incorrectly to Albertus are *Tractatus Secretorum, De Philosophorum Lapide, Compositum de Compositis,*[50] and others.

---

[49] Cf. Kopp, *Beiträge zur Gesch. der Chemie*, III, p. 76; Berthelot, *op. cit.*, I, 290. In the extensive and excellent discussion of Albertus Magnus in Thorndike's *History of Magic and Experimental Science*, the author says of this essay: "Of these various treatises in alchemy ascribed to Albert we shall now consider in more detail the one which has been included in editions of his works, and which is perhaps the most likely of any of them to be genuine."

[50] Cf. Kopp, *Die Alchemie*, p. 17.

Though in the works of Albertus we find nothing of chemical theory or data that is not to be found in earlier writings, yet it is evident that he possessed a wide knowledge of many chemical facts and of the ideas about chemical subjects which were prevalent in his period. He presents this knowledge with a clearness and directness that characterizes him as one of the ablest thinkers and writers of his century. His accounts are not always free from errors of commission or omission, because he was not himself a practical or operating chemist. On the other hand, this very clarity of expression—free from intentional secrecy or mystification—must have given his works an important value in helping to lay a foundation for sensible and sane chemical points of view, in a time when, according to many writers of those times, fraud, charlatanry and imposture in alchemy were very prevalent.

We find in Albertus a general knowledge of many specific facts and operations of chemistry. He knew of the operations of distillation and sublimation and of the apparatus used in these operations, of the purification of gold and silver by cementation and by the use of lead. He knew that quicksilver may be successively distilled without loss of weight; that cinnabar can be produced by the union of quicksilver and sulphur; that wine, when heated gives off an inflammable substance which he calls an oil (oleum) "supernatant" and "inflammable."[51] He describes many metals, minerals, salts and other substances, without, however, adding any facts of interest not comprised in the authorities which precede him.

"Sal armoniacum" is with Albertus, as with ancient writers, classed as a variety of common salt, though he refers to a salt of which he has heard, that is prepared from human urine, chiefly of young boys, prepared by the operations of alchemy, by sublimation and distillation. As he characterises this salt no further, it leaves a doubt as

---

[51] Albertus Magnus, *Mineralium*, Lib. III, Tract. II, Cap. I, and *Meteororum, IV*.

to whether he considered this as essentially different from common salt, though Arabian writers had previously made the distinction clear. No reference is made to saltpeter, in the authentic writings of Albertus, though it is mentioned in the works falsely attributed to him but of later origin. A characterization of flame as a burning smoke is considered by Kopp[52] as worthy of recognition for his time. This definition, however, we have seen in Vincent of Beauvais[53] and by him attributed to the *Liber de Naturis Rerum*. E. von Lippmann says that this definition occurs also in Aristotle and in Galen.[54] The character of the descriptions by Albertus can best be understood by a few typical examples.

In a discussion of the nature and mixture of bronze or brass,[55] after discussing the nature of brass, its origin from mercury and sulphur and its colors, quite after the conventional Arabian philosophy of the metals, he is speaking of the manufacture of brass (aurichalcum) from copper ores and zinc ores, (called by him calamina, as previously by Theophilus the Monk, and not "cadmia") :

"Those who operate much in copper in our region, namely in Paris or Cologne and in other places where I have been and have seen them work, convert copper into brass by the powder of a stone called calamina. And when this stone evaporates there still remains a dark brilliancy turning slightly to the appearance of gold. But that it be rendered paler and thus more like the yellowness of gold they mix with it a little tin by reason of which the brass loses much of the ductility of the copper. And those who wish to deceive and to produce a brilliancy like gold retain the stone (calamina) so that it remains longer in the brass in the fire (or furnace) not quickly vaporizing from the brass. It is [thus] retained by *oleum vitri* (liquified glass), for fragments of glass are powdered and sprinkled in the pot (testa) upon the brass after the calamina is introduced,

---

[52] *Beiträge zus Geschichte der Chemie,* III, p. 84.
[53] *See ante,* p. 241.
[54] E. von Lippmann, *Entstehung und Ausbreitung der Alchemie,* p. 492.
[55] *Mineralium,* Liber IV, Tract. I, Cap. VI.

and then the glass so added swims upon the brass and does not allow the stone and its virtue to evaporate, but turns the vapor of the stone back into the brass, and thus the brass is long and strongly purged and the feculent matters in it are burned away. Finally the oleum vitri vaporizes also and then vaporizes the virtue of the stone, but the brass is made much more brilliant than it would be without it.[56] He who desires to simulate gold still more completely repeats these operations of heating (optesim) and purging of the melted glass frequently and mixes with the brass silver instead of tin and thus it is made so red and yellow that many believe it to be gold itself when, in truth, it is still a kind of bronze (or brass, 'aes')."

While Albertus does not deny the possibility of the conversion of one metal into another in nature, he is very sceptical as to the alchemistic claims of such transmutation. In discussing the theory, which he opposes, that every metal contains every other metal, for instance, he says, after asserting that gold said to be produced from lead is not true gold although it may be something very similar to it:

"Besides we have never found an alchemist so-called operating generally (in toto) but that he rather colors with a yellow elixir into an appearance of gold, and with a white elixir colors to the resemblance of silver, seeking that the color may remain while in the fire and may penetrate the whole metal, just as a spirit (spiritualis substantia) is introduced into medicines, and in this manner of working it is possible to produce a yellow color, the substance of the metal remaining. And here again it is not to be maintained that several kinds of metals are contained in one another. It is from this and similar things that is demolished the dictum of those who say that any kind of metal you please is contained in another." [57]

Book II, Tract. II, of the *Mineralium* contains an alphabetically arranged description of precious stones and other

---

[56] Albertus here seems to accept the Aristotelian concept of the function of the zinc ore, that it has only changed the color without remaining as a constituent. Cf. this manuscript, p. 128.

[57] *Mineralium*, Liber III, Tract. I, Cap. VIII.

minerals. That this is an elaboration of the Arabian work on stones, falsely attributed to Aristotle, has been shown by V. Rose, one particular source being that of Arnoldus Saxo.[58] The description of the stone *magnes,* for instance, is evidently condensed from that description in Aristotle's *De Lapidibus* as given in the text of Spanish origin published by Rose. It describes nearly all the miraculous properties there ascribed to it.[59]

"Magnesia, which some call Magnesium, is a black stone which the glass makers frequently use. This stone distils and flows in great and strong fire but not otherwise; and then mixed with glass it removes substance and purifies the glass (ad puritatem vitri deducit substantiam)."

There can be little doubt that black oxide of manganese is the mineral here referred to though the description is not definite.

"Marchasita or Marchasida, as some call it, is a stone in substance and there are many species, wherefore it takes the color of any metal whatsoever and is thus called silver or gold marchasita and so of others. The metal that gives it color does not distil from it by itself, but vaporizes in the fire and thus there is left a useless ash. And this stone is known among the alchemists and is found in many places."

The name "marchasita" was generally applied to metallic sulphides such as iron and copper pyrites, and other sulphides of metallic luster, though taken by itself the above description gives little basis for such identification.

"Nitrum also approaches the solidity of a stone, but is somewhat pale and transparent and its power to dissolve and to attract is proved. It has value (as a remedy) for jaundice and is of the class of salts."

Nitrum, which at the time of Albertus, as with the ancients, meant sodium carbonate—or potassium carbonate (as obtained from the ashes of plants)—is elsewhere by him described more at length.[60]

---

[58] *See* ante, p. 205.
[59] *See* ante, p. 208.
[60] *Mineralium* Liber V, Cap. VII.

"Nitrum is thus called from the island Nitrea where it was first found. The Arabs call it baurac. It is a kind of salt less known than sal gemma (rock salt) transparent but in thin plates. It is roasted in the fire and then, all superfluous aqueous substance being given off, it is burned to a high degree of dryness ('efficitur siccum magis combustum'), and the salt itself is rendered sharper. The varieties are distinguished according to the localities where it is formed. With us it is found of three kinds (tripliciter) namely Armenian, African, and German, which latter is found abundantly in the place called Goselaria (Goslar in Hanover at the foot of the Harz Mountains). Rain falling on a mountain which is full of copper minerals is collected and conducted a hundred paces into a pit which the diggers have made. This water is seen to be converted into *nitrum* which nevertheless is thought by the inhabitants to be rock salt (sal-gemma) but I have proved by sight and touch that it is *nitrum*. It exists in a hollow of the mountain in the manner and form in which ice is formed on roofs by water dripping from them in time of freezing cold, and this is not laminated but rounded. The relation of African nitrum to other species of nitrum is the relation of nitrum to salt (?) (Comparatio etiam nitri Africani ad ceteras species nitri est comparatio nitri ad salem.)

"The foam, (spuma) of all nitrum, sometimes called flos nitri (flower of nitrum) is of more subtle substance and virtue than nitrum itself; that spuma is best, of which the color resembles marble, and which is very friable.

"All nitrum is warm and dry and therefore the applications of it are such that it is *inscissivum* (cutting or disintegrating?) *lavativum* (cleansing?) *excoriativum* (caustic?) and *corrosivum* (corrosive) and especially the African which is sharper than the others."

That carbonate of soda (nitrum) occurred in the form of stalactites in mountain caves in Goslar is doubtless an error, and the "inhabitants" were probably more nearly correct than Albertus who judged by sight and touch (visu et tactu), though both may have been at fault. The term "sal-nitri" or "sal-nitrum" meaning our niter did not come into use until early in the fourteenth century (for example pseudo-Lullus) but the designation of simply "nitrum" for our niter was first employed about the end of the sixteenth century.[61]

Here is Albertus' description of "tuchia" which was usually an impure sublimate of zinc oxide mixed often with more or less of other metallic oxides, dust from the flues or domes of bronze or brass furnaces.

"Tuchia which has frequent use in the transmutation of metals, is an artificial and not a natural mixture, for tuchia is made from the smoke which rises and is solidified by adhering to hard bodies, when brass is purified from the stones (minerals) and tin which are in it. But the best kind is from that which is sublimed from that (that is, re-sublimed), and then that which in such sublimation remains at the bottom is *climia*,[62] which is called by some *succudus*. There are many kinds of tuchia, as it occurs white, yellow and turning toward red. When tuchia is washed there remains in the bottom a sort of black sediment of tuchia. This is something called by some *Tuchia Irida*. But the difference between succudus and tuchia is as we have stated, namely, because tuchia is sublimed and succudus is what remains at the bottom unsublimed. The best is volatile and white, then the yellow, and then the red; the fresh is considered better than old. All tuchia is cold and dry and that which is washed is considered better in those operations[63] (that is, in above mentioned transmutation of the metals)."

From the above extracts and from his writings in general concerning chemical subjects it seems clear that Albertus neither claims nor possesses any special experience, his qualifications being those of an intelligent student of the literature of the subject and of a man of good powers of observation and of broad information and high scholarly ability. If, however, we were to accept as authentic the *Libellus de Alchimia* attributed to him, and included

---

[61] Kopp, *Geschichte der Chemie* III, p. 221; Rulandus, *Lexicon Alchemiae*, 1612, p. 346.

[62] Climia according to Rulandus *Lexicon* is a kind of "cathimia of brass, a smoke adhering to the upper parts of furnaces."

[63] *Mineralium*, Liber V, Cap. VIII.

as authentic even in the Paris edition of his works,[64] this judgment would have to be modified. For in this work the author, after statements of the variety and untruth of the many books on the art of alchemy, continues that nevertheless he had not despaired but had studied more deeply into the arts of alchemy, its decoctions, sublimations, solutions, distillations, cerations, calcinations and solidifications—

"whence I have found the transmutation into gold and silver to be possible; that [the metal] is far better than any natural, in every examination and malleation. . . . But I, the least of the philosophers, intend to describe to my associates and friends the true art easy and infallible; nevertheless so that seeing they shall not see, and hearing they shall not understand. And I beseech and adjure you by the creator of the world that this book be concealed from all foolish persons (insapientibus)."

As to the body of this brief treatise, it is so conventional a repetition of Arabian chemistry and so similar in style to a great number of fourteenth century alchemical works published under the pseudonyms of Albertus Magnus, Roger Bacon, Raymond Lully, Hermes, etc., that there can not be a reasonable doubt of its fraudulent authorship— even if it did not contain as already noted references to writers of later date and refer to substances as sal-petrae not known to Albertus or to his contemporary, Vincent of Beauvais.

Of the great value of the works of Albertus Magnus in helping to spread knowledge of the chemistry of his time there can be no doubt. With his elder colleague Vincent and his younger contemporary, Roger Bacon, he was assisting in distributing and popularizing among the educated classes the theories and facts of chemistry as then understood, a service which ultimately, though not immediately, was to help lay the foundation of a more productive interest in chemical thought.

---

[64] 1898, Vol. 37, p. 545 *ff.*

Roger Bacon, the great English theologian and philosopher of the thirteenth century was born about 1214, dying probably in 1292. We know that he studied at Oxford under Robert Grosseteste (or Greathead) an able Franciscan scholar, whom Bacon held in high esteem, and from whose inspiration he acquired a profound interest in mathematics and optics. In about 1240 Bacon went to Paris and there spent a large part of his later life. He acquired there much celebrity by his teaching and is said to have prepared popular elementary treatises for students. In Paris Bacon came in contact with many of the prominent scholars of the time. Not long after his removal to Paris he joined the Franciscan order.

For the history of science in the middle ages Roger Bacon is a more interesting personality than Vincent or Albertus, for while the latter were mainly recorders and interpreters of the natural science of this time, Bacon was more passionately interested in the accomplishments of scientific discoveries and aims. He possessed the fervor of a missionary in presenting the claims of science to the attention of his contemporaries, and an imagination which enabled him to look beyond the state of experimental science in his own time to a future of greater possibilities. It is evident that he was a zealous student of several branches of science especially of mathematics, physics (notably of optics), astronomy and the chemistry of his time.

"During the twenty years," he wrote in 1267, "that I have spent in the study of wisdom after abandoning conventional methods, I have spent more than 2000 libra on secret books and various experiments, and on languages and instruments and astronomical tables, etc." The Paris libra was about one third of a pound sterling; a considerable sum in his time for the Franciscan monk.

Bacon can hardly be called a great discoverer or a very productive experimenter. His points of view were those of his predecessors. But his was of the class of minds that make great teachers; he was an earnest student—satisfied only to obtain his information from original authorities, and filled with the desire to impart his points of view to others. He did not believe that all truth lay in the ancient and accepted authorities. *Quanto juniores tanto perspiciores*—the later the authorities the clearer they are —was not the spirit of the conservative middle ages. It is not surprising therefore that Bacon was at times a severe critic of his contemporaries and that with his reform spirit he should come into difficulties with his order and the church. In 1271 Bacon wrote a *Compendium Studii Philosophiae* in which he expressed his views on certain subjects. "In no previous writing had the moral corruption of the church from the Court of Rome downward been so fiercely stigmatized: the whole clergy is given up to pride, luxury and avarice."[65] His teachings of his doctrines of science evidently attracted attention for in June 1266 he received a request from Pope Clement IV to transmit secretly to him copies of his writings regardless of any conflicting regulations of the Franciscan order in Paris. From this request it might be inferred that the influence of his teachings was suspected of questionable orthodoxy. In response to this request Bacon composed his three greatest works, the *Opus Majus, Opus Minus* and *Opus Tertium*. The first two of these were forwarded to the Pope in 1268. It is doubtful whether the *Opus Tertium* was ever received, as Clement IV died Nov. 29, 1268.

What influence, if any, these expositions of his ideas on many sciences may have had upon the Pope, is not known. Certain it is that Roger Bacon's troubles were not thereby ended for in 1277 he was tried and condemned by the Minister General of the order of Franciscans to imprisonment on account of suspect innovations (novitates suspectas). Just what is meant by imprisonment is not made clear, whether actual bodily confinement, or as suggested by Prof. Walsh, only "enforced retirement,"[66] but at any

---

[65] Bridges, *Roger Bacon—The Opus Majus, Introduction.*
[66] See the interesting sketch of life and work of Roger Bacon in Walsh, J. J., *Catholic Churchmen in Science*, 3d ser., 1917.

rate he does not seem to be credited with any literary or scientific activity from 1277 until 1292 when he wrote his *Compendium Studiae Theologiae* and in this year also it is recorded that he was buried at Greyfriars, the Franciscan church at Oxford.

That which distinguishes Roger Bacon from other scholars of natural science of his century is not that in general he possessed more advanced knowledge or insight into the sciences. That this was the case in his more special field of optics may well be true. In other fields of science which interested him, however, he seems to have depended upon the same authorities as those of Vincent, Albertus or Bartholomaeus and to have granted them his confidence at times to an even greater degree than his contemporaries. The great distinction of Roger Bacon lay in the fact that in the domains of physics, mechanics and chemistry he had a living interest and enthusiasm for the practical achievements accomplished by science and beheld the vision of greater things to follow. In the field of chemical activities he was a keen student of the accepted authorities of the time, and, in at least one particular, his readings had been in a line which were unknown to Albert or Vincent. For he has heard of, if he has not seen, various contrivances for fires and explosives such as we have seen in the *Book of Fires* of Marcus Graecus, which neither of his slightly older contemporaries seems to have known. It may be recalled that the earliest copy of the manuscript of the *Book of Fires* is from Roger Bacon's period.[67] In another connection this feature of Bacon's knowledge has been previously discussed.[68] For the value of experiment in science he held great enthusiasm and advocated it with zeal. His arguments were logical and numerous though destined to fall upon sterile soil during his own time so far as any response can be noted, though we may believe that not in vain was this bread thrown upon the waters.

---

[67] *See* ante, p. 196.
[68] *See* ante, p. 199 *ff.*

Sciences of nature Bacon has classified into perspective (optics); astronomy (operative and judicial); the science of weights (heavy and light); alchemy; agriculture; medicine, and experimental science.[69] This notion of experimental science as a separate branch of science was a distinctly original idea with Bacon and the object of much consideration in his works.

"The things specially and strictly assumed as belonging to nature are those in which is the principle of motion and rest, as in the parts of the elements which are fire, earth and water, and in all things made from them which are inanimate as metals, stones, salts and sulphurs, pigments and colors such as minium and cerusa and lapis lazuli which is azurium, and Grecian green and things of that sort generated in the belly of the earth."[70]

Experimental science seems to Bacon a separate science operating in and through the other sciences. Experience and experiment, says Bacon, are necessary to establish confidence in truth. Nothing is established by argument and logic unless supported or confirmed by experiment. The function of experiment is verification and experiment attains to truth not to be reached by other sciences. Nature must be studied at first hand. In the *Opus Majus* Bacon also has emphasized certain causes which have hindered the progress of true philosophy among the Latin writers. The first of these is dependence upon the example of slight or unworthy authorities, the second the undue weight of established custom, third the power of public opinion, and fourth, ostentatious pretense to wisdom and efforts to conceal ignorance. This is assumed by our superiors—this is the popular opinion—therefore it must be accepted.

With respect to Bacon's experimental science F. H. Bridges[71] has well said:

"Last among the series of the Natural Sciences comes that which Bacon denotes as *Scientia Experimentalis*. The

---

[69] Bacon, *Communium Naturalium*, Steele ed., Liber I, p. 5.
[69] Bacon, *loc. cit.*, p. 2.
[71] Introduction to *Opus Majus*, Oxford Press, 1897, I, LXXVIII.

sample of it, for it can hardly be regarded as more than a sample, given in the sixth section of the *Opus Majus* indicates that it was connected in Bacon's mind with no special department of research, but was a general method used for the double purpose of controlling results already reached by mathematical procedure and of stimulating new researches in fields not as yet opened to inquiry.

In some respects this is the most original part of his work. Not that experiment was a new thing. Experiments without number had been made by man from the time of his first appearance on the planet. The Greeks towards the end of their marvellous scientific career had begun to use experiment in their investigations of natural truth. Galen had applied it in his researches into the nervous system; Ptolemy had arrived by its means at his remarkable discovery of the refraction of light. The Arab astronomers, far more skilful mechanicians than the Greeks, had constructed extremely elaborate apparatus for the same purpose, and also to verify the equality of the angles of incidence and reflection. But no one before Bacon had abstracted the method of experiment from the concrete problem, and had seen its bearing and importance as a universal method of research. Implicitly men of science had begun to recognize the value of experiment. What Bacon did was to make the recognition explicit.''

That the earnest exhortations of Bacon as to the importance and value of experiment fell on unfertile soil we may infer from the observation of Mr. Bridges in discussing the various known manuscripts of Bacon's work, that this sixth section of the *Opus Majus,* namely *Scientia Experimentalis,* appears to have been seldom copied. Kopp also in his *Roger Bacon*[72] remarks:

''What Bacon in respect to the method of the investigation of nature in general perceived and expressed has long been undervalued; what he has given us of particularities or announcements of discoveries is often overestimated.''

A. G. Little lists thirty-six titles of works by Roger Bacon and as many more doubtful or spurious.[73] Among

---

[72] *Beiträge zur Geschichte der Chemie,* III. p. 90.
[73] *Roger Bacon Essays,* collected and ed. by A. G. Little, Oxford, 1914.

the latter are several which deal with alchemy. Among those accepted by Little as original are only three or four specifically dealing with alchemical subjects and the authenticity of these has been questioned by competent critics. There are, however, in his works of unchallenged authenticity many passages which deal with chemical subjects. Tradition credits Roger Bacon with being a student and practitioner of alchemy and magic, though whether only on the basis of these works falsely credited to him it is difficult to decide. From his well authenticated writings it is certain that he was an interested and careful student of the literature of the subject and he takes many occasions to express his knowledge of and belief in the past accomplishments of chemistry (alchemy) and his faith in the importance of future possibilities. In these illustrations and references, however, Bacon nowhere claims such knowledge on the basis of his own personal experience with chemical manipulation, and usually quotes the authority for his statements. The claim that Bacon was the discoverer of gunpowder has been already discussed in another connection.[74]

Bacon's explanation of the meaning and significance of alchemy is characteristic of his point of view.[75]

"There is another science which treats of the generation of things from the elements and of all inanimate things and of simple and composite humors, of common stones, gems, marbles, of gold and other metals, of sulphurs and salts and pigments, of lapis lazuli (that is, azurium) and minium and other colors, of oils and burning bitumens and other things without limit, concerning which we have nothing in the books of Aristotle. Nor do the natural philosophizers (philosophantes) know of these, nor the entire crowd of Latin writers. And because this science is not known to the generality of students it necessarily follows that they are ignorant of all that depends upon it concerning natural things, namely of the generation of

[74] See ante, p. 199.
[75] Roger Bacon, Opus Tertium (in his Opera Quaedam Hactenus Inedita, ed. by J. S. Brewer, London, 1859, I, pp. 39, 41).

animate things, of plants, and animals and men, for being ignorant of what comes before, they are necessarily ignorant of what follows. For the generation of men and of brutes and plants is from the elements and the humors (or 'waters') and is related to the generation of inanimate things. Whence on account of their ignorance of that science it is not possible to know common natural philosophy, nor theoretical medicine, not only because natural philosophy and theoretical medicine are necessary for the practice but because all medicinal simples from inanimate things are obtained from that science which I have touched upon, as is made clear in the second book on medicine by Avicenna who enumerates the medicinal simples, and as is evidenced by other authors. Of these medicines neither the names nor their meanings can be understood except through this science, and this is theoretical (speculativa) alchemy which theorizes about all inanimate things and the entire generation of things from the elements.

"But there is another alchemy, operative and practical, which teaches how to make the noble metals, and colors and many other things better or more abundantly by art (artificium) than they are made in nature. And science of this kind is greater than all those preceding because it produces greater utilities. For not only can it yield wealth (expensas) and very many other things for the public welfare (rei publicae) but it also teaches how to discover such things as are capable of prolonging human life for much longer periods than can be accomplished by nature. For we die far earlier than we ought and this on account of defective regulation of health from youth up, and because also our fathers give us a corrupted constitution (complexionem) on account of the same defects in their own regulation of health, whence old age and death come more quickly and before the term which God has set for us. Therefore this Science has special utilities of that nature; while nevertheless it confirms theoretical alchemy through its works and therefore confirms natural philosophy and medicine, and this is plain from the books of the physicians. For these authors teach how to sublime, distil and resolve their medicines, and by many other methods according to the operations of that science, as is clear in health-giving waters, oils and many other things. Whence Galenus in his *Liber Dinamidiarum* instructs physicians how to make Calcecuminon, which physicians [nowadays] just as they know not how to make it so they know not how to name it. And Avicenna teaches in the first book of *Medicine* how to prove by the works of alchemy that it is not alone blood that nourishes as *Galenus* thought, but the other humors also; but this no physician knows either to understand or to perform, and similarly with very many things.

"Hence this duplex science of alchemy (that is, theoretical and practical) is unknown to nearly all men. For throughout the world many are working to make metals and colors and other things, yet extremely few know how rightly to make colors, or profitably, and scarcely any one knows how to make metals, and still fewer are they who know how to make preparations which are useful in prolonging life. And they also are few who know how to distil well, and to sublime and calcine and to resolve and do any of those works of art of that kind by which all inanimate things are certified (certificantur) and through which are confirmed theoretical alchemy, natural philosophy, and medicine.

"Hence there are not three among those Latin writers who have devoted themselves to the knowledge of theoretical alchemy, as it is alone possible to be known without the operations of practical alchemy, namely, according to that which those books and authors teach who have proved it through their own practice. There is but one who is competent and most skilled in all those things.

"Because so few know these things they do not deem it worth while to communicate them to others nor to associate with others, since they consider as asses and lunatics other men who are subject to the quibbles of law and those sophisms of artists (artistarum), which have debased philosophy and medicine and theology. Moreover the operations of that science are difficult and most expensive, for which reason those who know well the art of operating are not able to operate; and the books on that science are so secreted, that a man can scarcely find them, whilst they may be nevertheless more numerous than in any other department (facultate), and which also by reason of their multitude cost much.''

This description of alchemy well illustrates the fact that it is not so much the theoretical aspect as the practical value of the work of alchemy that commands his interest. Credulity toward the claims of alchemists to be able to produce gold from base metals, and prepare elixirs for long life was almost universal at the time though the impostures of most of those who professed to possess the arts were well recognized by Bacon as by his contemporaries Albertus and Vincent. Both elsewhere in the *Opus Majus,* and in the *Opus Tertium,* Bacon refers to the making of gold of a superpurity by Alchemy, as well as to the medicine that will prolong life. In the *Opus Tertium* he says:[76]

''Similarly in the domain of alchemy. For the natural grades of gold in the belly of the earth are twenty-four, but by art they can be multiplied indefinitely. But all works on alchemy do not teach of these grades nor in what manner the seventeen kinds (modi) of gold are compounded from these. For that whole art is scarcely able to make gold of the twenty-four grades from these, and neither can Nature in the belly of the earth—and yet these are in the domain of alchemy. But then comes the experimentor and investigates these twenty-four grades of gold and resolves the seventeen kinds (here *species* in place of *modi*) and is able to make as many more than twenty-four as he wishes, which neither the art of alchemy nor nature in the belly of the earth are able to accomplish and the medicine which the experimentor prepares for this is the greatest of secrets. . . . For that is what removes all corruption of baser metal and converts it into gold, and that is what takes away the corruptions of the human constitution so that life may be sufficiently prolonged.''

It is interesting in the above to see how Bacon endeavors to discriminate between the domain and powers of alchemy

---

[76] A. G. Little, *Part of Opus Tertium,* Aberdeen, 1912, p. 46.

as a science, from that of experimental science, though operating with the same subjects. A passage in the *Opus Majus* discusses this same topic rather more at length but to the same general effect. He is here a little more specific as to twenty-four grades of gold:

"When twenty-four grades are found in a mass of gold, this is the best gold that can be produced in nature—when there are twenty-four grades of gold and one part of or grade of silver, then this is a poorer gold than the former, and so proceeds the diminution of the grades of gold up to sixteen or until there are eight grades of gold mixed with silver."

And of the "medicine" he there says:

"For that medicine which could remove all impurities and corruptions from baser metal so that it could become the purest silver and gold is considered by the wise to be able to remove the corruptions of the human body to such a degree that it could prolong life through many ages (secula)."

A description of the manufacture of brass is given in the *Opus Minus*.[77] The description is introduced in connection with a discussion of the errors of writers resulting from their ignorance of languages, and consequently of the real significance of terms used.

"For it is unknown to nearly everybody how cuprum, aes, electrum and orichalcum, called by error aurichalcum, should be properly called. It is thought by nearly all that these are different kinds of metals, though this is not true. For aes, and orichalcum and electrum are made from copper (cuprum). The metal that is first smelted and purified from earthly impurities is really copper and so it should be called. But although into copper, melted and purified, powdered yellow calamina is sprinkled, yet it does not contain much of the powder, but as the copper is made a little harder and more yellow it is then called aes. Calamina is a certain vein of earth and is of many kinds but I refer here to the yellow. If considerably more of this powder is

---

[77] Roger Bacon, *Opera Quaedam Hactenus Inedita*, ed. by J. S. Brewer, London, 1859, Vol. I, p. 385.

added it makes it still harder and more deeply colored.
Thus is produced orichalcum, and since in all the books of
the Bible of greater preeminence and in other books of
medicine and of the saints *orichalcum* is found, *aurichalcum* is nothing, but called thus by moderns in error. . . .
If yet more calamina is added then is produced electrum,
though that is better made from yellow tucia as Avicenna
teaches in Book V.[78] Tucia is a certain vein of earth and
is of several varieties, but yellow tucia is here proper. In
certain regions of the earth they add tucia to copper to
obtain electrum but in others they use calamina.

"Because electrum is made with the bellows (blast)
therefore by the force of the bellows much tucia is blown
away; wherefore that which is subtle escapes and there remains a hard material which renders the electrum hard.
Hence it is harder than orichalcum which is made without
the bellows.

"Though electrum is generally thus made, it is possible
for it to be made far more beautiful and noble by means
of certain things, opposites of tucia and calamina, such as
the roots of uruscus [?] and the fig tree (ficus) and other
things, provided due skill is used (dummodo debitum artificium praebeatur). This electrum is good for astronomical
instruments and many other valuable uses. Though electrum is thus made, nevertheless, as says Servius (on
Vergil), threefold are the varieties of electrum, one is
from copper as just explained, another is a mixture of
certain proportion of gold and silver, and the third is a
gem stone. All authors, as Isidorus and others accept this
diversity from Servius. Pliny,[79] nevertheless, follows with
another kind of electrum. He says that this is collected
in the *glosaphis* islands (glaesaria, Pliny says) between
Germany and England, called electricae by reason of their
abundance of electrum. It is produced from the juice of a
pine tree, distilling from that into the sea as these trees are
in compidine alnei marini, and this juice is solidified into
a solid and translucent substance by the action of the sun
and sea. Pliny states that electrum is known to be

---

[78] This significance of Electrum is also met in Vincent of Beauvais, VIII,
36, ("hoc aurichalcum frequentis scripturae vocatur electrum").

[79] *Natutalis Philosophiae,* XXXVII, 11.

formed from this juice from the odor, since the juice and the electrum have a similar odor.''

It will be noted that in speaking of the use of calamina in brass, Roger Bacon seems to have the same opinion as Albertus, based upon Aristotle, that it is mainly a color effect that is produced by the calamina which is apparently thought to be itself again volatilized. The nature of these alloys and their compositions were but dimly understood by the writers of this period.

In the fragment of the *Opus Tertium* discovered by Professor Duhem there is contained a short treatise on the enigmas and keys (claves) of alchemy. The intention of this is to give to the Pope a brief account of the terms used in alchemy and of their significance so that, as he concludes, by these, ''with other things I have written, it is possible for Your Wisdom wisely to make use of them and to detect every impostor.'' This work is of interest in manifesting the care with which Roger Bacon has studied his authorities.

The work begins with an introduction in which he refers to the extracts on this subject in his other works principally the *Opus Majus,* and that in these he has hesitated to speak clearly of these things mainly because of the undesirability of spreading information on this subject to those who are not wise. Then follows:

''The Explanation of the Enigmas of Alchemy.

''Therefore the general explanation of the Enigmas is here necessary. Hence the philosophers explain what are bodies, spirits, planets, stones and many other things. Bodies are those which do not flee from the fire nor volatilize in smoke, as metals, stones, strictly taken (proprie sumpti) and other solids.

''Those things which flee from fire are called spirits, as mercury, sulphur, sal ammoniac, orpiment, which is arsenicum.

''The planets are seven, according to Avicenna in the first book 'de Anima,' that is in the major science of alchemy. For lead is called Saturn; tin, Jupiter; iron, Mars; gold,

Sol (the sun) ; copper, Venus; quicksilver, Mercury; silver, Luna (the moon).

"In whatsoever manner it is found written in books differently is the fault of the writer or translator or a mystification. For sometimes it is found that bronze (aes), is compared to Mars, but this is false. For bronze is nothing but copper colored by the powder of calamina, and similarly brass (aurichalcum) and electrum are made from copper and the same powder or the powder of tucia, as I have stated in the Second Book (*Opus Minus*).

"And quicksilver is called aurum vivum [quick gold] as Avicenna often misuses this word.

"Gold is sometimes designated also as stone or body of the river Iberus [the Ebro], of the Pactolus, or of the Tagus or some other, because grains of gold are found in these.

"Because the Hybernici [Irish] are named from the Iberus ('Hyberus') in the kingdom of Castile since they lived there for 300 years after they had departed from Egypt on the death of Pharoah in the Red Sea, and before the King of England had given them the island of Hybernia, as certain histories relate, therefore gold is called corpus Hybernicum, or stone (lapis) Hybernicus or something similar. . . . Silver is also called margarita [pearl] on account of its white color and is called unio, because margarita and unio are the same, as Solinus informs us in the book *De Mirabilibus Mundi*. For margarita is called "unio" because never more than one at a time is generated in the marine shell. For shells naturally open to receive the dew of heaven, and a single drop of dew received, (the shell) shuts again and by its power solidifies the drop into a margarita or unio.

"Silver is also called Anglia because silver abounds there. Similarly also, a less red gold is called Anglia because it is found there. And good gold is called Hispania, or Apulia or Polonia or any other region where good gold abounds.

"Rubificare [to redden] is to make gold, and albificare [to whiten] is to make silver. To convert Saturn into Sol, or into Hispania, or Apulia, or Palonia is to make gold from lead. To convert Venus into Luna or into Anglia is to make silver from copper, because gold has to be made from lead, and silver from copper. By medicine, or laxative medicine, is called that which projected into liquefied lead converts it into gold, and converts copper into silver, and this is called *elixir* in all the books. That is called the greater work (opus majus) when gold is made, the lesser (minus) when silver is made. Also that is called the *minus opus* when one pound of medicine converts ten or so up to 100 pounds of base metal into a nobler: and *majus opus,* when the medicine is so powerful that one pound converts two hundred or a thousand or a thousand thousand of baser into nobler metal. That such a medicine may be possible Avicenna and all others attest.'' [80]

"Concerning the Keys of Alchemy.

"The operations of that art are called keys (claves) which are performed according to the precepts of this science in order that the medicine may be had which is called *elixir*. Those Keys are purification [another manuscript says *putrefaction*], distillation, ablution, grinding, roasting, calcination, mortification, sublimation, proportion, incineration [another reading is *inceration,* softening],[81] decomposition (separating '*resolutio*'), solidification, fixation, cleansing (mundificatio), liquefaction, projection. These operations are known to all skilled in this science and their books are full of these. And very many alchemists perform these works but do not know how to elicit the chief object of them. This arrangement (*ordo*) of the operations is according to the execution but not according to intention of the profession (*artificii*). As to this mystification I have adduced the authorities of Aristotle in the first book of the *De Anima,* and the sixteenth of the *De Animalibus,* and Avicenna in the first *Liber Physicorum,* sixth of the *Metaphysica* and elsewhere, all of whom explain that what is first in intention is last in execution and vice versa as is evident to any wise man. . . .''

These examples will suffice to illustrate the scope and

---

[80] Little, *Part of Opus Tertium*, pp. 83, 84.

[81] *Inceratio* is the mixture of a liquid with a dry substance by gentle combination to the consistency of wax. *Lexicon Alchemiae,* Rulandus, 1612, A.D.

character of the chemical knowledge and ideas of Bacon. It is evident that his works contain no ideas or facts not generally known to the literature of his time, but that he is well informed and has carefully studied the authorities which he quite generally quotes for the authorization of his statements.

The *Speculum Alchemiae* or *Mirror of Alchemy* attributed to Roger Bacon is a short treatise in seven chapters treating of the composition and origin of the metals. It contains only the conventional Arabian theories of mercury and sulphur as the constituents of metals, with obscure metaphysical discussions of the origin of mercury and sulphur and vague allusions to transmutation and the red and white elixirs and their projections. Judging from its contents, this work might have been written as well in the twelfth century as in the more probable fourteenth. There is nothing in it that is characteristic of Roger Bacon's style or ideas, nor that distinguishes it from many unimportant alchemical lucubrations of anonymous writers of the thirteenth to the sixteenth centuries.[82]

The work is listed by A. G. Little among those of doubtful authenticity. Professor M. M. P. Muir[83] says:

"The directions for making the philosopher's egg given in the *Mirror of Alchemy* and the *Secrets of Nature and Art* closely resemble those contained in ordinary alchemical writings. There is in them the vague talk, the haziness, the thinking in images of words rather than in images of things which are the marks of most books on practical alchemy."

E. v. Lippmann considers the alchemistic works attributed to Bacon, *Brever Breviarium, Tractatus Trium Vertorium, Speculum Alchemiae*—as clearly pseudepigrapha.[84] Yet is is upon these books that Bacon's reputation as a

---

[82] Texts consulted by the writer are—the Latin text in Zetzner's *Theatrum Chemicum*, 1602, II, pp. 433–442, and the English text, *The Mirror of Alchemy*, London, 1597.

[83] *Roger Bacon Essays*, collected and ed. by A. G. Little, Oxford 1914, p. 301.

[84] E. v. Lippmann, *op. cit.*, pp. 493, 494.

practical alchemist was mainly based.

Considering the relation of these three great scholars of the thirteenth century, the Frenchman Vincent of Beauvais, the German Albert von Bollstedt, and the Englishman Roger Bacon, to the development of chemical knowledge, it appears then that no one of them contributed anything of importance either of facts or theories to the knowledge of their predecessors.   It would nevertheless be a grave error for this reason to underestimate the importance of their influence on the development of chemistry or of science in general, for by their extensive summarizing of the authorities existing in their time and by the weight of their authority they did much to make accessible the accumulations of the knowledge of the past, and to revivify and popularize the study of science.   And this was indeed largely due to the reintroduction of Aristotle's natural science to the western world, and to the rehabilitation of his authority.   "The triumphal progress of Aristotle is one of the marvels of man's mental history," says E. Withington.[85]   *The Physica* and *Metaphysica* of the pagan philosopher who taught the eternity of the Universe, the mortality of the soul, and the nonintervention of the Deity in the fate of the world or the affairs of men, were promptly and naturally condemned by the Church in 1209, 1215 and later.   Yet in less than a century, the greatest of Catholic theologians had converted them into a bulwark of orthodoxy, and the greatest Catholic poet had given their author the immortal title: Master of those who know.

It was, however, to take time before the new impulse to science was to be perceived in new contributions to chemical thought, unless indeed we may attribute to this influence the work of the unknown author who chose to write under the name of Geber and thus conceal his identity. To distinguish him from the Arabian alchemist, Djaber, he is generally alluded to as pseudo-Geber.

---

[85] *Roger Bacon Essays*, collected and ed. by A. G. Little, Oxford, 1914, p. 340.

CHAPTER VII

CHEMISTRY OF THE FOURTEENTH AND FIFTEENTH CENTURIES

Considering the intellectual awakening of the thirteenth century, and the revival of interest in the natural sciences, as shown in works of the encyclopedists and other writers, and the influence of new universities, it would seem reasonable to anticipate that the fourteenth and fifteenth centuries should have exhibited a marked advance in chemical thought and discovery. On the contrary these centuries exhibit very little which would justify such expectations. There were indeed causes operative which help to explain why the field of chemistry was comparatively sterile of productive activity.

From the statement of the thirteenth century encyclopedists, and from Arabian writers also, we know that there existed much imposture and charlatanry among writers on alchemy, with their assumptions and claims as to gold making and the elixir of life. Concerning the dates or authorship of such alchemical writings we rarely have specific or definite information. Works of this character were not generally issued except under precautions to conceal the identity of the writer.

That substantial reasons existed for such precautions we know from contemporary records. We have already referred to the close supervision and censorship exercised by the church even upon the natural science of Aristotle, in the early part of the thirteenth century. It is natural that the activities of the alchemists who claimed to make gold and to prolong life indefinitely, often associating these claims with magical invocations and mystic charms, should

273

have been closely censored and, so far as possible, suppressed.

Thus in 1317 Pope John XXII issued a decree against the alchemists:

"Alchemies are here prohibited and those who practise them or procure their being done are punished. They must forfeit to the public treasury for the benefit of the poor as much genuine gold and silver as they have manufactured of the false or adultered metal. If they have not sufficient means for this, the penalty may be changed to another at the discretion of the judge, and they shall be considered criminals. If they are clerics, they shall be deprived of any benefices that they hold and be declared incapable of holding others," etc.[1]

In Barcelona in 1323, Hervé Nedelic, General of the Dominican Friars, pronounced the penalty of excommunication against all clericals who should apply themselves to the study of alchemy or should not within eight days burn all books of that character which might be in their hands. Haureau[2] considers this as circumstantial evidence that the alchemical treatises attributed to St. Thomas Aquinas were not yet issued, a conclusion in harmony with all known facts, for no allusions to any of these works are known until much later.

It was not only the church which viewed with suspicion the activities of the alchemists. In 1380 Charles V of France proscribed the prosecution of alchemy throughout the kingdom and even forbade the possession of instruments and furnaces for alchemical operations. In England in 1404 Henry IV promulgated an edict against the practice of alchemy,[3] and in 1418 the greater council of Venice directed an edict against the alchemists. Many other instances might be cited of attempts by clerical or

---

[1] Full text of this decree in Latin and in English translation has been published by J. J. Walsh, *The Popes and Science*, London, 1912, pp. 125–126. A French translation is in L. Figuier, *L'Alchimie et les Alchimists*, Paris, 1860, p. 140.

[2] B. Haureau, *Histoire Litteraire de la France*, XXXIV, Paris, 1914, p. 312.

[3] L. Figuier, *op. cit.*, pp. 140, 141.

civic authorities to suppress the counterfeiters and supposed gold makers.

On the other hand, many rulers and nobles, believing in the possibility of transmutation, were tempted by the hope of gain to encourage and protect impostors who claimed to be able to supply unlimited wealth by occult means, and the efforts to suppress alchemical activities were notoriously ineffective, for the numbers of the alchemists and of alchemical writers seems to have increased rather than to have diminished.

Not only with authorities who were concerned with the protection of the stability of state coinage and currency from the feared debasement by false gold, but with the cultivated classes quite generally the alchemists were held in evil repute. Dante (about 1300) in his *Divina Commedia* pictures them in the tortures of the deepest regions of the Inferno; Petrarch (in 1366) satirizes their deceptions; and Chaucer in his *Canterbury Tales* (about 1388) voices the low estimate in which the alchemists were held.

It may readily be conceived that the conditions in the fourteenth and fifteenth centuries were not such as to make the field of chemical activities, other than the technical arts, attractive to men of scholarly inclinations nor to enlist the services of really able men. On the other hand a great number of men of mediocre ability were attracted by the very mystery and obscurity of the forbidden science to dabble in it, and others, who saw their opportunity to profit by the reputation of wonder workers, found in the popular belief in the reality of these mystical arts a fertile soil for their operations.

In these conditions may be found the reason for the sterility of these centuries in chemical literature of real merit. Very many treatises were written on the philosophy and practice of alchemy but they were almost all issued either anonymously or pseudonymously, as the authors did not wish to incur the penalties incurred by those who were suspected of practicing a forbidden art.

One writer only who can be credited to the end of the thirteenth or the beginning of the fourteenth century is a notable exception to the general mediocrity of the chemical writers of the period. This unknown person was probably a Spaniard versed in the Arabian chemistry, who wrote under the name of Geber.

We have already seen that besides the authentic works of the Arabian Djaber, there were certain Latin works of later origin long credited as translations of Arabian manuscripts of Djaber, which were in all probability original works by a different person.

When Professor H. Kopp in 1874[4] completed his elaborate studies into the personality and works of Geber, he expressed grave doubts as to the genuineness of the Latin texts attributed to the Arabian alchemist. He saw no relation between any Arabian texts known to him and the alleged translations into Latin. He called attention to the fact that the few allusions to Geber in thirteenth century writers were not to any of the well-known Latin works of the so-called Geber, nor indeed did they bear any resemblance to them. Kopp submitted the Latin works to an Arabic scholar to see if perchance there existed any internal evidence that these works were translations from Arabic originals, but this expert, Professor G. Weil, could find no traces of such evidence. Kopp noted also that the earliest references to the Latin works were in writings attributed to Arnaldus de Villanova (not earlier than 1310) and to Lullus (about the middle of the fourteenth century). He could find no manuscripts of the Latin Geber of a period earlier than the fourteenth century.[5] Nevertheless as Kopp knew that in the libraries of Europe there existed Arabian manuscripts attributed to Djaber the contents of which had not been investigated, with characteristic caution he hesitated to declare the alleged translations as spurious and tentatively discussed these writings

---

[4] *Beiträge zur Geschichte der Chemie*, Vol. III, pp. 13–54.
[5] The earliest now known manuscript of the Latin *Summa* is that of Munich, attributed to the latter part of the thirteenth century.

as of an original Geber or Djaber of the eighth or ninth centuries.

When M. Berthelot, however, in his exhaustive studies of Syriac and Arabian manuscripts, the results of which are comprised in his *La Chimie au Moyen Age,* three volumes, 1893, and in his *Archéologie et Histoire des Sciences,* Paris, 1906, had been able to compare the contents of many of these Arabian manuscripts, it became clear to him that there was nothing whatever in these documents that bore any resemblance to the Latin Geber's writings. His researches established the justice of Kopp's doubts, and apparently proved that the writings, which under the authority of Geber so widely influenced the chemists of the fourteenth and later centuries, were not expressive of views and knowledge of the eighth or ninth centuries, but of the close of the thirteenth or the fourteenth. This means that the later writer might even have had at his disposal such manuscripts as those of Vincent de Beauvais, Albertus Magnus, Roger Bacon, Avicenna, Rhazes, and the works falsely attributed to the two latter and so much used by Vincent and Albertus.

Naturally also it follows that the Arabian Djaber has been credited by historians of chemistry with a knowledge of chemistry and more especially with a definiteness of description of processes and manipulations to which he is not entitled. It also follows that the Latin Geber is to be considered as the inheritor of the accumulated results of Arabian alchemists, and possibly also of the popular summaries of that knowledge as presented in the thirteenth century by the great encyclopedists of the period. The works of Geber are extensively cited by Petrus Bonus (1330), and it may be assumed that they were first issued not far from 1300 A. D. If we were to accept as correct the report that Villanova wrote alchemical works while with King Robert at Naples, whither he went in 1309, and that among these works the *Novum Lumen* is correctly ascribed to Villanova, which is very doubtful, the citation of Geber in that work would be evidence that his works were then extant.

The works of Geber demand consideration because they exerted an almost epochal influence upon later chemists, and not without justice. Though his notions upon matter, upon the constitution of metals, upon transmutation, etc., are entirely those of his predecessors and the majority of the more important chemical facts known to him be found in the writings of Greek, Arabian, and Latin writers before his time, yet his method of presentation of his subjects is so essentially modern as compared with preceding writers that he could not fail to attract attention. In the first place it is at once manifest that the author is a man of practical experience in the manipulations of chemistry and not a mere compiler or editor of authorities. In the second place he is animated with the desire to explain experimental methods and apparatus so clearly that others may profit by his experience. His presentation is, moreover, orderly and systematic, clear and concise, contrasting sharply with the obscure style, vague descriptions, and confusing disorder in the writings of earlier alchemists, whether Greeks or Arabians.

The generally credited works of Geber are not numerous nor voluminous. They are four in number: *Summa Perfectionis Magisterii; De Investigatione Perfectionis; De Inventione Veritatis; Liber Fornacum.*[6] The works entitled *Testamentum Geberi regis Indiae,* and *Alchimia Geberi* are according to Berthelot and Darmstaedter manifestly of more modern origin.[7]

The long credited belief that Geber was an Arabian, expressed in phrases such as "Geber Arabis," "Geberis regis Persarum," "König der Araber," seems to be without foundation, as these appellations are interpolations of not earlier than the fifteenth century. The earlier manuscripts

---

[6] The editions accessible to the writer are: the Latin text entitled *Alchemiae Gebri Arabis Philosophi,* etc. Bern, 1545; *The Works of Geber,* translated into English by Richard Russell, London, 1678; *Die Alchemie des Geber,* Ernst Darmstaedter, Berlin, 1922.

[7] Berthelot, *La Chimie au Moyen Age,* I, p. 343.

of the *Summa* contain no such intimations and Geber's writings are noticeable in that he cites no authority, Greek or Arabian, but quotes from the "ancient philosophers" or in similar vague terms.[8] Kopp[9] states that the announcement that Geber was a Spaniard was first made about 1581. This ignores the authority of Petrus Bonus, the author of the *Pretiosa Margarita Novella*, dated 1330 at Pola, who alludes to Geber more than once as Geber Hispanus.

It is of interest that the title of Geber's principal work, elaborated in the printed edition to *Summa Perfectionis Magisterii, etc.*, is given in the earlier manuscripts by the word *Summa* alone. Petrus Bonus (1330) often referring to Geber, also uses only the title *Summa*. The statement made by Darmstaedter that no manuscripts of the three other works are known earlier than the first printed works, is in harmony with the fact that Petrus Bonus seems to know only the *Summa*. It may be possible, therefore, that the other works credited to Geber may be elaborations by later writers.

So far as present knowledge authorizes, we may assume that Geber was a European chemist, probably a Spaniard, who wrote largely from his own experience as a practical chemist and metallurgist, and that his theoretical views upon alchemy were those of the thirteenth century, which were largely the result of Arabian development. No Arabian originals are known which might have been translated by him nor which present so advanced a knowledge of chemical processes. On the other hand he makes no claim to originality, and seems to have endeavored to give a clear description of the practice of his time.[10]

---

[8] The recent work by Dr. Ernst Darmstaedter *Die Alchemie des Geber übersetzt und erklärt*, Berlin, 1922, presents well the latest knowledge on Geber and his works.

[9] *Op. cit.*, III, p. 20.

[10] E. J. Holmgard (in *Nature*, February 10, 1923) criticises the purely negative evidences presented by Berthelot, that the Latin Geber is not a translation of the Arabian Djaber, and expresses the hope that investigation of as yet unexplored Arabian manuscripts may prove the contrary, though he says that up to the present he has not found any Arabic works which can be considered as the originals of the Latin treatises.

Geber takes for granted that the supreme aim of the science is the removing of imperfections of metals so that they shall become "perfect." He accepts the existence of the philosopher's stone and of the elixirs, red and white, that in these elixirs red and white "there is no other thing than quicksilver and sulphur," "and because all metallic bodies are compounded of quicksilver and sulphur—pure or impure—accidentally (superficially) and not in their first nature, therefore by convenient preparation it is possible to take away such impurity." "The natural principles of the metals are three; sulphur, arsenic and quicksilver." . . . "Sulphur is a fatness of the earth thickened until it be hardened and made dry, and when it is hardened it is called sulphur." . . . "Arsenic is a subtle matter like to sulphur therefore it need not be otherwise defined than sulphur." . . . "Quicksilver is a viscous water united in the bowels of the earth with white subtle earth until the moist is tempered with the dry." Expressions such as these illustrate how completely the author is dependent upon the conventional chemical philosophy of the Arabian alchemists.

An extended discussion in scholastic style in the *Summa Perfectionis* concerning the various reasons why men—"sophists and ignorant men"—deny the truth and validity of the art, and his confutation of these reasons, does nothing to advance the knowledge of chemistry.

The work on "The Investigation of Verity or Perfection" contains descriptions of the preparations for coloring the baser metals white or yellow. These he calls medicines "white and red according to the nature and properties of the body (metal) to be transmuted." As might be supposed they are methods for staining or alloying of copper, lead and tin to present colors resembling gold or silver—though the writer claims that the applications of these medicines effect a real transmutation—"and it will be a medicine tincturing every metal and mercury itself into a true Sol (gold) or better."

While Geber in his theories is entirely bound by the conventions of his period, it is very evident that it is the experimental work that is his principal interest, and though he considers all these as steps having ultimate bearing on the problem of transmutation, yet the operations of chemistry are with him not dependent for their interest on this possible consummation.

The book on the Investigation of Perfection treats of the preparation and purification of substances or reagents which are useful for the perfecting of the metals, and the work is expressly intended to be an introduction to the main work, the *Summa*. It is confined to directions for purifying salt, alkali, sal ammoniac, alums, copperas, and similar salts, and to obtaining the metals in the form of solutions. These directions are invariably perfectly clear, consistent and practical, for example:

"Purification (Mundatio) of common salt.

"Common salt is purified in this way. First it is ignited, then dissolved in ordinary warm water, the solution filtered, the filtrate solidified by a gentle fire in a glazed dish. The solidified material, when calcined for a day and night with moderate fire, you may consider as sufficiently purified."

"Purification of sal alcali.

"Sal alcali is purified like common salt; and it is sagimen vitri.[11] First it is ground and entirely dissolved in ordinary warm water, afterwards filtered and solidified and calcined with gentle fire."

"Purification of Alum.

"First of glacial alum. Many kinds can be used without any purification. Nevertheless it is purified in this manner. It is placed in an alembic and thus the whole humidity extracted; which is of much value in this art. The residue (feces) remaining in the bottom of the vessel is either dissolved upon the stone in some moist place, or extracted with water, or reserved."

It is worthy of note that the distillate from alum which

---

[11] Sagimen vitri is sodium carbonate according to Thomas Thomson, *History of Chemistry*, London, 1830, p. 124.

he says is of much use in the art would be a solution of sulphuric acid.

"Purification of sharp vinegar.

"Vinegar (acetum) and all that kind of sharp or sour substances are rendered subtle and purified, and their virtue or effect improved by distillation."

It is important to note that Geber as well as the Arab chemists uses the terms for vinegar and sharp waters without attempting to discriminate as to their specific character. In the *Summa* Geber states that complete solution of substance is effected by the use of acute, sharp and saline waters having no feces (solid residues), as vinegar, sour grapes, very sour pears, pomegranates and such like, distilled. That such vegetable acids do not constitute all that he means by sharp waters or corrosive waters, is made clear by the following from his book *De Inventione Veritatis*.

"And first as to our solvent waters of which we have made mention in our *Summa* where we have spoken of solution by the sharpness of waters.

"Take first one pound of vitriol of Cyprus and a half pound of salpeter and a quarter pound of laminated alum and extract the water at the red heat of the alembic, for the solvent power is great, and make use of it in the forementioned chapters: it will be made much sharper if you dissolve with that a fourth part of sal ammoniac, because it then dissolves gold, sulphur and silver."

The distillate here described as obtained from the retort at redness would be a mixture of sulphuric and nitric acids, and by the addition of the ammonium chloride, hydrochloric acid. The solvent action of this acute or sharp water makes much more comprehensible the chemistry of many processes described, than if we assumed that the vegetable acids were the only ones used. It is probable that this is by no means Geber's invention, but he is perhaps the first who describes the preparation so clearly and comprehensibly.

The "preparation" of metallic bodies consists in general in first heating the crude "bodies" (metals or metallic ores) to expel all humidity and then in burning off all sulphur or other substances which can be removed by ignition in the air ("calcination"), continuing the ignition until the ore or metal itself is converted to a dry powder, and in treating the material thus obtained with these sharp or corrosive waters until solution is obtained. Easily fusible metals as tin and lead, after roasting at moderate temperature, are reduced in a perforated crucible contained in another crucible under protection of a layer of melted glass, the reduced metal flowing out through the apertures. This is again roasted and treated again with sharp waters till dissolved.

Each metal differs in detail in this manipulation according to its properties. The preparation of Venus (copper) will illustrate his method of description.

"Venus is prepared by this method. A layer of common salt is placed in a crucible and above it thin plates of copper, and above this a layer of salt and above this other plates and so on continuously until the vessel is filled, then covered and luted. It is then placed in a furnace of calcination for a natural day. Then it is taken out and that which has been calcined is scraped off and the plates replaced with fresh salt. And thus it is calcined repeatedly until all the plates shall have been consumed or corroded by the action of the salt and fire, because the salt corrodes the superfluous humidity and combustible sulphurity, and the fire elevates the volatile and inflammable substance with due proportion. It is then rubbed to the finest powder and washed with vinegar until the water running from it is free from blackness. (Probably the vegetable acids in use contain some tannin. The blackness would result from iron as an impurity). Again moisten it with fresh salt and vinegar, and grind, and after grinding place it in the calcination furnace in an open vessel and let it stand for three natural days. Then it is removed and ground well and fine and washed with vinegar well and long until cleansed and purged of all impurity. It is well dried in the sun, then half its weight of sal ammoniac added, ground well and long until it shall become an impalpable substance. Then placed in the open air (sub divo) or in a bath of manure (in resolutionis fimo) until whatever is subtle shall be dissolved, renewing the sal ammoniac, if necessary, until all becomes water (that is, solution). Honor this water which we have called the *water of fixed sulphur,* with which the elixir is tinctured, to infinity. These directions suffice for the preparation of Venus."

The *Summa* gives detailed descriptions of the processes for distillation, sublimation, calcination and for the preparation of various chemical substances. Details of directions for construction of the furnaces are given with much minuteness and throughout it is evident that the writer is himself thoroughly familiar with the processes. The general characteristics of the metals, the readiness with which they form alloys, or with which they unite with sulphur, are well described, though these facts may also be found scattered through writings of earlier Greek or Arabian authorities.

It is of interest to note that in his description of lead, Geber mentions that in calcination it does not preserve its proper weight but is changed to a new weight. He ventures no explanation however as to the cause of this phenomenon. A later chemist, Eck of Sulzbach, supposed to have written about 1490, whose work *Clavis Philosophorum* was printed in the *Theatrum Chemicum,* Vol. IV, states more specifically,

"Six pounds of mercury and silver amalgamated, heated in four different vessels for eight days showed an increase of weight of three pounds. This augmentation comes from the union of a spirit with the metallic body" (spiritus unitur corpori).[12]

Directions for constructing a water bath are clear though the device is rather crude.

"In a pan place hay or wool three fingers deep, cover the

---

[12] Cf. Hoefer, *Histoire de la Chimie,* 2d ed., I, p. 471; also Kopp, *op. cit.,* III, p. 119.

retort (concurbitum) with the same almost as high as the neck of the alembic and upon this lay many small sticks or weighty stones which by their weight may depress the hay, or like material. Pour in water until the pan is full, then place fire under it until all is distilled off.''

The book on furnaces, *De Fornacibus,* is a concise description of furnaces and apparatus for particular purposes, illustrated with drawings to explain their arrangement. Just what were these illustrations in the original manuscript is not known. In the printed edition of 1545 they are evidently elaborated into finished engravings characteristic of the period, but references in the text to figures show that drawings were also present in the manuscript. These engravings in the printed work are mainly, though not entirely, duplicates of illustrations in the *Summa,* as though the work were intended to furnish an abbreviated manual of furnaces and appliances for different operations.

It is not necessary to claim for the unknown writer of the pseudo-Geber works any original contributions either to the development of chemical philosophy or to advances in chemical practice, in order to explain the great influence which he exerted on his successors for two or three centuries. The fact that he presented to his world a manual of the general chemical practice of his time, so clear and concise as almost to make an epoch in chemical literature is sufficient to account for the great stimulus which he exerted. Indeed it is not too much to assert that, as a manual and guide to the ordinary operations of chemistry—distillation, sublimations and furnace operations generally—and to many accessory operations with metals, no later publication is known which rivals his before the sixteenth century.

As to the personality of the pseudo-Geber we know nothing. Petrus Bonus, erudite Italian writer on alchemy of 1330, the earliest writer to quote Geber extensively, calls him Geber Hispanus and there is no reason for supposing that he is not right in it. The facts that he draws upon

the traditional Arabian chemistry and writes in Latin
are quite in accord with the assumption that he is a
Spanish chemist.

With the alchemists of later centuries no names after
Geber had greater veneration as masters of the mystic
art than those of Arnaldus of Villanova and Raymundus
Lullus (Lull or Lully).

Arnald of Villanova was a physician of high reputation
in the latter part of the thirteenth century. He was born
as variously stated in 1235 or 1248. There has been much
dispute as to which of several towns named Villanova was
his birthplace, but according to evidence presented by his
biographer, Bartholomé Haureau,[13] he was a Spaniard
from Catalonia, and probably also sometime a resident of
Valencia. He had studied at Naples and from Arabian
medical masters in Valencia, knew Arabic, and his medical
doctrines were largely founded on Rhazes and Avicenna.
In 1285 he was called to the court of Pedro III, King of
Aragon. In 1300 he claimed Montpelier as his residence
and is named among masters of medicine at that new
medical university. While in Paris in 1299 on a mission
of a business nature for the eldest son of Pedro III of
Aragon, he was arrested on charge of heretical doctrines
and prophecies. There followed a long contest, the final
outcome being that he submitted to the Pope the book
which had been condemned, and that this was finally returned to him, absolving him from the charge. As he had
however many enemies in Paris, personal as well as clerical,
he left the city for a time, though in 1306 we hear of him
again in Paris. In 1308 he was with Pope Clement at
Avignon, and a little later he went to Sicily at the solici-
tation of King Frederick, to the court at Catania. In 1309
he was at the court of King Robert, at Naples, where he
was said to have written alchemical works. He met his
death in 1311 or 1312 by shipwreck while on his way to

---

[13] Bartholomé Haureau, *Histoire Littéraire de la France*, 1881, XXVIII,
pp. 26–126.

answer a summons to attend Pope Clement V then suffering from a painful malady.

Arnald evidently enjoyed a great reputation as a practical physician. He is also said to have achieved this reputation largely through his use of chemical medicines. He was interested in magic and alchemy and it is traditionally stated that in the presence of the familiars of Pope Clement V at his court at Avignon, he had turned plates of copper to gold. In 1317, or about five years after his death, thirteen small books of Arnald were condemned by the Inquisition on account of fifteen heretical propositions. The titles of these books listed by Haureau are all of a theological character and none alchemical. The works themselves are no longer extant.

Works upon alchemy attributed to Arnald, *Thesaurus Thesaurorum et Rosarius Philosophorum,* and *Novum Lumen,* were much later listed by the Archbishop Sandoval of Toledo,[14] among proscribed works.[15] The texts of these treatises are contained in the collected works of Arnaldus, for example in the Basel edition of 1585.

Notwithstanding the reputation of Arnaldus as interested in alchemy, there is much doubt as to the authorship of alchemical works attributed to him. Schmieder lists twenty alchemical treatises attributed to Arnaldus. It is quite certain that these were not all by the same author nor all of the same period.

Haureau doubts that Arnaldus was the author of any of these. That five years after his death no such works seem to have been known to the censors who proscribed other writings by him is in itself ground for doubt. That any works under his name were unknown to a writer on alchemy in 1330, Petrus Bonus[16] who cites elaborately all authors known to him, is circumstantial evidence in the same direction. Either these writings, if authentic, were kept secret during his life and for years after his death,

---

14 Archbishop Sandoval of Toledo was of the early seventeenth century.
15 Haureau, *Histoire Littéraire de la France,* XXVIII, *loc. cit.*
16 *See post,* p. 293.

or they are like nearly all writings on alchemy which pretend to personal achievements and claim to instruct in the art of transmutation, written by some impostor who seeks shelter under the name of a prominent scholar deceased.

H. Haeser[17] says Arnald's philosophical works were destroyed by the Inquisition; that the alchemical works upon which his reputation in great part rests are in all probability fraudulent and emanate perhaps from an Arnald who lived at Montpelier at the beginning of the fifteenth century.

The alchemical writings attributed to Arnaldus are characterized by the obscurity and charlatanry found in most of the anonymous alchemists of that century, treating of the transmutation of the metals, the red and white elixirs and their preparation, the philosopher's stone, etc. The reasoning, as is characteristic of this class of works, is analogical and weak.

The following is a typical illustration of his attempt to establish that transmutation is reasonable and possible. It is from the *Flos Florum,* one of the articles in his collected works.

"Ice or snow is converted by the action of heat into water. Therefore it was first water then snow or ice. But all metals can be converted into quicksilver, therefore they were first quicksilver. The method of converting them into quicksilver I shall teach below. But it being granted that a metal can be converted into quicksilver, there is refuted the opinion of those who assert that it is not possible for spirits (*spiritus,* that is volatile substances) and other materials to be transmuted into the elements and into the nature of metals, unless first reduced to their primal matter. This reduction to their primal matter is easy as I shall show below. Therefore the transmutation of metals is possible and easy. In the same way it can be shown you that the multiplication of metals is possible; for everything that is born and grows is multiplied, as is clear with plants and trees. For from one seed a thousand seeds are procreated,

---

[17] *Lehrbuch der Geschichte der Medizin,* 1875, I, p. 722.

from one tree proceed infinite shoots from which are produced various and an infinite number of trees, and thus
their number is increased and they multiply. But metals
are born in the earth and grow, therefore augmentation and
multiplication in these is possible even to infinity,'' etc.[18]

The experimental methods which he gives for accomplishing the desired objects are complicated and contain
nothing of interest when they are at all comprehensible.
His philosophy of matter and its changes is the conventional Arabian theory.

The most popular of his treatises in the fifteenth century and later was probably his *Thesaurus Thesaurorum
et Rosarium Philosophorum* (Treasure of Treasures and
Rose Garden of the Philosophers). It consists of two
parts, the first in ten brief chapters gives the conventional
Greek-Arabian doctrine of the origin and constitution of
metals, of sulphur, mercury, and the philosopher's stone,
and transmutation. The second part of thirty-two chapters
contains seemingly specific directions for operations for the
preparation and purification of substances supposed to be
necessary for the preparation of the elixirs and the philosopher's stone. As Professor Thomas Thomson pertinently
remarks,[19]

''Perhaps the most curious of all these works is the
Rosarium which is intended as a complete compend of all
the alchemy of his time. The first part on the theory of
the art is plain enough; but the second part on the practice,
which is subdivided into thirty-two chapters, and which
professes to teach the art of making the philosopher's
stone is in many places unintelligible to me.''

Hoefer[20] thus summarizes his judgment on Arnald's
work:

''To summarize, the works of Arnald of Villanova are almost insignificant, because they contain not a single observed fact of which the discovery is due to the author,
whom we do not believe we have judged severely enough.''

[18] Arnaldus, *Opera*, Basel, 1585, pp. 2044, 2045.
[19] *History of Chemistry*, 1832, I, p. 42.
[20] *Histoire de la Chimie*, Paris, 1842, I, p. 394.

Later critics have not seen reasons to modify the judgment of these earlier historians, that no new fact or theory
in chemistry is traceable to Villanova.

That, as a physician, his reputation may well have been
deserved; and that he, in his practice, made successful uses
of chemically prepared substances—as alcohol, arsenious
oxide and mercury preparations—more generally than was
customary among his contemporaries is also to his credit,
though the chemical facts contained in these papers present
nothing new. It is evident that he possessed a very exaggerated notion of the medicinal value of alcohol which he
calls aqua vini or aqua vitae. The aqua auri or water of
gold apparently contained no gold but was a yellow colored
solution containing alcohol and rosemary—to which he
attributed great curative value.

Among his medical treatises, two articles on poisons and
on wines manifest a comprehensive knowledge, though they
contain no new facts, and indeed draw largely, directly or
indirectly, from Pliny and Dioscorides.

The name Lullus ranks with Geber and Arnaldus de
Villanova high in the estimation of the alchemists of the
fifteenth and sixteenth centuries. We have here, however,
another illustration of a name respected for learning and
piety being used by later writers as a shield for their forbidden activities. The real Raymundus Lullus was of
Spanish nationality, born at Palma in Majorca in 1235,
shortly after the conquest of the Balearic Isles from the
Musselmen. He became a member of the Minorite friars,
was a prolific writer on theology, philosophy, logic, and
originated a system of graphic classification of syllogisms
which attracted much attention. He was widely traveled,
known at Paris, Rome, Naples, in Cyprus, and Armenia.
His great passion was to convert the Mohammedans to
Christianity, in which mission he encountered hardships
and imprisonment. He was stoned to death in Bugia, Al-
giers in 1315 while laboring in this cause.

There were doubts in the earlier centuries as to whether

the alchemical works attributed to Lullus were actually written by him, and the historians Kopp and Hoefer both freely voice their incredulity of the authenticity of such works. The thorough and elaborate investigations of B. Haureau on the biology and works attributed to Lullus[21] and the later investigations of Berthelot[22] establish beyond reasonable doubt that none of these alchemical writings is his own, but all are the production of Spanish or south of France writers, written at times much later than Lullus. It is here not practicable to discuss in detail the evidences advanced to prove the falsity of the eighty works on alchemy printed or in manuscript, the titles of which are given by Haureau among the more than 300 works on all subjects which he there discusses.

Suffice it to say, that in two bibliographical lists of his writings composed, one in 1311, the other in 1314, which are published by Haureau, no such work is included, that no manuscript copy of any such work attributed to Lullus has been found of date anterior to the fifteenth century, and that many of his most popular and frequently printed works profess to have been written in 1330 or 1332. In all probability, however, even these dates are falsified and the works themselves of later origin. Haureau considers all of these alchemical pseudo-Lullus works as not earlier than the fifteenth century.

It is also worthy of note that when in 1386 to 1394 certain works of the real Lullus were suspected and condemned upon the basis of heterodox theological expressions, there is no reference to any of these alchemical works, which would themselves at that time have given adequate cause for condemnation. This alone excites a fair presumption that no alchemical works attributed to him were then known. If we therefore ascribe to a pseudo-Lullus these alchemical writings, it is with the probability that more than one writer masqueraded under that name, and none of

---

[21] B. Haureau, *Histoire Littéraire de la France*, 1885, XXIX, pp. 1–386.
[22] Berthelot, *op. cit.*, I, p. 351 *ff.*

these works is, in all probability, earlier than nearly a century after the death of the real Lullus.

As to the general character of the best known of these works of the pseudo-Lullus, it is difficult for us to understand the high repute in which they were held. Considering the period in which they were actually written, they contain remarkably few facts which were not known to writers previously discussed.

The *Testamentum,* in two parts, *Theorica* and *Practica,* seems to be one of the earlier works, as it is often referred to in other treatises assuming to be by the same writer. The *Theorica* is well characterized by Hoefer[23] as "A tissue of generalities and speculative notions for the most part devoid of sense." Of the *Practica* the same author says: "One would search in vain for clear and positive experiments."

Professor Thomas Thomson[24] says:

"I have attempted several times to read over the works of Raymund Lully, particularly his *Last Will and Testament,* which is considered the most important of them all. But they are all so obscure and filled with such unintelligible jargon that I have found it impossible to understand them."

Gmelin[25] characterizes Lullus as the weakest (schwächste) of the great medieval authorities from Albert the Great to Arnald of Villanova, crediting him nevertheless with certain observations of chemical nature; the greater part of these however, as Hoefer later observed, were not new.

A work which is ascribed to Lullus but which as Haureau has noted makes no claim to be by any author of that name but was issued anonymously and arbitrarily attributed to Lullus by editors or publishers, is called *Experimenta* and dated 1330. It contains statements that the author received information as to at least two of the experiments described, from his friend Arnald of Villanova at Naples

---

[23] Hoefer, *op. cit.,* Paris, 1842, I, p. 400.
[24] Thomson, *op. cit.,* I, p. 40.
[25] Gmelin, J. J., *Geschichte der Chemie,* 1797, I, p. 70 *ff*.

(experiments number 19 and 23). He does not allude to any treatises written by Arnaldus, nor indeed is there any evidence that the date given for the work is authentic. It is apparently upon this shadowy foundation that is based the statement that Lullus was associated with Arnaldus at Naples.

These experiments, thirty-four in number, are all very circumstantially described but nevertheless are not very instructive. The term "aqua fortis" is used for nitric acid, the preparation of which however, though not under this name, was described by the pseudo-Geber. Whether this is the first appearance of the name "aqua fortis" is hard to decide on account of the uncertain date of this and of other fourteeenth or fifteenth century works which contain it. The preparation of a concentrated and purified syrup (oleum) of potassium carbonate from the ignition of tartarum (argol from wine) is given with elaborate and partly useless experimental detail, but that had been given by pseudo-Geber, intelligibly and much more concisely. The author of *Experimenta* describes a more concentrated alcohol than early descriptions previously noted. He directs to take aqua vitae of the highest strength, such that it burns a linen cloth, and to again put it through the alembic. It may be recalled that earlier descriptions cited describe the properties of the product as such that it will not burn the cloth or the finger upon which it burns, evidently therefore dilute.

An important writer of this period is Petrus Bonus, who is known through a book which bears the title: "*Petrus Bonus of Ferrara, Physicus. Introduction to the Arts of Alchemy.* Composed in 1330 in the City of Pola in Istria. A Precious New Pearl (Pretiosa Margarita Novella)." [26] The date 1330 is repeated at the end of the work, though 1339 is stated in the preceding paragraph as the date of completion of the work.

This work of Bonus is an elaborate and learned treat-

---

[26] Manget, Liber III, Sect. I, subsection I, pp. 1–80.

ise on the philosophy of alchemy. He treats in scholastic fashion the subject in its various aspects, stating at great length for instance the arguments that the art of alchemy is not true, and with similar elaboration the reasons for believing it to be true. Though prolix it gives a very good account of the fourteenth century philosophy of alchemy. It was much prized by later alchemists, being often published and was translated into German.

He cites authorities profusely, and this is of importance from the fact that Petrus Bonus seems to have been a writer whose personality and date are generally accepted as genuine. The work bears all the character of an earnest and honest treatise. Authors whom he cites, he cites very frequently. Thus the works of (pseudo-) Geber, written probably about 1300, are very often quoted, and apparently this is the latest authority he knows. There is no citation in his lengthy work, which is confined strictly to alchemy, of any treatise on this subject by Albertus Magnus, Roger Bacon, Thomas Aquinas, Arnaldus of Villanova nor Raymond Lullus. It is impossible that he should have cited Lullus in 1330, because, as we have seen, this pseudo-Lullus literature is certainly none of it earlier, and probably all of it considerably later.

The omission of the other names is significant, as works of alchemical nature attributed to those men were at later dates very much esteemed, on account of the high reputation of the men as scholars; and it seems safe to infer that if works like the *Libellus de Alchemia,* etc., attributed to Albertus, or the *Speculum, Alchemiae, Breve Breviarium de Dono Dei, De Arte Chemiae,* credited to Roger Bacon, or the various alchemical works credited to Arnaldus of Villanova, were then extant, that so conscientious a student of authorities as Petrus Bonus would not have been likely to have omitted them.

In the case of the work attributed to Albertus Magnus, as we have already noted[27] it bears evidence of much later

----

[27] *See* post, pp. 359, 360.

origin so that no critic longer considers it as original. As to the strictly alchemical works of Roger Bacon, modern criticism tends more and more to consider all these as pseudonymous and the fact that they were unknown to Petrus Bonus strengthens the theory that they were written in the fourteenth rather than in the thirteenth century. As to Arnaldus, we have already noted that his alchemical writings were not included in the bibliographies of the writings in 1310 and 1311, and were also not included in the list of books which, some five years after his death, were condemned by the Inquisition censors. Had any of these writings been really authentic, and extant in 1330, it seems probable that Peter Bonus, so familiar with the works of the Spaniard Geber, would not have been ignorant of the writings of the Spaniard Arnald of Villanova, so eminent as a physician and for some time a resident of Italy. As Arnaldus is known to have died in 1311 or 1312, we may infer that this omission confirms the assumption that if any of the treatises on alchemy were really written by him they were kept secret until some years after his death.

There is another alchemical writer called The Monk of Ferrara whom Lenglet du Fresnoy (1742) considers to have written about 1280, or at latest at the beginning of the fourteenth century, because he quotes Geber, Morienus and the Turba, but says not a word of Arnaldus of Villanova or of Lullus. Schmieder also (1832) speaks of this alchemist and attributes to him a date of about 1200, because he mentions neither Albertus Magnus, Roger Bacon, Arnaldus nor Lullus. So far as these considerations are concerned there is nothing that would necessitate placing the works of this writer earlier than the time of Peter Bonus, or in the first half of the fourteenth century, as according to the best evidence we now possess, the works of alchemical character attributed to all these authors were not earlier than the early part of the fourteenth century and some of them much later. That he cannot have been much earlier than Peter Bonus is evidenced by the fact that he does cite the pseudo-Geber.

The authors cited, and most of them frequently, by Peter Bonus comprise nearly all that are prominently quoted by writers who are known to have written before his time and who deal with the philosophy of alchemy. Thus among the ancients, Plato, Democritus (pseudo), Empedocles, Aristotle, "Philosophus," Galen, "Hermes," Porphyrus are cited. Of medieval writers, (pseudo-)Aristotle, Avicenna, (pseudo-)Avicenna, (pseudo-)Rhazes, Morienus, Senior (Zadith), and Mesue; also cited is the *Turba Philosophorum* (a twelfth century composition), and the numerous personages therein contained, Albumazar, Alphidius, Averroes, Hamec, Thebit, and Calid. Among the works and authors mentioned by him, the works of pseudo-Geber are apparently the only ones that were not known to Vincent of Beauvais and Albertus Magnus or other thirteenth century writers.

Among alchemical writings of the fourteenth and fifteenth centuries, in so far as present evidences exist, must be included the ten titles given by Schmieder, ascribed to Albertus Magnus, twelve to Thomas Aquinas, twenty-five to Raymund Lully, fifteen to Roger Bacon, (Professor Thomson cites eighteen), and twenty or more to Arnald of Villanova, all of which are in great part, if not entirely, pseudonymous. All of these works however are so lacking in originality and valuable contents that the reputations of those men, all of them justly prominent on account of their authentic works, gain rather than lose by being relieved of the responsibilities for the alchemical works ascribed to them.

This period was prolific in alchemical writings by many anonymous and pseudonymous writers or by persons whose dates and personalities are more or less vague and doubtful. Prominent among these are Johanus de Rupescissa, about 1350; Richard Ortulanus, about 1350; Nicholas Flamellus (Flamel), 1330–1413 (?); Bernhard of Treviso, 1406–

1490; John Cremer, who claims to have met Lullus in Italy in 1330! !; Geo. Ripley, 1415–1490; Thomas Norton, English writer who first appears in literature about 1600, and whose popular *Ordinal* is dated or misdated 1477; Philip Ulsted, who was teaching medicine at Freiburg in Breslau in 1500. Hortulanus, the alleged translator into Latin of a famous but purely mystical writing called the *Tabula Smaragdina,* supposed to have been written by the legendary Hermes, was believed by Schmieder and other early historians to have been of the eleventh or twelfth century, but Berthelot asserts that he wrote about 1350.[28] Two writers long credited by historians as belonging to the fourteenth or fifteenth century, who wrote under the pseudonyms of Basil Valentine and John or Isaac Hollandus, are known to be of the close of the sixteenth and the beginning of the seventeenth century, and therefore have no place in this chapter.

Of the works of all these writers, there is nothing that advances to any material extent the knowledge of chemical facts or thought, however they may have appealed to those who cultivated the philosophy of alchemy as such. Very many of these works enlarge upon the Arabian theories of matter and its changes without contributing anything new. Very many of them also are filled with extravagant claims and boasts as to what the authors have experienced or accomplished in prolonging human life or turning masses of baser metals into gold and silver.

When the art of printing with movable types had advanced so that printing books became easy, about 1500 A.D., quantities of these alchemistical writings were collected and published, either singly or in small or great collections. Among the more important of these collections are the following:

Artis Auriferae quam Chemiam Vocant.

Basel, 2 volumes, 1572, a 3d volume, 1610.

Theatrum Chemicum, Zetzner.

3 volumes, 1602, 2d ed. 6 volumes, 1613–1661.

---

[28] Berthelot, *op. cit.,* I, p. 234.

Bibliotheca Chemica Curiosa. J. J. Manget.
   2 volumes folio, Cologne, 1702.
Theatrum Chemicum Britannicum. Ashmole.
   London, 1652.
Bibliotheca Chemica. F. Rothscholzen.
   1719.
Lenglet du Fresnoy, in his *Histoire de la Philosophie
Hermétique* (Paris, 1742, 3 volumes), lists the titles of
nearly a thousand treatises upon alchemy, in print.

   In spite of the sterility of chemical literature, it should
not be inferred that no progress in chemical arts or practice was made during this period. The workers in the
practical arts of chemistry were not writing for the public,
but nevertheless were not inactive. Such chemical industries as the making of glass, and coloring of glass, paper
making, pigments, and metallurgy, were progressing
steadily, though for information concerning the processes,
we are indebted to works of a following century, when such
books as George Agricola's *De Re Metallica,* Biringuccio's
Pyrotechnia, and similar works of less importance made
their appearance.

   There was evidently in medical practice a considerable
tendency to make available for medicinal uses the preparations of chemistry. That effort was more or less manifested in the *Materia Medica* of Dioscorides and Pliny,
and we have already alluded to the use of new chemical
remedies by Italian physicians and by Arnaldus of Villanova.

   The *Distilling Book (Liber Distillandi),* published by
Hieronymus Brunschwygk in Strassburg in 1500, describes
a phase of application of chemical methods to medical
practice. The special purpose of the book was to apply
the methods of distillation with steam to separating the
active principles of medicinal agents from the nonessential
matter. These medicinal agents were largely plants or
herbs, but many other substances were evidently considered
of similar value, and the distilled "waters" of ants, frogs,

oxblood, flies, and a great variety of similarly strange remedies are described. These distilled waters were quite
radical innovations upon conventional medieval pharmacology.

The work of Brunschwygk had many successors developing the same kind of medicines.

The distilled waters of Brunschwygk's descriptions have
left little trace in pharmacology, but the attempt to utilize
chemical methods in the preparation of remedies which his
work illustrates was not without influence in helping to
pave the way for the more intimate connection of chemistry
and medicine brought about by Paracelsus and his followers in the sixteenth and seventeenth centuries.[29]

The real importance of the movement illustrated by the
distilled waters depends upon the recognition that drugs
and other medicinal agents depend for their efficiency upon
pure principles, "spirits" or "quintessences," and that
these principles may be extracted by the methods of chemistry.

---

[29] For a fuller discussion of the *Liber Distillandi* and its influence see the
writer's article ''Chemistry in Medicine in the Fifteenth Century,'' *Scientific
Monthly,* 1918, p. 167 *ff*.

CHAPTER VIII

THE PROGRESSIVE SIXTEENTH CENTURY

The centuries from the period of Geber to the beginning of the sixteenth century were, as we have seen, not distinguished by noteworthy advances in chemistry and that partly by reason of the bad name in which the alchemists were held. All kinds of chemical activities were under suspicion and there was little encouragement for the cultivation of chemical philosophy, or for venturing outside the practice of the technical arts.

But those centuries were marked, however, by events and by tendencies that were preparing better conditions for scientific speculation and progress. The foundation in all European countries of new universities so importantly inaugurated in the twelfth and thirteenth centuries continued actively in the fourteenth and fifteenth. In the fourteenth century over twenty new foundations were made; Italy led with Rome, Perugia, Treviso, Pisa, Florence, Pavia, Ferrara; France added Avignon, Cahors, Grenoble, Orange; Spain added Lerida, Perpignan, Huesca; and in other countries were founded the universities of Prague, Vienna, Erfurt, Heidelberg, Cologne, Cracow, Buda and Fünfkirchen. The fifteenth century excelled the fourteenth in the number of new foundations, more than thirty being recorded: in Italy, Turin and Catania; in France, Aix, Poitiers, Caen, Bordeaux, Valence, Nantes, Bourges and Besançon; in Spain, Barcelona, Saragossa, Valencia, Alcala, Palma in the Isle of Majorca; in the German Empire, then including the Netherlands, Leipsic, Würzburg, Rostock, Louvain, Treves, Greifswald, Freiburg im Breisgau, Basel, Ingolstadt, Mainz, Tübingen. Great Britain added

St. Andrews, Glasgow, and Aberdeen; and there were also founded in this century the universities of Upsala in Sweden, Copenhagen in Denmark, and Pressburg in Hungary.

While the natural sciences then found little place in the curricula of these universities, at least in any form which we recognize as science teaching—and chemistry, youngest of the sciences, least of all—yet gradually conditions were changing, the thoughts and experiences of men were widening and gradually also the problems of natural sciences were finding their way into university thought. Even chemistry, through the door of medicine, became a live subject in the universities long before it was recognized in any formal way as a subject worthy of university teaching.

The discovery of printing by means of movable metal type in the latter half of the fifteenth century was a factor hardly less influential than the universities, making accessible to a vastly larger public in the form of printed books and pamphlets, material hitherto only accessible in laboriously and expensively copied manuscripts.

When at the beginning of the sixteenth century the spirit of unrest in theologic matters culminated in the Protestant Reformation, and the censorship of the ecclesiastical authority was relaxed, a multitude of alchemical writings which had circulated surreptitiously were printed and circulated freely. The secrecy and mystery which had surrounded them in the past gave them an interest and importance which most of them would doubtless never have received except for the previous censorship.

The capture of Constantinople by the Turks in 1454 and the breaking up of the Byzantine Empire resulted in the scattering of the Greek scientists and made more available to Europe their accumulated manuscripts and scientific knowledge.

The discovery of America and of the ocean route to India (1498), were opening new centers of trade and commerce.

All these influences were stimulating to new thoughts

and wider interests, and we find the first half of the sixteenth century marked by many great names in all lines of thought. Such are, for instance, Michael Angelo, Leonardo da Vinci, Rafael, Machiavelli, Ariosto, Martin Luther, Columbus, Thomas More, Erasmus, Copernicus, Rabelais, Melancthon, Vesalius, Cardanus, and the list might be greatly extended. In the field of chemical activity the sixteenth century is marked by four great names, Theophrastus von Hohenheim or Paracelsus, (1493–1541), Vannucio Biringuccio, whose Pyrotechnia was published 1540, George Bauer or Agricola (1494–1555), and Bernard Palissy (1499–1589).

Before considering the work and influence of these men, certain anonymously printed works of German origin and important to an understanding of the progress of metallurgy are to be noted. These are small hand books for the use of miners and mining chemists or assayers, which were first printed about 1500 or possibly even before that. They were frequently reprinted throughout the century, and at various places, usually with slight changes, under the titles of *Ein Nützlich Bergbüchlein* and *Probierbüchlein*. The first named is a little book dealing with the occurance of metals in the mines, general descriptions of ore-veins, etc. This work contains little of importance with relation to chemistry.[1]

Descriptions of the nature and origin of the metals follow the conventional Arabian philosophy of the generation of the metals from sulphur and mercury in various degrees of purity and various degrees of combination and their relation to the seven planets. It is worthy of note, however, that bismuth (Wismüth) is mentioned in relation to its occurrence with silver veins, probably the first mention

---

[1] For a full discussion of these early booklets, see the appendix devoted to these early sources in Hoover's translation of Agricola's *De Re Metallica*. Their content and relation to Agricola's work are also considered in the voluminous and valuable historical footnotes in that work. For the opportunity of examining the early edition here cited the writer is under obligations to Mr. Hoover's valuable private collection of early works on mining and metallurgy.

of that metal. Agricola discusses it more at length in his
Bermannus (1530).

The *Probierbüchlein*, on the other hand, is a treatise very
important for the history of the development of mineral
chemistry. It will be recalled that the work of Theophilus
the Monk (ascribed to the latter part of the twelfth century) contained much detailed and accurate information
concerning the methods of separating gold and silver from
other metals and from one another by so-called cementation
processes, but contained no reference to methods depending on the use of the mineral acids. Geber (about 1300)
gives us our first definite information concerning the preparation and use of the strong mineral acids in the treatment of metals and ores. The *Probierbüchlein* reveals the
use of nitric acid and aqua regia in the systematic parting
of the metals as developed into a well conventionalized
system. As has been justly said:

"This is the first written work on assaying, and it displays that art already full-grown, so far as concerns gold
and silver, and to some extent copper and lead; for if we
eliminate the words dependent on the atomic theory from
modern works on dry assaying, there has been but very
minor progress." [2]

Hoover lists twenty-one editions of the *Probierbüchlein*
from the earliest about 1510, to 1782, though this list makes
no claim to completeness. The earliest edition known is
without date or place, but estimated at the British Museum
as probably printed at Augsburg in 1510. It is this edition from which the following illustrations are drawn.

It is of interest to note that the manufacture of various
balances for the laboratory seems to have been well developed. The directions say:

"First order a good and accurate Cologne or Nürnberg
assay balance with a long beam which is adapted and
proper to lift the silver button (Korn). Take care that you
lift nothing heavy with it, for by that the balance will be
lamed and [weigh] false.

---

[2] Hoover, *op. cit.*, note p. 614.

For a second you should have a one-way balance that is stronger, with which you weigh copper and ore for the assay hundredweight.

For the third you should have a balance for weighing added material (Zusatz) and lead, which carries 23 or 24 Lot (Lot is a half ounce). It must be quite strong so that a *Marck* (half a pound) can be weighed on it."

A later edition of the *Probierbüchlein* (1533) gives woodcut illustrations of these three balances, as does also Agricola in his great work.

The assay weights are described, and are on the same principle as the modern assay ton weights, except that the standard is not a ton but the hundredweight, the *centumpondium*. In the making of the weights (of brass) this hundredweight is not standardized but taken of any convenient weight and the fractional weights carefully made to exact parts of the large one.

The use of a fine grained black stone, the "touchstone," for determining the relative proportions of gold and silver in coins or other alloys is of very ancient origin. Theophrastus describes it, and Pliny, though describing the process inaccurately, applying it to the ore (vena) instead of to metals themselves, says that persons of experience can tell in a moment the proportions of gold and silver or copper "their accuracy being so marvellous that they are never mistaken." [3]

The *Probierbüchlein* contains elaborate directions for making sets of standard touch needles for comparison with alloys to be tested by their streak on the touchstone. These sets are composed of alloys of silver and gold, silver and copper, gold and copper, and gold, silver and copper. Each set numbers, usually, from twenty to thirty needles. Agricola's *De Re Metallica* incorporates these in his more systematic account.

To make cupels for assaying it is directed to "take horse bones burned and pulverized, and wood ashes, well washed,

---

[3] *See ante*, p. 60.

with an equal part of the bone ash, moisten these and strike cupels, which are good. On the newly made cupels sift through a very fine sieve on the deepest part bone ashes from calf's head or fish bone or pike's head to thickness of a poppy leaf, and then give it a blow with the stamp. This gives good cupels. Let them dry well and the older such cupels are the better they become. Sprinkle burned and powdered pike bone on the cupels when you wish to test an ore.''

The following recipes will illustrate the character of the very modern-sounding directions of this little book at the beginning of the sixteenth century.

"To test ore by sal alcali.

"All metals or ores can be melted and tested in a small sample, however infusible.

"Take for one pound of ore, or what you wish to melt, two pounds granulated lead, five *Lot* (2½ oz) salt, five *Lot* "sal alcali," a lye made from willow ashes and quick lime, five *Lot* corpus mortuus, that is the mud or residue from parting water, five *Lot* argol (tartar) and heat in a Viennese crucible, and cover it so that nothing unclean may fall into it, and let it fuse in the blast to a regulus (König) which then test.''

The test referred to was by burning off the lead and determining by the touchstone or by wet analysis the composition of the metal.

"To separate silver from gold.

"Take one part of silver which contains gold, one part *Spiessglas* (antimony sulphide), one part copper, one part lead, and fuse together in a crucible. When melted pour into a crucible containing powdered sulphur, and as soon as poured in cover it with a soft clay (laymen) so that the vapor cannot escape. Then let it cool and you will find your gold in a regulus. Place this on a dish and submit it to the blast (verblass).

"To reduce silver to a powder and again to silver.

"Dissolve in aqua fortis; take the resulting water and pour into bad [impure?] water which is warm or salty, and the silver settles as a powder. Let it settle well, pour

off the water and dry the silver powder (silver chloride).
"To make silver again from it.

"Take the powder, place it on a cupel (testa), and add
to it the powder from the residue of the *aqua regia* and
add lead, and subject it to the blast when there is enough
lead so that it encloses the powder. Otherwise it would be
blown away. Blast it till it "blickt" (flashes).
"To separate gold from silver.

"Beat the silver, in which you suppose gold to be contained, very thin, cut it in small pieces and lay it in strong
water and set it in a gentle fire until warm and as long
as it gives off bubbles. Then take it and pour off the water
into a copper dish and let it stand and cool. The silver
then settles in the copper dish. Let the silver dry on the
copper dish after the water is poured off, and melt the
silver in a crucible. Then take the gold from the glass
flask and fuse that to a lump."

The gold in the silver remains undissolved by the nitric
acid and the silver solution was evidently decanted or
filtered from this before pouring into the copper dish.

An edition entitled *Bergwerck und Probierbüchlein* published in 1533 at Frankfort am Main, republishes verbatim
a large part of the above noted 1527 *Bergbüchlein,* and verbatim also a large part of the 1510 (?) *Probierbüchlein,*
but adds a considerable amount of new matter on solution
and separation of the metals, on the polishing of gems,
preparation of excellent waters for separation and solution of ores, and on precautions against the evil effects of
poisonous metallic vapors.

The following direction for making strong water to dissolve all metals is more specific than that given by the
pseudo-Geber, but essentially the same process:

"Take one pound plumous alum, one pound vitriol, and
one pound saltpeter, pulverize well, put in a glass, set over
it a glass flask (alembic), cement the joints with lutum
sapientiae, which is made from one part strong potter's
earth, two parts well sifted ashes, one part sand, mixed
to a dough with a little water. Spread it on and let it dry
before putting on the fire. Then distil with gentle fire

until the first water is over, that is until the water begins
to be yellow or till it is colored. Then receive it in another
vessel. As soon as the yellow changes to red, there comes
over the strongest water. Receive that in a special vessel
and wait with all care until it is quite clarified. Let it
stay until all is distilled off. This last must take place with
a good strong fire. Then you will have the right water
that dissolves all things. Stopper it well so that no odor
or strength escapes. You can keep it in a thick strong
glass two days.

"If you wish to make it twice as strong; take one part
alum, one part green vitriol, one part saltpeter, one part
tucia. Pulverize and distil as above. It has indescribable
strength."

Plumous (feathery) alum is a variety mentioned also by
pseudo-Geber. The ancient and medieval chemists mention many varieties of alum—some of them being really
vitriols—but at this time all varieties are apparently varieties of alum proper, or at least sulphates of aluminium.
The three fractions distilled in the above process were
manifestly, first a dilute nitric acid, second a strong nitric
acid with some sulphuric, and third largely sulphuric with
some nitric. The use of tucia in the second recipe is of
doubtful advantage. Tucia was a crude zinc oxide, and
its addition would seem to have no other influence, if any,
than to hold back some sulphuric acid from distilling over.
The *Probierbüchlein* contains many other recipes for
strong waters, some of them containing salt or ammonium
chloride and yielding aqua regia. Not all the ingredients
added, however, are of any real significance in the process.

The formation of silver amalgam is described by dissolving silver in aqua fortis and then:

"Take the dissolved water, set in warm ashes and place
in the shade in a warm place. It will solidify to a hard
stone. Set in a bath of horse-dung so hot that the hand
can scarcely be held in it and in six weeks it will be a clear
water. Set it aside again as before and you will have the
Philosopher's Stone upon which the art depends.

"To project this medicine upon a quicksilver:

"Take 70 ounces quicksilver put in the furnace and blast it until the quicksilver is hot, then throw upon it one ounce of this stone and it melts like butter, penetrates all the parts (Glieder) of the quicksilver and turns it into fine silver which stands all the tests."

There is here a touch of alchemical pretention, in the interpretation of the result of this experiment, which may give a solid silver amalgam, but does not solidify the entire mass of quicksilver, when used in these proportions.

Warnings and precautions against the danger of poisonous gases from charcoal fires, lead and mercury fumes, and from the strong waters are given. Workers are advised to work in the open air, to cover the mouth, and it is advisable "some say" to eat garlic before and after the work!

These little books give many detailed directions as to apparatus and furnaces used in preparation of ores, separation of the metals, and other processes relating to metallurgy and assaying. They are extremely interesting as evidencing a well established technique which doubtless had an uninterrupted development from ancient times, and of which the book of Theophilus the Monk in the twelfth century and Geber at the end of the thirteenth century are illustrations of well defined stages. All these processes were to find their most complete summing up in the great work on mining and assaying of George Agricola.

In 1493 Theophrastus Bombastus von Hohenheim called Paracelsus was born in the village of Einsiedeln in Switzerland. He was a man of eccentric personality who was destined to exert an epochal influence on medicine and chemistry. His father, Wilhelm Bombast von Hohenheim, was a practising physician in Einsiedeln, was married to a woman in the service of the "Gotteshaus unserer lieben Frau zur Einsiedeln," and Theophrastus seems to have been an only child. When Theophrastus was nine years old the family moved to Villach, in Carinthia, a mining region and seat of a mining school founded by the Fuggers of Augsburg. It is very evident that his youth, passed in this mining region, and largely given to instruction in medicine by his father, afforded him many opportunities to become acquainted with the operation and facts of mining chemistry. It also appears that in early manhood he passed the better part of a year in the laboratory of a certain Sigmund Füger in Schwatz. Paracelsus alludes with gratitude to the instruction of Füger and to his experience with Füger's helpers. Throughout all his writings, whether on medicine, surgery or natural science or occult science, Paracelsus constantly draws for illustration or example upon his chemical knowledge and experience. Primarily Paracelsus was a physician. His medical education was probably inconsecutive, and it is not known where he received his degree of doctor. His adversaries later disputed his right to the title, a matter which he dismissed disdainfully. In his earliest medical works he writes his title as doctor, and alludes in 1527, in one of his defenses against his critics, to his doctor's oath, asking to whom this oath was taken whether to the apothecaries or to the sick.

From about 1518 to 1525, Theophrastus served a large part of the time as army surgeon in the Danish wars, the Netherlands and in the Neapolitan wars, returning from his travels to German territory at the age of about 32 years, with experience which qualified him to make a distinct impression as a practising physician. He was by that time a man of marked individuality, great self-confidence, strongly influenced by the spirit of revolt from traditional authority characteristic of the period of the Revolution, and imbued with the mission to free the practice of medicine from the domination of the traditional doctrines of Galen and Avicenna, and to further the founding of medicine upon independent observation and experience. And to chemistry he looked as an important factor in the new development of medical practices. Having attracted attention by successful treatment of prominent patients, he received in 1526 appointment as city physician of Basel, and was ex-officio professor in the university. Here his violent opposition to the accepted authorities, his unconventional practices and his aggressive temper brought him speedily into conflict with the medical faculty and profession, resulting in his abandonment of his position and his flight from Basel within a year.

From this time on his life was one of wandering and frequent hardship, in continual warfare with the conservative medical profession, while attracting many radically inclined adherents. He wrote voluminously but during his life time only a few of his more important medical or surgical works were published. The opposition of the faculties more than once blocked his plans of publication. He died in 1541 in his forty-eighth year, in the city of Salzburg in Austria.

The movement he had inaugurated gained rather than lost momentum by his death. Works he had published passed through several editions and copies of his manuscripts were jealously treasured by admirers. Several such collections of his manuscripts were known. About twenty-five years after his death there began almost a rage for works by Paracelsus, and several publishers vied in the publication of works not hitherto printed. In 1589 a publisher of Basel—Johann Huser—published his complete medical and philosophical works, in eight octavo volumes, these being reprinted in 1603 and 1616, while his surgical works were collected by the same publisher and issued in 1603 and again in 1618.

The chief contentions of Paracelsus; that the medical men ought not to be satisfied with leaning on the dicta of the ancients, but should use their own observations and experience unbiased by inherited dogmas; that to chemistry medicine should look for a fundamental support for medical practice; and that chemists or alchemists should seek a productive field for their activity in preparing new medicinal agents, appealed more and more to the younger medical and chemical generations of the progressive-minded.

The mass of writings published by Huser was from various sources. Some were from manuscripts treasured by former students, many from well known collectors of Paracelsus literature, some written up from lecture notes, some of dubious origin and some obviously spurious and so recognized by Huser, who nevertheless included them in the collection because they had appeared under the name of Paracelsus. The larger part perhaps were published by Huser from the original handwriting, though here it may be that they were at times only copied by Paracelsus from other writings for his own use. Certain it is that there is great variety in style and substance, and there is much uncertainty as to the authenticity of many of the works attributed to him. The works which deal more specifically with chemical subjects were printed from about 1567 on—the *Archidoxa, Von Natürlichen Dingen, De Natura Rerum, Von Metallen, De Mineralibus, De Cementis Metallorum,* etc. As, however, none of these writings has been found in any previous author, and none of the original manuscripts appear to be still in existence it must be assumed that these works are from the pen of Paracelsus, though uncertainty exists as to the degree of elaboration or interpolation at the hands of various editors, which may have taken place in some of these writings.

During the seventeenth and eighteenth and well into the nineteenth century, when it was assumed that the writings of the pseudo-Basil Valentine and the Hollandus, father and son, were of the fifteenth century, Paracelsus was supposed to have drawn principally upon these authors for many of his chemical facts and theories. Now, however, that modern research has shown that all of these works are of later origin than the publication of the works of Paracelsus the relation is reversed.

From the works which include his more specific chemical information, such as the above mentioned, it appears that Paracelsus possessed wide information on the chemistry of his time. His descriptions of processes and operations are numerous and various. They are often carelessly edited, often incomplete and explained by fantastic and unconventional theories.

Many of his operations on ores and metals are very manifestly derived from such books or manuscripts as are illustrated in the *Probierbüchlein*. Illustrations of this style of description are the following:

"Of the separation of the elements in metals.

"The separation of the elements from metals is a process in which you should provide yourself with good apparatus, and with experienced manipulation and workmanship. First make an aqua fortis thus: take of alum, vitriol, sal-nitri, equal parts, distil to a strong aqua fortis, return that to the residue and distil a second time in a glass flask. Dissolve in this silver and afterwards dissolve in it sal ammoniac. After this is done take the metal in thin plates and dissolve it in the water. When that has taken place separate it in the water bath (balneo maris), pour it over again until an oil is found at the bottom; from gold almost brown, from silver almost bluish, from iron red to almost black, from mercury quite white, from lead lead-colored, from copper quite green, from tin, yellow.

It is indeed true that not all metals are converted to an oil unless they have been first prepared. So mercury should be sublimed, lead calcined, copper converted to flowers (that is, oxidized), but gold and silver yield easily to it." [4]

This description is less clearly given than similar descriptions in the *Probierbüchlein* though practically the same. It is, however, evidently not free from errors, apparently due to too hasty condensation. It continues in an effort to explain how eventually the metals are separated into their constituents air, earth and water, quite after the style of fourteenth century philosophizing.

"To separate the elements from marcasites.

"Take any marcasite you please, wismuth, talc, or kobalt, garnets and things similar, one pound; add one pound saltpeter, crush and rub them together, heat in a flask

---

[4] Archidoxa, Liber III, Opera I, p. 792.

with alembic, and keep the water that passes over, and crush that which remains on the bottom and lay it in aqua fortis so that it dissolves to a water. After that add the water before collected and distil it all to an oil as previously described for metals. And by the same process separate the elements from one another. The margarita aurea is to be understood as like gold, argentea marcasita as like silver, wismuth as like lead, zinetus like copper, talc like tin, kobalt like iron, and this will suffice for the complete separation of the kinds of marcasites."[5]

"Wismuth" we have seen was mentioned in the *Probierbüchlein,* Kobalt is also there mentioned, though there is much doubt as to just what these terms meant. Kobalt seems to have meant generally a troublesome mineral to the miners. Paracelsus, in the book *De Mineralibus,* describes it thus:

"There is a metal from *Koboleten,* this metal can be poured, flows like zink, has a particular black color, blacker than lead and iron, but with no luster or metallic appearance, can be beaten and hammered but not so much that it might be used for anything."

Even Agricola does not make clear just what this Kobalt is, but says it is a kind of cadmia, which usually means a zinc oxide or zinc ore.

In a work *The Transmutation of Metals; on Cements,* Paracelsus describes the composition of some half dozen mixtures for separating gold from silver, or gold from copper or from silver and copper by cementation. These mixtures consist essentially of antimony sulphide, common salt, brickdust, sal tartar or argol, vitriol, alum, sometimes niter, flos aeris or copper oxide, and other constituents. The mixtures and their application are not very different from, (though not identical with), the many similar cementation processes in the *Probierbüchlein,* nor from those given by Agricola. The descriptions are less careful and complete as given in those sources. If published as originally written by Paracelsus, they bear evidence that he was

---

[5] Archidoxa, Liber, III, *op. cit.,* p. 793.

familiar in a general way with the laboratory methods in the mining laboratories.

A work entitled *De Natura Rerum* bears, in its dedication the date of 1537 at Villach. It was first printed in 1573 and Professor Sudhoff, the greatest authority on the writings of Paracelsus, doubts the authenticity of all portions of the nine books comprising the work.[6] This work contains much of chemical facts and processes and is of interest for the chemistry of the time.

In general this treatise accepts the customary Arabian theory of the origin and generation of the metals, with the exception that in addition to the origin from sulphur and mercury he introduces a third constituent, salt (sal). With the Greek alchemists and their Arabian followers, he believes in the gradual growth of the metals, and the ripening in the earth of imperfect into perfect metals. Also he credits the power of alchemy to so mature the imperfect metals and minerals.

"Namely in all ores in which the immature metal exists it can be brought to ripeness by the skillful devices of the alchemist. So also may all marcasites, garnets, zinc, cobalt, talc, cachimia, wismuth, antimonium, etc., which all contain immatured gold and silver be so matured until they resemble the best gold and silver ores, by cohobation."

Cohobation was the repeated treatment by liquid agents, by repeated pouring on and drawing off, or by distillation.

The character of specific chemical actions as described may be illustrated by his discussion of the mortification of metals (from *mors* "death," a term much used by the early alchemists for any process which seemed to deprive metals of their life or spirit. In general it corresponded to any process which we should call direct or indirect oxidation).

"Iron: Take steel beaten to thin sheet, ignite it and quench in strong wine vinegar. Perform the ignition and quenching so often that the vinegar is a fine red, and when you have enough of it pour it all together and distil

6 Prof. Karl Sudhoff, *Hohenheims Literarische Hinterlassenschaft*, Atti del Congresso Internationale di Scienze Storiche, Roma, 1903, XII,

off the moisture of the vinegar and bring to a dry powder. That gives a noble crocus martis (ferric oxide). There is another process to make crocus martis which is in some respects better and prepared with less trouble and expense. This method is that the thin-beaten steel plate is stratified and reverberated with equal parts of sulphur and tartarum (argol). This gives an extremely beautiful crocus which is removed from the steel plate. Likewise you should know that any iron or steel sheet wetted with aqua fortis also gives a fine crocus. Also with oil of vitriol (oleum vitrioli), with water of salt, water of alum, with water of sal-ammoniac, water of saltpeter, or with sublimed mercury. All of these *kill* (mortify) the iron, destroy and consume it and convert it to a crocus.''

For use in medicine he directs to use only the first two preparations, though others are used in chemistry. He explains further the preparation of the waters above employed which are merely water solutions of salts named. The mortification of copper, lead, mercury and other metals is similarly described, with products such as verdigris, white lead, corrosive sublimate, or the various oxides, etc. In some cases, as with gold, the processes are elaborate but the fancied results are not capable of rational interpretation.[7]

As with mortification, so Paracelsus deals at length with the ''resuscitation of natural things.'' Processes here described with respect to the metals are any processes by which the metals are reduced from their compounds to the metallic state. Thus mercury can be ''resuscitated'' from cinnabar, or from mercury precipitate by rubbing to a fine powder, mixing with egg albumen and soaps, making into balls the size of hazel nuts and heating in an earthen flask with a perforated iron plate luted to the neck, heating in a strong fire and distilling ''per descensum'' into cold water.

---

[7] The reader who is desirous of obtaining a more adequate idea of the extent and character of Paracelsus' chemistry may consult the English translation by A. E. Waite, *The Hermetic and Alchemical Writings of Paracelsus,* 2 volumes, London, 1894. The work suffers from being translated, not from the originals in German, but from the Latin version, itself very faulty.

"Thus you have again quicksilver." The following descriptions of the separation of gold and silver, and silver from copper, closely resemble similar processes in the *Probier-büchlein,* but the descriptions are less complete.

"The separation of metals in aqua fortis, aqua regia, and other similar strong corrosive waters is thus: that the metal which has another mixed with it is taken, and in thin sheets or small granules, placed in a parting flask (scheyd-kolben) and common aqua fortis poured on it in sufficient quantity. Let them then work upon one another until the metal is entirely dissolved to a clear water. If it is a silver which contains gold, the silver will then be all dissolved to a water and the gold be calcined and settle on the bottom like a black sand. And thus are the two metals gold and silver separated. But to separate the silver alone from the aqua fortis without distillation, and to precipitate it like a sand, and from solution to attain calcination, you should throw into the solution copper in sheets. Soon the silver will sink in the water or precipitate and settle like white snow to the bottom of the glass and the copper begin to be consumed.

"The separation of silver and copper in a common aqua fortis is as follows: in the same manner as above given, silver containing copper or copper containing silver, in thin sheets or small granules is placed in a glass flask and aqua fortis poured on until enough is added. The silver will then be calcined and settle on the bottom as a white calx, but the copper be dissolved to a transparent water. If now this water be poured off through a glass funnel from the silver calx into another glass, the copper dissolved in the water may be precipitated by a foul common rain water or running water or warm salt water and settle like a sand on the bottom of the flask."

It will be noted that in the separation of gold from silver the aqua fortis used must be free from hydrochloric acid, while in the separation of silver from copper the aqua fortis must have contained hydrochloric acid to have separated the silver as chloride. Paracelsus does not discriminate however. So also the description of the foul waters used

for precipitating copper from its solution is very inadequate. Such waters must have contained either sulphides
or alkaline carbonates.

"The separation of hidden gold from every other metal is
effected by extraction in aqua regia, for this water attacks
to dissolve no other metal than fine gold alone." This is
a loose and incorrect statement.

In the body of Paracelsan literature printed between
1536 and 1580, there occurs a great mass of chemical detail,
and much chemical philosophy. Assuming in the lack of
positive evidence to the contrary that all this is really by
Paracelsus, which is doubtless not the fact, yet there is
apparently nothing in the specific facts noted by Paracelsus
that would justify the conclusion that he was a real investigator or a discoverer of any important facts in chemistry.
His influence upon the development of chemistry is not to
be accounted for by his chemical discoveries. It is perhaps true that in the great number of chemical data and
processes contained in his works and works attributed to
him, there are none which were not a part of the common
knowledge of the chemists of his day, and may not be found
in multitudinous chemical and alchemical writings previous
to his time. His great prominence was due largely to his
vigorous personality and to his radical tendencies.

It is also true that many of the statements are inaccurate.
A notable exception is his characterization of the metal
zinc. This name first appears in his writings, though doubtless here also he is citing a name which was in use in some
mining region, though not in general use. That the metal
itself had been prepared before his time is beyond doubt,
though descriptions of it are not clear. As however, cadmia or calamina, ores of zinc, were used in the making of
brass from about the first or second century before Christ,
and remembering the very easy reduction of such ores to
the metallic state, it is inconceivable that it should not have
been prepared, though not recognized as a distinct metal.
Neither the *Bergbuch* nor the *Probierbüchlein* mentions it.

Agricola does not mention it in his *De Re Metallica* nor in
his other works until the revised edition of his *Bermannus*
printed in 1558. Paracelsus, however, mentions it in his
Chronica des Landts Körnten (Carinthia), dated August
24, 1538, a work of unchallenged authenticity, for he placed
it in the hands of the authorities of the archduchy of
Carinthia in the expectation that it would be published
and in their archives it remained until surrendered to Huser
for publication in the collected works. In this work dealing
with the natural and mineral resources of Carinthia, he
says:

"There are also many kinds of mines in this land, more
than in others; at Bleyberg a wonderful lead ore which
has not only supplied Germany but Pannonia, Turkey and
Italy. Similarly there are iron ores at Hütenberg and
vicinity richly endowed with a particularly excellent steel.
Also much alum ore which is mined and utilized. Also
vitriol ore of high grade. Also gold ore and wash gold
(placer gold) excellent in quality which is found notably at St. Paternioms. Also the ore *Zincken* which is not
elsewhere found in Europe, a very strange metal much
stranger than others. It also has excellent cinnabar ore
which is not without quicksilver." [8]

In *De Mineralibus* he says: [9]

"A metal is that which can be made into an instrument
by man. Such namely are gold, silver, iron, copper, lead,
tin; for these are generally known as metals. Now there
are some metals which are not recognized in the writings
of the ancient philosophers nor commonly recognized as
such and yet are metals: as Zincken, Kobaltet which may
be hammered and forged in the fire."

And again:

"There is also another metal called Zincken. . . .
This is not generally known; it is in this sense a metal of
a special kind, and from another seed. Yet many metals
alloy with it. This metal is itself fusible for it is from
three fusible elements (meaning doubtless, sulphur, mer-

[8] Paracelsus, *Opera*, folio, 1616, Strassburg, I, p. 251.
[9] Paracelsus, *Opera*, folio, II, p. 134.

cury, salt), but it possesses no malleability but only fusibility. Its color is different from the colors of others, so that it is not like other metals in growth. And it is such a metal that its primal matter is not yet known to me. For it is nearly as strange in its properties as quicksilver. It admits of no admixture and does not submit to metallic manufactures but stands by itself." [10]

In his *Bergkrankheiten,* or diseases pertaining to mining, he also refers to the vapors from zinc as injurious along with mercury, etc. [11]

Again in the *Philosophia* he refers to zinc in connection with a fanciful theory of the origin of the metals, in which he says:

" . . . Zincken which is a metal and yet none, also Wismuth and the like which fuse and to some extent are malleable. And yet although they are somewhat related to the metals through their fusibility, they are only bastards of metals. That is, like them and yet unlike. Zincken is for the most part a bastard of copper, and Wismuth of tin."

This characterization of zinc and bismuth as bastard metals, finds its analogue in a later century in the designation of these and other substances, as halfmetals or semimetals as for example by Boerhaave in the beginning of the eighteenth century.

The chemical philosophy of Paracelsus as comprised in the works attributed to him is in general thoroughly medieval. Based upon the traditional speculations of his predecessors, but elaborated in fanciful extensions by his own imagination, full of occult and superstitious notions current in his period, it did not tend to add clarity or rationality to chemical theory in general.

In one particular, however, Paracelsus contributed a theoretical concept which exerted a dominating influence on the theory of following centuries. This was the doctrine known as the *tria prima,* the idea that all matter from

[10] Paracelsus, *Opera,* folio, II, p. 137.
[11] Paracelsus, *Opera,* folio, I, p. 656.

metals to man is made up of three principles sulphur, mercury and salt. This notion was a development of the Greek-Arabian theory of the constitution of metals and other matter from sulphur and mercury. This theory with medieval philosophers generally implied that mercury and sulphur were the first substances to be generated in the earth from the elements fire, water, earth and air. By sulphur they had come to understand the constituent which is combustible, while mercury was understood as the mother of all the metals and liquidity was its characteristic property, càusing fusibility in metals which are generated from it and contain it. Sulphurs and mercuries differed, however, in their grades of purity, earthy admixtures, and the degree of digestion affecting their purity.

From another point of view matter was considered by the medieval philosophers as composed of body, spirit and soul. Body was that which gave solidity and permanence, spirit was that which fled from the fire or was volatile. Soul was not very intelligibly defined, and not so generally adopted. Paracelsus crystallized these vague theories into a more tangible form by assuming that all matter is made up of three primal substances, sulphur, mercury and salt. To these three constituents he ascribed more definite functions than had previously been recognized. Sulphur was the combustible principle, mercury that which imparts fusibility, liquidity and volatility, salt that which is nonvolatile and incombustible. This idea he developed extensively in very many of his works. Thus in the *Paramirum*:

"Three are the substances which give body (or substance) to everything: that is every body consists of three things. The names of these three things are sulphur, mercury, and salt. When these three are combined then we have what we call a body, and nothing is added to them except life and what depends upon it. . . . Now to understand the affair, take first (for example) wood. That is a body. Now let it burn, that which there burns is sulphur; that which vaporizes is mercury, that which turns to ashes is salt. . . . That which thus burns is sulphur, nothing burns but sulphur; that which fumes is mercury, nothing sublimes which is not mercury; that which turns to ashes is salt, nothing turns to ashes which is not salt." [12]

In the *De Generatio Rerum Naturalium* he relates his three principles to the theory of body, spirit and soul:

" . . . you should know all seven metals originate from three materials, namely, from mercury, sulphur and salt, though with different colors. Therefore Hermes has not said incorrectly that all seven metals are born and composed from three substances, similarly also the tinctures and the philosophers' stone. He calls these three substances, spirit, soul and body. But he has not indicated how this is to be understood nor what he means by it. Although he may perhaps have known, yet he has not thought (to say) it. I therefore do not say that he has erred, but only kept silent. But that it be rightly understood what the three different substances are that he calls spirit, soul and body, you should know that they mean not other than the three principia, that is mercury, sulphur and salt, out of which all seven metals originate. Mercury is the spirit, (spiritus), sulphur is the soul (anima), salt, the body (corpus)." [13]

Paracelsus does not, as he indicates in various places, consider his sulphur, mercury and salt as merely the common mercury and sulphur, but just as the earlier alchemists considered their sulphur and mercury as an idealized "mercury of the philosophers," etc., so he has a similarly generalized concept of the three principles. Thus in his *De Mineralibus,* where he discusses his three principles at considerable length, he says:

"For as many as there are kinds of fruits—so many kinds there are of sulphur, salt, and so many of mercury. A different sulphur is in gold, another in silver, another in iron, another in lead, tin, etc. Also a different one in sapphire, another in the emerald, another in the ruby, chrysolite, amethyst, magnets, etc. Also another in stones, flint, salts, springwaters (fontibus), etc. And not only so many

[12] Paracelsus, *Opera,* folio, I, p. 884.
[13] Paracelsus, *Opera,* folio, I, pp. 26, 27.

kinds of sulphur but also as many kinds of salt, different ones in metals, different ones in gems, stones, others in salts, in vitriol, in alum. Similarly with mercuries, a different one in the metals, another in gems, and as often as there is a species there is a different mercury. And yet they are but three things. Of one nature is sulphur, of one nature sal, of one nature mercury. And further they are still more divided, as there is not merely one kind of gold but many kinds of gold, just as there is not only one kind of pear or apple but many kinds. Therefore there are just as many different kinds of sulphurs of gold, salts of gold, mercuries of gold." [14]

This theory of the tria prima which is reiterated and discussed very extensively in numerous treatises of Paracelsus, made a strong appeal to the public of his own and later centuries. It indeed almost completely dominated chemical theory and philosophy until the rise of the theory of phlogiston. It was adopted by the authors of the later works ascribed to Basil Valentine and Johann and Isaac Hollandus, and so long as these works were believed to have been written in the fifteenth century, Paracelsus was naturally supposed to have acquired this concept from the works of those writers.

It is interesting to note the introduction of the Greek "chaos" by Paracelsus as a generalized expression for all aerial matter. For instance, discussing the four Aristotelian elements in their relation to the components of the human organism:

"They are born from the elements . . . as namely out of the element terra (earth), its species, and out of the element aqua (water), its species, out of the element ignis (fire), its species, out of the element chaos, its species." [15]

"The elements in man remain indestructible. As they come to him so they pass from him. What he has received from the earth goes back to earth and remains such so long as heaven and earth stand; what he has in him that is water becomes water again and no one can prevent it, his

---

[14] Paracelsus, *Opera*, folio, II, p. 132.
[15] Paracelsus, *Opera*, folio, I, p. 269.   *Labyrinthus Medicorum.*

chaos goes again into the air (Luft), his fire to the heat of the sun." [16]

"Thus all superfluous waters run into their element called the sea (mare); whatever is terrestrial returns to its element called earth (terra); what is igneous into the element fire (ignis); and what is aerial runs into its element *chaos*." [17]

The term *gas* as a generalization for aerial fluids was first suggested by van Helmont (1577–1644), himself very familiar with the works of Paracelsus and to some extent a champion of his views. He tells us that he derives this word from the Greek *chaos*,[18] and it is more than probable that it was the use of the word by Paracelsus in this sense that suggested the word gas to van Helmont.

Much more important than any specific chemical advances due to Paracelsus was his influence in attracting the attention of physicians and chemists to the importance of chemistry in the development of medicine in connection with his campaign against the blind worship of traditional authorities. In his life-long and intense struggle against the conservatism of the medical faculties and profession, he constantly emphasized the duty of the physicians to depend upon experiment and independent observation rather than on the dogmatic medicine of Galen and Avicenna, and emphasized the great value of new medicines derived from the development of chemistry.

He possessed a breadth of view as to the field of chemistry and its possibilities and stimulated chemists to seek a more important field for their activities than the search for gold making or the philosopher's stone. Not that he disbelieved in the possibility or reality of transmutation. On the contrary it received full attention and credence from him in his chemical philosophy. His estimate of the place of

---

[16] Paracelsus, *Opera*, folio, *Chirurgische Bücher*, p. 378. *Von Offenen Schaden.*

[17] Paracelsus, *Opera*, folio, I, p. 291, *Das Buch von den tartarischen Krankheiten.*

[18] J. B. van Helmont, *Opera omnia*, Frankfort, 1682, p. 69. *See* also Franz Strunz, *J. B. van Helmont*, Leipzig and Vienna, 1907; and E. O. von Lippmann, *Chemiker Zeitung*, XXXIV, p. 1.

chemistry in medicine is illustrated in the following from the *Paragranum*:[19]

"Now further as to the third foundation on which medicine stands, which is alchemy. When the physician is not skilled and experienced to the highest and greatest degree in this foundation all his art is in vain. For nature is so subtle and so keen in her matters that she will not be used without great art. For she yields nothing that is perfected in its natural state, but man must perfect it. This perfecting is called alchemy. For the baker is an alchemist when he bakes bread, the vine grower when he makes wine, the weaver when he makes cloth. Therefore whatever grows in nature useful to man, whoever brings it to the point to which it was intended by nature, he is an alchemist."

His notions of the functions of the animal organism are colored by his chemical ideas. Thus discussing the effect of poisoning from food, he holds that all food contains wholesome and unwholesome constituents, and he conceives that there is in the stomach of a presiding "Archaeus" whose business is to sort out the wholesome from the poisonous. "The body was given to us without poison and there is no poison in it; but that which we must give the body for its food contains poison." So long as the Archaeus performs his function our food is wholesome and the body thrives. Should from any cause the Archaeus become incapacitated from performing his functions properly, the separation of food and poison is incomplete and we suffer from the effects of the poisons. This Archaeus Paracelsus calls an alchemist because his functions are analogous to those of the alchemist in his laboratory.[20] His appeal to substitute medical chemistry for conventional alchemical aims, he voices frequently, for example:

"Many have said of alchemy that it is for making gold and silver. But here such is not the aim but to consider only what virtue and power may lie in medicines."[21]

---

[19] Paracelsus, *Opera*, folio, I, p. 219.
[20] Paracelsus, *Opera*, folio, I, p. 9 *ff.*, *Paramirum eus veneni.*
[21] Paracelsus, *Opera*, folio, I, p. 149, *Fragmenta medica.*

"Not as they say—alchemy is to make gold, to make silver: here the purpose is to make *arcana* and to direct them against disease." [22]

Another persistent feature in the campaign for reform in medicine, was his often emphasized conviction of the necessity of experiment and experience to the physician as against looking for all knowledge to the traditional authorities. It will be recalled that the value of experiment in science had been earnestly preached by Roger Bacon though his logic fell on unappreciative ears in the thirteenth century. The sixteenth century, however, found minds more responsive to the appeal of Paracelsus.

"For in experiments neither theory nor other arguments are applicable, but they are to be considered as their own expressions. Therefore we admonish every one who reads this not to reject the methods of experiment but, according as his power permits, to follow it out without prejudice. For every experiment is like a weapon which must be used according to its peculiar power, as a spear to thrust, a club to strike, so also is it with experiments. And as a club is not to be used for thrust nor a spear for hewing (zum hauen) just as little can any experiment be changed from its kind and nature. Therefore the highest aim is for one to recognize in any experiment its powers and in what form it is to be employed. To employ experiments needs an experienced man, sure of his thrust and blow that he may use and master them according to their nature." [23]

Again when defending himself against charges of medical opponents that he does not know at once, when he comes to a patient, what is the matter:

"For obscure diseases cannot be at once recognized as colors are. With colors we can see what is black, green, blue, etc. . . . what the eyes can see can be judged quickly, but what is hidden from the eyes it is vain to grasp as if it were visible. Take for instance the miner: be he as

---

[22] Paracelsus, *Opera*, folio, I, p. 220, *Paragranum*.
[23] Paracelsus, *Opera*, folio, *Chirurgische Bücher*, p. 301, *Von Frantzösischen Blättern*, etc.

able, experienced and skilful as may be, when he sees for the first time an ore he cannot know what it contains, what it will yield, nor how it is to be treated, roasted, fused, ignited or burned. He must first run tests and trials and see whither these lead. . . . Thus it is with obscure and tedious diseases, that so hasty judgments cannot be made, though the humoral (Galenic) physicians do this." [24]

The introduction into medical practice of many chemical preparations not recognized in the medical profession created an issue between Paracelsus and his adherents on the one side and the medical faculties on the other, that increased with time and eventually resulted in the great struggle between the Paracelsan and anti-Paracelsan medicine which waged for a century and more with considerable success eventually for the Paracelsists.

It was however not Paracelsus who first introduced the use of salts of the metals and similar products of chemistry into medical practice. The *Materia Medica* of Dioscorides and the *Natural History* of Pliny bear evidence that such substances were much used in their time. But in the middle ages, their use was more limited and conventionalized. To be sure the use of chemical medicines was being slowly extended principally through the initiative of Italian and Spanish practitioners, before Paracelsus, and it is possible that in Italy Paracelsus received this impulse.

He used, however, many chemical medicines not usual in his time, and this gave occasion for severe complaints from the regular medical schools. Paracelsus offers a special defense against his opponents who accuse him of using poisons in medicine. He challenges their ability to define what is poison, for all things even food and drink may be poison if used in excess, and many customary medicines are poisons, even fatal poisons, when used in greater than the proper doses. "You know," he says, "that mercury is a poison, yet you use it to smear the sick." Cinnabar and the sublimate they also used yet they blamed him for

---

[24] Paracelsus, *Opera*, folio, I, p. 262, *Die Siebente Defension.*

using vitriol because it is a poison. "Nothing is poison," he says, "that benefits the patient, only that is to be considered as poison which injures him." [25]

It is because of his appeals to a wakening unrest in the sixteenth century that Paracelsus owes his reputation as a reformer in medicine and as giving a fresh impetus to chemistry. As Professor Thomas Thomson long ago said:

"It is from the time of Paracelsus that the true commencement of chemical investigations is to be dated. Not that Paracelsus or his followers undertook any regular or successful investigation, but Paracelsus shook the medical throne of Galen and Avicenna to its very foundation: he roused the latent energies of the human mind, which had for so long a period remained torpid; he freed medical men from those trammels and put an end to that despotism which had existed for five centuries. He pointed out the importance of chemical medicines and of chemical investigations to the physician. This led many laborious men to turn their attention to the subject. Those metals which were considered as likely to afford useful medicines, mercury for example and antimony, were exposed to the action of an infinite number of reagents and a prodigious collection of new products obtained and introduced into medicine. Some of these were better, and some worse, than the preparations formerly employed; but all of them led to an increase of the stock of chemical knowledge, which now began to accumulate with considerable rapidity." [26]

The influence of the Paracelsan literature was, however, by no means entirely in the direction of progress. By 1600 there had appeared in print no less than two hundred and fifty titles, as listed by Dr. Karl Sudhoff in his great bibliography of Paracelsus, and by 1658, the date of the last Latin version of his works, this number had increased to about three hundred and ninety. This great number included new editions, reprints, and many works of dubious authenticity as well as obvious forgeries. In the mass of writings attributed to him—as in many of those of undisputed genuine-

---

[25] Paracelsus, *Opera*, folio, I, pp. 256, 257, *Die Dritte Defension*.
[26] Thomas Thomson, *The History of Chemistry*, London, 1830, I, p. 140.

ness—there is so much of credulity, superstition, mysticism, and obscurity that mystics and charlatans also found in his works much food for encouragement. Thus his own better influence was in part retarded by his medieval heritage.[27]

Of Vannucio Biringuccio's antecedents and life little seems to be known except that he was a citizen of Siena in Italy and what may be gathered from his single publication *De la Pirotechnia,* first printed in Venice in 1540. Under Pirotechnia he explains that he treats fully of all kinds of minerals and concerning the examination, fusion and working, of the metals and similar things. It is indeed the first systematic text on the arts of mining and metallurgy, anticipating the *De Re Metallica* of Agricola by sixteen years. In the treatment of mining and mining engineering it is not to be compared in detail and completeness with the latter work. In the assaying of ores, separation of the metals from the ores, and chemical processes generally his treatment compares favorably and is in some matters more complete. He treats also of subjects not covered by Agricola's work, as the casting of bells and cannon, of gunpowders, mines, artificial fires and fireworks and contemporary devices of chemical warfare. The work is excellently illustrated, by many woodcuts of apparatus of many kinds, though not so elaborately as is Agricola's work on mining.

Much of the material of the chemistry of the metals indeed is such as is described in the earlier *Probierbüchlein* and was the common knowledge of the miners and metallurgists of the day. Biringuccio nowhere claims any originality for his information, though he manifestly was familiar with much of it through his experience. His book was written not like Agricola's in the language of scholarship—the Latin—but in Italian, and from this fact it is inferred

---

[27] For a more general account of the life and work of Paracelsus see the author's work: *Theophrastus Bombastus von Hohenheim, called Paracelsus, His Personality and Influence as Physician, Chemist, and Reformer.* The Open Court Pub. Co., Chicago and London, 1920.

that he was not a man of conventional university training. However that may be, the descriptions of operations are generally clear and comprehensible. That its merit was appreciated is evident from editions and translations that succeeded the first publication.

The first book of the *Pirotechnia* treats of the ores of the six major metals, gold, silver, copper, lead, tin and iron, their location, surface indications, and the methods of recovering the metals from their ores. His style may be illustrated by his description of recovering gold from river sands. It may be recalled that this method is briefly given by Theophilus Presbyter in the twelfth century.[28]

"As before mentioned there is still found some (gold) in the sands of several rivers, as in Spain in those of the Tagus, in Thrace in the Ebro, in Asia in the Pactolus, in India in the Ganges, and in several rivers in Hungary, Bohemia, and Laslifia; in Italy in the sands of the Ticino, Adda, and Po; not in all sands of their beds but only in certain special places where in bends there are gravel beds, upon which the water during high water leaves sandy loams in which the gold is mixed in the form of minute flakes, resembling flaxseed. Now in wintertime as soon as the floods have subsided these are heaped above the bed of the river so that in case of other high water they will not be washed away. Then in summer time by patient and ingenious methods the prospectors wash the sands to get rid of worthless material, by using tables of poplar, elm or white walnut or other fibrous and tough lumber, the surfaces of which are bored with hollows by the saw or other tools. Upon these throughout their length is thrown with a concave shovel the sand with abundance of water. By this means the gold that is in the sand being heavier, enters the hollows and remains, being thus caught and separated from the sand. Then when any is seen to be thus caught, it is carefully recovered and finally placed in a wooden vessel, resembling the vessel used to wash sweepings, cut with many grooves in the middle; and again they wash it as much as they can to clean it better. Finally they amal-

---

[28] *See* ante, p. 224.

gamate it with mercury and strain it through a bag or submit it to the alembic (distil). There remains the gold, the
mercury being evaporated, like a sand in the bottom of the
vessel, which is fused with a little borax or saltpeter or
black soap and reduced in volume, cast in bars or other
shape as desired.''

The second book is devoted to what he calls the minor
minerals, under which he includes mercury and its ores,
sulphur, antimony, marcasites containing metals, vitriol,
alum, arsenic, orpiment, and realgar, common salt and other
salts, calamine, zaffre, ocher, Armenian bole, emery, borax,
lapis lazuli, rock crystal and glass.

His treatment of antimony is of especial interest in
view of the fact that as the much more complete description by the pseudo-Basil Valentine has been so long believed to be earlier instead of more than a half a century
later, this description by Biringuccio has been ignored by
historians.

Biringuccio begins, following the conventional Greek-
Arabian theory of the development of the metals, by expressing his judgment that antimony is a substance intended by nature to form a metal eventually, but, arrested
in its development, containing an excess of hot and dry
material, and insufficiently digested, it is, like mercury,
that anomaly and monster among metals, much like the
true metals. He notes its light and brilliant color, its heavy
weight and metallic appearance. It is whiter and has more
luster than silver but is more brittle than glass. Alchemistic philosophers are greatly interested in it for they claim
that an oil made from it tinctures silver permanently to
gold. He has himself seen a blood colored liquid in the
form of an oil, and the person who showed it to him claimed
that it permanently tinctured silver to gold color, but he
himself had not seen any silver so tinted by that person nor
any other.

''This mineral is only found, like other metallic ores,
in the mountains, and is extracted by various means. I
know that some is found in Italy in several places and from

Germany there is brought to Venice some melted in bars for use in bell-making, for they find that by mixing a certain amount of it with the metal it augments the sound. It is also used by the makers of pewter vases, and by the makers of mirrors of glass which give the appearance of metal. I understand also that it serves as a medicine for surgeons in the treatment of abscesses and incurable ulcers, and that, by it, corruption and dead flesh is removed and nature is assisted in the healing. It also serves in making several yellow colors for painting pottery and for coloring enamels, glass and similar things which are desired to be fused at a yellow heat. There are quite a number of mines of antimony in the province of Siena, one near the city of Masse, and another large one near another city called Sovana, and this one experienced investigators claim to be the best that is known. There are also some in the province of Santa Fiora, near a place called Selvena, and not only in these places I have mentioned but in many others where, because it is a mineral not bearing gold or other important perfect metals, it is held of no importance, and this that I have told you is all I know about antimony."

In the third book are described the testing of gold and silver, cupels and cupellation, the preparation, roasting, etc., of the ores of the various metals preparatory to smelting, furnaces for smelting and methods of smelting the ores, the separation of metals from one another. These descriptions are in general similar to those given in the small German manuals previously described.

In the fourth book he describes the preparation of and use of aqua fortis in parting, and processes of cementation by sulphur, or by antimony.

The fifth book treats of alloys of gold and silver, silver and copper, and other alloys of copper, of lead and of tin.

The sixth book is a lengthy treatise on casting of cannon, bells, and other objects, and for making the molds for such castings, the metals, furnaces and appliances. Those who have read the memoirs of Benvenuto Cellini (1500-1577) will remember his description of such a casting, and the works of Cellini and contemporary artists are evidence of the high degree of skill attained by Italian metal workers in this sixteenth century.

Books seven and eight deal with many methods of work pertaining to metals and their fusions, reverberatory furnaces, bellows and special methods of casting particular objects.

The ninth book is devoted to arts connected with chemistry, as distillation and sublimation, the arts of coining and the manufacture of jewelry, mirrors of metal, etc. In this book appears a description of "methods for extracting all gold or silver from the waste of the mines or sweepings of the mint." It is of especial interest because it describes, apparently for the first time, recovery of silver by the method of amalgamation, a process first apparently utilized on a large scale by the Spaniards in America later in that century. He says:

"Great consideration is due to the inventor of the short method of extracting gold and silver from the sweepings from all the trades which handle gold or silver, and also any of the substances left by smelters in the waste, as also from some minerals without the labor of fusion, by the use of mercury. For this there is first built a large walled structure of stone or lumber like a mortar, within which there is arranged a grindstone made to turn like a millstone. In this is placed the material containing gold, which must first be well ground in a mortar, washed and dried. In the machine mentioned it is reground while moistened with vinegar or water, in which there is dissolved sublimate, verdigris and common salt. Upon this material is placed a large quantity of mercury so that it covers it. It is then stirred and mixed for an hour or two by means of turning the grindstone by hand or horsepower, remembering that the more the mercury and material are rubbed together in the machine the more perfectly will the mercury extract the metal from the material. Finally the mercury is separated by means of a sieve or by washing from the powdered earth. It is then distilled in an alembic or passed through a bag, there remaining the gold, silver, copper or that metal which has been caught in the machine by the grinding. And for this secret, desiring to know it, I gave to him who taught me a ring with a diamond worth twenty-five ducats, and he also exacted from me the right to an eighth part of anything I should receive from operating it. And this I wish to say not that you should reimburse me for teaching you but so that you should appreciate it the more.'' [29]

In this book also Biringuccio discusses the subject of alchemy in general, and while not attempting to dispute the possible reality of the art, he yet inveighs strongly against the many vain efforts and consequent waste of time and money, and especially against the prevalent frauds and deceptions carried on by those who pretend to change the baser metals into gold and silver. He manifestly doubts that the art has any foundation in ancient times:

''Because there is not found a single ancient writer of history, in Greek, Latin or any other language, who ever mentions it. Neither is there any mention among the approved and great philosophers, as Aristotle, Plato or others like them, who have the means of knowing possible things.''

The tenth and last book is devoted to various substances and devices in warfare. Gunpowders are treated at length and careful discrimination made in their composition for various purposes. Thus for heavy artillery, a slower burning powder of three parts saltpeter to two of charcoal and one of sulphur; for medium sized artillery five of saltpeter, one and a half charcoal and one of sulphur; for harquebuses ten of saltpeter, one of charcoal and one of sulphur, or thirteen and a half of saltpeter, two of charcoal and one and a half of sulphur. These proportions are for heavy guns, like those now in use for blasting powders, while the proportions for light weapons are not far from modern black powder for rifle or shotgun use.

Various projectiles and mixtures for use in warfare—Greek fires, grenades, mines and countermines,—are de-

---

[29] Biringuccio, *De la Pirotechnia*, Libro IX, Cap. XI, p. 142.

scribed and illustrated. Marcus Graecus (Gracchus) is cited in connection with some of these "artificial fires."

The description of sources and methods of obtaining and purifying saltpeter is the earliest complete account of the preparation of that salt. It is apparently the basis of the somewhat condensed description in Agricola's *De Re Metallica*. Biringuccio says that saltpeter (sal nitro) is a compound of several substances, extracted by fire and water from dry earth containing manure, or from the efflorescence produced on new walls in damp places, or from the mouldy earth which is found in tombs or in unoccupied caves where rains cannot enter. After describing its physical properties and its explosive character, as in gunpowder, he continues:

"The best saltpeter is obtained from animal manure converted to earth in stables, or from latrines which have not been used for a long time and above all from pigpens. This manure must be converted to earth by time and entirely dried and powdered. Vats are then filled alternately by layers of this earth about four digits deep, and layers one digit deep of a mixture of two parts of quicklime and three parts of ashes from bitter oak. The vats are filled to about four digits or half an arm's length from the top and then filled with water. The water seeps through this earth, dissolves the saltpeter and trickles through holes in the bottom of the vats into conduits which carry it to other vats. This water is now tasted and if it is sharp and strongly salty it is good, otherwise it must again be passed over the same or other earths containing saltpeter. This process is continued until practically all of the saltpeter is dissolved. The water is then placed in copper kettles on furnaces and slowly boiled to about one third its original volume, then drawn off and put into a strong covered cask and allowed to settle until clear. The clear water is then drawn off and again evaporated by the same process as before. In order that water shall not foam and overflow and thus waste much good material, a measure is made of three quarts of soda or of ashes of bitter oak, or oak, or olive, and one quart of lime, and for every hundred pounds of water there is dissolved four pounds of alum of Rocca (or rock alum).

"A glass or two of this water is added whenever you see that it threatens to rise and form a foam. The saltpeter solution is boiled until it becomes clear and of bluish color indicating that most of the water has been evaporated. It is then drawn off and placed in casks and allowed to solidify. It is then placed in wooden casks and allowed to stay three or four days, and then decanted, either by inclining the vessel or by holes in the bottom. The decanted water is saved and reboiled. The solidified saltpeter is then chiseled out and washed with its own solution then placed on tables to dry thoroughly."

The purification of the saltpeter is effected by two methods which are briefly as follows:

First method.—For every barrel of water placed in a copper kettle is added four to six glasses of the clarifying solution above described. Then there is added as much crude saltpeter as will easily dissolve on boiling. After the first boiling when scum forms, it is drawn off and passed through a sand filter consisting of a cask with a single small hole in the bottom. The bottom of the cask is covered four digits deep with washed river sand and over this is placed a cloth. The filtered solution is then passed through the same process as in the first instance adding some of the clarifying solution when boiling.

Second method.—An iron or copper vessel is filled with saltpeter and securely covered. It is then placed in the middle of a good charcoal fire until the saltpeter is melted. When well fused, finely powdered sulphur is placed on the fused material and burned completely until the saltpeter remains clear and clean. When this occurs the vessel is removed from the fire and allowed to cool. The saltpeter is then found to be in a solid white mass resembling marble, and at the bottom of the vessel there remains all the earthy matter.

The foregoing must suffice to give an illustrative though inadequate impression of this earliest text book of a modern type upon the chemical technology of metallurgy. His contemporary, Agricola, in the preface to his *De Re Metallica,* thus alludes to Biringuccio's work:

"Recently Vannucio Biringuccio of Sienna, a wise man experienced in many matters, wrote in vernacular Italian on the subject of the melting, separating and alloying of metals. He touched briefly on the methods of smelting certain ores, and explained more fully the methods of making certain juices (that is salts); by reading his directions I have refreshed my memory of those things which I myself saw in Italy. As for many matters on which I write, he did not touch upon them at all, or touched but lightly. This book was given me by Franciscus Badoarius, a Patrician of Venice, and a man of wisdom and repute; this he had promised that he would do, when in the previous year he was at Marienburg, having been sent by the Venetians as an ambassador to King Ferdinand."

Georgius Agricola, (this being the latinized name of Georg Bauer), was born in Saxony in 1494. He received his degree of A. B. at the University of Leipzig when about twenty-four years of age. He taught Greek and Latin in Zwicken for a time. In 1522 he was a lecturer in the University of Leipzig and in 1524 went to Italy to pursue his studies for his profession of physician. In 1526 he returned to Saxony and in 1527 became city physician (stadt-artzt) in the mining town of Joachimsthal. Later (1533) he occupied a similar position in Chemnitz where he remained till his death in 1555.

Born and residing nearly all his life in a great mining district, he took the deepest interest in the problems and practices of mining, mineralogy, metallurgy and assaying, and his interest soon found expression in published works. In 1530 his first work on these subjects appeared, *Bermannus.* This book is in the form of conversation between three friends on matters relating to mineralogy. It deals largely with the names of Saxon minerals and the corresponding nomenclature of the ancients. To this work has often been credited the first mention of bismuth, but as we have seen this substance was first mentioned in the *Nützliches Bergbüchlein.* A brief treatise on the weights and measures of the Greeks and Romans appeared in 1532. Several other minor publications were issued, of no interest from the point of view of chemistry. In 1546 appeared a work on the "nature of fossils," which is the earliest important attempt to classify minerals. The basis of his classification is naturally the physical properties: fusibility, solubility, color, odor, taste, etc. In Agricola's time no other basis was possible, for except as to the ores of metals, and some metallic salts, there existed no knowledge of the chemical composition of rocks, minerals and salts.

Agricola divides the minerals into: 1. earths, such as clays, chalks, ochres, etc. 2. stones, properly so called, gems, semiprecious stones. 3. solidified juices, (succi concreti), salt, alum, vitriols, saltpeter, etc. This is an application of the theory of the ancients that these are derived from solidified waters. 4. rocks, such as marble, serpentine, alabaster, limestone, etc., hard and not friable like the earths. 5. metals. 6. compounds, or mixtures, under which head he classes various ores of the metals, from which he recognizes that simpler constituents, as the metals, may be obtained. The fundamental basis of this classification Agricola explains in the following manner.[30]

"Mineral bodies are solidified from particles of the same substance, such as pure gold each particle of which is gold, or they are of different substances such as lumps which consist of earth, stone and metal; these latter may be separated into earth, stone, and metal, and therefore the first is not a mixture while the last is called a mixture. The first are again divided into simple and compound minerals. The simple minerals are of four classes, namely, earths, solidified juices, stones, and metals, while the mineral compounds are of many sorts, as I shall explain later.

"Earth is a simple mineral body which may be kneaded in the hand when moistened, or from which lute is made when it has been wetted. Earth, properly so called, is

---

[30] Hoover, *op. cit.,* pp. 1, 2, footnote.

found in veins or veinlets, or frequently on the surface in fields or meadows. This definition is a general one. The harder earth, although moistened with water, does not at once become lute, but does turn into lute if it remains in water for some time. There are many species of earths some of which have names but others are unnamed.

"Solidified juices are dry and somewhat hard (subdurus) mineral bodies, which when moistened with water do not soften but liquefy instead; or if they do soften, they differ greatly from the earths by their unctuousness (pingue) or by the material of which they consist. Although occasionally they have the hardness of stone, yet because they preserve the form and nature which they had when less hard, they can easily be distinguished from the stones. The juices are divided into 'meagre' and unctuous (macer et pinguis). The meagre juices, since they originate from three different substances, are of three species. They are formed from a liquid mixed with a mineral compound. To the first species belong salt and nitrum (soda); to the second, chrysocolla, verdigris, iron rust, and azure; to the third, vitriol, alum, and an acrid juice which is unnamed. The first two of these latter are obtained from pyrites, which are numbered amongst the compound minerals. The third of these comes from cadmia.[31] To the unctuous juices belong these species: sulphur, bitumen, realgar and orpiment. Vitriol and alum although they are somewhat unctuous, do not burn, and they differ in their origin from the unctuous juices, for the latter are forced out of the earth by heat, whereas the former are produced when pyrites is softened by moisture."

Of stones he accepts the classification of writers "on natural subjects" into four classes:

"The first of these has no name of its own but is called in common parlance 'stone.' To this class belong lodestone, jasper (or bloodstone) and aetites (geodes). The second class comprises hard stones, either pellucid or ornamental with very beautiful and varied colors which sparkle marvellously; they are called gems. The third comprises those

---

[31] The Hoovers suggest, probably correctly, that this "unnamed" substance is zinc sulphate.

which are only brilliant after they have been polished, and are usually called marble. The fourth are called rocks. They are found in quarries, from which they are hewn out for use in building and they are cut into various shapes. None of the rocks show color or take a polish.

"Metal is a mineral body, by nature either liquid or somewhat hard. The latter may be melted by the heat of the fire, but when it has cooled down again, and lost all heat, it becomes hard again and resumes its proper form. In this respect it differs from the stone which melts in the fire, for although the latter regains its hardness yet it loses its pristine form and properties. Traditionally there are six different kinds of metals, namely, gold, silver, copper, iron, tin, and lead. There are really others for quicksilver is a metal although the alchemists disagree with us on this subject, and bismuth is also. The ancient Greek writers seem to have been ignorant of bismuth, wherefore Ammonius rightly states that there are many species of metals, animals, and plants which are unknown to us. Stibium (antimony), when smelted in the crucible and refined has as much right to be regarded as a proper metal as is accorded to lead by writers. If when smelted a certain portion be added to tin, a booksellers' alloy is produced from which the type is made that is used by those who print books on paper. Each metal has its own form which it preserves when separated from those metals which are mixed with it. Therefore neither electrum nor stannum is of itself a real metal, but rather an alloy of two metals. Electrum is an alloy of gold and silver, stannum of lead and silver. And yet if silver be parted from electrum then gold remains and not electrum, if silver be taken from stannum, then lead remains and not stannum.[32]

"Whether brass, however, is found as a native metal or not cannot be ascertained with any surety. We only know of the artificial brass, which consists of copper tinted with the colour of the mineral calamine, and yet if any should be dug up it would be a proper metal. Black and white copper seem to be different from the red kind. Metal there-

---

[32] It will be recalled that with the ancients and into the middle ages the word "stannum" was generally used for an alloy of lead and tin or other alloys of lead, but not as at present for tin itself.

fore is by nature either solid, as I have stated, or fluid as in the unique case of quicksilver.''

It will be recalled that the idea that copper obtains merely a color and not substance from the addition of cadmia, or calamine, the zinc ore, was the idea of Aristotle, repeated also by Albertus Magnus, and other medieval writers.

The above quotations illustrate the independence in thought and expression of Agricola in discussing the nature and origin of metals and minerals. He is by no means free from the concepts and ideas of his predecessors, but he does not merely reiterate the common phrases of the origins of these things from air, water and earth or the mercury-sulphur hypothesis of the origin of metals and minerals. While not disputing these theories, he places the emphasis upon the facts determined by observation and experiment, and for his time is unusually independent in his judgments, based upon his experience and upon the current manuals for mining and assaying as well as on earlier literature.

An interesting feature of the *De Natura Fossilium* is his attempt to latinize and systematize the nomenclature of many metallic ores and minerals whose names existed to a considerable extent in the German vernacular, and which had no equivalent in Greek or Latin usage. For the formation of these designations he employed to a great extent the German names. Thus for example silver ores and minerals are designated as ''argentum rude'' (crude), with a specific suffix which should characterize them on the basis of their more obvious physical property—especially color. He mentions eleven such silver minerals, beginning with pure silver, argentum purum; argentum rude, meaning silver minerals in general; and then argentum rude plumbei coloris, lead-colored crude silver (silver glance); argentum rude rubrum, red crude silver (''Rot gold ertz'' in German); argentum rude rubrum translucidum (''Durchsichtig rod gulden ertz'' or ruby silver); argentum rude album (''Weis rod gulden ertz'' or white silver ore); and

similarly for liver-colored—jecoris, yellow—luteum, cinera-
ceum—ash-colored or gray, nigrum—black, and purpur-
eum—purple. In his description of minerals in particular, when dependent upon the literature he is often indefinite, as were usually the descriptions in the mineralogies
of Arabian source, as we have already seen. In the field of
his own experience he is often very clear, as for instance:

"Lead-coloured rudie silver is called by the Germans
from the word glass (glasertz) not from lead. Indeed it
has the colour of the latter or of galena (plumbago), but
not of glass nor is it transparent like glass, which one might
indeed expect, had the name been correctly derived. This
mineral is so like galena in colour, although it is darker,
that one who is not experienced in minerals is unable to
distinguish between the two at sight, but in substance they
differ greatly from one another. Nature has made this
kind of silver out of a little earth and much silver. Whereas
galena consists of stone and lead containing some silver.
But the distinction between them can be easily determined,
for galena may be ground to powder in a mortar with a
pestle but this treatment flattens out this kind of rudis silver. Also galena, when struck by a mallet or bitten, or
hacked with a knife, splits and breaks to pieces; whereas
this silver is malleable under the hammer, may be dented
by the teeth and cut with a knife." [33]

The great work upon which the reputation of Agricola
mainly rests is his *De Re Metallica* first printed at Basel
in 1555, in Latin. A German translation was published in
1557, a second Latin edition in 1561, and an Italian translation in 1563. On the appearance of this work—a stately
folio of some 600 pages—it was evidently at once recognized
as a work of first-rate importance. No English translation was published until 1912, when Mr. and Mrs. H. C.
Hoover issued their scholarly translation enriched with a
mass of notes relating to the history and development of
mining and metallurgy.

The *De Re Metallica* is a work which gives very clear,

---

[33] Hoover, *op. cit.*, notes pp. 108, 109.

complete and detailed accounts of mining geology, mining engineering and working, as it existed in Germany in the fifteenth and sixteenth centuries, with very full descriptions of the smelting of ores and of the assaying and analysis of ores and alloys as then developed. The work is systematic in treatment and profusely illustrated with excellent woodcuts of all machinery, tools, and processes which lend themselves to pictorial illustration. The work is divided into twelve books. The first is devoted to a review of classical writers who have written of mines and mining, and to a consideration of the importance and dignity of the mining profession.

"Certainly, if mining is a shameful and discreditable employment for a gentleman because slaves once worked mines, then agriculture also will not be a very creditable employment, because slaves once cultivated the fields, and even to-day do so among the Turks; nor will architecture be considered honest, because some slaves have been found skilful in that profession; nor medicine because not a few doctors have been slaves."

One is reminded that this is the time when Agrippa, as previously cited, quotes the proverb, "All alchemists are either physicians or soap boilers." The dignity of the chemical arts was indeed to be established by the works of just such men as Agricola and Biringuccio.

The second book is devoted to the general discussion of mines, their location, ownership, indication, outcrops, and like matters.

The third to the sixth books deal with mining operations, veins, stringers, surveying, administration, machinery, tools, etc.

In Book Seven, the author discusses very fully the various methods of assaying ores. The methods given are much the same as previously given in the *Probierbüchlein* and in Biringuccio's *Pyrotechnia,* but described in greater detail and with more systematic explanations. Furnaces, crucibles, scorifiers and tools are described and illustrated. The methods of constructing the various sets of touch needles for determining the composition of alloys by the touchstone are given in all detail.

Book Eight is devoted to the preparation of ores for smelting—sorting, crushing, grinding, sifting, washing, roasting. The use of quicksilver for recovering gold is described, but the recovery of silver by amalgamation is not referred to though this process had been described by Biringuccio. Whether silver amalgamation was independently discovered by the Spaniards in Mexico about 1565 or 1570, or was introduced there from the experience of Europe is not known.

Book Nine describes the various processes and machinery for the smelting of ores. Gold, silver, copper, iron, lead, tin, antimony, quicksilver and bismuth are included in these descriptions.

Book Ten deals with the making of the mineral acids used in assaying and in "parting" operations. "Aqua valens" is the term which Agricola employs indiscriminately for the acids or mixtures of acids, ignoring the terms "aqua fortis" or "aqua regia" then already introduced by previous writers. His description of the materials used for preparation would indicate that a considerable variety of strength and composition of these acids were in use. He describes ten recipes for the materials to be subjected to distillation in the furnace.

The first consists of one libra of vitriol and as much salt with a third of a libra of spring water. On distillation this would yield at gentle heat hydrochloric acid only, by forced heating eventually some sulphuric acid also. The second recipe is two librae of vitriol, one of saltpeter with water, as much as will pass away while the vitriol is being reduced to powder by the fire. This mixture gives at first nitric acid, more or less dilute, and in the later stages of the distillation, mixed with some sulphuric acid. The third consists of four librae of vitriol, two and a half of saltpeter, half a libra of alum with water. The fourth to the eighth are mixtures of the same general nature, of somewhat varying proportions but with no essential difference. The ninth contains two librae of brick dust, one of vitriol, one of saltpeter, a handful of salt and three quarters of a libra of water. This mixture would yield an aqua regia of concentration increasing with the progress of the distillation. The tenth mixture consists of three librae of saltpeter, two of stones which are liquefiable in the furnace with the third degree of fire, half a libra each of verdigris, of stibium (antimony sulphide), iron scales and filings and asbestos, with one and one third librae of spring water. On distillation to dryness the product would seem to be nitric acid containing much nitrous acid and at the end of the distillation sulphuric acid also would pass over. The fuming nitric acid thus obtained had probably special applications in the laboratory.

The methods of parting gold from silver or silver from gold are those already comprised in Theophilus Presbyter, Geber, Biringuccio, and the *Probierbüchlein,* by cementation with salt, with sulphur, sulphide of antimony, and by the use of aqua fortis. Agricola includes a cementation with saltpeter not mentioned by these earlier authors. The separation of gold and silver from lead or lead ores by cupellation is treated in great detail. This is a very ancient process; Diodorus Siculus and Pliny refer to it and it is described more or less completely by Theophilus, Geber, and later writers.

Book Eleven is mainly devoted to an elaborate description of the "liquation" process of separating silver from copper. This is the method by which an alloy of copper with lead is heated in a reducing atmosphere to such a temperature that the lead melts and largely separates out carrying the silver with it to a considerable extent. Frequent repetition of the process makes for efficiency. Book Twelve deals with the sources and preparation of "solidified juices" by which Agricola means soluble salts. He begins with common salt from sea water and from salt springs or mines, describing in great detail methods and appliances.

Soda (nitrum) is also briefly described. Saltpeter, its origin and production from earth in which it has rested many years, and from exudations from stone walls of wine cellars and dark places is described much as by Biringuccio, from whose work it appears to have been somewhat condensed. Then follows the manufacture of alum, vitriol, sulphur, bitumen and finally of glass of which he gives a clear and interesting description concluding:

"The glass-makers make divers things such as goblets, cups, ewers, flasks, dishes, plates, panes of glass, animals, trees and ships, all of which excellent and wonderful works I saw when I spent two whole years in Venice some time ago. Especially at the time of the Feast of the Ascension they were on sale at Morano, where are located the most celebrated glass works. These I saw on other occasions and when, for a certain reason, I visited Andrea Naugerio, in his house which he had there, and conversed with him and Francisco Asulano." [34]

The *De Re Metallica* is clearly the greatest treatise upon a chemical industry which is known to the history of chemistry up to or during the sixteenth century. It cannot claim to have introduced any great chemical discovery nor any new idea of importance into chemical thinking. On the other hand it is the product of a man of broad information, of scholarly training and taste, of excellent judgment and sound common sense devoting his wide knowledge and experience to compiling a work which should give to the interested public as clear and complete as possible an account of the profession of mining, metallurgy and accessory arts and sciences.

To the scholarly and chemical world of his time this work of Agricola made no great appeal, for the great interest of that time lay in the struggle against conservatism in medical chemistry among the physicians, or in the more or less transcendental chemical philosophy of the alchemists. But among miners and mining chemists, the work of Agricola took at once a standing which left it on a pedestal

[34] Hoover, *op. cit.*, p. 592.

unattained in its field and unparalleled in any other field
of chemical technology for more than a century.

The development of technical chemistry in the sixteenth
century was marked in France by the labors and writings
of Bernard Palissy (1510–1589). Palissy was a man without classical training, could not read Greek or Latin and
like Biringuccio and Paracelsus wrote in the vernacular.
Palissy's place in the history of chemistry is due not so
much to any new facts or theories that he developed, as to
his influence by precept and example in advancing the importance of experimentation and independent observation
over the reliance upon authorities. Becoming early interested in the problems connected with pottery and especially
with enamels on pottery he devoted his life to the solution
of these problems struggling with indomitable courage
against long years of discouragement and disappointment,
until he succeeded in the development of a characteristic
art in pottery that France has been proud to cherish among
her early art treasures. Palissy has told the story of his
first incentive to work on enamels. He had been shown a
cup fashioned and enameled in great beauty.

"Without regard to my having no knowledge of clays,
I set myself to seek enamels like a man who gropes in darkness. Without having heard how enamels were made, I
crushed all materials which I thought might make something, and having crushed and ground them, I purchased a
quantity of earthen pots and after breaking them in pieces,
I put the materials that I had ground upon these, and having marked them I put aside in writing the medicaments
(drogues) that I had used upon each of them for memory:
then having made a furnace after my notion I set to bake
the said pieces, to see if my medicaments could make some
color of white, for I sought no other color than white, for I
had heard say that the white was the foundation of all
other enamels.

"But because I had never seen earth baked nor knew at
what heat the said enamel ought to melt, it was impossible
for me to do anything by this means, if at any time my
doings had been good, because sometimes the thing was

heated too strongly and at other times too little, and when the said materials had been too little baked or overburned I could not judge of the reason why I made nothing good but laid the blame on the materials.''

The story of Palissy's difficulties that brought him continual disappointments during fifteen or sixteen years of struggle in which he felt his efforts and his means wasted, is an interesting one. Minor successes in pottery which helped to recoup his losses were no satisfaction to him so long as the main aim of his search remained unachieved.

''When I had invented the means of making my rustic pieces,[35] I was in greater trouble and weariness than before. For having made a number of rustic basins and having had them baked my enamels were found to be some beautiful and well fused, others badly fused, others over-burned, for the reason that they were fusible at different degrees; the green of the lizards had been burned before the color of the serpents was melted; also the colors of serpents, crawfish, turtles, crabs were melted before the white had received any beauty.

''All these faults have caused me such labor and sadness of spirit, that before I had succeeded in making my enamels of the same degree of fusibility I thought I should enter the gates of the tomb.''

The final result of the labors and sacrifices of Palissy was the achievement of the enameled pottery which made his reputation. It is said that it was not superior nor even equal to similar Italian pottery of that period, but it was his own achievement.

Palissy wrote several works, of which the most important is a book on pottery—*Des Terres d'Argile,* in which he records his experience and methods of making and decorating pottery. He wrote a work on salts in which he classifies as salts, couperose (green vitriol) saltpeter, alum, borax, sublimate, rock salt, tartar, and sal ammoniac. He emphasizes the occurrence of salts in plants and animals, and the importance of salts to agriculture. He even be-

---

[35] Pottery with colored glazes, representing vessels with animals or figures in relief.

lieved that manures were useful only on account of their salt contents, and that soils become infertile through the gradual loss of their salts. In a treatise on marl (*De la Marne*), he discusses the improvement of the soil by its application, though its use was already in vogue. The Romans indeed used among other manures plaster, marga, and the Greeks leucargillos, or "white clay" which was probably the same.[36]

With respect to the aims of the alchemists he expresses himself in his *Treatise on the Metals and Alchemy*. He cherishes no illusions as to the possibility of the making of real gold or silver, and asserts that their pretended gold and silver can easily be shown by cupellation to be false. Nevertheless he says,

"Let them go on, that saves them from greater vices, since they have the means to try these things. As to the physicians, in following alchemy they will learn to know nature, and that will be of service to them in their art and in doing it they will recognize the impossibility of the business."

By exerting his influence to encourage experiment, research and independent thought, as against scholasticism, the blind faith in authority and superstition, and by his own example as an indomitable and successful investigator, Palissy materially contributed to the advancement of chemical science, and did much to dignify the labors of the chemist.

Palissy was a Protestant in religion and survived the persecutions of the Huguenots even through the St. Bartholomew's massacres, perhaps, as has been said, through the favor of the queen-mother Catherine of Medici, for he was employed in decorating by his art the royal castles and grounds. In 1589 however, being then about eighty years of age, it is related by D'Aubigné that his death was demanded as a heretic, and the king, Henry III, visited him to see if he might not persuade him to renounce his errors.

---

[36] Cf. Hoefer, *Histoire de la Chimie*, 2d ed., I, pp. 188, 189.

As however he would not renounce his faith, he was not released and died in prison.

Giovanni Baptista Porta (1537?-1615), a Neopolitan scholar of ability who had devoted great attention to the study of natural and physical science, even visiting France, Spain and Germany to perfect his knowledge, deserves notice here by reason of the publication of his *Magia Naturalis*—(Natural Magic). This work first published in 1558 in three books was later, in 1584, expanded to twenty books comprehended in one volume. In this form the book had a great vogue, being translated from the original Latin into the principal European languages, and republished in the Latin edition in many places for a hundred years. It is in fact a work on popular science including books on many subjects of natural science, cosmology, geology, optics, plant products, medicines, poisons, cooking, etc. Included are books on transmutation of the metals, not however confining transmutation to the alchemistical signification but including chemical changes generally; distillation, artificial gems, the magnet and its properties; cosmetics used by women, fires, gunpowders, Greek fires including preparations of Marcus Graecus, (whom he, like Biringuccio, calls Marcus Gracchus); on invisible and clandestine writing.

In the treatment of these subjects Porta includes statements of the ancients from the time of Theophrastus and Aristotle, as well as the contemporary knowledge of his own time, not always with any critical discrimination between the ancient interpretations and the more modern facts. Thus under the heading "To change stibium into lead" [37] he says, "if you frequently heat and burn stibium which the chemists call *regulus* you will burn it into lead, because we see it noted by Dioscorides saying, 'Stibium if heated somewhat further is turned into lead.'" The chemists of Porta's time by the regulus of antimony or stibium, meant metallic antimony, and no longer considered

[37] Joh. B. Portae, *Magia Naturalis*, Amsterodami, 1664, p. 245.

it as lead as did Dioscorides and Pliny. The book on imitation gems is of interest, including the coloring of glass by metallic compounds, burned copper for the aqua marine, manganese for the amethyst, zaffre (cobalt) for the sapphire, copper and iron for the emerald, etc. So also the making of enamels and their coloring for pottery are described in this book, this art in Italy being further advanced at this time than elsewhere except in China.

It should be remembered that few in Porta's time were free from credulity toward many marvels and superstitions which were inherited from the past and Porta's work shows that he was no exception, as much of the marvellous is found in his writings. On the whole, however, his information is definite and practical and his work is as good as could be expected of one not himself a practical experimenter or investigator, but a conscientious and scholarly student of the literature, ancient and contemporary. His directions and recipes on a great variety of applications of chemistry are sufficiently definite and detailed to be of service in stimulating experimentation and all in all, the work must have been of considerable influence in disseminating interesting and useful chemical information.

Porta published in 1608 at Rome a work on distillation, its methods, apparatus and applications, which is of interest as giving a more comprehensive view of the applications of distillation in the sixteenth century than is found in any other work of the period. Methods and apparatus for distillation had been described from very early times, by Zosimus, pseudo-Geber, Brunschwyk, Biringuccio, Agricola and many others for particular applications.

This treatise of Porta's, which is very different in plan and content from the book on distillation in his earlier work, is divided into nine books, dealing successively with the kinds of distillation, the methods and apparatus for distillation in general, furnaces, retorts, condensers, etc.; with the preparation of distilled perfumed waters, from roses, violets, myrtle, lavender, jasmine, lilies, etc.; with

volatile oils of roses, myrtle, cloves, lemon, absinthe, jasmine, lavender, mint, (mentha), salvia, chamomile, anise,
laurel, cypress, angelica, cinnamon, pepper, cardamom, etc.;
oils distilled from resins, mastic, benzoin, styrax, ammoniac, opponax, turpentine, camphor, etc., and from woods,
guaiacum, juniper, aloes, and aspalatum. The seventh
book deals with the distillation of strong waters, "aquae
validae," he calls them. These are the corrosive mineral
acids in the variety described in the German *Probierbüchlein,* and in the works of many writers following pseudo-
Geber. He includes among them the "oil of bricks," oleum
de lateribus, obtained by distilling olive oil from hot bricks
as given in the manuscripts of Marcus Graecus[38]. Its virtues are adapted he says to tense nerves, cold abscesses and
to cold distillations. ("vires tensis nervis, frigidis aposte-
matibus, ac frigidis distillationibus.") This distillation of
alcohol from wine, and the preparation of certain oils of
animal origin are also given, musk, civet, beaver, scorpion,
etc.

The methods of obtaining all these oils and waters, and
very often also the quantitative yields obtainable are given.
Altogether it is an illuminating exposition of the scope of
application of distillation in the sixteenth century.

The works of these practical chemists of the sixteenth
century manifest a more serious appreciation of the dignity
and importance of chemistry in its relation to the practical
arts, and had a great stimulating influence on all chemical
workers. It will be noticed however that with the exception of Paracelsus these men were not greatly interested
in the problems of chemical philosophy. To the extent that
they refer to chemical theory they accept the conventional
Aristotelian or Arabian concepts. Paracelsus by the impression made by his three principles indeed did much to
shatter the blind faith in the ancient theories and to pave
the way for later constructive speculation. In so far as
chemical theory is concerned the sixteenth century marks

---

[38] *See* ante, p. 197.

the decay of the old rather than the birth of a distinctly new philosophy.

It must not be thought that the period of superstition, charlatanism and alchemy had yet passed away. Illustrations of this we shall see in the next chapter.

CHAPTER IX

CHEMICAL CURRENTS IN THE SIXTEENTH CENTURY

Though the works of the four last-mentioned men were in their domains the most far-reaching and permanent influences of the sixteenth century, they by no means summarize the chemical activities of the period. Several factors were influential in determining the current of thought and interest. Doubtless the most dominant motive was mainly excited by Paracelsus, though not entirely due to him, the revolt against the absolute authority of Galen and Avicenna in medical theory and the campaign for the extension of the use of the so-called chemical medicines. This resulted in the bitter and intense struggle between partisans of Paracelsus and his new medicines, and the conservatives of the medical profession and especially of the university medical faculties, who vigorously resisted those encroachments. As Agrippa puts it, at just this period all chemists were either "physicians or soap boilers." It was generally true that a large part of the chemical thinkers were also physicians, and the chemical and medical scholars were generally involved in this warfare, which occupied the center of the stage for more than a century after Paracelsus.

Among the more prominent supporters of Paracelsus were Michael Toxites, physician at Hagenau, who published a commentary on Paracelsus under the title of *Testamentum Paracelsi* in 1574 at Strassburg, and Gerhard Dorn, physician at Frankfurt, author of various works relating to alchemy and an enthusiastic adherent of the doctrines of Paracelsus who published in 1567 the *Clavis Totius Philosophiae Chymisticae,* in the introduction of which he acknowledges his indebtedness "first to God and then to the doctor and our preceptor Theophrastus Paracelsus easily chief of physicians and philosophers."[1] Adam von Bodenstein (1528–1577), a pupil of Paracelsus, who lectured as Professor of Medicine at Basel on the Paracelsan system of medicine followed Paracelsus in his mystical notions as well as in his practice, and is credited with a work *De Lapide Philosophorum,* and a commentary on the *Rosarium* attributed to Arnaldus de Villanova. G. Dorn was a pupil of Adam von Bodenstein, to whom his *Clavis* was dedicated.[2]

Alexander von Suchten, of Danzig, also a student at Basel, was an advocate of Paracelsus and interested in chemistry and alchemy. He wrote a *Clavis Alchemiae* and a work *De Secretis Antimonii Liber,* first published in Basel in 1575 and said to have been translated from German into Latin. It is of interest to note that both the above works were published in German in 1604 at Leipzig by Johann Thölden, the supposed author as well as publisher of the *Triumph Wagen Antimonii,* 1604, and other earlier and later literature of the mythical Basilius Valentinus.[3] The works of Alex. von Suchten were published in many later editions during this century.

Oswald Crollius, or Croll, (1580–1609) was another influential advocate of Paracelsus, and a contributor to the chemical remedies. His *Bascilica Chemica,* Frankfurt, 1608, often republished, was his most popular work. It contained an exposition of the teachings of Paracelsus, a treatise on materia medica in which he emphasizes the chemical medicines, and a treatise on the doctrine of Signatures, a subject also treated in the Paracelsan literature, and which assumes that medicinal plants or other sources

---

[1] Gerhardus Dorn, *Clavis Totius,* etc. Lugduni, MDLXVIII, p. 3.

[2] Schmieder, *Geschichte der Alchemie,* p. 276, 321.

[3] Prof. John Ferguson in his *Bibliotheca Chemica,* II, p. 415, lists an edition of Von Suchten's work, "*Antimonii Mysteria Gemina,* Alexander von Suchten, das ist von den grossen Geheimnissen des Antimonii, etc., durch Johann Thölden Hessum, Leipzig, 1604." The first edition of the *Triumph Wagen des Antimonii of* "Basilius Valentinus" was published in the same year and place by Johann Thölden.

of medicine bear some symbol or sign of their value for medicine in their color, shape or other visible sign, by which God intends that they shall become known to those expert and wise in the interpretation of these signs.

Croll is credited with being the first to mention the explosive fulminate of gold and with having given the name of *luna cornea,* horn silver, to the fused chloride of silver. Kopp also credits to him the first announcement of the acid from amber (succinic acid) "flos succinii."

Leonhard Thurneysser (1530–1596) was one of the most noted and notorius adherents of Paracelsus. Son of a goldsmith of Basel, he was first distinguished for having sold to a Jew gilded bars of lead as pure gold, as a result of which he was obliged to flee from Basel. He visited England and France. In 1552 he joined the army in Brandenburg, but a year later abandoned that to take up his earlier trade of goldsmith, which he seems to have pursued in various German cities for a few years, finally turning to mining. In 1560 his success was such that he was patronized by the Archduke Ferdinand of Austria, who sent him at his expense on an extensive investigating journey to Scotland, Spain, Portugal, Egypt, Palestine, Greece, Hungary and other countries. During all this experience he evidently acquired considerable knowledge of medicine as well as of the mining and metallurgical arts, and in 1569 we find him appointed as court physician to the Elector of Brandenburg. About this time he became an advocate of the Paracelsan medical doctrine, and published several works of chemical and medical character. Eventually on account of swindling operations he was forced to leave Berlin (1584). He then went to Italy and operated as alchemist pretending to be able to make gold, eventually returning to Germany and dying in great poverty at Cologne 1596. At one time in Berlin he had amassed considerable wealth and displayed it with ostentation. He let it be understood that it was acquired by transmutation, but it was acquired doubtless by his chemical and medical practice, aided by the arts of the charlatan and impostor. As Kopp[4] says, not a single useful experiment is found in his works, his entire accomplishment being a paraphrasing of the ideas of Paracelsus. Even these were not clearly understood by Thurneysser, nor correctly rendered.

Ferguson says of him:

"He was endowed with quickness and obviously a powerful memory; but he tried to pass as a man of science, a learned physician, and an accurate scholar, when in reality he was a man of action, with a gift for organizing and commercial advertisement. At the present day he might have been a successful manufacturing chemist, able to turn his raw material into gold without the red elixir."[5]

An enthusiastic advocate of Paracelsan ideas was Joseph Duchesne, better known under his Latin appellation of Quercetanus (1521–1609). He was born in Gascony, studied in Germany, and in France was attached as physician to the court of Henry IV. He was an extreme partizan of the chemical medicines of Paracelsus and added others of his own initiative. His position at court protected him from the hostility of the medical profession, then generally opposed to the new remedies, though his arrogance and many fantastic notions served to make him many enemies in the profession.

As a chemist he contributed nothing of note. Hoefer cites a passage from his treatise in *Materia Medica,* in which he says that saltpeter, (sal petrae) "contains a spirit which is of the nature of air and which nevertheless cannot sustain flame, but is rather opposed to it." Though this description would apply to nitrogen, yet as the above statement is accompanied by no further elucidation it seems a rather strained interpretation that nitrogen might have been isolated from saltpeter by Quercetanus.[6]

A later French physician and better chemist than Quercetanus, was Turquet de Mayerne (1573–1655), a well

---

[4] Kopp, *Geschichte der Chemie,* I, p. 109.
[5] Ferguson, *Bibliotheca Chemica,* II, p. 453.
[6] Hoefer, *Histoire de la Chimie,* 2d ed., II, p. 25.

educated physician and Professor of Chemistry at the University of Paris. Though not rejecting the Galenic medicine he nevertheless was an advocate of the new chemical medicines, preparations of mercury and antimony, acetate of potassium, benzoic acid, copper and iron sulphates, etc. In 1603 the medical faculty of Paris condemned him, for this reason, to be deprived of the decorations of the school and the academic privileges and forbade all true physicians to have any relations with him.[7] As a result of this decree De Mayerne left France, and spent the remainder of his life in England where he was court physician to James I and later to James II. He died at Chelsea in 1655.

Turquet de Mayerne was held in high estimation as a physician. His medical works containing much chemistry were published from about 1604 on. His complete works were published in England by Dr. Joseph Brown in 1701. Turquet is credited with some notable observations in chemistry, with the preparation of the black sulphide of mercury by rubbing together mercury and melted sulphur, with the preparation of benzoic acid from benzoin by volatilization with a paper cone for condensation. He has also been considered as the first to recognize that by the action of sulphuric acid upon iron an evil smelling and inflammable air is evolved, though whether to Turquet or to Robert Boyle this discovery is due, is a question not yet settled beyond doubt. The problem of priority in the observation of evolution of this gas and of its inflammability, as recorded in the history of chemistry, is interesting. F. Hoefer, in his *Histoire de la Chimie*,[8] gives Paracelsus credit for the earliest observation as follows:

"The effervescence which manifests itself when water and oil of vitriol (sulphuric acid) are brought into contact with a metal such as iron had not escaped this observing spirit. He knew that in this operation there is given off an air like a 'wind' (Luft erhebt sich und bricht herfür wie ein wind) and that this air separates from water of which

[7] Cf. Hoefer, *op. cit.*, p. 239, for Latin text of this decree.
[8] Hoefer, *Histoire de la Chimie*, 1st ed. 1843, II, p. 16.

it is an element. Paracelsus had glimpsed the truth without retaining it, etc.''

So also, H. Kopp, two years later,[9] evidently depending upon Hoefer says, speaking of hydrogen:

"The older alchemists seem to have had no knowledge of this gas, even Basilius Valentinus in the fifteenth century [really the seventeenth] who repeatedly describes the solution of iron in sulphuric acid does not with any word mention the kind of air which is developed. Paracelsus, in the century following [really preceeding] first called attention to it. His *Archidoxa* contain the description of how iron is dissolved in dilute sulphuric acid with the observation "Luft erhebt sich und bricht herfür wie ein wind.''

R. Jagnaux[10] cites Hoefer as to the first observation of hydrogen by Paracelsus, quoting also the above German phrase.

Hermann Schelenz,[11] also speaking of hydrogen, refers to Paracelsus and Thölden as having had it in their hands. Thölden is the accepted author of the Basilius Valentinus literature.

In 1875, Herman Kopp,[12] discussing the discovery of the composition of water, again refers to this subject. After asserting that nowhere does Basilius Valentinus allude to any evolution of gas or air in connection with the described preparation of iron vitriol from iron and oil of vitriol, he says:

"That Paracelsus mentions it has, indeed, been asserted. Hoefer says in his *Histoire de la Chimie*, III, 1st ed., p. 15, 2d ed., p. 12. 'The effervescence which,' etc. [as above quoted]. I have, therefore, in my *Geschichte der Chemie*, III. *Theil*, S. 260, also stated that Paracelsus had called attention to the evolution of air on the solution of iron in dilute sulphuric acid. The edition of the works of Paracelsus to which Hoefer refers[13] I cannot now consult, but in

9 Kopp, *Geschichte der Chemie*, 1845, Bd. III, p. 260.
10 *Histoire de la Chimie*, Paris, 1891, I, p. 385.
11 *Geschichte der Pharmazie*, Berlin, 1904, p. 560.
12 Kopp, *Beiträge zur Geschichte der Chemie*, Pt. III, p. 241, note 10.
13 Huser's 1st ed., 1589, *Archidoxis*, VI, p. 12.

Huser's edition of Strassburg, 1616, I could find nothing
to substantiate this statement. In Book III of the *Archidoxa,*
on the separation of the elements, is stated," (here Kopp
quotes the German text which contains the above quoted
phrase and then says) "as to action between sulphuric acid
and a metal there occurs nothing whatsoever about these
substances." [14]

The text of the passage under discussion[15] is as follows:

"So merck dass die Elementen in der Scheidung ge-
funden werden gleich in der Gestalt und Form wie sie an
den wesentlichen Elementen seind. Dann der Lufft erzeiget
sich gleich dem Lufft und ist nicht zu befassen, als etliche
in ihren Gemüttern vermeinen; Auss der Ursachen dass in
dem Instrument der Scheydung der Lufft sich erhebt und
herfür bricht gleich wie ein Wind, und etwan mit Wasser
aussfehret, etwan Erdtrich, etwan Fewer. Dann ein
sondery wunderbarliche Auffhebung ist im Lufft. Als
wann auss dem wesentlichen Element Wasser soll der Lufft
gescheiden werden als dann geschicht durch das Sieden:
Und so bald es seudt so scheidet sich der Lufft vom Wasser und nimpt mit sich die leichtist Substanz vom Wasser:
Und so viel das Wasser gemindert wirdt also nach seiner
Proportion und Quantitet wirdt auch gemindert der
Lufft."

This may be translated:

"Note, therefore, that the elements are found in their
separation (Scheydung) the same in shape and form as
they exist in the essential elements. For air shows itself
like air and is not to be grasped (or confined), as some
in their minds imagine. For the reason that in the apparatus for 'parting' (or separating) the air rises and breaks
forth like a wind, and sometimes passes off with water,
sometimes with earth, sometimes with fire. For such a
special wonderful lifting power exists in air; as when from
the essential element water, air is to be separated, that

---

[14] The Strassburg 1616 edition is the second reprint of Huser's edition of
1589. Though less carefully edited than the first edition, its text differs in
no essentials from the 1589 print, as shown in the critical bibliography of
the works of Paracelsus by Dr. Karl Sudhoff *"Versuch einer Kritik der Echtheit der Paracelsischen Schriften,* 2 vols., Berlin, 1894-1899.

[15] *Paracelsus Opera,* Huser, Strassburg, 1616, Vol. 1, *Archidoxa.* Lib. III,
p. 791.

takes place by boiling, and as soon as it boils, the air separates from the water and carries with it the lightest substance of the water; and as much as the water is diminished
just so much in its proportion and quantity the air is
diminished.''

The foregoing passage is from a treatise on the separation of the elements, (meaning the four Aristotelian elements) from their complexes. The whole discussion is
obscure and metaphysical. The interpretation of this
passage is none too easy.

If we assume that Paracelsus here means by ''Instrument der Scheydung'' the operation of parting in assaying, a common process in his time and elsewhere described
by him, that process consisted in the solution of alloys of
silver or gold with other metals by aqua fortis (nitric
acid) and the effervescence he refers to would be caused
by nitrogen oxides, not hydrogen.

If, on the other hand, he means only the separation of
air from water by boiling, as illustrated in the latter of
the above sentences, then it may be conjectured that he
only observes the conversion of water into vapor (air) in
boiling. In no case is there any justification for Hoefer's
conclusion. Paracelsus and ''Basilius Valentinus'' may
therefore both be eliminated as early observers of the formation of the gas now known as hydrogen.

Eliminating Paracelsus and ''Basilius'' or Thölden, the
credit of recording the first observation of the air evolved
from iron and sulphuric acid and of its inflammable character seems to lie between Turquet de Mayerne and Robert
Boyle. The passage upon which rests the claim of the
former occurs in his *Pharmacopoea*. The date of the first
appearance of this work is doubtful. It is included in his
Opera Medica, edited by Joseph Browne in London, copies
being apparently variously dated 1700, 1701, and 1703. On
citations from this publication are based all notices thus
far recorded of the description of this gas. Among the
lists of the publications of Turquet there seems to be no
earlier publication of the *Pharmacopoea* recorded, except

that R. Jagnaux states that Albrecht von Haller, a writer
of the latter half of the eighteenth century says that Turquet's *Pharmacopoea* was issued with his *Medicamentorum Formulae* in 1640. Jagnaux, however, states that,
unfortunately, he has not succeeded thus far in verifying
this publication.[16] Should such a publication be confirmed
and the same passage be found in that work, the priority
of notice would unquestionably be established. Even could
it be shown or accepted that the text of the works as published by Browne were all by Turquet and not later added
to, the priority would still be his, as he died in 1655, some
years before Boyle's observations. That this edition was
published as written, Browne specifically states, according
to Jagnaux. In the account of the life and work of Mayerne in the Dictionary of National Biography, Vol.
XXXVII, London and New York, 1894, it is stated: "On
June 25, 1616, he was elected fellow in the College of Physicians of London and in 1618 wrote the dedication to the
King of the first pharmacopoea published by the College."
One is tempted to wonder whether the pharmacopoea published as Turquet's work was not this work to which he
wrote the dedication and whether the work itself was not
revised after the death of Turquet. The *Nouvelle Biographie Générale*, Tome 34, Paris, 1861, lists among his publica-
lications the *Medicamentorum Formulae,* London, 1640,
but makes no mention of the *Pharmacopoea* in that connection.

Kopp also, who, in his *Geschichte der Chemie,* credits
Turquet de Mayerne with the first notice of inflammability
of the gas on the basis of the *Pharmacopoea* which he variously ascribes to "about 1600," [17] and "about 1650," [18] in
his later work[19] says of the *Pharmacopoea* of Turquet de
Mayerne, that he knows no other edition than that in the
collected works edited by J. Browne, London, 1703, but re-

[16] Jagnaux, *Histoire,* Paris, 1891, I, p. 386.
[17] Kopp, *Geschichte der Chemie,* II, p. 114.
[18] Kopp, *op. cit.,* III, p. 178.
[19] *Beiträge zur Geschichte der Chemie,* III, p. 242, note 11,

marks that Turquet de Mayerne knew in the first half of the seventeenth century that a product of the action of dilute sulphuric acid upon iron was of disagreeable odor and inflammable but that his observation was doubtless first published in the beginning of the eighteenth century.[20]

This statement in the *Pharmacopoea* is very clear. It is published in the original Latin by Kopp in the above mentioned passage in the *Beiträge,* and in French translation by Jagnaux.[21] Translated the statement reads:

"I have taken 8 ounces of iron filings, and in a deep glass cup (concha) I have added successively 8 ounces of oil of vitriol and a little later I have added an equal quantity of warm water. There was produced an enormous agitation, a great ebullition, and a *meteorism* of matter easily quieted by stirring with a rod. There is also raised a most fetid sulphurous vapor, very noxious to the brain, which (as happened to me, not without danger) if brought near a candle takes fire, on account of which this operation should be made in the open air or under a chimney."

The priority of De Mayerne in this matter then depends upon whether this observation was really written by him or was a later addition to the Pharmacopoea of the London College of Physicians, to the first edition of which he is said to have written the dedication.

In 1670 (or 1672) Robert Boyle published his *New Experiments touching the Relation between Flame and Air,* in which he says:

"Having provided a saline spirit (this was hydrochloric acid) which, by an uncommon way of preparation, was made exceeding sharp and piercing, we put into a phial capable of containing three or four ounces of water, a convenient quantity of filings of steel. This metalline powder being moistened in the phial with a little of the menstruum, was afterwards drenched with more, whereupon the mixture grew very hot, and belched up copious and stinking

---

[20] The published catalogues of the British Museum and of the U. S. Surgeon General's libraries contain no edition of the *Pharmacopoea* earlier than 1700.

[21] Kopp, *loc. cit.*

fumes, which, whether they consisted altogether of the volatile sulphur of the Mars, (iron) or of metalline steams participating of a sulphureous nature, and joined with the saline exhalations of the menstruum, is not necessary here to be discussed. But whencesoever this stinking smoke proceeded, so inflammable it was that upon the approach of a candle to it, it would readily enough take fire; and burn with a blueish and somewhat greenish flame at the mouth of the phial for a good while together; and that, though with little light, yet with more strength than one would easily suspect." [22]

Unless, therefore, it can be shown that the statement of the evolution of hydrogen gas by Turquet was really by him, this description by Boyle, written some sixteen years after the death of Turquet, but about thirty years before the known appearance of Turquet's *Pharmacopoea,* seems to be the first announcement. That both of these writers attempt to explain the vapors of fumes as of sulphurous nature is accounted for by the fact that chemists of the time were thinking in terms of the concept of sulphur as the combustible constituent of matter.

Returning from this digression to the chemists of the sixteenth century, and first to the progress of the campaign for chemistry in medicine, which was the most prominent feature of chemical activity of the century, the work and influence of Libavius cannot be ignored.

Andreas Libau, better known under his latinized name, Libavius, was born at Halle about 1540, and was from 1588 to 1591 professor of history and poetry at the University of Jena; later city physician and director of the gymnasium or secondary school at Rothenburg; in 1607 and until his death, in 1616, director of the gymnasium at Koburg. Broadly trained, somewhat conservative by nature but endowed with an independence of judgment none too common for his time, Libavius, in the latter part of his life, became interested in chemistry and the new chemical medi-

---

[22] The Works of Robert Boyle (Birch ed.) London, 1744, III, pp. 255, 256.

cine campaign inaugurated by Paracelsus. In this campaign he occupied a position between the enthusiasts who followed Paracelsus in his vagaries as well as in his reforms, and the antagonists of the Paracelsan reforms, who rejected all the new medicines as dangerous innovations. Thus in his printed works we find Libavius, while a Paracelsus follower in the campaign for chemical remedies, to which he contributed much himself, severely criticizing the extravagances and vagaries of Paracelsus and his followers, and at other times opposing Erastus and other antagonists of the Paracelsan movement in chemical medicines. The independence of his attitude is well evidenced by the fact that many later writers classify Libavius as a Paracelsus follower or antagonist, according to their own predilections or prejudices.

The chemical work of Libavius was generally upon the preparations of chemistry with reference to their uses in medicine. The tetra-chloride of tin, which he prepared by heating corrosive sublimate with tin and which he called "liquor or spirit of mercury sublimate," was long known as *spiritus fumans Libavii*. He is credited with the description of the glass of antimony,[23] an observation previously credited to Basilius Valentinus, in the *Currus Triumphalis Antimonii,* a work which, however, first appeared a few years later than the *Alchemia* of Libavius. He is further credited with the first recorded observation of the blue color produced in ammonia by copper, and with the first preparation of sulphuric acid by the action of sulphur and saltpeter. Ammonium sulphate is said to have first been prepared by him and to have had later extensive use in medicine. His work published under the title of *Alchemia* is characterized by Kopp[24] as the first real text book on chemistry. This work was divided into two parts, the first, *Enchiria,* on the methods of operating, or manipulations, the second, *Chymia,* on the preparation of substances which

---

[23] In his *Alchymia,* a work first printed in 1595.
[24] Kopp, *op. cit.,* III, in article on "Libavius," pp. 145–150.

are made by these methods. It appears to be a summary of chemical methods and operations of the time. In the theory of chemistry it does not appear that he presents anything new of importance. He presents Paracelsus's three principles, salt, sulphur, and mercury, in one place, but elsewhere discusses the composition of the metals according to earlier concepts of Geber and the Arabians, in which mercury and sulphur alone are alluded to, evidencing that he is here a recorder of the ideas of others rather than himself a contributing thinker. With respect to the possibility of transmutation of the base metals into precious metals, Libavius not only admits its possibility but records in his works various methods of carrying out such operations, doubtless here also as a recorder of the current chemical literature of the period, rather than from any experience of his own in matters so foreign to his own field of experimentation.

The orthodox medical profession, adherents of the medical theories of Galen and Avicenna, naturally combatted energetically and violently the new tendencies. Naturally also they were often not deeply interested in chemistry, and did not leave a deep impress on the positive accomplishments of that science. Too often also, while doing good service in criticizing the weaknesses, extravagances, and impositions of Paracelsus and his followers, they depended more on the *argumentum ad hominem,* on personal abuse and ridicule, than on the presentation of facts and the logic of facts, to influence the thought of the time. The most violent of the early critics of Paracelsus perhaps was Erastus (his name was Thomas Lieber) (1523–1583), professor of medicine in Heidelberg and later in Basel. Erastus criticized the salt, mercury, sulphur theory of Paracelsus, discredited the efficacy of the cures he claimed to have made in the use of his new medicines, and upheld the validity of the older Galenic system as against the new. He was supported by very many conservative medical professors and practitioners, such as Dissenius, a prominent physician and writer on materia medica; the learned Konrad Gesner; and many others whose works belong to the history of medicine but have little of interest for the development of chemistry.

Although the campaign for and against the chemical innovations in the theory and practice of medicine was the most notable feature in chemical activity, there were other influences making toward progress or toward reaction in chemistry.

The extension of travel and discovery in the Americas, Asia and Africa, with their discovery of new plants, animals and other observations of nature, gave a new impetus to the study of the natural sciences, and tended to weaken the authority of the natural history of Aristotle and his imitators. With the exploitation of new civilizations in Mexico and Peru, and opening of new sources of knowledge by such travellers as Magellan and Sir Francis Drake, old systems of natural science proved inadequate, and new and independent points of view became more numerous. A new spirit of observation and criticism inspired many strong and original thinkers. Such was that universal genius Leonardo da Vinci, artist, scientist, engineer and inventor. Konrad Gesner, Swiss scientist (1516–1565) was also a man of great versatility, writing on zoölogy, minerals, botany, medicine, and pharmacy as well as on philology and philosophy. As says Professor Ferguson: ''There is no more notable man in the history of learning and of science in the sixteenth century than Gesner.'' Even in chemistry Gesner was not without influence for his *De Secretis Remediis Liber, etc.,* a compendious work on distilled waters, oils, resins, and on distillation processes in general, passed through many reprints and translations and served as the basis of similar works by other writers.[25]

Cardanus, Stevinus, Tycho Brahe, Galileo, Vesalius, Co-

---

[25] A work entitled ''Quatre Livres de Médicine et de la Philosophie chimique, faits Francois par M. Jean Liebaut, Dijonnois, Docteur Médecin à Paris, Rouen, MVIC'' (preface dated 1573) is a translation from this work of Gesner's.

pernicus, are names that evidence the forward looking tendencies of the sixteenth century. On the other hand a strong reactionary tendency toward mysticism and superstition in natural philosophy and toward the revivification of alchemical notions and aspirations was operative in the sixteenth and seventeenth centuries, emanating from a revival of neoplatonism and the Cabbala in Italy toward the end of the fifteenth century. The Cabbala was a transcendental philosophy of nature, supposed to have originated among Hebrew Alexandrian neoplatonists, and was in the first instance a mystical interpretation of the scriptures. It assumes the magical power of words, signs, and numbers, and the possibility through the knowledge of this power to foresee and influence future events. It recognized the power of amulets, magic formulae, conjurations of spirits and other supernatural agencies. Giovanni Pico della Mirandola (1463–1494) and Marsilius Ficinus of the Florentine Academy were active propagandists of the Cabbala, and in Germany, Reuchlin (1455–1522) Trithemius (1462–1516), and Agrippa von Nettesheim (1480–1535). Trithemius is mentioned by Paracelsus as one of his valued teachers and from him and possibly also through Agrippa's influence, Paracelsus became a believer in this magical or occult philosophy, as is evidenced in many of the treatises written by him or ascribed to him.

Giovanni Pico della Mirandola also wrote a treatise *De Auro,* in which he testifies to having several times witnessed the making of gold. Marsilius Ficinus is credited with a work *De Arte Chimica,* in which theosophical and chemical notions are mingled. Through all these influences, and not the least through Paracelsus, minds mystically inclined, and they were very many in those times, were often turned to alchemy with its mysteries, rather than to the saner aspects of chemical research.

For the history of chemistry these occult philosophers are without importance, though they were very prominent in the time when alchemy was a live issue. Such were, among a multitude of less prominence: Denis Zacchaire (1519–1556), Gaston Claves (or Dulco), Sebastian Siebenfreund, John Dee (1526–1608), Edward Kelly (1555–1597), Heinrich Khunrath (ca. 1560–1601), Michael Mayer (ca. 1568–1622), Alexander Sethon, or Setonius (? –1604), Robert Fludd (1574–1637), Michael Sendivogius (1566–1646). Some of these later alchemists were simply mystics, others credulous fanatics, some simply charlatans and confidence operators. The story of their careers sometimes ends in assassination, and sometimes in legal execution, oftener in obscurity; but their works have left little if any permanent influence unless it be that they have served as encouragement for such mystic or theosophic cults as the Rosicrucians or their modern successors.

The works published under the names of Johann Isaac Hollandus and Isaac Hollandus deserve consideration here, not on account of any intrinsic value, but because of the place they have held in the history of chemistry. From the early years of the seventeenth century until quite recently they were generally, though not universally, believed to have been written in the fifteenth century. Even Kopp and Hoefer accept this literature as of the fifteenth century, the former evidently with some uncertainty. B. G. Penotus (1608) states that the works of Is. Hollandus are based upon Paracelsus.[26] T. Bergman, also, in his *Opuscula Physica et Chemica* (1779–1788) places Isaac Hollandus at the beginning of the seventeenth century.[27] Hoefer remarks that the works of Hollandus so resemble those of Basilius Valentinus that they perhaps are by the same author.[28] The latter works are also now known to be of the beginning of the seventeenth century, but there is no evidence that they are by the same author. Even Prof. John Ferguson, in his *Bibliotheca Chemica* (1906) is uncertain as to the period and authorship of the Hollandus

---

[26] Schubert & Sudhoff, *Paracelsus Forschungen*, Frankfurt, 1887, Pt. I, p. 76.

[27] The English translation (Edinburgh 1791) of Bergmann's Essays, says by error seventh instead of seventeenth, 3, p. 123.

[28] Hoefer, *Histoire de la Chimie*, I, p. 478.

literature. More recent researches into the extant literature, printed and manuscript, have established, beyond
reasonable doubt, that the works attributed to the supposed
father and son Hollandus are post-Paracelsan. Prof. Karl
Sudhoff, who as early as 1887 claimed that the Hollandus
literature rests on Paracelsus,[29] in a personal communication to the present writer in 1913, stated that after examination in recent decades of thousands of manuscripts
there is no possible room for doubting that all the Hollandus and Basilius literature is post-Paracelsan. More
recently still another authority in alchemical literature,
E. von Lippmann,[30] has discussed the question of the period
and authorship of the Hollandus literature, and gives an
extensive list of writers on alchemy and alchemists whose
works of the fifteenth, sixteenth and early seventeenth centuries he has examined without finding any reference to
works of Hollandus. The first mention of Johann Isaac
Hollandus he finds in a work printed in 1582, falsely attributed to Paracelsus, the *Centum Quindecim Curationes
Experimentaque*. Libavius (1597) alludes to him also and
first voices the accusation that Paracelsus plagiarized the
idea of the three principles from Hollandus, a theory
hailed with enthusiasm by the many anti-Paracelsus writers of the seventeenth century. Von Lippmann also is convinced that the Hollandus literature is post-Paracelsan and
that it depends on Paracelsus for any contents of essential
value. Other scholars of early chemistry, as H. J. Holgen,[31]
and Paul Diergart,[32] confirm this conclusion.

The first recorded publication of any work by one of
these authors was in 1572 at Prague; Joh. Isaac Hollandus,
*Liber de Minerale Lapide et Vera Metamorphosi Metallo-
rum*. The *Opus Vegetabile et Animale,* by the same author
was published in 1582. Other works were printed up to
1659. No original manuscript is known, though W. P. Jo-

---

[29] Schubert and Sudhoff, *loc. cit.*
[30] E. von Lippmann, *Chemiker Zeitung,* 1916. Vol. 40, p. 605; 1919, Vol. 43, pp. 265 *ff* and 286 *ff.*
[31] H. J. Holgen, *Chemiker Zeitung,* 1917, Vol. 41, p. 643.
[32] Paul Diergart, *Chemiker Zeitung,* 1919, Vol. 43, p. 201.

rissen cites a manuscript copy of a work which bears the date of 1567. It will be remembered that Paracelsus died in 1541. Jorissen also still clings to the possibility of the early date of the Hollandus works, but presents no evidence in support of that hypothesis.[33]

The mystery which puzzled the early historians as to the personality of the two supposed Hollandus,' Johann Isaac, and Isaac, is not yet convincingly solved. Von Lippmann calls attention[34] to certain circumstantial evidence found in Ben Jonson's play *The Alchemist*. In this play, first staged and printed in 1610, referring to the charlatan and pretended alchemist who fills the title role, it is said:

"*Face*.   Will he win at cards too?

"*Sub*.   The spirits of dead Holland, living Isaac, you'd swear were in him, such a vigorous luck as cannot be resisted."

Wharton, the eighteenth century editor of Jonson's works, remarks on this passage, "The poet alludes to the two famous chemists, Isaac and John Isaac Hollandus, who flourished about that time and wrote several treatises on alchemy." Ben Jonson is also known to have himself spent some time in Holland previous to 1610.

Antonio Neri, who wrote a treatise on glassmaking, first printed in Italian in 1612, refers to "This method of imitating gems which I received (or obtained) from Isaac Hollandus when I was in Flanders." Neri's sojourn in Flanders was about 1609. This statement does not in itself necessitate the interpretation of personal contact between Neri and Isaac Hollandus, although von Lippmann calls attention to the fact that Neri, in his work, has not the habit of citing written works as authorities, and that the published works of Hollandus contain no such matter as Neri here describes.

Von Lippmann also calls attention to the reference to Hollandus by Sir Francis Bacon, (1561–1626), as presumptive evidence that one of that name was still living. Bacon,

<hr>

[33] W. P. Jorissen, *Chemiker Zeitung*, 1919, Vol. 43, p. 105.
[34] v. Lippmann, *Chemiker Zeitung*, Vol. 40, p. 605; and Vol. 43, p. 265.

after speaking in strong disapproval of the influence of
Paracelsus and his many radical adherents, says, "Such
a one is Isaac Hollandus and by far the greater part of
the crowd of chemists."[35] That this circumstantial evidence as to the personality and period of the authors of
these works is not conclusive must be admitted, and
this has been emphasized by Jorissen[36] and also by
H. Schelenz.[37] These writers, however, present no positive evidence as to the pre-Paracelsan period of their
authorship. The fact that the "Hollandus" writers
cite no authorities of the sixteenth century is an argument of no weight if we consider that the writings
were expressly intended to convey the belief that they
were more ancient than Paracelsus and his contemporaries, which is apparently the fact. Von Lippmann presents many items[38] in the writings themselves that indicate
the improbability of their early date. Holgen quotes from
the *Opus Vegetabile,* attributed to J. I. Hollandus:[39] "Take
the best sugar of the Island of Madeira which is very
hard," and cites Reese,[40] as stating that sugar from Madeira first came to Amsterdam in the early sixteenth century. Whatever the facts may be as to the authors, it may be
taken as established beyond reasonable question that the
Hollandus literature is of the latter third of the sixteenth
and the early part of the seventeenth century.

As to the character of these works: *Hand der Philosophen, Opuscula Alchimica, Opus Saturni, Opera Vegeta-
bilia, Opus Minerale, Von der Cabala, De Lapide Philo-
sophica,* etc., it may be said that they contain nothing that
distinguishes them from a great mass of contemporaneous
alchemistical literature.[41]

[35] F. Bacon in *De Interpretatione Naturae Sententiae,* "Talis est Is. Hol-
londus et turbae chemistarum pars longe maxima."
[36] *Loc. cit.*
[37] H. Schelnz, *Zeitschrift fur Angewandte Chemie,* 1917, p. 195.
[38] v. Lippmann, *Chemiker Zeitung,* 1919, p. 265 *ff* and p. 286 *ff.*
[39] Amsterdam edition of 1659, p. 82.
[40] *De Suckerhandel von Amsterdam,* Haag, 1908.
[41] The editions of works of the Hollandus, Joh. Isaac and Isaac, accessible
to the writer are *Die Hand der Philosophen, etc.,* Frankfurt, Götzen,
MDCLXIVI (1667?); and *Sammlung unterschiedlicher bewährter chymischer
Schriften* [etc.], Wien, 1773.

Although the literature which appeared under the name of the alleged Benedictine monk, "Basilius Valentinus" is now generally conceded to have been written at the close of the sixteenth or the beginning of the seventeenth century, its relation to the history of chemistry is so similar in many ways to that of the Hollandus literature that it may be best considered in this connection.

The earliest publications under that name were published by Johann Thölde (or Thölden), of Hesse, himself a chemist, part owner of salt works in Franckenhausen in Thuringia, and a councillor (Raths Kämmerer) of that town. He was also author in his own name of a work on salts (*Haligraphia*) in 1603. The principal works of "Basilius" were issued by Thölden as follows: *De Microcosmia, von der Welt im Kleinen,* Eisleben, 1602. *Vom Grossen Stein der Uhralten Weisen,* Zerbst, 1602. *Tractat von Natürlichen und Uebernatürlichen Dingen,* Eisleben, 1603. *De Occulta Philosophia,* 1603. *Triumph Wagen Antimonii,* Leipzig, 1604.

These works Thölden claimed were translated into German with great labor from original Latin manuscripts. It does not appear that Thölde ever gave any information as to the source of these alleged manuscripts, nor were the original manuscripts ever placed in evidence. The works attracted great attention, and were frequently republished, commentated, and translated into Latin and other languages. Other works were also published by various persons and ascribed to Basilius Valentinus.

The interesting fact was soon noticed that there was a strange similarity of many ideas, points of view, and even of modes of expression between this Basilius Valentinus and Paracelsus. Such were these resemblances that it was a reasonable assumption that one of these writers was dependent on the other for many facts and ideas. In the state of opinion and feeling toward Paracelsus at the beginning of the seventeenth century, it was natural that the orthodox medical faculties and practitioners should prefer to believe Paracelsus the borrower, rather than the newly discovered Basilius. The problem as to the personality of Basilius and the period of the literature which was attributed to him was much contested in the seventeenth century. The works themselves gave no definite information for identification, and the statements that the supposed author was a brother of the Benedictine order, that he was a native of the upper Rhine region, and had traveled in the Netherlands, England and Spain, was all that the works themselves indicated.

In 1675 Gudenus, in his history of Erfurt, stated that in 1413 a monk named Basilius Valentinus lived in St. Peter's cloister in Erfurt, a man deeply versed in medicine and natural science. This very definite information, though unsupported by any evidence to substantiate the statement, was evidently largely accepted as answering the doubts. To be sure it was soon recognized that the alleged date 1413 must be an error, because the works of Basilius Valentinus were found to refer to the use of antimony in metal type used in printing, a use known to be not earlier than the latter half of the fifteenth century; and they also contained references to the disease of syphilis under the name of *morbus gallicus,* which name was first used about the close of the fifteenth century. Elaborate search into the records of the Dominican monasteries in Germany and the records at Rome revealed no Dominican member of that name. At a somewhat later period the statement appeared and became generally accredited that in 1515 the Emperor Maximilian I had instituted a search to establish the existence and identity of the alleged Basilius Valentinus, though with negative results. The importance of this rumor consisted in this, that if Basil Valentine was known in 1515, he was evidently pre-Paracelsan. Prof. Kopp, who in his *History* in 1843 credits and repeats this rumor, in his *Beiträge* in 1875 calls attention to the baselessness of the statement, and states that, of the many manuscripts which he has consulted in the principal collections of Europe, there is none that refers to Basilius that is with any probability earlier than the seventeenth century.

However, the seventeenth and later centuries generally, while doubting the authenticity of the personality of Basilius Valentinus, accepted the assumption that the literature under his name was really written in the fifteenth century. Paracelsus was therefore suspected or believed to have had access to some copy of these works, and this explained the similarity of ideas and expressions. To be sure, there were skeptical critics, as Vincent Placcius, an early bibliographer who asserted that the real name of Valentinus was Thölden.[42] To this conclusion also came the anonymous author of the *Beytrag zur Geschichte der Hohen Chemie,* 1785.[43] Older skeptics as to the early origin of the Basilius Valentinus literature did not, however, prevent the general acceptance of the fifteenth century period for these works. Thus Gmelin, in his carefully compiled and conscientiously edited *Geschichte der Chemie,* 1797, accepts that period for the writing of the works, although dubious as to the alleged personality of the author. Kopp also in his *Geschichte der Chemie,* 1843–1847, accepts the fifteenth century as the probable date of these writings, though in his later *Beiträge* Kopp presents, very circumstantially, evidences for doubting that conclusion and for believing that the works are really of the seventeenth century. He hesitates, however, to attribute their authorship to Thölden, seeing no reason why this chemist should have wished to deceive the public. In his latest work, *Die Alchemie,* 1886,[44] Kopp hesitates no longer and, in view of all that he then had been able to learn, states that the reasonable interpretation of the situation is that Thölden must be considered as the author as well as publisher of the Basilius literature which he issued. Hoefer,[45] also states that the evidence is that there was

---

[42] Cf. A. E. Waite, *The Triumphal Chariot of Antimoy,* by Basilius Valentinus, p. xv, citing Placcius' *Theatrum Anonymorum et Pseudonymorum, Hamburg,* 1708.

[43] John Ferguson, *Bibliotheca Chemica,* II, p. 446.

[44] Kopp, *Die Alchemie,* pp. 29–32.

[45] *Histoire de la Chimie,* 1st ed., 1842, 1, pp. 453–454 and 2d ed., 1866, Vol. I.

no Benedictine monk of that name, and that the pseudonymous author belonged at the end of the fifteenth century
or perhaps even later. He states that none of these works
was printed before 1602–1604, and refers to certain manuscripts of the seventeenth century—French translations of
certain treatises. Yet he, like Gmelin and Kopp, accepts the
pre-Paracelsan character of the Basilius works and his history is written accordingly. With the impetus given by these
three important authorities on early chemical history, the
Basilius literature has in the later and brief histories of
chemistry generally been treated as pre-Paracelsan.

Since Kopp expressed his conviction that Thölden, from
1602 on, must be held as responsible for the authorship as
well as the publication of the Basilius works, the researches
of many scholars interested in the early history of chemistry, medicine, and pharmacy, have served only to confirm
the conclusion of Kopp; and the question may now be considered as settled beyond reasonable doubt that all the
facts and ideas contained in the literature of Basilius
Valentinus were compiled after all the works of Paracelsus,
Biringuccio, Agricola, Porta, Konrad Gesner, and many
lesser compilers and writers were in print. From this
viewpoint there is little if anything of importance, even
in the *Triumphal Chariot of Antimony,* that is not antici-
pated in these other writers. It is worthy of mention also
that this latter work was issued by Thölden in the same year
in which he published an edition of Alexander von
Suchtens[46] *De Secretis Antimonii.* The author of the *Beytrag* (1785) suggests that the *Triumphal Chariot of Antimony* may possibly have been compiled from this work.[47]

However this may be, and whatever sources besides
Paracelsus Thölde may have utilized, there is no doubt
but that his treatise brought together into one volume the
facts of the chemistry of antimony and its combinations,
and its uses in medicine in a form that made his book the

[46] *See* ante, p. 354.
[47] Ferguson, *Bibliotheca Chemica,* II, p. 417.

standard work on that subject for many decades. The work is, in so far as its chemistry is concerned, clear and comprehensible for its time. Its philosophy is medieval indeed, except that the *tria prima* of Paracelsus are utilized formally though without the interpretation of those terms so frequently emphasized by Paracelsus. Thus "Basilius" says:[48]

"Woe unto you, who neither understand nor care to understand my words! If you knew the meaning of fixation and volatility, and of the separation of pure and impure, you would cease from your foolish occupations and follow me alone. It is I, Antimony, that speak to you. In me you find mercury, sulphur, and salt, the great principles of health. Mercury is in the regulus, sulphur in the red color, and salt in the black earth which remains. Whoever can separate these, and then re-unite and fix them by art, without the poison, may truly call himself blessed; for he has the Stone, which is called fire, and in the Stone, which can be composed out of Antimony, he has the means of perfect health and temporal subsistence."

In his bitter and contemptuous arraignments of the conventional physicians, he imitates Paracelsus, so that it is not surprising that, if the seventeenth century accepted the Basilius literature as of the fifteenth century, it should also have concluded that Paracelsus was the imitator, having had access to some unknown copy of this early author's work. This long accepted theory, however, may be considered as finally abandoned, for all modern historians[49] who have studied into the literature of this period agree upon the post-Paracelsan character and on the fraudulent intent of the writer in ascribing to his alleged

---

[48] Waites' translation, p. 89.

[49] Authors who may be cited as expressing these convictions on the question are for example: Kopp, *Die Alchemie*, 1886, pp. 29-32; M. Berthelot, *Introduction a l'étude de la Chimie*, 1889, pp. 279, 280; Ferguson, *Bibliotheca Chemica*, 1906, I, p. 81, and II, pp. 445, 446; H. Schelenz, *Geschichte der Pharmazie*, 1904, p. 480; F. Dannemann, *Die Naturwissenschaften in ihrer Entwickelung und in ihren Zusammenhang*, 1910, I, p. 343; J. Campbell Brown, *A History of Chemistry*, 1913, p. 196; E. von Lippmann, *Entstehung und Ausbreitung der Alchemie*, 1919, p. 640; Karl Sudhoff, (*see ante*, p. 369).

Basilius Valentinus an earlier date than the dates of publication.

Briefly characterizing the contributions of the sixteenth century to the development of chemistry, we recognize the appearance of a new spirit of appreciation of the dignity and importance of the science and the decadence of the veneration for ancient and traditional doctrines and authorities, which so characterized preceding centuries. Many able and independent thinkers and workers contributed valuable additions to the literature of chemistry. Numerous experimental additions to chemical knowledge were made and many compilers and editors gave wide circulation to these advances in knowledge. None of these discoveries, to be sure, can be considered as epoch-making, but they were preparing the way and providing the material for future constructive developments. These advances in experimental chemistry were mainly in practical lines, in chemical processes and preparations and in their application to the chemical arts or to the arts of medicine and pharmacy. Great advances in the philosophy of chemistry we do not find, but in the newly established and more liberal attitude of thought toward traditional authority and ancient dogma, new ideas were not so universally felt to be necessarily dangerous heresies, merely because they were new.

The one important theoretical advance is the notion of the three Paracelsan principles constituting substances, mercury, sulphur, and salt, replacing in interest, to a great extent, the Platonic-Aristotelian concept of the four elements, and the more mystical Greek-Arabian concept of sulphur and mercury as the constituents of metals. Unquestionably, the appeal of the *tria prima* to the chemists of the period lay in its more comprehensible relation to experimental observation. Mercury, as the embodiment of whatever was merely volatile in the heat, sulphur of what burned away, and salt as the constituent which was fixed and nonvolatile and noncombustible, was a concept the

justification of which was to be found in experiment and
experience rather than in inherited dogmas.

The importance of the wide acceptance of the three principles lay not in any permanent value this theory possessed,
but in that this acceptance was a distinct break with ancient
authorities and appealed to experience for its justification,
and opened the way for further development on the basis
of wider future experience.

CHAPTER X

THE SEVENTEENTH CENTURY

The seventeenth century is marked by an increase in chemical experimentation and by a still greater independence of thought. Though the ancient authorities and theories found many stout defenders, there were many chemists who ventured new explanations of phenomena on the basis of an increasing knowledge of chemical facts and of observations. The main current of chemical thought and activity in the first half of the century was in the domain of their application in medicine and pharmacy, though metallurgy and other practical arts were not neglected. The most important of the chemical writers of that period were physicians, as Angelus Sala, Daniel Sennert, J. B. van Helmont, Sylvius de le Boë, Otto Tachenius, Werner Rolfinck, and others of less importance. J. R. Glauber was distinctively a metallurgist, though his activities also extended to chemical medicines. Robert Boyle, whose chemical publications appeared from 1660 on, is credited with being the first chemist of the century to study chemistry for its own sake, and not as an accessory to medicine or any chemical art.

Angelus Sala, born at Vicenza, went to Germany when young and passed his life there. He practised medicine first in Dresden, and later in Bavaria and Austria. Sala was interested in chemistry and an able experimenter. His works were published in 1647 by F. Beyer. He seems to have been a man of conservative judgment, free from vanity, which was rather the exception in chemical writers of his period. He criticized both Paracelsists and Galenists. Sala is credited with a number of notable observations and discoveries in chemistry, though it is difficult to know with certainty at this period whether a particular one of them is first discovered by him. Sala seems to be the first who prepared sal ammoniac synthetically. "If you place together one part of sal volatile from urine, with a proper proportion of spirit of salt, you will obtain a product resembling in all respects ordinary sal ammoniac."

Sala tried to prove that the precipitation of copper from vitriol solution by metallic iron was not, as was supposed by many, due to a transmutation of iron into copper but was due to the separation of copper present in the vitriol. He recommended lime and albumen from eggs for refining of sugar, promoted the use in medicine of the fused silver-nitrate (lunar caustic) and noted that oil of vitriol, or as he called it "spirit of sulphur," was produced by burning sulphur in moist air under a bell jar. Lemery improved this process by the addition of saltpeter (4 lbs. of sulphur to 4 ounces of saltpeter) and thus began the commercial manufacture of sulphuric acid, which had previously been obtained by distillation of vitriols or alums. Ward in England established a factory on this principle, and when in 1746 Roebuck and Garbill replaced the glass jar by lead-lined chambers, the price of sulphuric acid was reduced to perhaps a very small fraction of what it was before this development began.

Sala was also an important champion of the introduction of the chemical medicines. Sala's description of "fermentation," as an intimate movement of elementary particles which tend to group themselves in a different order to make new compounds, is evidence of a concept doubtless derived from the atomic theory of the Greeks, and differs from the concept of chemical action in the nineteenth century mainly by lacking qualitative and quantitative definition.

Daniel Sennert (1572–1637) of Breslau, a celebrated teacher of medicine at Wittenberg, was a follower of Paracelsus in the campaign for the chemical medicines, though independent in his judgment, so that he criticized Paracelsus and many of his followers in many things, especially for his belief in the existence of a universal medicine or

Alkahest. He also blamed the Galenists for resisting the progress of medicine by their obstinate conservatism. Robert Boyle manifestly considered Sennert one of the chief exponents of the theory of the "three principles" and cites him in the *Sceptical Chymist*.

Johann Baptista Van Helmont (1577–1644), born in Brussels, was the most prominent chemist of the first half of the seventeenth century. He came of a noble family, was educated in the conventional classical course at the University of Louvain, though he refused to accept the degree of Master of Arts on the ground that he was not qualified for that degree. He also attended courses in magic and mystical philosophy conducted by Jesuit teachers, and began the study of theology. An interest in natural science together with a missionary and unselfish impulse to the service of his fellows determined him to follow medicine as a profession, and in 1599 he took his doctor's degree at Louvain.

As a student of medicine he was strongly influenced by the works of Paracelsus, not only by his progressive ideas, but also by his transcendental and mystical philosophy. Van Helmont resembled Paracelsus, however, too much in his disregard of traditional authority to be a blind follower of Paracelsus. While he accepted some of the latter's most characteristic ideas, as the "Archaeus" presiding over functions of digestion, etc., he rejected some of his more prominent theories as, for example, the three principles of matter.

As chemist and as physician Van Helmont held a high place. He visited London in 1604–1605 and was received with honor, returning to Vilvorde near Brussels where he resided until his death in 1644. His complete works were first published by his son, Franciscus Mermurius Van Helmont, in 1648, and were often reprinted and translated.[1]

The chemistry of Van Helmont was largely developed with reference to physiological or medical functions, but not exclusively. His ideas of matter and its changes were

---

[1] The edition of his works accessible to the writer is that of Frankfurt, 1682.

largely original. For instance the Aristotelian theory of
the four elements as well as the Paracelsan concept of the
three principles were alike rejected by Van Helmont.[2] Of
the latter he says they are new inventions against the
truth of nature and fact. They are not primary constituents but are produced by the agency of fire and are hence
new entities which were previously nonexistent. Instead
of the four elements of Aristotle, he assumes that there
are two primitive elements, *air* and *water*. Of these two,
water, he says, is the more active, because from it all other
substances, except air, are produced; and into it all other
substances, excepting air, may be changed.

His reasons for the belief that water can be changed
into all other forms of matter, except air, are based upon
his own experiments and observations rather than upon
the authority of Thales, though it is not impossible that he
was influenced by the thought of that Greek philosopher.
Van Helmont calls attention to the fact that a great number
of substances, mineral, animal, and vegetable, yield water
on distillation or ignition, and he assumes that they are
partly converted into water. His widely cited experiment
upon the willow tree was his most impressive argument.

Van Helmont placed two hundred pounds of carefully
dried earth in an earthen pot, and planted in it a five-
pound willow. The pot was covered with a perforated plate
of tinned iron to guard against loss or gain of weight by
dust, etc. The pot was supplied with nothing but water,
either rain water or distilled water. After five years, he
removed the willow, weighed it again, finding one hundred
sixty-nine pounds and three ounces. The earth was dried
and again weighed and was found to have lost but two
ounces. Van Helmont concluded that one hundred and
sixty-four pounds of willow tree had been produced from
pure water.

If we recall that at that time there was no knowledge or
suspicion of the presence of carbon dioxide or of nitrogen

---

[2] *Opera Omnia*, 1682, p. 101.

compounds in the atmosphere, and that nothing was known of their relation to vegetation, and again if we consider the large number of substances obtained by the distillation of wood, we cannot regard Van Helmont's conclusion as anything but a reasonable deduction from the facts as he knew them. Furthermore, his conclusion was confirmed from certain facts of which he knew but had not personally experimented upon. Such was the often repeated account of certain springs which have the power of converting wood or charcoal into stone, a process usually interpreted at that time as a kind of transmutation. As charcoal is producible from water alone, and as charcoal can be changed to stone, this proved to Van Helmont that the stone also is materially water.[3]   Also the fact that fishes spend their lives in the water and obtain their development by things occurring in the water is interpreted by Van Helmont to mean that they, like his willow tree, are also ultimately produced from water.

Van Helmont experimented also with chemical processes in which various gases are produced and was the inventor of the term *gas* to distinguish these substances from ordinary air or from easily condensible vapors. Especially was our carbon dioxide, which he called *gas silvestre* or *spiritus silvestris,* the object of his attention. We have already noted that he derived this word from *chaos,* a term used by Paracelsus as a sort of generalized term for air.[4] Van Helmont burned sixty-two pounds of charcoal and found there was left one pound of ash. The other sixty-one pounds had disappeared as an invisible spirit. "This spirit, hitherto unknown, I call by a new name *gas,* which cannot be confined in a vessel nor reduced to a visible body, unless its seed be first destroyed."[5] And again he says, "Therefore with the privilege of a paradox and needing a name I have called this vapor *gas,* not very different from the *chaos* of the ancient secrets." He recognized that this gas

---

[3] *Op. cit.,* p. 104, 105.
[4] *See* ante, p. 323.
[5] *Op. cit.,* p. 102.

is heavier far than air, but more "subtle" than the vapor of water.[6]

Gas silvestre, he also found, was produced not only from burning charcoal, alcohol and other substances of organic origin, but by fermentation of wine and beer, by the action of acids or of distilled vinegar upon shells of crabs (lapides cancrorum), and occurs in some springs and subterranean caves.   The action of aqua fortis upon silver, of heat upon saltpeter, the burning of sulphur and the action of sal ammoniac and aqua fortis, all produce *gas silvestre.* Though he notes differences of odor or color in some of these products, he does not seem to consider it necessary to give them different names, they are all *gas silvestre.*   This is not very surprising for there was as yet no notion of their composition nor of any relation of odor or color to composition.

Van Helmont distinguished clearly between the uncondensible vapors that he calls gas, and those which are easily condensible, or are substances vaporized by heat but condensible in the cold to their original state.   He recognizes, as did the metallurgists at the time, the persistence of metals in their preparations or solutions.   He states that silver dissolved in parting water, though invisible is yet present in its previous essence, just as salt dissolved in water remains salt and can be recovered unchanged.   He also asserts that when glass is made from sand and alkali, the sand even in the fusion remains as such, being merely enveloped in the transparent glass.[7]

Van Helmont uses the terms acid and alkali, and refers to the effervescence of alkali with acid in the production of this *gas silvestre,* and uses the term saturation in a way that indicates some comprehension of limiting conditions. He devised (apparently about 1620) the term '*sal salsum*' to distinguish from *sal acidum* and *sal alkali* that which is now commonly called a neutral salt.

---

[6] *Op. cit.,* p. 69.
[7] Cf. Strunz, *F. J. B. Van Helmont,* Leipzig, and Wien, 1907, for an interesting study of his points of view and his work.

His notion of the cause of chemical action is quite mystical.   He supposes a "ferment," a formless and unsubstantial something implanted by the divine will in all substances, to be the thing which determines what the action and the products shall be.   The various functions of the body, for instance, take place under the initiative of the ferment, and under the guidance or direction of the Archaeus, a sort of resident spirit, the concept of the Archaeus being derived, somewhat modified, from the Archaeus of Paracelsus.[8]

Much of Van Helmont's theory and speculation is mystical and difficult to understand.   In the words of Professor Thos. Thomson: "The system of Van Helmont has for its basis the opinion of the spiritualists.   He arranged even the influence of evil genii, the efforts of sorcerers, and the power of magicians among the causes which produce diseases."

Toward the marvelous he was certainly credulous, and was sometimes thus led to endorse the facts of transmutation of the metals.   He relates for example that in 1618 he had received from an adept one fourth of a grain of a powder with which he himself had changed eight ounces of mercury into pure gold.[9]

Van Helmont's chemical experiments and his chemical theories exerted a powerful influence on the chemists of his century.   No chemist is cited more frequently nor with higher respect.   Yet, his theory of the two elements, air and water, did not, with many, replace the four Aristotelian elements, nor the three principles, though the latter had by this time been frequently elaborated into five, sulphur, mercury, salt (the active principles), and phlegm (water) and earth (the passive principles).   The suggestion of the rational and desirable term *gas* which he used, was ignored by his early successors.   Boyle, Boerhaave, and Priestley used instead the terms "artificial air," "factitious air,"

[8] *See* ante, p. 324.
[9] *Op. cit.,* "Vita Aeterna," p. 697 b.   *See* Kopp, *Alchemie,* I. Tl., p. 82.

or "different kinds of air," and it remained for Lavoisier and Macquer, a hundred and fifty years later, to reintroduce Van Helmont's convenient word *gas* to comprehend that class of bodies.

Johann Rudolph Glauber (1604–1670) was a German chemist, born in Karlstadt, who also shared the esteem of the seventeenth century, second only, perhaps, to Van Helmont. He was a man of very different training and experience from his elder Netherlands contemporary. He lacked classical training necessary at that time to the student of the chemical or medical literature. He wrote his many works in German, though later they were translated into Latin, and into French and English. He was an active chemical worker, and his experience in the field of the metallurgist and assayer is summarized in his really important work (for his time) on *New Philosophical Furnaces.*[10] This is a well organized book on the construction of various furnaces, illustrated with woodcuts of furnaces and accessory apparatus, and is an extensive treatise not only on furnaces, but also on the various methods of distillation and on the various kinds of "spirits, oils, and flowers" (that is distillates solidified to powders on cooling) of animal, vegetable or mineral sources, and on their uses in chemistry, medicine, and other arts. For the well described observations and many new experiments described here, Glauber well deserves to be remembered.

Glauber wrote many other works, and his *Opera Omnia Chymica* were published in 1658 in Amsterdam and in the same year in Frankfort.[11]

Much practical information of chemical value is contained in many of these works. Next in importance perhaps to the *Furni* was his treatise on the *Welfare of Germany—Des Teutschlands Wahlfahrt,* in which he discusses the natural resources of Germany. This work is a powerful appeal to German chemists and manufacturers to

---

10 *Furni Novi Philosophici*, Amsterdam, 1651.
11 Johannis Rudolphi Glauberi, Philosophi & Medici Celeberrimi, *Opera Chymica*, 2 Vols., Franchfurti am Main, 1658, 1659.

develop their natural chemical resources, and to become thereby less dependent upon Italy and France for many raw materials. This work was published in six parts.[12]

The chemical philosophy of Glauber is much the same as that of Paracelsus, whom he esteemed highly, and of whose works (published in German) he was a student. Of the constitution of matter, he says:

"The principles of vegetables are water, salt, and sulphur, from which also the metals are derived, not from running mercury, as many of you think, for that mercury is a special metal and from these same three principles as other metals and vegetables, namely, from water, salt, and sulphur, which are found on decomposing (Anatomisirung) them."[13]

This substitution of "water" for Paracelsus' "mercury," finds its analogy in the practice of other contemporary chemists in substituting the term "spiritus" for "mercury" to represent the principle of volatility.

That Glauber, in spite of his many valuable improvements in metallurgy and other branches of practical chemistry, and his many clearer descriptions of processes,[14] was something of the charlatan, is quite evident. The wonderful and absurd claims he makes for the virtues of his *sal-mirabile,* and the quarrels he had with his contemporaries on account of the exaggerated values he assumed for the secret remedies he sold, make it evident that he was not free from practices very common at his time, and not unknown to-day. The name "Glauber's salt," still much in use, especially in medicine, as applied to crystallized sodium sulphate, is a reminder of the great virtues which Glauber assigned to his *sal-mirabile* or wonderful salt. This *sal-mirabile* is discussed at great length in his treatise on *De Natura Salium,* and in *Miraculum Mundi.* He does not claim that the discovery of it is his own, but

---

[12] Only the first two are in *Opera Chymica* of 1659, the others being issued in Amsterdam between 1659 and 1661.

[13] "De Natura Salium," *Opera Chymica,* p. 452.

[14] Gmelin, *Geschichte der Chemie,* I, pp. 625–657, records a large number of his observations.

believes it to be the rediscovery of the *sal-enixum* of Paracelsus, for which also the properties of universal solvent
and medicine were claimed. His general description of the
properties, uses, and preparation of his *sal-mirabile* are
written much in the style of the modern vendor of secret
nostrums;[15] and it is by no means clear that he intends to
describe its preparation and properties so clearly as to be
understood by his competitors. Here are the directions
as given:[16]

"It should be known that my *sal-mirabile* may be separated and prepared from all common salts, but from some
more easily than from others. For not only common cooking salt, but also saltpeter, alum, and vitriol can yield it.
But because alum and vitriol possess many sulphureous
and mineral qualities which are troublesome to separate,
and saltpeter is burning and volatile, therefore we had better leave these salts alone and prepare our *sal-mirabile*
only from common cooking or kitchen salt, separate from
it its earthiness by the aid of fire and water, and use it to
the honor of God and the service of our neighbor as we
know or can: and first:

"Concerning the external form, color, taste, and odor
of the *sal-mirabile*.

"This salt when well prepared, appears like frozen water
or ice, crystallizing much like saltpeter, quite clear and
transparent, melting easily on the tongue like ice; in taste
not sharp, but peculiarly saltish and somewhat astringent,
not decrepitating like common salt when laid on glowing
charcoal, nor inflaming like saltpeter, but may be ignited
without giving off odor, which takes place with no other
salt."

These properties of the *sal-mirabile* agree with those of
sodium sulphate, though the description of the preparation
of the salt is vague—from common salt by aid of fire and
water. Nevertheless, all later writers identify his *sal-
mirabile* with the salt now known as sodium sulphate. The
powers that Glauber attributes to this salt are absurdly

---

[15] *Opera Chymica*, I, pp. 495–502.
[16] Glauber, *op. cit.*, I, p. 495.

exaggerated. He devotes great space to the enumeration of its powers and virtues. Professor Thomson says:[17]

"In the treatise called *Miraculium Mundi* his chief object is to write a panegyric on sulphate of soda, of which he was the discoverer, and to which he gave the name of *sal-mirabile*. The high terms in which he speaks of this innocent salt are highly amusing, and serve well to show the spirit of the age, and the dreams which still continued to haunt the most laborious and sober minded chemists."

Though Glauber's writings on chemical philosophy followed the obscure, medieval transcendentalism of previous centuries, and though he elaborately advertised the remedies he dispensed, nevertheless, as a practical chemist, and as a careful and reliable recorder of the results of the experiments of himself and others, Glauber set a new landmark in technical chemistry, and insured for himself a deserved place in the history of the arts of chemistry.

Glauber practised chemistry and medicine in many cities of Germany, Austria, and Switzerland,—Salzburg, Vienna, Basel, Frankfort, and Cologne. In 1648, he removed to Amsterdam, where he spent the remainder of his life, dying in 1670.[18] In Amsterdam, his first book—on the furnace—had been printed for the first time in 1651.

The two most prominent representatives in the middle of the seventeenth century of the iatro-chemical impetus so vigorously inaugurated by Paracelsus and his followers, and so strongly developed by the efforts of Libavius, Sala, Glauber, and Van Helmont, and others, are, perhaps, Franciscus Sylvius de le Boë (1614–1672) and his enthusiastic supporter, Otto Tachenius (ca. 1620–1690). Both were primarily physicians, but experienced in chemistry and both inclined to make the theory and largely also the practice of medicine depend upon chemical analogies.

Sylvius was born in Hanau, whither his Netherlands parents had taken refuge during disturbances in their home country. He received his first schooling at Sedan and

[17] *History of Chemistry*, I, 229.
[18] Ferguson, *Bibliotheca Chemica*, I, p. 329, for reasons for 1670 as against usually cited date of 1668.

at Leyden, eventually receiving his doctor's degree at Basel. As a practising physician, he first resided at Hanau, later at Leyden, finally settling in Amsterdam, where he achieved a high reputation as a skilful and careful physician and a scientist. This reputation finally brought him the position of professor of medicine at Leyden, in 1658, a position he occupied with great prestige until his death.

The writings of Sylvius were first published between 1659 and 1674, all on medical subjects primarily, unless we except his brief treatise on Chemical Medicines,[19] which is practically confined to the various medicinal compounds of antimony—"flowers," "liver," "regulus," "glass," antimony diaphoreticum, butter of antimony, the latter made by distilling crude antimony (that is sulphide) with mercury sublimate (that is mercuric chloride). These compounds were, however, all known by 1600 and well summarized in pseudo-Basilius's (Thölden's) *Currus Triumphalis Antimonii*.

Sylvius was profoundly influenced by Van Helmont in his theories of the chemical functions of the organism, and the authority of his position and reputation gave much weight to his chemical speculations. He was also a well informed chemist for his time.

His tendency was the same as that of nearly all medical chemists of his period—to accept a plausible analogy instead of waiting for more basis in facts for his conclusions. Especially notable was his attempt to make the chemical function of the body depend on action between acids and alkalies. So for instance he said that in the right auricle and ventricle of the heart, the blood in its circulation meets the blood charged with bile. The mixture of these two effervesces on contact like iron and oil of vitriol. This is the source of animal heat. The function of respiration he concludes is to temper the heat produced by this effervescence, and expiration from the lungs carries away the vapors produced by the effervescense.

---

[19] Sylvius, *Opera Medica*, Venice, 1696, pp. 576, 577.

Diseases generally are, in his view, due to some super-acidity or superalkalinity, acids are generally the causes of stomach disease because alkaline medicines are more frequently the remedies. The plague is caused by sal-volatile, because its injection into the veins causes symptoms similar to the plague. Acid remedies are therefore the best remedies.

Otto Tachenius, younger partisan of the medical phil-osophy of Sylvius, was born in Herford, Westphalia, and studied the trade of apothecary at Lemgo. Driven thence because of some theft,[20] he served as apothecary's assistant in Kiel, Danzig, and other German cities, then going in 1644 to Italy, and there studying medicine, eventually taking his degree of M. D. at Padua, and remaining in Venice where he was still living in 1699.[21]

Among works on medicine, the most interesting from the chemical point of view are his *Hippocrates Chymicus* (1668) and his *Hippocraticae Medicinae Clavis* (1668), both republished in many editions and in English transla-tion. While the immediate aim of these volumes was to prove that the medical philosophy of Hippocrates really meant nothing essentially different from the then prevalent chemical medical theories (an object as may be imagined, only attained in a purely scholastic sense, if at all), yet they are a treatise on the chemical philosophy of medicine and upon chemical medicines. His philosophy is similar to that of Sylvius, especially in the relative importance of acids and alkalies. Indeed, his statements are even more extravagant than are those of Sylvius. Thus while Tache-nius, not unreasonably, says that "all salts are composed of an acid and an alkali," [22] yet he says also "But we for our greater knowledge and light call these two Hippocratic principles acid and alkali, because from these two universal principles are made all things in the universe," [23] and yet

[20] Kopp, *Geschichte der Chemie*, I, p. 140.
[21] Ferguson, *Bibliotheca Chemica*, II, p. 424.
[22] Otto Tachenius, *Hippocrates Chimicus*, 3d ed. Lugd. Bat. 1671, p. 8.
[23] Otto Tachenius, *Hippocraticae Med. Clavis*, 3d ed. Lugd. Bat. 1671, p. 2.

again he says, "After showing above that there is nothing
in the universe but alkali and acid, from which nature com-
poses all things," [24] etc.   This theory Tachenius was pro-
pounding in Italy as Sylvius was doing in Holland.   Tachenius appears to have been well versed in the chemical
knowledge of his time.   We may note his statement that
lead gains one tenth in weight when roasted to red heat
and is reduced to its previous weight upon reduction,[25] a
very accurate statement for his time.   He accounts for this
increase by the absorption of "acids" from the fuel or
wood ("acidis lignis").

The latter half of the seventeenth century is marked by
the activity of a considerable number of able investigators
and writers on chemistry, notable among whom are Nicolas
Le Febre (or Le Febure), (?-1674); Christopher Glaser
(died about 1670–1673); Robert Boyle (1627–1691); Thomas
Willis (1621–1675); Johann Kunkel (1630–1702); Johann
J. Becher (1635–1682); John Mayow (1645–1679); Nicolas
Lemery (1645–1715); and Wilhelm Homberg (1652–1715).
All these men contributed to the increase of knowledge of
the facts of chemistry by their researches and publications,
which appeared from about 1660 to the close of the
century.

We may note in general a more rational discussion of
chemical problems, and, while correct solutions were often
lacking, thinkers were less dominated than their predecessors by the extravagant and imaginative conceptions
of the past.   In this period also were founded the influential learned scientific societies, the "Academia del Cimenti"
of Florence, founded in 1657; the "Academia Naturae
Curiosorum" of Vienna, 1652; the Royal Society of Great
Britain, 1662 (formed by the association of two local societies of Oxford and of London); and in 1666, from a
similar amalgamation of local societies, was established in
Paris the "Académie Royale des Sciences."   The influence
of these societies, where scholars could exchange and discuss

---

[24] Tachenius, *loc. cit.*, p. 42.
[25] Tachenius, *Hippocrates Chemicus*, 3d ed. p. 167.

their knowledge and speculation, is not easy to overestimate.

No one of the chemists of this period exerted so profound an influence upon the development of chemistry toward a real science as did Robert Boyle. The particular circumstances which conspired to give him the place were, in the first instance, as Kopp emphasizes, that he was "the first chemist whose efforts are employed primarily in the noble impulse to investigate nature." In other words the facts of nature interested him rather than their applications to medicine or any of the arts. More important was his mental attitude, unique in his time, toward the solution of the problems he studied. He approached these problems singularly unbiased by previous authorities or speculations, and was able to preserve the attitude of really scientific skepticism toward generally accepted theories. These qualities, with his excellent preliminary education, his untiring energy, his modesty, made effective by ample means and leisure for experiment, his lack of dogmatism, and the respectful consideration which he gave to the views of opponents, gave him a unique place in his generation.

Robert Boyle, seventh son and fourteenth child of Sir Richard Boyle, Earl of Cork and Lord High Treasurer of Ireland, was born in Lismore in the province of Munster, January 25, 1627. He tells us that he "was born in condition that neither was high enough to prove a temptation to laziness nor low enough to discourage him from aspiring." His early education was careful and thorough. He spent four years at Eton and later studied with private tutors, and at twelve years of age he was sent to Europe and remained there for six years, studying, with masters at Geneva and Florence, French and Italian, mathematics, geography, and physical accomplishments—fencing and dancing. In Florence, he tells us, he spent the time, spared from his language study, in reading modern history in Italian and "the new paradoxes of the great star-gazer Galileo, whose ingenious books, perhaps because they could not be so otherwise, were confuted by a decree from Rome."

Galileo, indeed, died while Boyle was in Florence, January 8, 1642.

Boyle's return was delayed until 1644 by the Irish Rebellion which embarrassed for a time the affairs of the English Lord Treasurer, his father, and determined Robert to go on to their English estate at Stallbridge, where he lived until 1650, applying himself devotedly to his researches into natural philosophy and chemistry. In 1654 he removed to Oxford, where he continued his scientific work and was associated with the framers of the Royal Society in 1662 of which he was President from 1680 until his death in 1691.

The scientific publications of Boyle began in 1660 with his extensive treatise on the *Spring of the Air,* in which he made use of an improvement on the air pump discovered by Otto von Guericke,—his "newly discovered pneumatic engine." This was a very important contribution to the physics of air, in the course of which he announced the generalization still called "Boyle's Law" and sometimes called "Marriott's Law," though Marriott announced it some seventeen years later.

In 1661, appeared *Certain Physiological Essays and other Tracts,* largely chemical, and the *Sceptical Chymist,* which doubtless was most influential of all his works upon chemical thought. This work was an elaborate analysis and criticism of the two then prevalent theories of the elementary composition of substances, the Peripatetic or Aristotelian theory of the four elements, air, fire, water, earth, and the Spagyric or Paracelsan concept of the three principles, mercury, sulphur, and salt, and of the variations of the latter theory which had arisen. Boyle was the first to challenge the validity of both these systems. He saw no reason, and asks to be shown any reason, for supposing that any four or three or five substances are the elements that enter into the composition of all matter. Though the first edition of the Sceptical Chymist was issued anonymously, the work attracted wide attention and the authorship soon became known. The second, also unsigned, edition was issued in 1679[26] with supplementary articles, though in the mean time the work had been translated into Latin and "reprinted many times" before this second edition appeared.

The influence of this work was epoch-making and did more than any other work of the century to arouse a truly critical spirit of scientific logic in chemical thinking. The book is written in the form of a discussion among a group of scientific friends, Carneades representing the sceptical chemist; Themistus, the exponent of the Aristotelian or Peripatetic four elements; Philoponus, the defender of the three Paracelsan principles, Eleutherius an independent and open-minded participant, and "I," the anonymous reporter of the conversation.[27]

Themistus first presents the customary arguments for the truth of the four elements, to which Carneades replies at length. In his summing up, for example, he says:

"I consider then [says Carneades] in the next place that there are divers bodies out of which Themistus will not prove in haste that there can be so many elements as four extracted by the Fire. And I should perchance trouble him if I should ask him what Peripatetic can show us (I say not all the four elements, for that would be too rigid a question, but) any one of them extracted out of gold by any degree of Fire whatsoever, etc.

"The next argument [continues Carneades] that I shall urge against Themistus's opinion shall be this. That as there are divers Bodies whose analysis by Fire cannot reduce them into so many heterogeneous substances or ingredients as four; so there are others which may be reduced into more, as the Blood (and divers other parts) of men and other animals, which yield when analyzed five distinct substances, Phlegm, Spirit, Oyle, Salt, and Earth," etc.[28]

The doctrine of the three principles is discussed much more elaborately

---

[26] The title page is dated 1680.

[27] *The Sceptical Chymist* is easily accessible in the form of a volume of the popular series entitled "Everyman's Library."

[28] Boyle, *Sceptical Chymist*, 2d ed., 1680, pp. 32–34.

"because the Chymical Hypothesis seeming to be much more countenanced by experience than the other, it will be expedient to insist chiefly upon the disproving of that. Especially since most of the Arguments that are employed against it may, by a little variation, be made to conclude at least as strongly against the less plausible, the Aristotelian Doctrine."

Carneades begins this discussion by stating four propositions as a preliminary basis of the discussion. These are of interest as formulating Boyle's hypothesis of the constitution of matter in general, and his notion of what should constitute an element. These propositions are as follows:[29]

"1. It seems not absurd to conceive that the first Production of mixt Bodies, the Universal Matter whereof they among other Parts of the Universe consisted, was actually divided into little Particles of several sizes and shapes variously moved.

"2. Neither is it possible that of these minute Particles divers of the smallest and neighboring ones were here and there associated into minute Masses or Clusters, and did by their Coalitions constitute great store of such little primary Concretions or Masses as were not easily dissipable into such Particles as composed them.

"3. I shall not peremptorily deny that from most of such mixt Bodies as partake either of Animal or Vegetable Nature, there may by the Help of the Fire be actually obtained a determinate number (whether, Three or Four or Five, or fewer or more) of Substances worthy of differing Denominations.

"4. It may likewise be granted, that those distinct Substances, which Concretes generally either afford or are made up of, may without very much Inconvenience be called the Elements or Principles of them."

It appears from the above that Boyle entertains the hypothesis of a universal matter, the concept of atoms of different shapes and sizes, and the possibility of existence of substances that might properly be called elements, though

---

[29] Boyle, *op. cit.*, pp. 36–46.

in his extended discussion of the problems he does not venture to assert that any known substance can safely be asserted to be such an element, though he knows, for example, no fact that would prove that gold, for instance, might not as well be called an element as anything else.

The atomic theory as originally conceived by Democritus and Epicurus, developed by Lucretius, and resurrected by Gassendi from about 1647 on, was doubtless the source from which Boyle derived his ideas on this subject, as he cites both Epicurus and Gassendi. Boyle, however, in the above proposition carefully avoids any dogmatic assertion of these hypotheses. It is plain, however, that these atoms or "corpuscles" as he calls them are a constant element of his thought. In part six (an appendix) to the *Sceptical Chymist,* he states more distinctly his definition of a chemical element. Carneades says:[30]

"And to prevent mistakes, I must advertize you, that I now mean by Elements, as those Chymists that speak plainest do by their Principles, certain Primitive and Simple, or perfectly unmingled bodies; which not being made of any other bodies, or of one another, are the Ingredients of which all those called perfectly mixed Bodies are immediately compounded, and into which they are ultimately resolved."

This definition was as accurate a definition as the knowledge of the time permitted; and was indeed the same as given by Lavoisier and by later chemists until the development of the phenomena of radioactivity afforded a more intimate concept of the nature of the element. Neither Boyle nor his contemporaries ventured to assert that any known substance was such an element, and the subsequent rise and acceptance of the Phlogiston Theory tended to postpone any recognition of the elementary character of even such metals as gold or silver, until that theory was abandoned.

It is not necessary here to summarize the mass of evi-

---

[30] Boyle, *op. cit.,* p. 354.

dence presented by Boyle—experiments, observations, and logical deduction—to show the lack of basis in fact or reason for the theory of three, or five, principles or elements. It will be sufficient to explain that theory in the form in which it was generally accepted in the latter half of the seventeenth century. This can best be seen in the series of text books on chemistry most popular and authoritative by Nicolas Le Febure (or Lefebre) (first edition 1660), Christofle Glaser (first edition 1663), and Nicolas Lémery (first edition 1675). These chemists occupied successively the position of Chemist at the Jardin des Plantes at Paris. All these authorities present essentially the same explanation of the theory in question. The presentation by Nicolas Lémery in his *Cours de Chymie* is the best and clearest. Lémery's work marked a distinct advance on any preceding works as a general text on chemistry. Lémery (1645–1715) was himself an able chemist, and he was free from dogmatism and egotism. His *Cours de Chymie* passed through some fourteen editions in Paris alone, through four editions in English, was translated into Latin, Italian, German, and Spanish, and was the most authoritative text in general chemistry for more than fifty years.

The theory of the principles of Le Febure, Glaser, and Lémery varies from the original Paracelsan theory in that it recognizes, beside the original three active principles, two passive principles. Lémery presents the theory thus:[31]

"The first principle that can be accepted in the composition of mixed bodies is a mineral spirit, which being distributed everywhere, produces various things according to the different matrices or pores of the earth in which it may be entangled: but as this is somewhat metaphysical, and as it is not subject to the senses, it is well to establish the sensible principles of it. I will report those in common use.

"As the Chemists in analyzing various mixtures, have

---

[31] The passages here are translations from the ninth Paris edition of the *Cours de Chymie*, 1701.

found five kinds of substances, they have concluded that
there are five principles of natural substances, water, spirit,
oil, salt, and earth. Of these five, three are active, spirit,
oil, and salt, and two passive, water and earth. Those are
called active because being in active motion they cause the
activity of the compound. The others are called passive
because being in repose they serve only to restrain the
vivacity of the active ones. Spirit, which is called Mercury is the first of the active principles which we obtain
in making the analysis of a compound.[32] This is a subtle
substance, slightly penetrating, which is in livelier agitation than any other principle. It is that which makes compounds grow in greater or less time according as it occurs
there in greater or less quantity: but also by its too violent
motion, it follows that bodies in which it abounds are more
subject to corruption: this is what is noticed in animals
and plants. On the contrary the greater number of minerals where it is present in small quantity seem incorruptible. It cannot be obtained pure from compounds, for
either it is mixed with a little oil which it carries with it,
and then is called volatile spirit, such as the spirits of
wine, of rosemary, of ginger, or else it is entangled in salts
which restrain its volatility, and then it may be called fixed
spirit, such as the acid spirits of vitriol, of alum, of salt,
etc.

"Oil, which is called sulphur, because it is inflammable,
is a substance mild, subtle, unctuous, which passes off after
the spirit. It is said to cause the variety of colors and
odors. According to its distribution in bodies it causes
their beauty or their ugliness; binding the other principles.
It also allays the sharpness of salts and by stopping the
pores of the compound, it prevents decay from seizing them
either from too much moisture or by the cold; this is why
some trees and plants which abound in oil last longer than
others in verdure, and resist entirely the severity of bad
weather. It is always recovered impure from compounds,
for it is either mixed with spirits, like the oils of rosemary
or lavendar which swim upon water or it is filled with salt
which it entangles in the distillation, as with the oils of

[32] Distillation is here meant.

box, guayacum, and cloves which are precipitated to the bottom of water because of their heaviness.

"Salt is the heaviest of the active principles, it is also commonly recovered last: this is an incisive and penetrating substance, which gives to a compound solidity and heaviness, it preserves it from decay and excites various flavors according as it is differently combined. . . . . .

"The water called Phlegm is the first of the passive principles: it passes over in the distillation before the spirits when these are fixed, or after them when they are volatile. It never passes off pure and there always remains some impression of the active principles. This it is that causes it to have, ordinarily, more detersive power than is possessed by natural water. It serves to dilute the active principles and to moderate their agitation.

"Earth, which is called *Caput Mortuum* or *Damnatum,* is the other passive principle, it can no more than the others be separated pure, for it always stubbornly retains some Spirits, and if, after being so far as possible deprived of these, it is left long exposed to the air, it takes them up anew."

That this theory contains very much that is not established as a scientific consequence of any known facts is evident, and Boyle's arguments to show upon what inadequate basis of fact and logic it was sustained were very impressive to open-minded readers. Boyle's discussions generally are very clear, illustrated with a wealth of examples from known facts and experimental evidence. His style, however, is often almost painfully prolix.

Another theory which at this period had been developed to an unwarranted extent, and which also became a target for Boyle's logical analysis, was that of acids and alkalies. We have seen the extent to which, under the authority of Van Helmont, Sylvius de le Boë, and Tachenius, these concepts had been carried. It may be recalled that the ancients did not differentiate sharply between the acid of vinegar, *acetum,* and other acid juices. The Arabic word alkali, was derived from Kali, the name of a plant (a glasswort) the ashes of which were leached to obtain the salt (carbonates of potassium and sodium) used in soap-making and glass-making. The application of the term had gradually been extended to mean any substance which effervesced with an acid, and finally it came to be understood that any effervescence was evidence of reaction between acid and alkali. Thus Sylvius states[33] that effervescence always shows the coming together of an acid and alkali; and Lémery states that an alkali may be recognized by the effervescence which occurs when an acid is poured upon it.

The seventeenth century concepts of acid and alkali are well given by Dr. Bertrand in his book devoted to that subject in 1683.[34] In this work of 359 small octavo pages he discusses very fully the current concepts referring especially to Van Helmont and Tachenius. He is by no means an extremist like Tachenius or Sylvius. His concepts do not differ essentially from those expressed more briefly by Lémery in his *Cours de Chymie.*

Bertrand explains that in endeavoring to define the "nature of these two salts, I shall not imitate the process of some who content themselves with saying in general that an acid is that which ferments [that is, effervesces] with an alkali, and that an alkali is that which absorbs the acid. These notions are too vague and obscure.

"I say that an acid is a liquid body composed of small firm and pointed particles, slightly resembling very fine and delicate needles. This idea accords exactly with all the actions of which we see acids to be capable. For by its particles of such a shape it excites a prickling when applied to the tongue, and is fitted to cause effervescence when mixed with certain bodies which it penetrates, and of which it violently disturbs the particles. Some of these it dissolves by disturbing and breaking up the tissues in penetrating their pores, and others it coagulates by becoming entangled in their branching and irregular particles, as occurs with milk. Moreover, as acids are not all entirely

---

[33] In his *Disputatio de Chyli Secretione,* 1659.
[34] *Réflections nouvelles sur l'acide et sur l'alcali,* par M. Bertrand, Docteur en Médecine Agrégé au Collège des Médicins de Marseille, Lyon, 1683.

alike, as their particles may have different sizes and points more or less fine, it should follow that they ought not to dissolve every sort of body indifferently, but only those the pores of which are accommodated to their shapes, and the textures of which cannot resist their force and activity. We see therefore that aqua fortis which dissolves silver cannot dissolve gold, and that distilled vinegar while it dissolves lead cannot act on mercury.

"Alkali, on the contrary, should be a solid earthy body the particles of which have between their junctions pores of different structure. It is for this reason it can be dissolved by an acid, and that it effervesces with it and blunts its points: that it cleans cloth and is capable as alkali of some other actions that experience teaches us to recognize. But it is only by reason of this particular contexture that it accomplishes these."

Bertrand does not agree with Tachenius and others that every substance contains an acid or an alkali, nor does he deem it necessary to assume that every body which ferments with an acid is necessarily an alkali or contains an alkali. There may be structural peculiarities of substance other than those pertaining to alkalis.

In 1676 Robert Boyle published a paper in which he criti-
other than those pertaining to alkalies.

"I cannot acquiesce, [says Boyle] in this hypothesis of alcali and acidum, in the latitude wherein I find it urged and applied by the admirers of it, as if it could be usefully substituted in the place of matter and motion.

"And first it seems precarious to affirm that in all bodies, or even in the sensible parts of all mixeds, acid and alcalizate parts are found: there not having been, that I know, any experimental induction made of particulars anything near numerous enough to make out so great an assertion. . . . Some spagyrists when they see aqua fortis dissolve filings of copper, conclude from thence that the acid spirits of the menstruum meet with an alcali upon which they work; which is but an unsafe way of arguing,

---

35 *Reflections upon the Hypotheses of Alkali and Acidum*, Opera, 1744, III, pp. 603-608.

since good spirits of urine, which they take to be a volatile alcali [that is ammonia or ammonium carbonate] and which will make a great conflict with aqua fortis, will, as I have elsewhere noted, dissolve filings of copper both readily enough and more genuinely than the acid liquid is wont to do . . . and yet if one should urge that quicksilver readily dissolves gold in amalgamation, he may expect to be told, according to their doctrine, that mercury has in it an occult acid, by which it performs the solution: whereas it seems much more probable that mercury has corpuscles of such a shape and size as fit them to insinuate themselves into the commensurate pores they meet with in gold, but make them unfit to enter readily the pores of iron to which nature has not made them congruous. . . . It seems a slight and not philosophical account of their nature (that is, of acids), to define an acid by its hostility to an alcali, which, they will say, is almost as if one should define a man by saying that he is an animal that is at enmity with the serpent, or a lion that he is a four footed beast that flies from a crowing cock.''

With respect to the phenomenon of effervescence as a sign of action between acids and alkalis, Boyle says:

''And as for the other grand way that chemists employ to distinguish acids and alcalies, namely by the heat commotion and bubbles that are excited upon their being put together, that may be no such certain sign as they presume, they having indeed a dependence upon particular contextures, and other mechanical affections, that chemists are not wont to take any notice of. For almost anything that is fitted variously and vehemently to agitate the minute parts of a body will produce heat in it, and so, though water be neither an acid nor an alcalizate liquid, yet it would quickly grow very hot, not only with a highly acid oil of vitriol, but (as I have more than once purposely tried and found) with the fiery alcalizate salt of tartar'' (that is, potassium carbonate).

Of the notions of sympathy and antipathy in connection with chemical actions he expresses himself:

''I am dissatisfied with the very fundamental notion of this doctrine, namely a supposed hostility between the tribe of acids and that of alkalies, accompanied, if you will have it so, with a friendship or sympathy with bodies belonging to the same tribe or family. For I look upon amity and enmity as affections of intelligent beings; and I have not yet found it explained by any, how those appetites can be placed in bodies inanimate and devoid of knowledge or of so much as sense,'' etc.

In his conclusion Boyle voices his scientific spirit in saying:

''Nor do I pretend by the past discourse, that questions one doctrine of the Chemists, to beget a general contempt for their notions, and much less of their experiments. For the operations of chemistry may be misapplied by the erroneous reasonings of the artists, without ceasing to be themselves things of great use, as being applicable as well to the discovery or confirmation of solid theories, as the production of new phenomena, and beneficial effects. And though I think that many notions of Paracelsus and Helmont and some other eminent Spagyrists are unsolid, and not worthy of the veneration that their admirers cherish for them, yet divers of the experiments which either are alleged to favour these notions or on other accounts are to be met with among the followers of these men, deserve the curiosity, if not the esteem, of the industrious enquirers into nature's mysteries.''

Just as Boyle in his *Sceptical Chymist* offers no scheme of elements to replace the discredited Aristotelian and Paracelsan scheme, so also here he presents no definitions or criteria of acid and alkali as satisfactory to his judgment and experience.

Theories of combustion, as they existed at the end of the seventeenth century and before the advent of the phlogiston theory of Stahl, may perhaps be advantageously considered here.

With the ancients, following Plato and Aristotle, burning was interpreted as the passing off of the element fire from its compounds. When the alchemistic notion of sulphur and mercury as constituents of metals and other substances became prevalent, combustion was understood to be the burning of the sulphur. Paracelsus in his extension of this theory to the three principles also says, "all that burns is sulphur." So also says Lémery in his text book (1675), "Sulphur is the only principle which takes fire."

That air was necessary to maintain combustion was a fact of common knowledge from ancient times, but the function of air in combustion other than to carry off the heat and "sulphurous vapors" seems to have received no attention from chemical philosophers before the sixteenth century. And the earliest speculations on this matter seem to have been excited by the fact that in the case of some metals, their burning or calcination was accompanied by a gain in weight. In the case of ordinary combustibles, the volatile and gaseous product escaped into the air, and the unburned residue was lighter than the original material. Why, on the other hand, should lead or tin or antimony gain in weight when fire or sulphur departed? Why should the calx be heavier than the metal? No methods were then known for collecting, isolating, and weighing the gaseous and volatile product of combustion, and it was assumed very naturally that the burning of these metals was exceptional in increasing the weight which existed before burning. Eck von Sulzbach, about 1490, seems to be the first who records the increase of weight of metals in calcination and he describes his experiments on mercury and quicksilver amalgams when calcined. Even the pseudo-Geber (about 1300) speaks of tin as acquiring weight in heating (in magisterio) and says that when obtaining silver from lead, the lead does not keep its own weight but is changed into a new weight.

Speculations as to the cause of this phenomenon are various and numerous in the sixteenth and seventeenth centuries. Thus Cardanus (1553) notices the increase in the weight of lead on calcination and attributes it to the loss of celestial fire. By the departure of this life giving principle or soul the metal becomes heavier, and the notion seems to be that the metal is buoyed up by the fire element. Later theorists, the phlogistic philosophers, thought of phlogiston, or the fire element, as having negative weight. Cardanus's contemporary, Scaliger, thought the gain in weight must be due to consumption or vanishing of the element air enclosed in the metal, and that its loss left the metal denser, confusing thus specific weight with absolute weight. Le Febure (1660) thought the increase due to some material of the light or heat of the flame. Tachenius (1666) attributed it to the fixation in the calx of acids from the flame of the fuel. He determined the gain in weight of lead roasted to minium very closely at one tenth the weight of the original lead and showed that by reduction the lead returned to its original weight. The French physicist and chemist Duclos experimented on the change in weight when antimony is oxidized in the heat of the burning glass, and attributed the gain to the absorption of sulphur from the air.

Christophle Glaser attributed the gain in weight to "corpuscles of fire" which are incorporated with the calx.[36]

Becher (1635–1682), whose ideas of combustion were later elaborated by Stahl into the phlogiston hypothesis, in 1669 discussed the gain in weight of metallic calxes, and pronounced the opinion that the only source of this must lie in the fixing of some fire material which was the only thing which could pass through the glass of the apparatus— and this material of the fire when fixed by the calx caused the gain in weight. This opinion being reinforced by such authorities as Robert Boyle and Nicolas Lémery was quite generally accepted. Boyle in 1673 published a series of tracts upon this subject under the titles of *New Experiments to make the Parts of Fire and Flame Stable and Ponderable, Additional Experiments about arresting and weighing of Igneous Corpuscles, A Discovery of the Perviousness of Glass to Ponderable Parts of Flame.*

In these treatises Boyle subjects many metals, calxes of

---

[36] *Traité de Chimie*, 4th ed., 1676, p. 109.

metals, and other substances to the action of heat in glass or metal, usually in the presence of more or less air, and in all cases finds a greater.or less increase of weight. In the case of calxes of the metals, this gain of weight which he finds, must have been due either to the fact that they were originally incompletely calcined or to their absorption of carbon dioxide, or sulphur dioxide from the fuel gases. Boyle also took fresh and well burned quicklime. Even that he found after two hours heating upon a cupel over a strong fire had increased in weight "somewhat beyond my expectation." Two drachmas had increased to two drachmas and twenty-nine grains. That at the temperature of heating of his cupel in the furnace, this calcium oxide had absorbed carbon dioxide, was at that time beyond the knowledge or conjecture of Boyle. He even found that two drachmas of good red coral, hermetically sealed up in a thin bubble of glass and heated upon kindled coals increased in weight by over three grains and a half. This fact is difficult to explain except on the basis of some error in experiment. Boyle also heated weighed amounts of tin and of mercury in sealed flasks so that no extraneous matter should enter during the heating, and still found small amounts of calx produced, and slight increases in weight. This was convincing to Boyle and his contemporaries, as proving that fire material was the source of the increased weight, because there was in none of their minds the notion that this might be caused by an essential part of the enclosed air.

The experiments and conclusions of Boyle appeared to have been convincing, and the gain in weight of metals when roasted was now very generally accepted as due to fixed fire substance. That this apparent proof was in part due to the very inaccuracies of some of Boyle's experiments, and to the misinterpretation of some others is evident. This acceptance, however, was an important obstacle in the way of a true concept of the function of the air in combustion. When we consider the great amount of experimentation upon phenomena of combustion, it seems strange that a clearer notion of the part played by air in that process was not reached earlier. Some speculations of thinkers seemed to lead toward such a consummation.

That universal genius, Leonardo da Vinci, (1451–1519), expressed his view of the relation of the air to combustion in a way that if followed up would seem to have led toward a correct solution. He said, ''The element fire consumes continuously the air as concerns that portion which nourishes it, and there would be formed a vacuum, if other air did not come to supply its place,'' and again,
''When a flame occurs there is started a current of air generated thereby. This draft serves to maintain and to increase the flame. The fire destroys without intermission the air which supports it and would produce a vacuum if other air could not come to supply it. So soon as the air is no longer in condition to sustain a flame, no earthly creature can live in it any more than can the flame.'' [37]

Leonardo utilized his conception by devising a lamp chimney to regulate the draft, but seems not to have discussed the problem further.

In 1630 there was printed a treatise by a French scholar, Dr. Jean Rey, upon the causes why tin and lead are augmented in weight when calcined. The answer to this question as given by Jean Rey is[38]
''that this increase of weight comes from the air, which has been condensed (spessi), made heavy (appesanti), and rendered somehow adhesive by the vehemence and long continued heat of the furnace, which air mixes with the calxes (frequent agitation aiding) and attaches to their more minute parts, not otherwise than water makes sand heavy by moistening and adhering to the smallest of its grains.''
The postulates that led Rey to this conclusion are in-

[37] Marie Herzfeld, *Leonardo da Vinci, der Denker, Forscher und Poet,* Jena, 1911. *See* also E. v. Lippmann, *Leonardo da Vinci als Gelehrter und Techniker,* Vortrag in 1899 in *Abhandlungen,* [etc.,] *zur Geschichte der Naturwissenschaften,* I, pp. 361, 362. Cf. D. H. Grothe, *Leonardo da Vince als Ingenieur und Philosoph,* 1874, p. 46.
[38] *Essais de Jean Rey.* Reimpression de l'édition de 1630. Publiée avec preface par Édouard Grimaux, Paris, 1896. Essays of Jean Ray, *Alembic Club Reprints,* No. 11, Edinburgh-London, 1895.

teresting. In the first place he accepts that all the four elements—air, fire, water, and earth, have positive weights —that is that all tend to approach the center of the earth. This is in opposition to the concept of negative weight held by some theorists of the time. He calls attention to the fact that tests with the balance may deceive in this, for you cannot weigh air in air, nor water in water, but you can show that air has weight by compressing it or rarefying it in a vessel before weighing. Another concept of Rey was, that as in nearly all distillations of what he calls homogeneous bodies, as turpentine, vitriol, wine, these are by the action of the fire separated into parts of varying densities, the parts longest subjected to heat and remaining in the retort longest being, as he thinks increased in density by the fire. Thus even water is acted upon as he thinks, a lighter distillate first passing over and subsequent fractions being ever heavier as the process continues. Distilled water is therefore more penetrating and subtle than ordinary water. So it is also with air, and consequently by long and intense heating of lead, tin, antimony, etc., in air, the air is constantly being rendered more dense and this air is what sticks to the particles of calx formed by heat from the metal and so increases its weight. To the question why one could not increase the weight indefinitely by the continued action of heat in the air, he replies by saying that there is a definite maximum of such absorption of air by the calx just as there is in the absorption of water by sand or flour, etc. Rey also discusses the various theories advanced by his predecessors and contemporaries to account for this gain in weight of some metals by calcination and shows why they are, from his point of view, inadequate, though his reasoning is not always scientific, nor conclusive.

This theory of Rey's, of course, did not explain the function of the air as now known, but it was an approach to the truth, in so far as it recognized air as the contributing source of the increase of weight instead of fire material absorbed by the calx, which was the explanation offered by Boyle and Lémery nearly half a century later.

Rey's work seems to have made no impression on his times. This is in part explained by very much in his treatise which shows a curious lack of understanding of elementary physics. His work was forgotten and his little book was extremely rare, when it was recovered from oblivion by Bayen in a communication to the Journal de Physique in 1775. This was after Lavoisier's proof in 1774 that the gain in weight of tin heated in a sealed vessel in a confined volume of air, was equal to the loss of weight by this air, and due to fixation of a part of the air by the metal forming the calx.[39]

Robert Hooke, in 1665, was seemingly the next writer to advance the theory of the function of the air in combustion. Hooke concludes that there is a certain substance in the air, which is similar to, if not the same as, a substance contained in saltpeter. This substance has the power to "dissolve" all combustibles when they are sufficiently heated. Fire may be caused by this solution, which is not merely a phenomenon of motion. The products of this may be aerial, liquid, or solid. In saltpeter this substance is so condensed that there is more of it in a given space than in the same space of air. Combustion in a limited air space ceases when the quantity of this substance in the space is saturated.[40] Hooke's promise to explain further this theory was never carried out. Hooke's explanation of combustion is at fault in his supposition of solution instead of combination, and his uncertainty as to what the substance is which in air or in saltpeter supports combustion.

So also in 1671 Thomas Willis proposed a theory of combustion. When a flame arises and is maintained there is need of continuous supply of air, not merely to prevent the flame being suffocated by vaporous effluvia, but to supply the nitrous food (pabulum nitrosum) necessary to the burn-

---

[39] See E. O. von Lippman. *Zur Würdigung Jean Reys;* 1910, *Abhandlungen,* [etc.] *zur Geschichte der Naturwissenschaften,* II, p. 292.
[40] Kopp, *op. cit.,* III, pp. 133, 134.

ing of anything which is supplied by the air, for every sublunary fire is fed by particular sulphurs from the combustible body, and nitrous particles (nitrosis) which everywhere abound in the air.[41]

This notion seems quite similar to Hooke's except that Willis appears to entertain the notion of a combination by the collision between the sulphureous particles of the combustibles and the nitrous particles of the air. It is interesting to note that Robert Hooke, Dr. Willis, and Robert Boyle were intimate friends and co-workers in Oxford and later in London, and were alike early members of the newly founded Royal Society. Thomas Birch, in his life of Boyle, for instance, referring to the air pump which Boyle made in 1558–1559 and which was perfected by Mr. Robert Hooke, says:

"Mr. Hooke, who was afterwards professor of geometry in Gresham College, and doctor of physic, then lived with Mr. Boyle, whom he assisted in chemistry, having been recommended to him by Dr. Willis, the physician whom he had before served in the same capacity."[42]

Boyle, who contributed so greatly to the physics of the air, and experimented much with various chemical actions in air, shares the same concept of the relation of air to combustion as Hooke and Willis. In his *Suspicions about the Hidden Realities of the Air* (1674), his ideas are expressed:

"I have often suspected that there may be in the air some yet more latent qualities or powers differing enough from all these [that is from gravity, elasticity, light refraction] and principally due to the substantial parts or ingredients, whereof it consists. . . . For this is not as many imagine a simple and elementary body, but a confused aggregate of effluviums from such differing bodies that though they all agree in constituting, by their minuteness and various motions, one great mass of fluid matter, yet there is scarce a more heterogeneous body in the world.

---

[41] Kopp, *op. cit.*, III, pp. 135, 136.
[42] Boyle's Works, Vol. 1, p. 33.

. . . The difficulty we find of keeping flame and fire
alive, though but a little time, without air, makes me sometimes prone to suspect that there may be dispersed through
the rest of the atmosphere some odd substance, either of a
solar, or astral, or some other exotic nature, on whose
account the air is so necessary to the subsistence of flame.
. . . And indeed it seems to deserve our wonder, what
that should be in the air, which enabling it to keep flame
alive, does yet, by being consumed or depraved, so suddenly render the air unfit to make flame subsist, and it
seems by the sudden wasting or spoiling of this fine substance, whatever it may be, that the bulk of it is but very
small in proportion to the air it impregnates with its virtue. . . . And this undestroyed springiness of the air
seems to make the necessity of fresh air to the life of hot
animals (that is warm-blooded animals) . . . suggest
a great suspicion of some vital substance, if I may so call
it, diffused through the air, whether it be a volatile nitre,
or [rather] some yet anonimous substance, sydereal or
subterranean, but not improbably of kin to that, which
I lately noted to be so necessary to the maintenance of
other flames."

The statement of Boyle that only a very small proportion of the bulk of the air is consumed, is easily explained
by the fact that he has used alcohol or other organic combustibles, so that the volume of oxygen consumed has been
replaced largely by the carbon dioxide and monoxide produced, and that only that variable volume has disappeared
produced by the oxidation of the hydrogen of the combustible. It will be noted that Boyle, in using the term volatile
nitre, recognizes like Hooke and Willis the similarity of
the action of saltpeter to the unknown substance in the
air.

The most important of these seventeenth century researches into the relation of the air to combustion was
published in the same year, 1674, in which Boyle published
the above observations. This was the work of John Mayow
(1643–1679) a young English physician, a fellow of All
Souls' College, Oxford University. He also became a fel-

low of the Royal Society, being nominated for that honor
by Robert Hooke in 1678. Mayow's publication comprised
five treatises on chemical and medical subjects, those bearing on the present topic being the first and second, entitled
De Sal-Nitro et Spiritu Nitro-Aereo and *De Respiratione.*
These works published in Latin were reprinted at the
Hague, 1681, and at Geneva, 1685, but appear, nevertheless,
to have failed to make the impression that they deserved
and were quite generally forgotten for nearly a century,
though Stephen Hales refers to Mayow in his *Vegetable
Statics* (1727). J. F. Gmelin mentions him and refers casually to his explanation of respiration in that the lungs
of animals draw in from the air a substance (Gmelin called
it "saltpeter") which passes over into the vital spirits
and gives warmth to the blood, but "without any experiments of his own." Though Gmelin cites the work above
mentioned, he seems to have seen only the earlier publication of 1668 by Mayow on *Respiration,* and not the
treatise *De Sal-Nitro.*[43] The later historians, Hoefer and
Kopp, however, recognize more fully the value of his work.
In recent years the work of Mayow has been issued in accessible form.[44]

Mayow was acquainted with the publications of Hooke,
Willis, and Boyle, all of whom he cites, the last frequently.
His own work may be considered as the final stage of the
development of the theory of a "nitrous" substance in the
atmosphere as the cause of combustion of sulphureous
(that is, combustible) materials, though he also fails to
understand the actual process taking place. Mayow, like
Hooke and Boyle, is impressed by the fact that the same
substance which enables saltpeter to burn combustibles
out of contact with the air, is the substance which main-
tains combustion in the air. His treatise begins with the

---

[43] Cf. J. T. Gmelin, *Geschichte der Chemie,* 1798, II, p. 112.
[44] Ostwald's *Klassiker der exakten Wissenschaften Nr. 125,* Untersuchungen
über den Salpeter und den salpetrigen Luftgeist, das Brennen und das
Athmen von John Mayow, Leipzig, 1901; *Alembic Club Reprints,* No. 17;
medico-physical works being a translation of *Tractatus Quinque Medico-
physici,* by John Mayow, LLD, M.D. (1674), Edinburgh and Chicago, 1908.

statement that air "is impregnated with a universal salt of a nitro-saline nature, that is to say, with a vital, igneous, and highly fermentative spirit." He first discusses extensively the properties of niter or saltpeter, and the conditions of its formation in the soil. He then gives experimental evidence to show that when niter is distilled it is separated into a volatile spirit which passes over into the receiver, and a "fixed niter" resembling sal alkali which remains in the retort. So also if the acid spirit of nitre is poured upon any alkali, sal nitrum is generated.[45] He next discusses the formation of niter in the soil, giving evidence to show that it is derived in part from an alkali contained in the earth while the more volatile part, "its acid spirit" has its origin in the air itself. He further concludes that not all the acid spirit of niter is obtained from the air, but some part of it only.

"With regard then to the aerial part of nitrous spirit, we maintain that it is nothing else than the igneo-aerial particles which are quite necessary for the production of any flame. Wherefore let me henceforth call the fiery particles which occur also in the air, nitro-aerial particles or nitro-aerial spirit. . . . As regards the sulphureous particles which are also indispensable for the production of fire, the necessity for them seems to arise merely from this, that they are naturally fit to throw nitro-aerial particles into a state of rapid and fiery commotion. . . . Nor should it be overlooked that antimony, calcined by the solar rays, is considerably increased in weight as has been ascertained by experiment. Indeed, we can scarcely imagine any other source for this increase of the antimony than the nitro-aerial and igneous particles fixed in it during calcination."

And considering the action of niter heated with antimony, he says:

"Clearly, then, the fixation of antimony appears to be caused, not so much by the removal of extraneous sulphur,

---

[45] It should be noted that "sal" (salt) was used at that time in a very general way to indicate not only substances we call salts, but also acids and bases as well.

as by the fixation in it of the nitro-aerial particles in which the flame of niter abounds.

"With regard to fire, it is to be noted that for the burning of things, it is necessary that nitro-aerial particles should either be already in the burning mass or be supplied from the air. Gunpowder burns very readily on account of the nitro-aerial particles it contains: plants burn partly from the nitro-aerial particles they contain, and partly from such as come from the air; but sulphureous matter, pure and simple, can only be ignited by nitro-aerial particles supplied by the air."

Mayow advances many speculations as to the functions of the nitro-aerial spirit, which show that he does not distinguish clearly between this spirit and the phenomena of heat generally, as in producing rigidity in bodies, and in affecting their elasticity, and that the elastic power of air is due to nitro-aerial spirit. He arrives at these conclusions on the basis of experiments described, but often misinterpreted.

The similarity of respiration to ordinary combustion Mayow clearly comprehended. He cites the previous observation of Lower that the venous blood becomes bright red by the air in the lungs. Mayow cites experiments to show that blood which has been kept some time in a glass vessel and is bright red only at the surface, when placed under the air pump, will at the surface effervesce gently and rise in bubbles, but fresh arterial blood on the other hand will, in vacuo, expand remarkably and rise in an almost infinite number of bubbles. Mayow considers that the nitro aerial spirit thus absorbed in the lungs by the blood plays the same part as in other combustions and this accounts for the heat of the animal body.

Mayow's experiments on burning substances over water in a fixed volume of air and similarly on the respiration of animals in a fixed volume of air are well devised. He notes that when alcohol or camphor is thus burned, that the air is diminished in volume and weight. His observations of the diminution of volume are complicated by the fact that he had burned material composed largely of carbon and hydrogen, and the carbon oxides replace a large and variable part of the oxygen consumed. He notes for instance, in one experiment, that the volume is reduced one thirtieth, and that respiration of an animal reduces the volume by one fourteenth. He notes also that the air left over when an animal or a lamp has expired in it "is possessed of no less elastic force than any other air." But this, he says, seems flatly to contradict what has been said on this matter, but his attempt to explain this contradiction is not clear or convincing.

Yet these experiments, observations, and ideas, of Mayow, on the existence and actions of his nitro-aerial spirit, foreshadowing clearly as they do the existence and behavior of the oxygen of the air, are far from the discovery and identification of oxygen. He apparently has no concept, for instance, that this spirit is a gas or that it forms any definite proportion of the volume of the air. He has no idea that it might be isolated. He seems to consider the nitro–aereo–spirit as excitable particles, which are capable of being set into violent motion by contact with sulphureous matter, and this motion is the cause of heat. Nor does his discussion of the gain in weight of antimony from heating in air necessarily conflict with Boyle's idea of the fixation of fire material, for Mayow seems to consider heat to be mainly due to the excited motion of his nitro-aereo spirit. Yet Mayow's experiments were so well directed and his reasoning so keen, that it seems in no way improbable that, had his life not been so early cut off, he might have been the one to discover the existence of oxygen gas and of its real function in combustion—a discovery that was to wait a hundred years after his time.

Both Boyle and Mayow were disciples of the mechanical or corpuscular theory of matter. Boyle seems to have been particularly influenced by Gassendi, though familiar also with Descartes' theory, while Mayow was a disciple of Descartes. While Gassendi, 1592–1658, was a follower of

Epicurus, and maintained the existence of indivisible atoms and the existence of vacuum, Descartes, 1596–1650, disbelieved in both indivisible atoms and in the possibility of a vacuum, assuming an ether to fill the spaces between other matter. Yet, in so far as their doctrines concerned the phenomena of chemistry and physics, Boyle considers the two doctrines for all practical purposes one philosophy.

"I esteemed that, notwithstanding these things wherein the atomists and the Cartesians differed, they might be thought to agree in the main, and their hypotheses might by a person of reconciling disposition be looked on as upon the matter, one philosophy. Which, because it explicates things by corpuscles, or minute bodies, may not very unfitly be called corpuscular." [46]

Both Boyle and Mayow, in attempting to visualize chemical and physical actions of bodies consider them as due to the properties and the coming together of different corpuscles, as Boyle calls them, or particles, according to Mayow. Boyle, however was not the first of the chemical philosophers of that period to think in terms of some sort of atoms or corpuscular hypothesis. Van Helmont (1577–1644), who was an opponent of the Aristotelian natural science, frequently uses the atomic hypothesis, though not with consistency nor very clearly. Daniel Sennert (1572–1627) considered all changes in bodies as due to different atoms participating. The ideas of van Helmont and Sennert seem to be derived from the ancient atomists rather than from the influence of Gassendi or Descartes.[47]

Two prominent names among the chemists of the seventeenth century were Johann Kunckel (or Kunkel) (1630–1703), and Johann Joachim Becher (1635–1682). Kunkel was born in Holstein near Rendsburg. He was at first apothecary, but soon become interested in the problem of alchemy, and, for a time, endeavored to realize the conversion of base metals into gold. He was encouraged by

---

[46] Boyle's Works, 1744, Vol. I, p. 228.
[47] Cf. Richard Ehrenfeld, *Grundriss einer Entwickelungs-Geschichte der Chemischen Atomistik,* Heidelberg, 1906, pp. 131-148.

various noble patrons in this endeavor. He was apparently honest, however, in this aim, and more than once unmasked the designs of impostors. It is evident that the possibility of transmutation was an abiding conviction, for he says, in his *Laboratorium Chymicum,* written late in life and published after his death, referring to alchemistic works which had been accomplished for the electorate of Saxony:

"Who can not see from this, that the Transmutatio Metallorum is a certain and true art, which certain ones out of gross ignorance deny and speak of in mockery, etc.?" But he did not believe all the assumed possibilities of the alchemists, who claimed to be able to transmute not only the metals but even to create living beings, since he says in the same work:

"There are in chemistry separations, combinations, purifications, but there are not transmutations. The egg hatches by the heat of the hen. With all our art, we cannot make an egg. We can destroy it and analyze it but that is all."

The influence of Kunckel on theoretical ideas was, however, small, and his interest was largely in practical chemistry, and it is on account of the many practical chemical facts and observations that his name achieved its prominence. None of these was at all epoch-making, though he wrote several works on chemical topics, which were much read in his time. Among these his *Ars Vitraria Experimentalis* (1679) was the one of most lasting value. This was a treatise on glass making and coloring, extending the earlier works of Neri and Merret.[48]

Kunckel attracted attention also as the discoverer of phosporus. This substance had really been prepared by a man named Brand. This coming to Kunckel's knowledge, he determined to obtain from him the process and went from Dresden to Hamburg to see him. Finding manifest disinclination to impart the secret, Kunckel wrote a note to a Dresden friend, a Mr. Krafft, telling him of the situation. Krafft, according to Kunckel's story, did not answer him, but journeyed to Hamburg and purchased the

---

[48] An Edition in French was issued by Baron d'Holbach, Paris, 1752.

secret from Brand for 200 thaler, on the condition of
Brand not revealing that fact to Kunckel, after which he
returned to Dresden as he had come, Brand telling him he
had not succeeded in repeating his previous work. Kunckel,
however, found out that urine was the source of the preparation and, by his own labors, prepared the substance.
He published a book upon phosphorus and its properties
in 1678, in which, however, he gives no details of its preparation.

About the same time, Robert Boyle had discovered phosphorus, and, in 1680 (September), he described its preparation in a paper deposited with the Royal Society but not
published until 1692. In the same year, 1680, however, he
published papers on the *Aerial Noctiluca* in which he speaks
much of the samples brought by Mr. Krafft to show King
Charles. Boyle met Krafft and states that Krafft gave
him no information as to its preparation other than that it
was derived from "somewhat that belonged to the body of
man." This information gave him a valuable clue to limit
his experiments to a few substances and a further hint he
received later from a stranger, "countryman, if I mistake not, of Mr. Krafft," who referred to the high degree
of heat necessary for the operation. The method which
Boyle used was to distil evaporated urine with about three
times its weight of fine sand at high temperature for several hours, condensing under water the eventually distilling
phosphorus. This is also the process by which Kunckel
prepared phosphorus as described by Homberg (in 1692)
who had seen the operation of Kunckel. Stahl relates that
Krafft, whom he knew, told him that he had described the
process to Boyle, but this statement is hardly credited as
against Boyle's specific statement to the contrary, considering the universal conscientiousness and professional courtesy and consideration of Boyle. Boyle makes no claim to
the discovery of phosphorus, and on the contrary says:

"I find the first invention is by some ascribed to the
above mentioned Mr. Krafft (thought I remember not, that

when he was here, he plainly asserted it to himself), by others attributed to an ancient chemist dwelling at Hamburgh, whose name, if I mistake not, is Mr. Branc [evidently Brand], and by others again, with great confidence, asserted to a famous German chymist in the court of Saxony, called Kunckelius, but as to which of these so noble an invention . . . is justly due, I neither am qualified nor desirous to judge.'' [49]

The process used by Kunckel and Boyle gives slight yield and phosphorous was an expensive product in their day. Only after Gahn or Scheele (about 1771) had shown that bones contain phosphoric acid, was discovered the process in which the syrupy liquid produced by removing a great part of the calcium by nitric or sulphuric acid is reduced by ignition with carbon. By this method phosphorus was obtained at a cost which removed it from the class of expensive rarities, and gave it wide industrial possibilities.

Johann Joachim Becher was born at Spire, 1635, the son of a Protestant minister. The Thirty Years War, which so devastated Germany, deprived him of property, and his father's death compelled him as a boy to earn his livelihood by teaching writing and reading. He was not systematically educated, but possessed a vivid imagination, a passion for chemical knowledge, and an ambition which soon brought him into prominence. In 1666 he was appointed professor of medicine at the University of Mainz (Mayence), and afterwards became court physician to the Elector of Bavaria at Munich; from there he went to Vienna as a member of the newly formed Commercial College. Becher had evidently an unfortunate disposition which soon lost him the favor of patrons, and made many enemies, and in 1678 he fled to Holland; in 1680 he appears at London where he died in 1682. Becher's fertile imagination, together with his unpractical character, caused him to suggest many plausible schemes of an industrial nature, which attracted more or less attention but apparently were

---

[49] Boyle's Works, 1744, IV, p. 21.

rarely realized. Thus he proposed to the States General of Holland, a project for utilizing the sand of the dunes to recover gold. He promised an income of a million thalers a year from the project, through a process involving the use of a million marks a day of silver. The States General were favorably impressed and agreed to pay him a royalty, but after preliminary tests, the scheme was abandoned as impracticable. Becher was a prolific writer. Gmelin records sixteen works of his authorship, covering a wide field of subjects, practical and theoretical, and his works attracted much attention in his time. They evidence much independence and imagination and show him to be a man of much native ability. He was, however, like many other self-made and self-educated men, unclear in his reasoning and his ideas were so often obscure and his various theoretical developments so inconsistent, that there is little that has left any impress on the history of chemistry. His place in the history of chemistry is due to his theory of the constitution of matter, which as interpreted and extended by Stahl and his followers, formed the basis of the phlogiston theory, an explanation of the processes of combustion, oxidation, and reduction, which dominated chemistry until Lavoisier.

Becher, in his earlier writings, adopts the *tria prima*—mercury, sulphur, and salt—as the composition of matter. Later he framed a new formulation which, however, is rather verbal than essentially new. Though not always clear or consistently expressed, his concept seems to be practically as follows:[50]

All earthly substances are compounds; there is no existing substance which is elementary. All mineral substances are composed ultimately of earth and water, but proximately of three earths: terra prima, fusible or stony; terra secunda, pinguis or fatty; terra tertia, fluid. The first of these earths he describes as resistent to fire and vitrifiable, the second is to the first as soul to body and imparts combustibility, the third imparts malleability, volatility, fusibility to its compounds.

---

[50] Cf. Kopp, *Beiträge zur Geschichte*, 1875, III, p. 203, *ff.*

It will be noticed that these are respectively the characteristics and influences attributed at that time by chemists to salt, sulphur, and mercury. Becher recognizes this
but objects to the latter terms because the actual substances
thus named are not elementary but themselves compounds.

These three earths or *terrae* are also the constituents
from which animal and vegetable substances are composed,
but their proximate constituents are more complex than
in the minerals.

Ordinary sulphur is composed of an acid and this *terra
secunda* or fatty earth, and it is the content of this earth
which makes any substance combustible. Combustion is the
separation of the burning substance by fire into hetero-
genous parts, but the fatty earth is not alone sufficient for
combustion, for saline parts must coöperate. From the
baser metals, a volatile part is driven off by fire. As to the
gain in weight of metals upon calcination, Becher attributes this, as do Boyle and Lémery, to absorption of fire material.

It is difficult to see in the characteristics and properties
of the three earths of Becher any substantial improvement
on the *tria prima* of Paracelsus and his successors, other
than the avoidance of the use of the three names which
were in common use in two different meanings. For the
three principles of that name, as chemists of that school
took great pains constantly to explain, were not the same
as the common substances so named. Nevertheless, the
new name *terra pinguis* or fatty earth for the older
sulphur, as the substance which departs in combustion, certainly gave the stimulus which incited Stahl and
his followers to develop the influential phlogistic hypothesis
and Becher thus played a not unimportant part in the history of chemical theory.

Despite the growing tendency toward real and practical
aims in chemistry, the seventeenth century is marked by
the vitality of traditional alchemical activity. Several
of the chemists who contributed to the expansion of

chemical knowledge still held belief in the reality of the transmutation of metals. On the other hand, there were a multitude of writers who may be classed as merely alchemists. Such for instance, were notably the Scotch alchemists, Alexander Setonius, Michael Sendivogius, Pierre Fabre, George Starkey, Joh. Friedrich Helvetius, and the pseudonymous Philaletha. The writings of earlier alchemists were also frequently republished. Compilations of alchemical writings, early and late, were issued by Elias Ashmole (Theatrum Chemicum Britannicum, London, 1652, a collection of 32 writings by English alchemists), and by J. J. Manget, (Bibliotheca Chemica Curiosa, 1702). Olaus Borrichius published a *Dissertatio de Ortu et Progressu Chemiae,* 1688, and *Conspectus Scriptorum Chemicorum,* 1697. He was professor of philology, poetry, and chemistry at Copenhagen and an ardent defender of the truth of alchemy.

The secret society of the Rosicrucians, which tradition says was originally established in the fifteenth century, was particularly active in the seventeenth century. This was an organization of mystics devoted to alchemy, cabalism, and theosophy. Its existence and the impression it produced on the popular imagination are evidences of the persistent appeal which mysticism and mystery exerted in this century.

CHAPTER XI

THE EIGHTEENTH CENTURY: THE RISE AND FALL OF THE PHLOGISTON THEORY

If Boyle was the first chemist, as Kopp believes, to prosecute the study of chemistry solely with the object of finding the truths of nature, the century following his death contained many followers of that ideal. To what extent that is due to the teachings and example of Boyle or to what extent it was a general tendency of which Boyle was an early and influential example, it would be hard to say. That Boyle's influence was great, we know from the almost universal tributes of admiration which the early eighteenth century elicited. Thomas Birch, the biographer of Boyle, says that his merit as a writer in natural philosophy and chemistry is universally acknowledged. Boerhaave, author of the most popular treatise on chemistry of the early eighteenth century is quoted by Birch as saying:[1]

"Mr. Boyle, the ornament of his age and country, succeeded to the genius and inquiries of the great Chancellor Verulam. Which of Mr. Boyle's writings shall I recommend? All of them. To him we owe the secret of the fire, air, water, animals, vegetables, fossils: so that from his works may be deduced the whole system of natural knowledge."

And the Italian natural philosopher, Francisco Redi, is cited as asserting that "he was the greatest man who ever was, and perhaps ever will be, for the discovery of natural causes."[2]

At all events the scientific spirit of Boyle found a fertile soil in the eighteenth century. Many influences conspired

---

[1] In his *Methodus discendi medicinam.*
[2] Redi's Works, IV, Florence, 1724.

to encourage scientific investigation and publications, notably, in chemistry, the publication of serials. Besides the publications of the Royal Society and the *Memoirs* of the Paris Academy, the *Journal de Physique* (founded 1778), *Annales de Chimie* (1789), Crell's *Chemische Journal* (1778), and the publications of the Berlin *Akademie der Wissenschaft* (1710) may be especially noted.

The teaching of chemistry in the universities of Europe was steadily acquiring a more important status. Instead of being mainly an appendix to medicine, it was given more and more by men who were primarily chemists, even though holding the degree of Doctor of Medicine, for the university courses in medicine were still the conventional courses for those who were interested in natural science. Many works on chemistry—texts and special treatises—theoretical and practical, appeared in the eighteenth century, evincing the rapidly growing importance of chemical science. Thus in the early part of the century may be mentioned the works of Stahl, F. Hoffmann, Boerhaave, Juncker, Neumann, and in the latter half of the century works by Marggraf, Macquer, Bergman, Scheele, Morveau, Black, Priestley, Cavendish, Berthollet, and Lavoisier.

The most influential development of chemical theory of the eighteenth century was the phlogistic theory which attempted to explain and to correlate the phenomena of combustion, oxidation and reduction in a relatively simple and comprehensive manner. The credit of founding this theory and of attracting the attention of chemists to it is due to Stahl.

Georg Ernest Stahl was born at Ansbach in 1666. He was educated as a physician at Jena and taught medicine there from 1683. Upon the foundation of the University of Halle, he was appointed professor of medicine at that university where he taught medicine and chemistry for twenty-two years. His especial interest in chemistry was shown here by the number of his students whom he inspired to chemical study. In 1716 he was called to Ber-

lin as Royal Court Physician, where he continued to pub-
lish chemical works until his death in 1734. Stahl is credited with many minor discoveries and rediscoveries of
chemical phenomena which found their place in the rapidly
growing body of chemical facts in his period and which
manifest his chemical knowledge and ability.

But the high place in history which is accorded to
Stahl is mainly due to his formulation of the properties and relation to chemical action of the supposed
phlogiston. The term "phlogiston," from the Greek
φλοξ, flame, was used, though rarely, by Becher to
designate his *terra pinguis* or sulphur principle, the
inflammable principle which was supposed to be given off
during any combustion process. Stahl, in his earlier works,
also used the word "phlogiston" seldom, more commonly
using the conventional terms "sulphureous principle,"
"fatty earth," or "principle of inflammability," though
in his later works he formally adopts the word "phlogiston" as best expressing the supposed substance. He says:

"From all these combined circumstances, I have judged
that no more fit name could be given to this material than
that of inflammable matter or principle. Indeed, as up to
the present time no one has been able to find or recognize
any portion of it except in combination, and no one consequently can give a definition of it nor any name after some
property which uniquely belongs to it, it seems to me
nothing is more reasonable than to name it after the general effects that it produces even in its final combinations,
that is why I give it the Greek name of 'phlogiston,'
phlogistic or inflammable." [3]

Stahl ascribes many properties to phlogiston which are
conjectural rather than demonstrated, as that of imparting
colors and odors to its combinations, and on account of the
properties of the many solid substances in which it occurs,
he approves Becher's characterization of it, that it is of
earthy nature, dry and well adapted to solid combination.

[3] *Traité du Soufre* translation in Paris, 1766, from Stahl's *Zufällige Gedanken und nützliche Bedenken über den Streit von dem sogenannten Sulphur,*
Halle, 1717, page 57 of the French translation.

Stahl gives Becher full credit for originating the theory of phlogiston.

Later students of the history of chemistry consider that Stahl has drawn a much more consistent hypothesis from his studies of Becher than that chemist himself possessed. Becher's modification of the older mercury, sulphur, salt hypothesis was little more than a change of names. But Stahl conceived Becher's idea of the *terra pinguis* or sulphureous principle to be that of a single substance, instead of the earlier notion that there were many sulphurs, mercuries, *et cetera*. In other words, he starts with the concept of a definite substance, the same in all its combinations, which existed in definite chemical union in various proportions with other substances. Possibly Becher's idea of common sulphur as being a combination of an acid and this *terra pinguis* or phlogiston was the starting point of Stahl's development as he also lays some stress on this idea. From the early belief that the metals also contained sulphur, which Becher extended to his *terra pinguis,* Stahl formulated his theory that the calcination of metals was analogous to the burning of sulphur or other combustibles, that the metals lost combined phlogiston and that the metals were themselves definite combinations of phlogiston and the resulting calx. When these substances, which are left by the escape of phlogiston, are heated with substances which contain much phlogiston, as oils or fats, sulphur or charcoal, they again combine with phlogiston and the original unburned substance is produced. Stahl reproduced sulphur from oil of vitriol by combining the latter with an alkali salt, heating this with charcoal and precipitating the sulphur with acetic acid. By this experiment he understood that he had reconveyed phlogiston from the charcoal to the acid and again reproduced the sulphur. So phosphoric acid—obtained from burning phosphorus—when heated with carbon recombined with phlogiston and the phosphorus was again produced. The reduction of calxes of the metals to the metallic state by heating with char-

coal was similarly interpreted as recombination of the
calx with phlogiston. Stahl was entirely cognizant of the
fact that some of the metals gained in weight by the process in which phlogiston was supposed to leave the metal,
but these facts bore no relation to the process in his mind,
nor in those of his followers, as phlogiston was known only
in combinations and nothing whatever was known of its
own intrinsic properties, or its possible relation to weighable matter. Its influence on the weight of other substances
came to be considered as diminishing rather than increasing this weight. Stahl also refers to the necessity of the
presence of air for reactions resulting in loss of phlogiston,
but expressly states that the air seems not to enter into
the combination.

Until it could be proved that the source of the gain in
weight of calcined metals was due to combination with a
definite and weighable constituent of the air, and that
this constituent of the air takes part in all cases of combustion and calcination, the phlogistic philosophy—as
elaborated by Stahl, his pupils, and adherents—offered to
chemists the first coherent and plausible explanation of all
those phenomena which are now known as oxidation and
reduction, direct and indirect.

Two serious obstacles to continuous progress were, however, inherent in this theory. The supposed phlogiston
could not be separated or isolated and weighed. It could
not be known whether it had a positive weight in combination, nor whether it could affect in any definite or determin-
able way the weight of other substances. It might even
have the effect of buoyancy or of diminishing the weight
of substances with which it was combined, and so long as
such ideas were held the weights as given by the balance
could not be depended upon to give the real quantitative
relations of chemical reactions.

The second obstacle this theory offered to chemical development lay in the fact that so long as this theory was
maintained, no identification of substances as elements was

possible. Boyle had given us a proper definition of an element, but so long as such oxidizable substances as phosphorus, sulphur, iron, zinc and carbon were considered as combinations of phlogiston with other substances, (namely, their oxides) and so long as the products of combustion, as we now know them, such as the oxides of phosphorus, sulphur, iron, etc., were considered as products of the loss of phlogiston, and therefore to that extent simpler or more nearly elementary than the combustibles from which they were produced, it is manifest that the elementary character of most of the now known elements could not have been recognized.

The importance of the phlogistic hypothesis in the development of chemical science as founded by Stahl and elaborated by the most able and prominent chemists of the century must, however, not by any means be underestimated. Although its fundamental basis was what we now regard as a mistaken idea, and although it is quite probable that in some respects its false concepts may have delayed the discovery of oxygen and of the function of air in combustion, nevertheless, it must be remembered that it was the first important generalization in chemistry correlating in a simple and comprehensive manner a great number of chemical actions and certain relations existing between a great variety of substances, and that it thus served to inspire an enthusiasm for research in a great body of able scholars whose results fell into place easily and more simply when Lavoisier and his co-workers elaborated the true theory.

The theory of phlogiston did not appeal to all of Stahl's contemporaries. Notably is this true of two of the most popular chemistry teachers of the time, Hermann Boerhaave at Leyden and Friedrich Hoffmann of Berlin. Both of these men accepted the idea of Becher's predecessors that combustion consisted in the loss of some substance, sulphur (sometimes called "phlogiston" by Hoffmann), oil (oleum) or "pabulum ignis" (the food of fire) by Boer-

haave. Both accept the idea that when sulphur burns the
inflammable substance leaves and the resulting acid is what
remains. Both of them reject the analogy of the action of
metals in calcination, Hoffmann believing that the calx is
a combination of the metal with an acid "sal acidum" from
the air, while Boerhaave thought that the calxes of metals
were not very different from the metals, and that the occasional gain in weight by calcination was due to the fixation of some foreign substance from the heating vessel but
not of fire material, which he considered imponderable.
Hoffmann thought that metal calxes heated with charcoal
gave up some material to the charcoal. In rejecting the
analogy between ordinary combustion and calcination,
however, they naturally saw no great difference between
the phlogiston theory of Stahl and the previous sulphur
theory of Paracelsus or its subsequent variations by later
theorists. For the same reason doubtless neither Hoffmann nor Boerhaave was interested in actively opposing
Stahl's theory.

Friedrich Hoffmann (1660–1742) as professor of medicine at Halle was influential in obtaining for Stahl his
professorship in that university. His friendly relations
with Stahl were, however, later disturbed by differences of
opinion on scientific subjects. Hoffmann was a broadly
trained scholar, a public-spirited and devoted officer of
the university, a constant correspondent with eminent
chemists of the time, and a member of several scientific
societies. He was a widely known and highly respected
physician, chemist and teacher, and published very many
works and papers on medicine and chemistry. The latter
were mainly on methods of analysis. Gmelin[4] cites the
titles of 122 books and papers pertaining to chemical analyses and descriptions of properties. Especially important
were his treatises on mineral waters, and the salt contents
of these, with methods for detecting the presence of the
various constituents. He introduced the use of a mixture

[4] *Geschichte der Chemie*, Vol. II, p. 179 *ff.*

of alcohol and ether in equal parts for quieting pain, which long carried the name of "anodyne liquor of Hoffmann." Hoffmann's works *Chymia rationalis et experimentalis* (1784) and his collected works under the title of *Opera omnia physico–medica* (published first at Geneva 1740–1760 in eleven folio volumes) were highly esteemed sources of medicine and chemistry in the eighteenth century.

Hermann Boerhaave (1668–1735) was born at Voorhout, near Leyden, and is called by Professor Thomas Thomson[5] "perhaps the most celebrated physician that ever existed, if we except Hippocrates." He received his degree of Doctor of Medicine at the University of Harderwyk in Holland in 1693. In 1702 he was appointed professor of medicine at Leyden, and later also was awarded professorships there in botany and in chemistry. The reputation of Boerhaave attracted a great body of students to Leyden and raised that university to an eminent position for the study of medicine and the natural sciences.

Boerhaave's lectures on chemistry excited wide attention. In 1724 there was published in Paris an apparently unauthorized edition of his chemistry, *Institutiones et Experimenta Chemiae,* which was translated with many notes by Peter Shaw and E. Chambers in 1727. This edition was so full of errors and perversions of his ideas that in 1732 Boerhaave published his *Elementa Chemiae* in two quarto volumes, on the history, science, and practical experiments of chemistry. This edition contains his manuscript (autograph) signature to a statement of the authenticity of the work and his repudiation of the responsibility for any earlier work. The *Elementa* soon became the most popular treatise on the chemistry of the period. The Latin edition, according to Hoefer,[6] passed through ten editions between 1732 and 1759 in Leyden, Paris, London, Basel, Leipzig, and Venice, and it was translated into German, French, and English in several editions.

---

[5] *History of Chemistry,* 1830, I, p. 209.
[6] Hoefer, *Histoire de la Chimie,* 2d ed. II, p. 368.

Professor Thomson says of the work that it "was undoubtedly the most learned and most luminous treatise on chemistry that the world had yet seen; it is nothing less than a complete collection of all the chemical facts and processes which were known in Boerhaave's time." That some of his supposed facts, resting on the authority of many previous writers, were not entirely to be depended upon is not a reproach, for all these facts could not be verified by his own experiments. Yet the work was a conservative summary of chemical facts and theories, free from all mysticism, and presented in orderly, dignified and comprehensive system.

This work "adopted in all schools," as says Hoefer, exerted a profound influence toward a sane and scientific attitude in the study of chemistry. It is worthy of note that Boerhaave, in his *Elementa,* makes no reference to Stahl or to his phlogistic theory, though he mentions Stahl among his list of authorities in the division of his book relating to the history of the science. When in the latter half of the century, this theory became the most important phase of chemical thought and became almost universally adopted, Boerhaave's work lost in popularity, being replaced by texts containing phlogistic philosophy, as Neumann, Macquer, and Bergman.

Boerhaave's many experimental researches described in his textbook or in the Philosophical Transactions of the Royal Society of London, or the Memoirs of the Academy of Sciences of Paris, show no discoveries that are in any sense epoch-making. By his experiments on the transmutation of metals he assisted materially in giving the death blow to the traditional belief, still more or less accepted by chemists of his time, that mercury was capable of being rendered a hard metal by long subjection to heat and that it was a constituent of other metals. He kept mercury for fifteen years at a warm temperature in an unsealed vessel, and for six months at high temperature in a sealed vessel, and distilled mercury five hundred times, without any material change being produced. So also he disproved the alchemical statement that mercury could be generated from lead by dissolving lead nitrate in water, precipitating with ammonium chloride, and digesting for some time with caustic potash or soda. This product, when distilled was supposed to yield mercury. Calcined sugar of lead treated with caustic lye was also supposed to give the same result. Boerhaave repeated these experiments extending the time of digestion, even to six months. That he obtained no mercury went far toward discrediting these lingering traditions of the alchemists.

An early disciple of Stahl in his phlogistic hypothesis was Dr. Johann Juncker (1683–1759), professor of medicine at Halle, who published in 1730 a *Conspectus Chemiae* expressly stated as "explained from the dogmas of Becher and Stahl." This is one of the best of the early treatises of chemistry on the phlogistic foundation. Another follower of Stahl was Caspar Neumann (1683–1737) who was first an apothecary, but his ambition not being satisfied by that profession, he was, by the favor of the King of Prussia, financed to traverse Holland, England, France, and Italy, where he formed connections with eminent chemists, and on his return was made professor of chemistry in the Medicinisch–Chirurgische Bildungsanstalt in Berlin. He was made a member of the Royal Society of London as also of the Berlin Academy. Neumann was particularly interested in his researches on the chemical analyses of various products, chiefly organic—camphor, wines, thyme oil, ethereal oil of ants, etc. His reputation as an able analyst was well deserved, though no very important discoveries or observations can be cited. As a lecturer he was very popular and after his death, his works were published in various editions; by Johann C. Zimmerman in Berlin 1740, two volumes quarto, second edition by Zimmerman, 1755–1756, a more extended edition by C. H. Kessel, Zullichau, four volumes quarto, 1749–1755, and a condensed edition of this in two volumes, 1755–1756. This condensed edition was translated into Dutch in 1766, French in 1781, and a further condensation was published, with later notes, in English by William Lewis in London, 1759, one volume quarto.

The general character of Neumann's chemistry is practical rather than theoretical. It describes plainly and in considerable detail the occurrences, properties and preparations of a large number of mineral, animal, and vegetable products, and the value which it must have possessed at that time as a condensed encyclopedia of chemical facts is manifest. Neumann apparently accepts the phlogiston hypothesis without reservation. In the discussion of metals, which he divides into perfect metals—gold and silver; imperfect metals—lead, copper, iron and tin; and semimetals (not malleable)—mercury, bismuth, zinc, antimony, arsenic, he has this to say under the head of imperfect metals:[7]

"These metals appear actually to contain an inflammable principle, which is burnt out in the calcination, and extracted from them by acids. Nitre, which deflagrates with and dissipates the inflammable principle wherever it is to be found, deflagrates with the imperfect metals, and thus occasions instantly the same change that fire alone would more slowly produce: Some of these metals emit visible flames by themselves.

"The phlogistic principle is the same in one metal as in another, in metals as in other bodies, in the mineral as in the vegetable and animal kingdoms. When metals, by the loss of their own pholgiston, have been changed into a calx or vitreous mass; the introduction of any other inflammable matter, from vegetables or animals, charcoal, resins, oils, fats, etc., instantly restores their metallic appearance, and all their pristine qualities.

"The calx differs greatly in different metals; it is on this that the distinguishing characters of each particular metal depend, the calx of one metal forming always with phlogiston no other than the same metal again. These calxes are

---

supposed to consist of a certain fixed vitrescible earth, and a more volatile principle called mercurial.[8]

"When metals have been but barely calcined, they have lost only their phlogiston, and are recoverable by the introduction of fresh phlogiston. A more thorough calcination, by long continuance of fire or by additions, dissipates a part of this mercurial principle; and as no method has been discovered of supplying it by art, the quantity of metal revivable will be proportionably less. The process is capable of being carried so far that no metal at all shall be recovered, and that the calx shall differ but little from mere earth.[9]

"The perfect metals of the foregoing class, though not resoluble by these operations into any dissimilar parts, are supposed, from analogy, to consist of the like principles."

This is an exceptionally concise and clear statement of the essentials of the phlogiston theory. In discussing sulphur, Neumann says:

"Experiment has fully evinced that sulphur is no other than the concentrated vitriolic acid combined with a small proportion of the phlogistic or inflammable principle, and to this combination alone, which is always one and the same except for adventitious admixtures, the more judicious chemists have wholly confined the name."

Neumann notes the gain in weight of lead and of zinc when calcined, but does not mention nor seem to see any bearing of these facts upon the phlogiston hypothesis.

Three prominent German chemists, each exerting much influence on his time, and all supporters of the phlogiston theory were: Johann Theodor Eller (1689–1760), Johann Heinrich Pott (1692–1777) and Andreas Sigismund Marggraf (1709–1782).

Eller belonged to a wealthy family, and received an excellent education; studying first jurisprudence at Jena,

---

[8] Here we have a vestige of the ancient belief that mercury is a constituent of the metals, as to which Boyle expresses his scepticism, and which Boerhaave combatted experimentally.

[9] Apparently an attempt to explain the fact that oxides of some metals when very strongly heated or when fused with certain vitrefiable impurities are with difficulty reduced or dissolved by acids.

later medicine and natural science at Halle, Leyden, and Amsterdam. He studied also at Paris and visited London, meeting many distinguished scholars of the time. In 1724 he received appointment as professor of anatomy at Berlin. In 1755 he was appointed privy councilor and first court physician by Frederick the Great, and in that year also was made director of the physics class of the Berlin Academy of Sciences. Eller's personal influence and his interest in chemistry were influential in obtaining from the government support for chemical and other scientific institutions. He issued, between 1745–1757, in the Berlin Academy publications, many papers which, after his death (1760) were published in collected form. None of his discoveries in chemistry are of more than minor importance. He noted that water saturated with one salt was capable still of dissolving other salts, and he determined the solubility of several salts with greater accuracy than had been previously accomplished. His theoretical discussions were not always logical, and did nothing to advance that branch of chemistry.

Pott was a native of Halberstadt, and was sent to Halle by his parents to study for the ministry, but, developing interest in medicine and especially in chemistry, he studied with Hoffmann and Stahl and devoted himself to chemistry. He made his residence in Berlin, and was elected to the Academy. After the death of Neumann (1777), Pott was appointed his successor in the professorship of chemistry in the Medicinisch-Chirurgische-Bildungsanstalt. He was a well-informed chemist, an energetic experimenter, and was very clear and straightforward in his descriptions. He was, on the other hand, of a contentious disposition, and his many disputes with other members of the Academy—as Eller, Marggraf, Brandes—often overstepped the bounds of courtesy. In 1761 his relations with his colleagues in the Academy were such that he severed his connection with it entirely.

Pott was a devoted worker, chiefly in the field of inorganic chemistry. He was especially active in the study of the action of high temperatures on minerals and mixtures of minerals. This feature of his work was doubtless due, in part at least, to the commission of the King of Prussia requesting him to find out the constituents of the porcelain made at Meissen, Saxony. Pott devised an improvement on a portable form of furnace originally invented by Becher. He thus obtained more effective blast. He gave great attention to devising more resistant compositions for crucibles. This enabled him to study high temperature reactions of minerals, especially at fusion temperatures, more efficiently than any of his predecessors. He often erred in the interpretation of his results, and paid little attention to analysis by wet methods. He investigated pyrolusite without discovering the manganese, describing it as a combination of an alkaline earth resembling alumina, with a combustible material. He proved that plumbago, or black lead, contained no lead, as was previously believed, but thought it perhaps the same as molybdenite.

As the result of his investigations, Pott believed that earths may be classified into four divisions, the alkali, or lime earths, the aluminous, the gypsumlike, and the vitreous or flinty earths. In his experiments on porcelain constituents, he is said to have executed three thousand separate fusions of single or mixed earths or minerals, at varying temperatures, and through this work to have aided materially in the development of the art of porcelain manufacture.

Pott was an adherent of the phlogiston theory, but added nothing of interest to theoretical chemistry. His extensive practical observations gave him a wide reputation in his time, and his works, collected by himself and by others, were published and translated into other languages, furnishing a mass of clearly described operations which contributed in an important way to the growing body of chemical phenomena.

The last of the important German phlogistonists was

Andreas Sigismund Marggraf, son of a Berlin pharmacist.
He studied his first chemistry in his father's business,
later studied chemistry under Neumann, and afterwards
studied at the University of Frankfort on the Oder, at
Strassburg, at Halle, and at the Mining School at Freiberg, returning to Berlin, where he became connected with
the Academie der Wissenschaften and devoted himself to
chemical research. He was made a director of the Academy
in 1760 and continued researches until his death in 1787.
Marggraf's work was contributed to the publications of
the Berlin Academy from 1747 to 1781, though many of the
more important papers were translated into German[10] and
published in two volumes in 1761 and 1767.

Marggraf's many contributions to chemical research are
characterized by painstaking thoroughness, and careful description. Like Boyle, he was, however, cautious in making theoretical deductions. Marggraf did much to extend
the use of the wet methods of analysis, as distinguished
from dry fusions and distillations, then the prevalent
methods of examination of various chemical substances.
His chief field of research was the salts and earthy minerals, but he did not confine his researches to these. He
first proved (1750) that gypsum consisted of lime and sulphuric acid. He investigated the properties of platinum,
"the new mineral body called platina del Pinto," which
had recently been described by English chemists.

As early as the sixteenth century some information had
arrived in Europe, from the Spanish gold miners in Central America, of the existence of an infusible metal which
gave trouble to the refiners of gold. Julius Scaliger (who
died 1558), in criticizing some views of Cardanus, and especially his definition of a metal as something that can be
melted and becomes hard when cooled, says that according
to that, mercury would be no metal, and besides in regions
between Mexico and Darien there are known to be fodinas

(mines or diggings) of a brass ("orichalci") which thus
far can not be melted by fire nor by any Spanish devices.
The first recorded introduction of platinum into Europe
was through Charles Wood, an assayer of Jamaica, who
had obtained it from Cartagena, Colombia, about 1741.
He gave some of the pieces to a Dr. W. Brownrigg who
presented them to the Royal Society of London in 1750.
On December 13, 1750, a brief description of this "semi-
metal" by Brownrigg was presented at the Royal Society
by Dr. W. Watson. Brownrigg stated that at Cartagena
some time previously five pounds had been purchased for
less than its weight of silver. A note by Enrico Mendez
da Costa was read by Watson at the same session, stating
that in 1742–1743 there were brought from Jamaica several bars supposed to be gold. These bars had the same
specific gravity as gold, or rather more, and were of like
color and grain. A piece of one of these counterfeit gold
bars sent to the mint for testing showed it to be twenty-
one carats and three grains "worse than standard."

Gold and platina alloys were said to be brittle and hard
and it was reported that it was impossible to separate the
gold from this alloy. The name platina del Pinto, by
which the metal was known, was derived from its general
resemblance in appearance to silver (plata), platina being
the diminutive of plata, and from the river Pinto, a principal source of the grains and nuggets in the gravel. In
1752 Scheffer, a Swedish chemist, presented before the
Stockholm Academy, a series of experiments on the metal,
emphasized its resemblance to gold, and called it white
(or blanc) gold. He suggested that its qualities would
make it suitable for construction of telescope tubes, being
very permanent in the air. In 1754, William Lewis presented a paper before the Royal Society in which he described a great variety of its reactions.

The discovery of a new metal created a real sensation
among chemists at the time, and many of the prominent
chemists obtained specimens of the crude platinum metal

and repeated and extended the examinations. Marggraf was one of these, and his researches were published in 1757 in the proceedings of the Berlin Academy, confirming results of Scheffer and Lewis, but adding little of positive value to their observations. The same may be said of the work of Macquer and of Baumé, issued in 1758 in the memoirs of the Paris Academy. The early investigations on platinum were made exceptionally difficult by the impossibility of fusion, and by the many impurities of the crude metal. That from Colombia was rarely of greater than 85 per cent purity, the principal impurities being iron and the then unknown rarer platinum metals—iridium, osmium, rhodium, and palladium. The labors of Buffon, Achard, Bergman, Knight, Wollaston, and others gradually wrought toward a greater purification of the metal and to methods of working it into wire foil, crucibles, etc., though it was not until well into the nineteenth century that these methods were sufficiently developed to bring platinum ware into common use. The statement by Lewis that the specific gravity of platina is from 18 to 19[11] (pure platinum is over 21), and of Marggraf that its weight is to gold as 18½ to 19 (the specific gravity of gold is about 19.3) are evidences of the impure state of the metals with which they were working.

Marggraf contributed importantly to the knowledge of phosphoric acid. He prepared the oxide of phosphorus, noting the increase of weight when phosphorus was burned,[12] and improved the process of making phosphorus, by reducing phosphoric acid by ignition with charcoal or soot. As a consistent phlogistonist, Marggraf naturally believed that what we call an oxide of phosphorus was produced by the loss of phlogiston from phosphorus and he interpreted the reduction of phosphoric acid by charcoal as the restoring of phlogiston from the charcoal. From serpentine, Marggraf separated the base magnesia and

---

[11] Hoefer, *op. cit.*, 2d ed. II, p. 361.
[12] Marggraf, *Chymische Schriften*, Berlin, I, p. 49.

recognized it as different from lime or alumina (thonerde). The name "magnesia" was, however, of later origin.

The most important of Marggraf's investigations, from the industrial point of view, was the demonstration in 1757 that the source of the sweetness of the juice of certain domestic vegetables was the same as the sugar from the sugar cane. Out of eight ounces of dried white beets (Weisser Mangold Würzeln) he obtained half an ounce of crystallized sugar (about 6.2 per cent) by drying and powdering the beets and exhausting with alcohol, filtering and evaporating to crystallization. From a half pound of dried red beets he obtained two and one half Quentchen.[13] As Marggraf estimated the dried beet as one quarter of the weight of the fresh beet,[14] the percentage of sugar he recovered was a small fraction of the real contents, but there was then no method known to determine the content of sugar in the juice as is now done with the polariscope.

Marggraf realized the importance of his discovery, and continued his experiments to show that sugar and syrup could be obtained from the beet by slight modification of the customary process of production of sugar from the cane. He says:

"From what has been related it is clear what domestic advantages may be drawn from these experiences, of which for example, I will only advance this: that the poor cultivator could well serve himself with this plant sugar or its syrup instead of the usual costly product,[15] if by help of inexpensive machines he pressed this juice from these plants, somewhat purified it, and reduced it to the consistency of a syrup. This would certainly be cleaner than the ordinary black sugar syrup [molasses] and there is no doubt the

---

[13] Marggraf, *Chymische Schriften*, 2d ed., 1767, Theil II, p. 74. Assuming the pound at 7219 grains and the Quentchen at 57.47 grains (old German weights) this would give about 4 per cent sugar obtained.

[14] Marggraf, *loc. cit.*, p. 86.

[15] Refined cane sugar in London, as cited by von Lippmann, cost eighty-three (83) marks per fifty kilo in 1750 and one hundred and fifty-three (153) marks per fifty kilo in 1805, or twenty (20) cents to thirty-five (35) cents per pound. Cf. Lippmann's *Kurzer Abriss der Geschichte des Zuckers.* In his *Abhandlungen und Vorträge zur Geschichte der Naturwissenschaften,* I, p. 273.

residues from the pressing could be usefully employed by the farmer.''

The attempt to utilize this discovery of Marggraf's, on a technical scale, was first made by a former pupil of Marggraf's, Franz Karl Achard. After years of experiment he established, with assistance from King Friedrich Wilhelm III of Prussia, a beet sugar factory at Cunern in Schlesien in 1802. This was the beginning of the beet sugar industry. The beet as a source of sugar achieved great importance during the partial blockade of Europe during the Napoleonic wars, and Napoleon stimulated the manufacture in France by liberal subsidies, so that the French factories were for a time the largest source, though Germany later achieved supremacy.[16]

Though the phlogistic theory was of German origin and though the most influential of German chemists were phlogiston supporters, adherence to the theory was by no means confined to the chemists of Germany. The most prominent chemists of all nations were followers of this theory. Such were in France, Pierre Joseph Macquer (1718–1784), Guyton de Morveau (1737–1816); in Sweden, Torbern Olaf Bergman (1735–1784), Karl Wilhelm Scheele (1742–1786); in Great Britain, Joseph Black (1728–1799), Henry Cavendish (1731–1810), Joseph Priestley (1728–1804), Richard Kirwan (1733–1812).

Macquer was not, indeed, the first to introduce the phlogistic theory into France. Several prominent chemists and teachers had adopted it in their philosophy. Such were Stephen Geoffroy (1672–1731), Duhamel de Morveau (1700–1781), and Guillaume François Rouelle (1703–1770). Yet, by common consent Macquer is considered the most prominent and most enthusiastic French advocate of the phlogistic philosophy. Macquer was born in Paris of Scotch ancestry, followers of the Stuarts who migrated to France on the expulsion of that dynasty. The original Scotch

---

[16] For a more detailed account of Achard and the foundation of the beet sugar industry *see* E. O. von Lippman, *Einige Worte zum Andenken Achard's.* In his *Abhandlungen,* I, p. 296.

form of the name seems not to be known.[17] Macquer early
acquired prominence in medicine and chemistry and was,
at the age of twenty-seven, elected member of the Academy.
As professor of chemistry at the Jardin des Plantes, he
occupied a position long held by a succession of the
most influential teachers of chemistry in France.

Macquer was an active worker and writer on chemical
subjects, and his contributions to the memoirs of the
Paris Academy are numerous. While no discovery of great
importance is credited to him, he contributed many investigations to the rapidly increasing mass of chemical facts in
this active period of chemical observations. He made valuable observations on the solubility of many oils and salts
in alcohol, the properties of platinum, on reactions of
Prussian Blue, on arsenic acid and the arsenates, upon the
manufacture of optical glass, and contributed by his supervision and encouragement to the manufacture of the Sèvres
porcelain ware. His works on chemistry were of far
greater influence than his experimental researches. His text
books, *Eléments de Chymie théoretique* (1709) and
Eléments de Chymie pratique (1751), were widely circulated in France and in other countries, and a new edition in
1775 *Eléments de la Théorie et de la Pratique de la Chymie,*
met with similar wide approval, being issued in many editions and translations. In 1766 he issued his famous *Dictionnaire de la Chymie,* a work in three volumes octavo,
which was reissued in 1778 expanded to four volumes. This
work was practically the first great encyclopedia of chemical knowledge and long held its prestige, being translated
into nearly all European languages. Macquer was the last
of the great French phlogistonists, as Marggraf was the
last of the great German phlogistonists. When Lavoisier's
work appeared, Macquer found it necessary to attempt to
reconcile the new facts with the phlogiston theory and
though his attempts were not satisfactory to himself in all
respects, he still believed that eventually these facts would

[17] Cf. William Thomson, *History of Chemistry,* 1, p. 295.

444 THE STORY OF EARLY CHEMISTRY

be found to be explained without sacrificing the theory which had been of so great service in chemical thinking. He died, still a phlogistonist, in 1784.

Sweden furnished two very able and influential chemists in Torbern Olaf Bergman (1735–1784) and Karl Wilhelm Scheele (1742–1786). Bergman was born in Katharinaberg, West Gothland. After his elementary schooling, he was sent in 1752 to the University of Upsala to prepare for law or the ministry, but soon developed a taste for mathematics, physics and chemistry. By industry he succeeded in fulfilling the desires of his relatives as to law studies while giving his main attention to natural science. By overwork his health was affected so that he was compelled to leave the university and return to his father's home, and to observe a careful régime of out door exercise. He utilized this time under the inspiration of the great Linnaeus, then teaching at Upsala, in making collections of plants and insects, and sent many new insects to Linnaeus by whom they were classified and named. He returned to the University after the restoration of his health and, released from the obligation to study law, he devoted considerable attention to natural history, and his first published paper was on the ovum of a species of leech. His work met the approval of Linnaeus and was printed in the memoirs of the Stockholm Academy in 1756. Bergman took his master's degree in 1758, his thesis being on Astronomical Interpolation. He soon received an assistant's position in the university and in 1761 was appointed adjunct in mathematics and physics.

When in 1767 Johann G. Wallerius, professor of chemistry at Upsala, resigned his chair, Bergman presented himself as a candidate for the vacancy and, not without spirited rivalry, was elected. This position he held till his death, though in 1776 Frederick the Great of Prussia made him a tempting offer to join the Academy of Sciences at Berlin. Notwithstanding Bergman's comparatively early death, in his forty-ninth year, he succeeded in putting forth such a variety and volume of excellent experimental work as to distinguish him among the greatest chemists of his century. It was against his declared principles to permit himself to carry speculation or theory beyond the logical bounds of deduction on the basis of clear and sound experimental evidence, and he kept within that limit about as well as was possible for a thoughtful and earnest student.

The investigations and opinions of Bergman were very important for his time, and it should be remembered that his was a period of great activity in chemistry, and there were many able investigators whose work was contemporaneous with his. The publications of the various scientific academies and societies, as well as of many journals, served to keep the chemical writers in touch with the general progress better than ever before. We can perhaps realize more fully the chemical atmosphere of this time, if we recollect that at about 1775 all the following distinguished investigators were in the prime of their working power: Bergman, Black, Scheele, Cavendish, Priestley, Kirwan, de Morveau, Klaproth, Berthollet, and Lavoisier. The high authority which Bergman achieved in his time was gained only by valuable contributions to chemistry in many lines.

In the introduction to the first volume of his *Opuscula,* Bergman lays down the principles of investigation which he had adopted as his guides. They are, categorically, as follows:[18]

I. "A. In investigating the principles of a body, we must not judge of them from a slight agreement with other known bodies, but they must be separated directly by analysis, and that analysis shall be confirmed by synthesis.

"B. Analysis should chiefly be conducted in the humid way. (He comments that the dry way may sometimes be useful, but oftener tends rather to confusion.)

---

[18] Cited from Dr. Cullen's translation, *Physical and Chemical Essays,* translated from the original Latin of Sir Torbern Bergman, London, 1784. I, p. xxx *ff.*

"C. Such experiments should be instituted as are adapted to the discovery of truth.

"D. Experiments should be made with the utmost possible accuracy.

"E. The experiments of others, particularly the more remarkable ones, should be candidly reviewed.

II. "(Of Causes)

'A. In the investigation of causes, we must begin by phenomena sufficiently varied, and well observed; and proceed in order from proximate causes to the more remote.

"B. A cause in whatever way indicated by phenomena, may for a while be assumed as true, and from it may be deduced the necessary consequences, which, being separately examined by suitable experiments, either confirm or overturn the position.

"C. Besides, the cause should, if possible, be so compared with the effect, that the exact relation may be discovered, even as to quantity.

"Finally, I aim at giving denominations to things, as agreeable to truth as possible."

Bergman comments on these various principles in an interesting and illustrative way, but the principles themselves even as categorically stated are an excellent program for the investigator of the unknown.

A very important series of investigations was carried out by Bergman, and published in 1778, upon the "Analysis of Waters," comprising natural waters, including hot and cold mineral waters, and on the artificial preparation of hot medicated waters. After a careful summary of the work of previous writers on various tests and reagents for detecting particular constituents, he discusses the various known constituents, the reagents used for detecting these, and thus elaborates for the first time a scheme for qualitative analysis of the many substances found in natural waters. The contents of natural waters, which he notes as either constant or occasional (translating the nomenclature into modern phraseology), are dissolved air, carbonic acid, "inflammable air" (hydrogen or hydrocarbons), potassium carbonate, sulphate or nitrate, sodium carbonate, sulphate

or chloride, ammonia "probably from putrid vegetable or
animal substance," barium chloride, calcium carbonate,
sulphate, nitrate, or chloride, magnesium carbonate, sulphate, nitrate or chloride, aluminum sulphate; and among
metals, iron as carbonate or sulphate or chloride, manganese
"has not yet been found except as chloride," copper as sulphate, arsenic rarely, organic matter, sometimes a sulphurous substance.

Reagents employed by Bergman, were: litmus, Brazil-
wood, turmeric solution, tincture of nutgalls (for iron),
"phlogisticated alkali" (for iron; this was potassium ferro-
cyanide), giving a blue with iron, red with copper, white
with manganese. Concentrated vitriolic acid (sulphuric)
immediately precipitates any "terra ponderosa" (baryta).
The "acid of sugar" (oxalic acid) is one of the most delicate tests known for lime; more slowly and less effectually
acts microcosmic salt. Aerated fixed alkali (potassium or
sodium carbonate) precipitates all earths and metals from
solution. Aerated volatile alkali (ammonium carbonate)
precipitates all earths and metals, but caustic volatile alkali (ammonium hydroxide) has no effect on lime or
baryta. This reagent produces a cloudiness in a very dilute
solution containing copper, which becomes an intense blue
solution with a super-abundance of the volatile alkali.
Limewater dropped into water containing any "aerial acid"
(carbonic acid) renders it instantly turbid. Salited terra
ponderosa (barium chloride) is of use in discovering the
smallest trace of vitriolic acid (sulphuric).

"Salited lime" (calcium chloride) is considered a useful test for fixed alkali, for the aerated lime (that is, carbonate) separates, "but this experiment is ambiguous because if vitriolated magnesia (magnesium sulphate) be
present, a double decomposition takes place and a gypsum
is formed."

"Nitrated silver" solution affords a most complete
method of detecting the smallest trace of marine acid (hydrochloric acid); he cautions, however, that sulphur com-

pounds present ("hepar") turn this white precipitate brown and black and that silver nitrate also gives a cloud with a solution of sal soda.

Mercury nitrate, corrosive sublimate, white arsenic, and lead acetate are occasionally used as reagents, and soap is described as an indication of hard waters, as when added to them "a decomposition takes place, the acid unites with alkali, and the oil is disengaged: such waters as these are generally called hard waters, and are unfit for washing cloaths, as also for boiling pulse and the harder kinds of flesh." [19] The list of qualitative reagents for the many constituents of natural waters comprises the principal reagents in use at present.

Bergman's scheme for the quantitative analysis of mineral waters evidences much knowledge and careful consideration of properties of the chemical constituents. On the other hand, his methods were not always capable of giving results of great accuracy. Gaseous contents were obtained by boiling in a retort a fixed volume of the water, collecting the gases in a graduated cylinder, correcting for the volume of air in the retort before the boiling, and determining the carbon dioxide by absorbing it by means of lime water.

The solid contents were obtained by evaporating to dryness and weighing. The various constituents were separated first by extraction with alcohol, thus dissolving chlorides and nitrates of calcium, magnesia, and barium if that were present, and sometimes ferric sulphate (dephlogisticated martial vitriol).

The residue from alcoholic digestion is then treated with a limited quantity of cold water (eight times its volume) and, after standing, filtered. The filtrate contains alkali salts, and sulphates of alkaline earths, and of metals. These he separated usually by their varying solubilities, identified by their crystallized form and other properties. The residue from the cold water extraction was then boiled

---

[19] Bergman's Essays, (Cullen), I, p. 139.

with a large quantity of water, which dissolved mainly sulphate of calcium (gypsum). The residue from hot-water extraction, Bergman says contains generally three ingredients, though sometimes more.

If iron is present, the dry residue is subjected for "several weeks" to sunlight which renders the iron insoluble in acetic acid, after which acetic acid dissolves the calcium and magnesium carbonates which are separated by dilute sulphuric acid, precipitating calcium and dissolving magnesia. The residue from acetic acid treatment consists of clay, silicious matter and iron. The clay and iron are dissolved by "marine" acid (hydrochloric) and the iron precipitated by caustic alkali (phlogisticated alkali) and the clay by alkali carbonate. The silicious matter may be identified by its complete solution with effervescence under the blowpipe with "mineral alkali" (sodium carbonate).

This outline of the general scheme of quantitative analysis is necessarily incomplete, but it can be readily seen that systematic as it is, it could not give very accurate results. Though Bergman was apparently considered in his own time the master of quantitative analysis, and his method was quite generally adopted as authoritative, yet he was not himself so accurate an analyst as some of his contemporaries. This was partly owing to his habit of weighing constituents in the form of their crystalline salts, a method which itself was capable in many instances of introducing errors. Some of Bergman's contemporaries exceeded him in accuracy of determination—even when following his own scheme of analysis. Klaproth improved on Bergman by heating constituents to dryness when possible before weighing, and thus obtained more accurate results in general.

Bergman published also a treatise on the analysis of several noted mineral waters of Europe—Seltzer, Spa, Pyrmont, Seydschutz, Aix-la-Chapelle, Medway, and various local water supplies of Upsala in Sweden—tabulating the results.

One very interesting research by Bergman was intended to throw light upon the much disputed differences in composition of cast iron, wrought iron and steel. For this purpose he collected very many samples from many sources. His first effort was to ascertain the relative purity of these samples of iron. As a phlogistonist, he believed that the purest sample of the metal should contain the most phlogiston, and he accepted the interpretation first announced by Cavendish that "inflammable air" (hydrogen) obtained by the action of hydrochloric or dilute sulphuric acid on certain metals was nearly pure phlogiston. He therefore dissolved equal weights of his irons in hydrochloric acid ("marine acid") and measured the volumes of hydrogen set free. He found that the average of his various kinds of iron gave volumes in the ratio of 50 for wrought iron, 48 for steel, and 40 for cast iron. The inference was therefore that their relative purities were in this ratio.

To confirm the results obtained by this method, Bergman utilized another process which the phlogistonists understood in this way: when a metal forms a calx (we say oxide) or when it forms a salt, it loses phlogiston. When a metal is precipitated in the metallic state from a solution of its salt, it regains phlogiston. Therefore when one metal replaces another from a neutral solution, when no effervescence takes place, the quantity of phlogiston given off by the dissolving metal will be proportional to the quantity of the metal precipitated. Bergman used neutral solution of silver salts, and added identical weights of different iron samples, of which he had already determined the relative volumes of inflammable air given off. When action was complete, he weighed the quantities of silver precipitated.

For instance, Bergman found that 66.7 pounds of silver were reduced by 19.5 pounds of Osterby iron and by 17.9 pounds of Grangen iron. These quantities therefore contained the same amounts of phlogiston. Or in equal weights of the three metals, if we assume for silver the amount of phlogiston (unweighable) to be 100, the other metals would have 342 for the Osterby iron, and 373 for the Grangen iron. But these two irons gave ratios of volumes of inflammable air of 48 to 51 volumes. The ratio of 48 to 51 is practically the same as that of 342 to 373. Another pair of iron samples, which had yielded inflammable air in ratios of 48 to 46 cubic inches, gave ratios of phlogiston by the other method of 347 to 333. These ratios are identical, and Bergman naturally assumed that the method of measuring the relative phlogiston contents by the relative volumes of inflammable air yielded by solution in hydrochloric acid was a reliable method, and therefore the relative purity of the various samples of iron could be thus determined.

Bergman pursued his investigation of iron samples to determine what substances, not iron, constituted their impurities. He examined the residues from the solution in acid of the weighed samples. His results showed:

IMPURITIES IN IRON

	Silicious Matter	Graphite or Plumbago
Crude cast irons..................	1.0 to 3.4	1.0 to 3.3
Steels...........................	0.3 to 0.9	0.2 to 0.8
Wrought irons...................	0.05 to 0.3	0.05 to 0.2

the rest being iron with varying manganese content. His method of determining manganese was imperfect and the quantitative results unreliable.[20]

It may be seen from this illustration how the phlogistic philosophy, before oxidation phenomena were understood,

[20] Bergman, *Analyse du Fer*, translation of M. Grignon, Paris, 1783, p. 58. Bergman states that ''Plumbago is a species of sulphur composed of an acid saturated with phlogiston.'' By ''species of sulphur,'' Bergman means a combustible substance, which was generally understood by phlogistonists as a combination of phlogiston and some acid, here carbonic. His friend Scheele had, in 1779, shown that plumbago (graphite), by ignition with salt-peter, was converted into fixed air (carbon dioxide), and concluded that it was a combination of fixed air and phlogiston.

and before there was any concept of atomic weights of the elements, was used to explain quantitative relations discovered by experience. It may also be understood how this hypothesis, mistaken though it was, yet came to obtain such a hold on the chemists of the period, that it was with difficulty that they could accommodate themselves to accept the simpler and more correct explanation of existing relations.

Bergman contributed to many important fields of chemical knowledge of his time. He added to the knowledge of crystallography in a treatise on the "Forms of Crystals" presented to the Royal Society of Upsala in 1773. He was also the first to attempt a serious classification of minerals on the basis of their chemical composition. The Swedish mineralogist, Cronstedt, had indeed, in 1758, attempted a classification on this basis, but the facts of chemical composition were then too limited for a satisfactoiy outcome. Bergman himself had, however, in the following quarter of a century, made so many analyses of minerals, and Wenzel, Kirwan, Scheele, and others, had so added to the material, that Bergman's classification was far in advance of Cronstedt's beginning.[21]   This classification of Bergman was superseded by the later work of Hauy on mineralogy, 1801.

The contribution of Bergman to the knowledge of carbonic acid or "aerial acid" will be alluded to in connection with the development of Pneumatic Chemistry, and his extensive work on Chemical Affinity will be referred to in connection with the history of early ideas on that subject.

Bergman died in 1784 at the age of about fifty years, having contributed so importantly to many fields of chemical knowledge as to have won the respect and admiration of the whole chemical world. Though Lavoisier's new interpretation of the phenomena of oxidation and reduction was already promulgated. Bergman died still a believer

---

[21]Bergman, *Sciagraphia, regni mineralis secundum principia proxima digesti,* 1782. Accessible to the writer through the French translation of M. Mongez; New edition by J. C. Delamethène, two volumes octavo, Paris, 1792.

in the phlogistic hypothesis, and at the time of his death none of the important phlogistonists was convinced of the superiority of the new explanation.

Carl Wilhelm Scheele (1742–1786) was born in Stralsund, Swedish Pomerania, on December 19, 1742. He was one of the younger sons of a family of eleven children of a Stralsund merchant. At the age of fourteen years, after a brief school experience, Carl was entered as apprentice with an apothecary in Gothenburg. Endowed by nature with the ability and enthusiasm for investigation, the boy was fortunate to find in Herr Bauch a sympathetic master and in the pharmacy many chemicals and some apparatus. He also had access here to the textbooks on chemistry of Caspar Neumann, Nicolas Lemery, and Herman Boerhaave, then the best texts extant. Works of Kunckel, and Stahl were also studied by him. The eight years that Scheele spent with Bauch were years of intense study and experiment, his work keeping him often late into the night performing experiments described in his texts or on his own initiative.

When Bauch disposed of his business, Scheele took a place with an apothecary in Malmö. His new master, Kjellstrom, also encouraged his zeal for study. Here he formed a useful friendship with Andreas Johann Retzius, afterward a professor in Stockholm. In 1768, Scheele removed to Stockholm, as assistant to another apothecary. While there in connection with Retzius who had also come to Stockholm, he worked on cream of tartar and discovered and isolated tartaric acid, the work giving rise to a paper presented by Retzius to the Academy of Stockholm and published in 1770, being the first published paper bearing Scheele's name. In this year, Scheele moved to Upsala, taking a position with an apothecary named Lokk. The five years of his residence in Upsala were of great importance to Scheele's development and reputation. Not the least important event was his meeting here with Bergman, then at the height of his fame and influence.

The story of his introduction to Bergman, as told by
Thomson and by Kopp, is very interesting. It seems that
Lokk, his employer, had noticed the curious fact that when
saltpeter was kept in fusion for some time, its properties
were changed. Although still neutral, yet, when distilled
vinegar was poured upon it, red fumes were given off,
whereas previous to the heating, vinegar (acetic acid) had
no such action. Lokk mentioned this curious fact to the
mineralogist Gahn, desiring an explanation. Gahn could
offer none, but related the fact to Bergman who also could
offer no suggestion. When Gahn called later at Lokk's
shop, he learned that Scheele had explained the fact by
stating that there were two "spirits of niter"; besides
the ordinary spirit (our nitric acid) there was another
related to it. By heating saltpeter, this other was formed;
the first acid possessed a greater affinity for the base than
did the acid of vinegar, while the second variety (our
nitrous acid) had a less affinity for the base than the vinegar and was consequently driven off by it, forming those
red fumes.[22] When this was reported by Gahn to Bergman, he expressed a desire to become acquainted with
Scheele. The acquaintance thus formed led to a life long
intimacy of the two distinguished chemists. Retzius afterward stated that their relations were such that it was
difficult to decide which of the two was the teacher and
which was the taught.

Bergman persuaded Scheele to undertake the chemical
investigation of the "black magnesia" (black oxide of manganese) which resulted in the discovery of many manganese compounds and the first preparation of chlorine.
Bergman also facilitated the publication of Scheele's most
celebrated work on *Air and Fire,* and wrote a lengthy introduction to the work. In 1775, Scheele was elected to
membership in the Royal Academy of Sciences, an honor
never before extended to a man with no higher academic
status than that of a student of Pharmacy. The position of

---

[22] Cf. Kopp, *Geschichte der Chemie,* I, p. 256.

Scheele as a chemical genius was now firmly established. If he had possessed an ambition for prominence, he could certainly have fulfilled his aim; but he was desirous only of opportunity to quietly pursue his studies.

In this year of 1775, he learned of a position as superintendent of a pharmacy at Köping made vacant by the death of the pharmacist Pohls, the business being inherited by the young widow. Scheele understood that the business was prosperous and that the widow had considerable property. He applied to the goverment for a license for the appointment, passed the required examination with distinction, and received the appointment. As a matter of fact, he found the business more or less financially burdened. He learned soon after that the widow contemplated the sale of the pharmacy to another, and his disappointment was expressed in his letters to his friends. This occasioned many invitations to Scheele. Bergman invited him to come to Upsala, Gahn to join him at Falun, and it was also suggested that he become Chemicus Regius (royal chemist) at Stockholm. It is stated that he also received an offer of a salaried position in Berlin. Meanwhile his reputation and personality so appealed to the citizens at Köping that permission was obtained for him to open an independent pharmacy, with the promise of adequate patronage. As a result, the contemplated sale of the pharmacy was given up, and in 1777 a contract was signed with the widow whereby the title of the pharmacy passed to Scheele. The remainder of his life was passed at Köping, and here much of his splendid work was done. It is related that he contemplated marriage with the widow of his predecessor, who had acted as his housekeeper, so soon as he should have accumulated some means of his own. At all events, shortly before his death (1786) he willed the property to her, and in his last illness, and but two days before his death, they were married.[23]

[23] Cf. Tilden, William A., *Famous Chemists, the Men and Their Work*, London, 1921, p. 53 *ff.*

Bergman had died two years before, and thus, as wrote Crell in 1787, "the world lost in less than two years two men, Bergman and Scheele, two chemists who were deeply beloved and mourned by all their contemporaries, and whose memories a grateful posterity will never cease to honor." [24]

The scientific work of Scheele was of such a character as to have attracted the admiration of his contemporaries and of all succeeding chemists. His thorough chemical preparation, the ingenuity and skill with which he designed his experiments, the care with which he confirmed his results by varying his methods, the clearness with which he described his proceedings, and the independence and scientific logic with which he interpreted his results place him among the most brilliant of investigators. He was a consistent disciple of the phlogistic philosophy until his death, and though cognizant of the experiments of Cavendish and Lavoisier which were destroying the basis of that theory, he did not accept the interpretation of these facts as made by Lavoisier. But these developments came to him at a time when his working powers were impaired by ill health and in the last years of his life. It is hard for the reader of Scheele's papers to believe that, had he continued to work, he would have long continued an adherent of this theory, for difficulties were occurring to him which he hoped later to explain without discarding the phlogiston hypothesis.

The work (*Air and Fire*) which Scheele had completed for printing, by 1775, but which was not printed until 1777, was undertaken to attempt to solve the problem of the constitution of fire. Scheele recognized that this problem was not to be solved unless the constitution of the air, in which combustions take place, was also known. His first effort therefore was to analyze the air, and his first step was to subject a confined volume of air to various substances which, as he would say, give off phlogiston readily

[24] *Chemische Annalen*, 1787, Band I, p. 192.

(or as we would say, take up oxygen readily) and to note
the effect on the volume and character of the residual air.
For this purpose, he employed "alkaline liver of sulphur"
(alkaline sulphides), cloths dipped in a solution of potassium carbonate, and submitted to fumes of burning sulphur
(therefore potassium sulphite), turpentine oil, iron vitriol
precipitated by caustic lye (that is, ferrous hydroxide),
sulphur, phosphorus, etc. In all these cases, he found the
volume of air reduced roughly by one fourth to one third
of its original volume. Reasoning from the point of view
that a combustible substance is composed of phlogiston
and some acid, he concludes that the air has an attraction
for phlogiston, that the combination with phlogiston is the
cause of the disappearance of the air, but as to whether
the phlogiston still exists in the remaining air or whether
the disappearing air has combined with, or become fixed
in, the liver of sulphur, oil, etc., these, he says, are questions
of importance.

He then proceeds to prepare by various methods this
constituent of the air which supports combustion and which
he calls "fire air" (that is, oxygen), distinguishing the remainder of the air by the name "spoiled air." He prepares fire air by distilling fuming nitric acid and absorbing
the acid distillates by slacked lime; by heating black oxide
of manganese and sulphuric acid, by distilling manganese
nitrate or saltpeter (the latter, he says, is the cheapest and
best method). He also obtained fire air from silver nitrate,
precipitated by potassium carbonate, washed and dried.
The aerial acid (he adopts Bergman's name for Black's
"fixed air" or carbon dioxide), also given off was removed
by slacked lime from the oxygen given off. Scheele took
the "fire air" obtained by these methods, mixed it with
two or three parts of the "spoiled air," and showed that
it acted in all respects like common air. Scheele says:[25]

"I have reported that I have found the spoiled air lighter

---

[25] Scheele, *Sämmtliche physische und chemische Werke: Chemische Abhandlung über Luft and Feuer*, Berlin, 1891, p. 115.

than the ordinary air. Must it not follow that the fire air is heavier than our air. Indeed I actually found that after I had accurately weighed as much fire air as occupied the space of twenty ounces of water, this was nearly two grains heavier than just as much ordinary air.

"These experiments therefore show that the fire air is just that air by means of which fire burns in common air, it is only here mixed with such an air as seems to have no attraction for the combustible, and this it is which causes some hindrance to the otherwise rapid and violent kindling. And indeed if the atmosphere consisted of fire air only, water would furnish poor service in extinguishing conflagrations."

It is difficult, from a modern point of view, to understand why Scheele was not led on, by this clear comprehension of the nature of common air and of its relation to combustion, to see the unnecessary character of the phlogiston hypothesis, or at least to accept promptly the suggestion of Lavoisier to that effect. We can better understand the weight of authority of that theory, however, when we remember that Bergman (who prefaced Scheele's book), Kirwan and Priestley, who read and commentated it, also saw no reasons for abandoning that hypothesis, though Kirwan indeed some years later appreciated the logic of the facts.

Priestley had discovered oxygen "dephlogisticated air" in 1774 and published his experiments in 1775. Scheele's manuscript was with the publisher in 1775, but it is generally accepted that each worked without any knowledge of the experience of the other. Priestley is entitled to the credit of original discovery by priority of publication, though Scheele's laboratory notes, published by Nordenskjold in 1892, give evidence that Scheele had really obtained oxygen as early as 1771. He then called it "aer vitriolicus." [26]

Scheele's idea of what takes place in combustion in air is that the combustible body loses phlogiston under the

---

[26] Muir, *History of Chemical Theory*, p. 40.

influence of more or less heat when some substance is present which can take up the phlogiston, *for the latter never exists uncombined.* The fire air takes up this phlogiston and disappears as visible volume, and the combination of fire air and phlogiston becomes heat or the material of heat and light. Of the nature of fire air itself, he says: "I consider the fire air as an elastic fluid, consisting of a general inelastic foundation or saline principle (principium salinum), of a certain though small quantity of phlogiston, and a certain quantity of water." This statement Scheele made in 1785 after there had come to his attention the experiments of Cavendish and of Lavoisier, showing that water is produced by the union of definite weights of inflammable air (hydrogen) and "pure air" (oxygen). Scheele repeated the experiments himself with carefully dried inflammable air and fire air, and verified the deposition of water, but this did not convince him of the correctness of the conclusion of Lavoisier. For inflammable air was for Scheele as for Cavendish nearly pure phlogiston, and fire air (oxygen) he thought contained water as a constituent. But 1785 was the year preceding the death of Scheele, and his health was poor and his working power seriously impaired.

Scheele was in his forty-fourth year when he died. The volume of his publications was small as compared with the number and value of his experimental results. He was distinctly an investigator, and all his publications were upon subjects of his research, and these were in many fields of chemistry. At the suggestion of Bergman, he undertook an investigation of the so-called "black magnesia" (black oxide of manganese), the results of which he published in 1774. This investigation is a model of systematic and well directed research of a substance of unknown composition, In the course of it he observed and recorded the principal properties and reactions of manganese compounds, including the chameleon solution, or permanganate solution. He did not indeed obtain the metal manganese itself, though

Bergman in that year announced, apparently on the basis of Scheele's work, that the "black magnesia" was the calx of a metal as difficult to fuse as platina. Gahn, the Swedish mineralogist, however, in the same year succeeded in obtaining manganese metal by improved furnace methods.

By the action of "marine acid" on the black oxide of manganese, Scheele obtained chlorine gas and described its principal characteristic properties. He called it dephlogisticated marine acid. The name was reasonable from his point of view, since "inflammable" air (hydrogen) was conceived to be chiefly phlogiston and the above action deprived marine acid of its hydrogen. Chlorine was not conceived to be elementary in its nature even by Lavoisier; Sir Humphry Davy, in 1810, was the discoverer of its elementary nature, and he it was who suggested the name "chlorine."

Scheele proved that plumbago, when ignited with saltpeter, was converted into fixed air (carbon dioxide) and assumed therefore that it was composed of that acid and phlogiston, that is, it was the same in composition as charcoal. It will be recalled that Pott had demonstrated that plumbago contained no lead (plumbum) as had been generally assumed by his predecessors.

Scheele first prepared prussic acid, and first separated the hydrofluoric acid from fluor spar. He obtained and studied molybdic acid, tungstic acid, arsenic acid, and a number of organic acids, lactic, citric, and malic. He isolated a "sugar substance" (glycerol) from fats and oils. The green pigment, the arsenite of copper, still bears the name of "Scheele's Green." In these and other researches, Scheele operated with such skill and intuition, and his descriptions were so clear and his deductions so convincing that he acquired the highest reputation as an investigator among all his contemporaries.

CHAPTER XII

THE DEVELOPMENT OF PNEUMATIC CHEMISTRY IN THE EIGHTEENTH CENTURY

It was the development of the chemistry of gases that contributed chiefly to the overthrow of the theory of phlogiston. Yet the men whose discoveries contributed most definitely to that end were all themselves phlogistonists, with the single exception of Lavoisier, who was himself less a discoverer than a clear interpreter of the results of others.

The most productive of English chemists of the latter half of the eighteenth century, Joseph Black (1728–1799), Henry Cavendish (1731–1810), Joseph Priestley (1733–1804), and Richard Kirwan (1733–1812), were all phlogistonists, although Black and Kirwan indeed ultimately acknowledged the force of Lavoisier's logic, after their own chemical work was over.

The researches which distinguish Black, Cavendish, and Priestley as chemists, were almost entirely on the preparation, properties, and reactions of gases. On account of the importance of the chemistry of gases or "pneumatic chemistry" in the development of chemical science, it will be worth while to follow chronologically the work and ideas of chemists on this subject, the researches and views of Van Helmont, Rey, Boyle, Hook, and Mayow having already been considered.

The first investigator after Mayow to devote any considerable attention to the subject, was an English clergyman, Stephen Hales, who was interested in problems connected with the development of plant life. In connection with this subject, he made many experiments. Hales had observed that the atmosphere seemed to play an important part in the life and growth of plants as well as of animals, and he sought to discover something of these relations. Hales was not a chemist, but he had read much of Boyle's and of Mayow's work, and was manifestly impressed with the fact that gaseous bodies were often fixed, that is, absorbed or combined in many substances, and that this fact might be of importance. Hales published two volumes of his investigations, *Vegetable Staticks,* in 1727, and *Statical Essays* or *Haemastaticks,* in 1733. These works contain chapters on "Analysis of Air," etc., which comprise his work on gases.

He starts from the point of view that distillation will disengage gases absorbed or fixed, and therefore he distils various substances, vegetable, animal, and mineral, to discover the nature and the quantities of air so fixed. For this purpose, he subjected to distillation such various substances as hog's blood, tallow, horn, oyster shell, oak wood, peas, mustard seed, tobacco, brandy, well-water, niter, pyrites, phosphorus, antimony, (that is, the sulphide), etc. The various gases and mixtures of gases thus developed were passed from the retort and collected over water and their volumes measured, and then allowed to stand, after which the diminution of volume due to absorption by the water was noted. Not only distillation but also fermentation and putrefaction changes were studied in the same way. Hales also obtained and measured gases produced by the action of acids on metals, of aqua-regia on gold and on "antimony," of nitric acid on iron and on "antimony" ("antimony" meaning then the sulphide of antimony), and of diluted oil of vitriol on iron filings, etc.

It is evident therefore that the gaseous products obtained by Hales comprised nearly all the common gases in the impure state, and mixed with other gases. Even oxygen was evidently obtained, as he found much "air" set free by distilling saltpeter and bone ash, although he did not distinguish it from other kinds of air.

As Hales was chiefly interested in finding out how much "air" was "fixed" in all these substances and whether these airs retained their elasticity or were more or less "fixed" by standing over water, he did not investigate the essential differences between the various gaseous mixtures he obtained. In fact, it would appear that as he was not a chemist, it did not occur to him that it was possible for him to distinguish between them. "Air" was probably to him, as it was to Boyle, the same substance containing, however, many kinds of impurities which imparted to it various differing properties, odors, colors, etc., such as Boyle called "effluvia."

Though Hales' work contributed no completed chemical discoveries, his conscientious observations were later a source of inspiration and interest to experimenters, and he was an oft-cited authority for later chemists.

The next important work upon the gases of the atmosphere was that of Dr. Joseph Black (1728–1799). Black was of Scotch extraction, though his father was born in Ireland, and himself at Bordeaux, where his father was established as a wine merchant. Black's elementary schooling was at Belfast; thereafter he attended the University of Glasgow as a student of medicine. Here he came under the inspiring influence of Dr. William Cullen, a professor of medicine and a lecturer on chemistry. Black was taken by Dr. Cullen as his assistant in chemistry in which capacity he served three years.

While Cullen himself was not an important original investigator, as a teacher he exerted an unusually inspiring influence on the development of interest in chemistry in Great Britain. Professor Thomas Thomson[1] says of Dr. Cullen, referring to his call in 1756 to the professorship of chemistry at Edinburgh:

"The appearance of Dr. Cullen in the College of Edinburgh constitutes a memorable era in the progress of that memorable school. Hitherto, chemistry, being reckoned of

---

[1]Thomson, *History of Chemistry*, I, p. 307.

little importance, had been attended by very few students. When Cullen began to lecture, it became a favorite study, almost all the students flocking to hear him, and the chemical class becoming immediately more numerous than any other in the college, anatomy alone excepted. The students in general spoke of the new professor with that rapturous ardor so natural to young men when highly pleased.''

It will be recalled that Cullen translated Bergman's Essays into English.

Black left Glasgow for Edinburgh in 1751 to complete his medical studies. He received his medical degree there in 1754, presenting for his thesis the results of his investigation upon magnesia, lime, and many other alkalies, and "fixed air," upon which his fame chiefly rests. The subject of this thesis was prompted by the differing opinions of physicians as to the actions of certain remedies then in use for alleviating the pains of urinary calculi, these being usually strong alkalies. The results achieved by Black far surpassed in chemical interest however their possible medical value, and it resulted that in 1756 Black was appointed professor of anatomy and chemistry at Glasgow, succeeding Dr. Cullen, who was in that year called to Edinburgh. When, in 1766, Dr. Cullen resigned the Chair of Chemistry at Edinburgh, Black was appointed as his successor. This position he held until his death. In the last years of his life, his health failed and he was compelled to limit his activities. His last lectures were given in 1796–1797, and he died in 1799. Dr. Thomson says of Black at Edinburgh, that his talent for communicating knowledge was not less eminent than his faculty of observation, and that his lectures were attended by an audience which continued increasing from year to year for more than thirty years.

It is well to remember that at the time Black undertook his investigations, the prevalent belief was that the alkaline carbonates, or "mild alkalies," were simple bodies, that when they were combined with phlogiston, they yielded the caustic alkalies. So when limestone was heated and yielded quicklime, it was supposed that the heat material or phlogiston combined with this material to form quicklime. Hales had also shown that chalk, when heated, yielded a considerable quantity of absorbed or fixed air. Van Helmont, too, had long before noted that a peculiar kind of air, which he called *gas sylvestre,* was given off by burning charcoal and by the action of acids on lime and other alkaline substances, although he did not clearly differentiate the gas from other gases set free by other chemical reactions. Black's work from 1752–1754 (printed in 1755[2]) was the first to establish clearly the relation of his fixed air (carbon dioxide) to the "mild" and "caustic" alkalies.

Black first experimented on "magnesia alba" (carbonate). He proved that magnesia is essentially different from lime. He heated "magnesia alba" (carbonate) to "such a temperature as is sufficient to melt copper" to see whether, at that temperature, it would yield a true quicklime. He noted that the magnesia alba lost about seven twelfths of its original weight. The calcined magnesia dissolved in acids without effervescence, and from the solutions he obtained the same salts as were produced by dissolving the magnesia alba in those acids. Black then heated in a retort a weighed quantity of mild magnesia (carbonate), and, as he found in the cooled distillate only a little water, he justly concluded that the loss of weight on heating was mainly due to the loss of air. He next calcined two drams (160 grains) of mild magnesia, dissolved the residue in sulphuric acid, and added "alkali" (by which he meant the carbonates of sodium or of potassium) and obtained 150 grains of a magnesia with the same properties as the original uncalcined material. He therefore concludes that the "air" which was "fixed" in the alkali had been driven out by the acid and had been attached to the magnesia, yielding again the mild magnesia.

---

2 *Experiments upon Magnesia Alba, Quicksilver, and some other alcaline substances,* 1755, being the chemical part of his Latin thesis printed in 1754.

Black repeated these experiments, using chalk instead of magnesia, and showed in the same way, and with fairly accurate quantitative data, that quicklime differed from chalk in the same way that his calcined magnesia differed from mild magnesia. Although it was not possible to drive off the fixed air from "alkalies," to form caustic alkalies, yet he recognized that a similar relation must exist between them, as between chalk and quicksilver.[3]

Black recognizes that his fixed air is not the same as ordinary air, as Hales appears to think, for in discussing the attraction which quicklime and its aqueous solution possess for fixed air, he says:

"Quicklime, therefore, does not attract air when in its most ordinary form, but is capable of being joined to one particular species only, which is dispersed through the atmosphere, either in the shape of an exceedingly subtile powder, or more probably in that of an elastic fluid. To this I have given the name of *fixed air,* and perhaps very improperly: but I thought it better to use a word already familiar in philosophy than to invent a new name, before we be more fully acquainted with the nature and properties of this substance, which will probably be the subject of my future inquiry."[4]

As to the real nature of fixed air, Black, in his manuscript notes, says: "With regard to its origin, when treating of inflammable substances and metals, I shall consider this more completely. I shall now only hint that it is a vital air, changed by some matter, seemingly the principle of inflammability," [that is phlogiston].[5] A contemporary of Black, Dr. Leslie, also says, "Dr. Black seems to consider fixed air as a particular modification of common air with the principle of inflammability."[6]

Black was an adherent of the phlogiston theory until after Lavoisier had published, in 1789, his *Elementary*

[3] *See* M. M. P. Muir, *History of Chemical Theories and Laws,* N. Y. and London, 1907, pp. 203–207.
[4] Cf. Wm. Ramsay, *The Gases of the Atmosphere,* London, 1896, p. 55.
[5] Wm. Ramsay, *loc. cit.,* pp. 59, 60.
[6] P. Dugud Leslie, *A Philosophical Inquiry into the Cause of Animal Heat,* London, 1778, p. 152.

Treatise on Chemistry, with the new chemical nomenclature based on the antiphlogistic philosophy. In a letter to Lavoisier in 1791, Black first acknowledges the superiority of the new point of view, although he says that for thirty years he has believed and taught the phlogistic theory.[7] Black is further distinguished by his discovery of the latent heat of melting, and of vaporization of water (1762), although a Swedish physicist, J. C. Wilcke, had also developed the idea of latent heat about the same time.[8]

David Macbride, a prominent surgeon of Dublin, was the next to contribute to the chemistry of gases. He published a work entitled *Experimental Essays* in 1764.[9] Macbride was especially interested in the fermentation processes in the animal body. Knowing that "fixed air" was an important product of these fermentations, he was led to investigate fixed air. His book consists of five essays, two of which, "On the nature and properties of fixed air," and "On the dissolvent power of quicklime," contain his contribution to the knowledge of fixed air.

Macbride was cognizant of the earlier work of Van Helmont and he recognized that his *gas sylvestre* was the same as *fixed air.* He also cites the term *gas subtile* of early chemists as a synonym, and he uses the simple word *gas* as synonymous.[10] Macbride also was thoroughly acquainted with the work of Hales and of Black, whose results he understands and thoroughly appreciates.

Macbride lays great stress on a supposed function of fixed air in acting as the immediate cause of cohesion in bodies either mineral or organic. This theory he accepts from the earlier speculations of Hales and of Haller. Hales had said:[11]

[7] Cf. Kahlbaum and Hoffman, *Ueber die Einführung der Lavoisier'schen Theorie in Deutschland,* Leipzig, 1897, p. 133.

[8] For a recent and comprehensive account of Black, *see* Sir William Ramsay, *The Life and Letters of Joseph Black, M.D.,* London, 1918.

[9] His book was translated into French by Dr. Abadie, and published in Paris in 1766. It is this translation upon which the present writer is dependent.

[10] "Afin d'éprouver les effets du gas, ou le vapeur qui se dégage dans le premier degré de fermentation." Macbride, *Abadie,* p. 319.

[11] Hales, *Vegetable Staticks,* London, 1727, I, page 314.

"The air is very instrumental in the production and growth of animals and vegetables, both by invigorating their several juices while in an elastic and active state, and also by greatly contributing in a fixed state to the union and firm connection of the several constituent parts of those bodies, viz. their water, salt, sulphur and earth." Macbride finds great justification for this idea in his own experiments, for after the distillation, ignition, or fermentation of substances which yield fixed air by these processes, they all lose their coherency. Macbride says:

"We shall see in what follows that the opinion of Hales and Haller is well founded and that the principle which is generally known as fixed air is the immediate cause of cohesion, since the preservation of the solidity and good condition of bodies depends upon that which prevents the flight of this air; for at the moment when it escapes and recovers its elasticity, we shall see that the other constituent parts, the terrestrial, the saline, the oily or inflammable, and the aqueous, being set in motion by that, commence immediately to exercise their different powers, attractive and repulsive, and enter into new combinations which first change and finally destroy the texture of the substances that they had previously composed, provided that this substance contains in it water enough to permit the intestinal [that is internal] movement by giving it the proper degree of fluidity."

Macbride also attributed to fixed air important antiseptic and antiscorbutic properties. This opinion of Macbride inspired Priestley's invention of water charged with fixed air or "soda water" as it came to be called.

Attributing such importance to the functions of fixed air, Macbride conceived it of importance to determine in his experiments the amount of fixed air set free, as distinguished from any other airs or mixtures of airs also produced. His method was well devised, though the apparatus, he says, was "the invention of Dr. Black, who communicated it to my very ingenious friend Dr. Hutchison, lecturer on chemistry in the University of Dublin."

This apparatus consisted of two bottles or jars, with a

glass tube connecting the necks of the bottles. The smaller of the bottles contained "volatile alkali spirit distilled over quicklime" (ammonia water freed from carbonate), while the larger bottle contained the material which evolved the fixed air—fermenting or putrefying material, or chemicals generating fixed air. This bottle was provided with a stoppered inlet, through which acids or other material could be added. Fixed air generated in this bottle passed over into the smaller vessel and was absorbed by the ammonia, forming the carbonate. When the evolution was complete, clear lime water was added and chalk was precipitated. The chalk was allowed to settle, filtered, and acid was added to set free the fixed air, which, measured, gave the quantity of fixed air given off by the fermentation or other reaction, as distinguished from any other gaseous products mixed with it.

Macbride made an interesting test to ascertain the carrier of fixed air in blood. He drew blood from a healthy person and separated the clear serum from the coagulum containing the red corpuscles. The clear serum, treated with clear lime water, he found yielded no precipitation of chalk on standing. The coagulum, however, gave a notable precipitation of chalk when so treated, and he rightly concludes that "the fixed air appears to be united to the red corpuscles and to that portion of the blood that M. Senac calls 'lympha coagulabilis.' "[12]

In 1766 appeared the first contribution of Henry Cavendish (1731–1810), that distinguished investigator and eccentric personality. Descended from a long line of English aristocracy, he was born at Nice, his mother having gone to that genial climatic region on account of her health. She died when he was but two years old. Little is known of his earlier years except that he attended school at Hackney in 1742 and that he entered St. Peter's College in Cambridge in 1745. He remained at Cambridge in regular attendance for the conventional four

---

[12] Macbride, *Abadie*, p. 354.

years, but did not take the degree. His biographer, Professor George Wilson, surmises that this may have been for the reason that he was reluctant "to submit to the stringent religious tests applied in his day to candidates for degrees." [13]

From the time of his leaving Cambridge until he joined the Royal Society in 1760, there seems to be no record of his activities. But in the Royal Society, where he formed his few associations, he was soon recognized for his scientific ability as well as for his strangely shy personality and eccentric behavior.

Dr. Thomas Thomson[14] relates that during his father's lifetime Henry Cavendish received an annuity of 500 pounds. After the death of his father and of other relatives, he became very wealthy, but as he had no extravagant tastes, he had little use for his large income. At the time of his death, he was the largest shareholder in the Bank of England, and his estate was estimated by Dr. Thomson at 1,300,000 pounds, and by Sir William A. Tilden[15] at about 1,500,000 pounds.

Biot says, in the *Biographie Universelle,* that he was the wealthiest of all scholars (savants) and probably also the most scholarly (savant) of all the wealthy. His wealth, however, meant little to him; he did not vary his methodic style of living and left to his bankers the investment of funds, stipulating only that he should not be bothered about it. He occasionally made gifts, often of generous amounts, to worthy objects, but apparently only when friends suggested the desirability of such action, and with little deliberation on his own part. An incident illustrative of this is given by Professor George Wilson on the authority of W. H. Pepys:

"At one time Mr. Cavendish had a large library in

---

[13] George Wilson, *Life of the Honorable Henry Cavendish,* etc., London, 1851, p. 181.
[14] *Op. cit.,* I, p. 336.
[15] Sir William A. Tilden, *Famous Chemists, the Men and Their Work.* London and New York, 1921.

London, which was in a bad state of arrangement. It was proposed to him to allow a gentleman, who was not very well off, to reside in the house, as being a clever man he would in return arrange the books, and render the library more useful for consultation, which Mr. Cavendish freely allowed. After this gentleman had resided there a considerable time, and had succeeded in classifying the books, he left to go to the country. Mr. Cavendish, dining one day at the Royal Society Club, some person present mentioned this gentleman's name, upon which Mr. Cavendish said, 'Ah! poor fellow: how does he do? How does he get on?' 'I fear very indifferently,' said this person. 'I am sorry for it,' said Mr. C. 'We had hopes you would have done something for him, sir.' 'Me, me, me, what could I do?' 'A little annuity for his life, he is not in the best of health.' 'Well, well, well, a check for ten thousand pounds, would that do?' 'Oh, sir, more than sufficient, more than sufficient.' "

Cavendish died in his seventy-ninth year after a brief illness, quietly and refusing all attention or attendance at his deathbed. His biographer, Dr. Wilson, offers his estimate of the character of Cavendish, in part, as follows:

"Morally it was a blank, and can be described only by a series of negations. He did not love; he did not hate; he did not hope; he did not fear; he did not worship as others do. He separated himself from his fellow men, and apparently from God. There was nothing earnest, enthusiastic, heroic or chivalrous in his nature, and as little was there anything mean, grovelling, or ignoble. He was almost passionless. All that needed for its apprehension more than the pure intellect, or required the exercise of fancy, imagination, affection or faith, was distasteful to Cavendish. An intellectual head thinking, a pair of wonderfully acute eyes observing, and a pair of very skilful hands experimenting or recording are all that I realize in reading his memorials. . . . Cavendish did not stand aloof from other men in proud or supercilious spirit, refusing to count them his fellows. He felt himself separated from them by a great gulf, which neither they nor he could bridge over, and across which it was vain to stretch hands or exchange greetings. A sense of isolation from his brethren made him shrink from their society and avoid their presence, but he did so as one conscious of an inferiority, not boasting of his excellence. . . . His theory of the universe seems to have been, that it consisted solely of multitudes of objects which could be weighed, numbered, and measured; and the vocation to which he considered himself called was to weigh, number, and measure as many of those objects as his alotted three score years and ten would permit." [16]

To whatever degree this estimate of Dr. Wilson may be true to the real Cavendish, it may be accepted as a faithful picture of the impression which Cavendish made by his personality upon the great majority of his acquaintances, but no one seems to have doubted his devotion to his ideals of scientific truth nor the consistency and honesty with which he pursued them.

The first publication by Cavendish was on *Factitious Airs,* three papers read before the Royal Society in 1766. The term "factitious air" was used in the same sense as by Boyle a century earlier. Cavendish says:

"By factitious air, I mean in general any kind of air which is contained in other bodies in an inelastic sense and is produced from thence by art. By fixed air, I mean that particular species of factitious air, which is separated from alcaline substances by solution in acids or by calcination; and to which Dr. Black has given that name in his treatise on quicklime."

The first of the three papers is on inflammable air, the second on fixed air, and the third on certain experiments on the air produced by fermentation and putrefaction, an examination to see whether they yield any other sort of air besides fixed air as shown by Dr. Macbride.

Inflammable air was first clearly noted by Boyle about

---

[16] Readers are referred for a comprehensive account of Cavendish's life and work to the above-noted life by Dr. Wilson, and especially to the *Scientific Papers of the Honorable Henry Cavendish, F. R. S.* Two volumes, Cambridge. 1921. Volume I contains his electrical papers with introduction by Clark Maxwell; Volume II contains his Chemical and Dynamical Essays wth an introduction by T. E. Thorpe.

1670, or possibly even earlier by Turquet de Mayerne,[17] and was well known by chemists of the time of Cavendish, although there was much confusion between different inflammable airs as to their nature, as they were sometimes hydrogen, sometimes carbon monoxide, and sometimes hydrocarbons. Cavendish, however, leaves no doubt of the kind he means, because he begins his paper by saying: "I know of only three metallic substances, namely zinc, iron, and tin, that generate inflammable air by solution in acids: and those only by solution in the diluted vitriolic acid (that is, sulphuric acid) or spirit of salt (hydrochloric acid)."

Cavendish found that at 30 inches barometer, and 50° Fahrenheit temperature, one ounce of iron gave 412 and one ounce of zinc gave 202 ounce measures. These volumes are approximately inversely proportional to the present atomic weights of these metals.[18]

Cavendish determined that from nitrous (nitric) acid, or concentrated oil of vitriol, no inflammable air was produced by these metals, also that from copper and "spirit of salt" (hydrochloric acid) there was nearly no action in the cold and that from hot acid no inflammable air was produced, but that the air that was then given off, lost its elasticity when in contact with water. This he notes as "remarkable enough to deserve mentioning." Evidently this was hydrochloric acid gas, though Cavendish does not examine it further than to describe its sudden absorption by the water.

Cavendish studied the inflammable air obtained by different acids on the metals, and found no difference between the properties of the gas from these sources. He showed this gas to be insoluble in water or alkalies, fixed or volatile. He found inflammable air to be about 10½ to 10⅞ times lighter than common air. The real value is about 14.4 times lighter, but Cavendish's method at this period of his work of weighing either common air or in-

---

[17] *See* ante, pp. 357–362.
[18] It will be remembered that in Cavendish's time there was as yet no concept of combining, or atomic, weights of the elements.

flammable air in a distended bladder, was incapable of giving accurate results. What Cavendish understands respecting the nature of this inflammable air he expresses thus:

"It seems likely from hence that when either of the above mentioned metallic substances (zinc, iron, tin) are dissolved in spirit of salt, or the diluted vitriolic acid, their phlogiston flies off, without having its nature changed by the acid, and forms the inflammable air; but that when they are dissolved in the nitrous acid, or united by heat to the vitriolic acid, their phlogiston unites to part of the acid used for their solution, and flies off with it in fumes, the phlogiston losing its inflammable property by the union."

This suggestion of Cavendish that inflammable air is phlogiston was accepted as the reasonable interpretation by nearly all his contemporaries, though in later years Cavendish saw reasons for believing that inflammable air was a combination of phlogiston and water, but this idea was not promulgated by him until 1784.

The paper on fixed air is an extension of the work of Black and Macbride in determining more carefully and quantitatively the properties and reactions of fixed air. He determined that water at 55 degrees Fahrenheit dissolved a little more than an equal volume of "the more soluble part of this air." He found that after boiling for fifteen minutes, all fixed air was expelled from the water solution. By the use of bladders for weighing, he found the specific gravity of fixed air at 1.57 heavier than common air. This result was much more accurate than his determination of the specific gravity of inflammable air, the correct value being 1.53. He determined the proportion of fixed air in marble at 408/1000 (instead of about 440/1000), and determined also the proportion of fixed air in other alkaline carbonates. In connection with this work, he notes an observation of Dr. Black that a solution of salt of tartar (potassium carbonate), exposed to the open air for a long time, formed some crystals which seemed to be the alkali united to more than its usual proportion of fixed air. To test this, Cavendish dissolved a weighed quantity of pearl-ash (potassium carbonate) in water in a bottle, to the open mouth of which was fixed a bladder kept full of fixed air by means of a tube from a generating bottle, which was supplied with "marble and spirit of salt." The bottle was agitated from time to time, and the crystals forming on the surface were thrown down and fresh solution exposed to the "air." These crystals were finally removed, dried on filter-paper, and analyzed. He found 42.3 per cent of fixed air (theory is 43.6 per cent). Previous experiments with pearlash yielded 28.7 per cent (theory is 31.8 per cent). Thus Black's surmise was proved justified, and the quantitative relation approximately determined.

The third investigation, on the air production by fermentation and putrefaction, was undertaken with a view of determining whether these processes yield any other air than the fixed air which Macbride had shown was given off. He therefor conducted fermentation experiments with sugar solution, and with fresh apple cider, and found that the gas given off was all fixed air, with properties identical with the fixed air from marble. The putrefaction experiments were conducted with "gravy broth" and with raw meat and water. The air given off was conducted into a bottle containing alkali (sope leys) and the unabsorbed gas, which was of considerable volume, was found to be inflammable and its specific gravity about one tenth of that of common air. He concludes that this air is the same as that from metals, though it seems a little heavier, and is "mixed with some air heavier than it, and which has in some degree the property of extinguishing flame like fixed air."

These experiments of Cavendish, carefully described and giving characterizations of fixed and inflammable air more specific and detailed than in any previous investigations, was of considerable volume, was inflammable and its spe-only theoretical suggestion made, that inflammable air was phlogiston, was also immediately adopted by Kirwan, Scheele, and other phlogistonists.

The next significant publication on air was by Daniel Rutherford in 1772 in his doctor's thesis. Rutherford was a pupil of Dr. Black and the subject was suggested by Black. Dr. Black had shown that fixed air could be separated from the air which no longer supported combustion and respiration, but other constituents of the air which no longer supported combustion were uninvestigated, and this was the problem he suggested for Rutherford.

Rutherford's experiments were devoted to completing combustion in a confined volume of air, and examining the residual air, after absorbing the fixed air by lime water. He found that it was not a simple thing to burn the air to complete saturation with phlogiston, as the current theory had it, or as we would now say, to complete combination of its oxygen. After a mouse died in the enclosed air, the residual air still supported the combustion of a candle, and after the candle was extinguished, lighted tinder would still smoulder a short time. Rutherford found that burning phosphorus was most efficient, and the fumes of the burning phosphorus could be absorbed by limewater. Though Rutherford does not appear to have investigated thoroughly the properties of this residual air, he calls it mephitic air and characterizes it as atmospheric air saturated with phlogiston.

To Rutherford is attributed the first isolation of the gas now called nitrogen. It is worthy of note in this connection that Cavendish left among his unpublished papers one describing this gas more specifically than did Rutherford. The manuscript in question bore a superscription by Cavendish "Communicated to Dr. Priestley," and Dr. Priestley himself refers to its contents in his account of *Experiments and Observations made in and before* 1772, the same year in which Rutherford's paper appeared. This paper by Cavendish was published by Mr. Harcourt in 1839 in the British Association's Papers (page 64 *ff.*).

Cavendish prepared the gas by passing atmospheric air repeatedly over red hot charcoal and removing by means of caustic potash the carbon dioxide (fixed air) formed. Cavendish says:

"The specific gravity of this air was found to differ very little from that of common air: of the two, it seemed rather lighter. It extinguished flame, and rendered common air unfit for making bodies burn, in the same way as fixed air but in a less degree, as a candle, which burned about 80 seconds in pure common air, and which went out immediately in common air mixed with 6/55 of fixed air, burnt about 26 seconds in common air mixed with the same portion of this *burnt air*." [19]

In 1774 Torbern Bergman presented his treatise on the atmospheric acid (Luftsaure or "Aerial acid") the most complete and systematic discussion of the sources, preparation, properties and combinations of carbon dioxide and carbonic acid. He begins by explaining that about 1770 he had informed his foreign correspondents of his ideas of the nature and properties of that elastic fluid, and cites Dr. Priestley who mentioned his ideas in the Philosophical transaction for 1772 and in a new edition of his work on airs had confirmed them by several fine experiments.

Bergman explains why he prefers the term "air acid" or aerial acid to the then usual name—fixed air. In the first place, because this is only one of several kinds of air which occur fixed, and in the second place, because it is at the same time a true acid and a constant constituent of the atmosphere. Fixed air, he says, is a true acid, because it possesses a distinctly acid taste; it reddens litmus ("turnsol"); it attacks caustic fixed alkalies, rendering them mild; a smaller quantity of this acid than of the stronger acids saturates these alkalies and renders them crystallizable and less soluble; it makes the volatile alkali (ammonia) more fixed, less odorous and penetrating and causes it to crystallize; when it just saturates quicklime, it deprives it of its solubility and acrimony and causes

[19] Dr. G. Wilson, *Life and Works of Henry Cavendish*, p. 28.

it to crystallize, but when in excess it renders it again soluble; it produces the same effect with terra ponderosa (baryta); it produces with magnesia a neutral crystallizable earthy salt; with iron, zinc, and manganese it forms salts which, when dissolved in water, redden the tincture of litmus, like all other salts of the metals.[20]

Bergman describes at length the preparation and properties of carbonic acid salts, with determinations by weight of the quantities of acid and base (these not always accurately, however). He also determined the relative "elective attractions" of the acid for different bases. His order of such affinities is as follows:

pure terra ponderosa	(baryta)
pure lime	(calcium oxide)
pure fixed vegetable alkali	(potassium hydroxide)
pure fixed mineral alkali	(sodium hydroxide)
pure magnesia	(magnesium oxide)
pure volatile alkali	(ammonium hydroxide)
zinc	
manganese	
iron	

This is a fairly correct order of the general stability of the corresponding carbonates. Bergman notes that this acid appears the weakest of all known acids and that the specific gravity is one and a half times that of air. Cavendish had announced it at 1.57.[21]

In discussing an experiment by Priestley—in which an electric spark passed through air confined over litmus solution in an inverted U-tube produced an acid reaction on the litmus (oxidation of nitrogen to nitrous acid)—Bergman makes this interesting statement:

"We now know that common air consists of three elastic fluids mixed together; viz., 1st of the aerial acid in its disengaged state, but in so small quantity that it alone cannot impart a visible redness to tincture of turnsol; 2nd of an air unfit for sustaining flame, or being subservient to

---

[20] *Bergman's Essays,* translated by William Cullen, I, p. 72, *ff.*
[21] *See* ante, p. 474.

respiration (this we may call vitiated air until we are better acquainted with its nature and properties) ; and 3rd of air indispensably necessary to flame, and animal life, which forms only about one fourth of common air, and which I call *pure air.*" [22]

This obvious reference to oxygen is of especial interest. It is established that Priestley had in August 1774 first prepared oxygen and had, in October, stated to Lavoisier and to others in the latter's laboratory in Paris his discovery of an extraordinary gas, which supported combustion to an unusual degree. This he had obtained by heating mercury precipitate. At this time he stated that he had given no name to the new gas. Lavoisier repeated the experiment in November 1774, and in February 1775 announced his discovery to the Academy of Sciences, calling the new air, *purer air* (air plus pur). Priestley's publication of his discovery of "dephlogisticated air" was in 1775. Bergman's treatise was delivered in 1774 at the Academy of Sciences of Upsala, though not printed until 1775. The question arises as to whether Bergman was drawing upon earlier knowledge of Scheele's discoveries or possibly had revised his manuscript for the printing in 1775. The expression "pure air" is not Scheele's, who called the gas "Feuer Luft," or "fire air." It is not Priestley's "dephlogisticated air." It is more like Lavoisier's "more pure air" or "very pure air." Scheele's best attempts to determine the proportion of his fire air in the atmosphere gave him about one fourth instead of one fifth, as Priestley's experiments showed.

No Englishman took a more prominent part in the discoveries in pneumatic chemistry than did Joseph Priestley. Without training in science, unfamiliar with the previous work of chemists in general, Priestley took up the study of chemistry as an amateur, but with great enthusiasm, a decided talent for experimental devices, and

[22] *Bergman's Essays,* translated by William Cullen, London, 1784, I, pp. 75, 76; also the same in French in *Opuscules chymiques et physiques de Bergman,* translated par M. de Morveau, Dijon, 1780, pp. 62, 63.

keen powers of observation; and he accomplished many
notable results.

"If," says Mr. Frederic Harrison "we choose one man
as a type of the intellectual energy of the eighteenth century, we could hardly find a better than Joseph Priestley,
though his was not the greatest mind of the century. His
versatility, eagerness, activity, and humanity; the immense
range of his curiosity in all things, physical, moral, or
social; his place in science, in theology, in philosophy, and
in politics; his peculiar relation to the Revolution, and the
pathetic story of his unmerited sufferings, may make him
the hero of the eighteenth century." [23]

Priestley was born at Fieldhead near Leeds, England,
on March 13 (old style), 1733. His family were Calvinists
and his schooling was directed toward the ministry. As he
early developed dissenting views, it was finally granted
him that he should be trained for a more liberal or less orthodox ministry at Daventry. He finished his formal course
of training of three years at twenty-two years of age. After some years of experience in the ministry, and in school
teaching, he was appointed teacher of classical languages
and polite literature at the Warrington Academy in 1761.
Here he remained until 1767, and his experience here was
of great importance to him in many ways. His teaching
was by no means confined to his nominal chair. Thorpe
says that there was practically no department of education
at the Academy in which at one time or another he was
not called upon to assist. Lectures on chemistry were
given at Warrington by Matthew Thorner, a Liver-
pool physician, who is believed to be the first to attract
Priestley's interest to chemistry, although Priestley apparently did nothing with it there.

He published an *Essay on Education* (1764) and conducted lectures on the *Study of History in General, History
of England,* and the *Present Constitution and Laws of Eng-*

[23] This quotation from Frederic Harrison serves to introduce the volume
of H. C. Bolton's *Scientific Correspondence of Joseph Priestley*, New York,
1892; and likewise the excellent work of Professor T. E. Thorpe, *Joseph
Priestly*, London and New York, 1906.

land. The first two were later published. He published a *Chart of Biography*—a tabulated compilation of eminent men of every age and profession, from which could be readily ascertained the relative periods and ages of the men at any time, the lengths of their lives, and so forth. For this accomplishment, he received from the University of Edinburgh the degree of Doctor of Laws. Occasional visits to London gave him opportunity of enlarging his acquaintance with eminent men. Here he became acquainted with Benjamin Franklin and formed an enduring friendship with him. Under the inspiration and at the suggestion of Franklin, Priestley wrote a *History and Present State of Electricity,* mainly a compilation from the *Philosophical Transactions,* though entailing much correspondence and some experimentation. This work met with general approval and passed through five editions during the author's lifetime. This publication secured his election to the Royal Society in 1766.

In 1767 Priestley accepted a call to preach at Mills Hill Chapel at Leeds. Here his position permitted him leisure to continue his scientific activities. He published in 1770 *A Familiar Introduction to the Theory and Practice of Perspective,* and in 1772, a *History and Present State of Discoveries Relating to Vision, Light and Colours.* In this year also (1772) appeared his first contribution to the chemistry of gases. Living next door to a brewery, he was stimulated to study the properties of the *fixed air* which lay over the surface of the liquid in the fermentation vats. When he removed his dwelling from that neighborhood, he continued his experiments with fixed air obtained from chalk and acid. Priestley added nothing of importance to the discoveries of Black, Macbride, Cavendish, or Bergman, with respect to fixed air, but he made an application of its use in 1772, which brought him the award of the Copley Medal in 1773. The basis of the award was thus described by Sir John Pringle, then President of the Royal Society:

"For having learned from Dr. Black that this fixed or mephitic air could, in great abundance, be procured from chalk by means of diluted spirits of vitriol; from Dr. Macbride, that this fluid was of a considerable antiseptic nature; from Dr. Cavendish, that it could in a large quantity be absorbed by water; and from Dr. Brownrigg that it was this very air which gave the briskness and chief virtues to the Spa and Pyrmont waters; Dr. Priestley, I say, so well instructed, conceived that common water impregnated with this fluid alone might be useful in medicine, particularly for sailors on long voyages, for curing or preventing the sea scurvy."

In 1772, Priestley accepted an offer of a position as librarian to Lord Shelburne, who had been Secretary of State for the Southern District with charge of the affairs of the American Colonies, under the ministry of Pitt. But, because of his conciliatory attitude towards the colonists, he had been, in 1768, relieved from this latter charge, and in the same year had resigned his office and was living in comparative retirement at his estate at Calne, though he was still active in the House of Lords. He was of scholarly tastes and desired a congenial companion as well as a librarian. Priestley was recommended by a mutual friend, Dr. Price, a well-known liberal, and, as Priestley had taken a prominent part in the support of the colonists' side of the controversies, he was doubtless for that reason more acceptable. The new position gave Priestley a much larger income, 250 pounds a year, with a residence at Calne in the summer and at London in winter, and with the assurance of 150 pounds annuity for life at the severance of their relations.

This situation Priestley held until 1780; and here he made his most important discoveries in chemistry, which were much appreciated and encouraged by Lord Shelbourne. Priestley's activity in political, educational, and theological propaganda was likewise continued, although the freedom with which he maintained his theological heresies produced an increasing unpopularity and eventually created somewhat strained relations between him and his patron. In 1780, therefore, their contract was terminated on Priestley's initiative, Priestley receiving regularly thereafter the promised annuity from Lord Shelburne.

In 1780, Priestley accepted the ministry of a dissenting congregation at Birmingham, where he found many sympathizers in his liberal views on political and religious matters, as well as an enthusiastic group of scientists in the celebrated Lunar Society, so called because it met monthly on the Monday evening nearest the full moon. Here he completed his six volumes on *Different Kinds of Air,* and produced a revised and condensed edition of the same work in three volumes, in 1784. His views on religious and on political questions were becoming more and more radical; a work of his on the history of the *Corruptions of Christianity* was received with a storm of hostile criticism from English and European Calvinists and Lutherans. In 1785, it was ordered to be burnt by the hangman at Dordrecht, in Holland. Priestley replied to his antagonists with a four volume work on the *History of Early Opinion Concerning Jesus Christ,* which only added to his unpopularity in orthodox religious sects and especially in the Established Church of England.

Conservative sentiment in England was also seriously disturbed, at this time, by the success of the American Revolution, and still more by the development of democratic spirit and the antichurch sentiment excited by the rise and progress of the French Revolution. As Priestley had favored the cause of the American colonists, so he was sympathetic with the ideals which dominated the rise and earlier development of the French revolutionary movement. The government party in England was aroused against Priestley, especially by his caustic reply to Edmund Burke's attack on the French Revolution in 1790. As Burke had been an outspoken advocate of the cause of the American colonists before the American Revolution, Priestley, who considered the principle of human liberty equally involved in both revolutions, arraigned Burke severely in a pamphlet dated January 1, 1791, in which he made a strong plea for the French Revolutionists. Government and church adherents, fearful of the influence in England of the revolution, were very indignant with Pristley, whom the great majority doubtless considered as a dangerous agitator.

At last, on July 14, 1791, the anniversary of the fall of the Bastile, a body of some eighty sympathizers having gathered for a celebration at a hotel in Birmingham, a mob assembled and stoned the hotel windows, though well after the adjournment of the meeting. (Priestley was not an attendant at this meeting.) Becoming more excited, the mob went to the New Meeting House, where Priestley preached, and burned all that was combustible in it. It then destroyed the Old Meeting House, and proceeding to Priestley's residence, the mob destroyed that and his laboratory, and other residences and meeting houses of unpopular dissenters. After three days of rioting, the arrival of dragoons put a stop to the activities of the mob. The King (George III) is quoted by Thorpe[24] from a letter to Secretary Dundas, approving the sending of the dragoons:

"Though I cannot but feel pleased that Priestley is the sufferer for the doctrines he and his party have instilled, and that the people see them in their true light, yet I cannot approve of their having employed such atrocious means of showing their discontent."

Priestley escaped personal injury by the mob, through the assistance of friends, and finally arrived in London. Here he endeavored to continue his ministry and other activities for some three years, and, though he had many offers of assistance from friends and admirers, public sentiment in general was so adverse that he gradually realized the futility of his efforts. He was assailed by the press

[24] Op. cit., p. 134.

and received many abusive communications. Edmund Burke attacked him on the floor of the House of Commons, and his fellows of the Royal Society were so generally unfriendly that he felt compelled to resign formally from that body. The facts that the French Academy of Science in July 30, 1791, addressed him a message of sympathy, and that the French Assembly in September, 1792, made him a citizen of France, and offered him a membership in the National Committee, were not calculated to increase his popularity in England. The courts eventually awarded him about 2500 pounds for the damage to his property at Birmingham, and he finally decided to emigrate to America, where his three sons were already established, and in April, 1794, he sailed for New York.

Here he was welcomed by many societies and individuals. He was offered the ministry of the Unitarian Church in New York, and was urged to take the professorship of chemistry in the University of Pennsylvania, but he finally decided to accept neither, and established himself at Northumberland, Pennsylvania, where he built a house and laboratory and spent the rest of his days. Here he completed his *History of the Church from the Fall of the Western Empire to the Reformation*. He wrote many theological papers, continued his chemical experiments, wrote two defenses of the phlogiston theory, the more elaborate on *Doctrine of Phlogiston Established and that of the Composition of Water Refuted*, printed at Northumberland in 1800, with a second edition at Philadelphia in 1803. He died in 1804 in his seventy-first year, and was buried in the Quaker cemetery at Northumberland.

The chemical work of Priestley which has given him so prominent a place in the history of chemical discovery was carried out between the years 1771–1777; and though his work and publications extended almost to the time of his death, yet in these later years he added little of importance. His chemical experimentation was indeed the recreation of a lifetime deeply engrossed in the duties of a preacher and theological writer. He was about thirty-eight years of age when he began his chemical activity; and, as before noted, he was possessed of no considerable previous training in chemical knowledge, or experimental methods. This may well account for the fact stated by Thorpe:[25]

"The contrast between Priestley, the social, political and theological reformer, always in advance of his times, receptive, fearless and insistent; and Priestley the man of science, timorous and halting when he might well be bold, conservative and orthodox when almost every other active worker was heterodox and progressive—is most striking."

The most productive years of his chemical discoveries were those spent with Lord Shelburne, when he was relieved from parochial responsibilities. Though Priestley entered upon his chemical researches with the preparation and the spirit of an amateur, his native ingenuity, the intense scientific curiosity he possessed, and his unquenchable enthusiasm enabled him to achieve very many important discoveries. The absolute frankness and, one might say, naiveté, with which he described his experiments and his interpretation of their significance rendered his writings readable and attractive. All that he did and thought was as a new world to him and he conveys that feeling to his readers. His attitude toward research, he states in the preface to the first volume of *Different Kinds of Air,* when he says:

"I do not think it at all degrading to the business of experimental philosophy, to compare it, as I often do, to the diversion of hunting, when it sometimes happens that those who have beat the ground the most, and are consequently the best acquainted with it, weary themselves without starting any game; when it may fall in the way of a mere passenger; so that there is but little room for boasting in the most successful termination of the chase."

Priestley's earliest important discovery was that of the gas which he called "nitrous air," now known as nitric

---

[25] T. E. Thorpe, *Joseph Priestly*, 1906, p. 168.

oxide. He tells us that he had been struck with Dr. Hales' account of an experiment, performed by him, in which an air, produced by the action of spirit of niter upon Walton pyrites, when mixed with common air "made a turbid red mixture and in which a part of the common air was absorbed." Priestley had never expected to see this interesting phenomenon, "supposing it to be peculiar to that particular mineral." Priestley, mentioning this to Mr. Cavendish in London in the spring of 1772, the latter suggested that other kinds of pyrites or even the metals themselves might answer as well, as probably the phenomenon depended on the spirit of niter. Acting on this suggestion, Priestley found that all the common metals gave, with spirit of niter (nitric acid), this peculiar kind of air, and that from all these metals the air was apparently the same.

The reaction between "nitrous air" and common air, he then studied in great detail. He collected the nitrous air over water and over mercury, and mixed it with common air in various proportions over water and over mercury. He soon established that the presence of a certain amount of water seemed to produce the greatest contraction of volume. He also found that the greatest amount of reduction in the volume of air so produced was one fifth, and that this reduction could be produced by about one volume of nitrous air to two of air. He then tested the behavior of nitrous air toward common air vitiated, or rendered impure, by combustion, putrefaction, or respiration, and thus found that the purer the air, the greater was the contraction in volume on addition of the fixed volume of nitrous air.

This discovery, that the "relative purity" of the air could be thus easily determined, attracted general attention, and more convenient forms of apparatus for measuring the purity of the air were soon proposed. One of the earliest was by Felix Fontana, professor of mathematics at Florence. Cavendish read a paper on a *New Eudiometer* before the Royal Society on January 16, 1783, which begins with these words: "Dr. Priestley's discovery of the method of determining the degree of phlogistication of air by means of nitrous air, has occasioned many instruments to be contrived for the more certain and commodious performance of this experiment; but that invented by the Abbé Fontana is by much the most accurate of any hitherto published." He then discusses in detail the relative merits and results of Fontana's and of his own apparatus. The word "eudiometer," now so commonly used for graduated apparatus for gas measurements, was thus first used to mean a measure of purity of the air. As the discovery of oxygen by Priestley was not made until August 1774, what was here meant by purity was the degree to which the air could support combustion or was respirable. Priestley had shown also that inflammable air and fixed air gave no reaction with his nitrous air. Priestley's determination of purity was no less important because no one yet knew that what they were really determining was the relative oxygen content of the airs tested.

Interested by Cavendish's observation upon the action of spirit of salt upon copper, in which he found no inflammable air produced, but an air which was extremely soluble in water, Priestley repeated this experiment but, as he had done with nitrous air, he collected this air also over mercury. He thus obtained a colorless gas very soluble in water. With lead, iron, tin, and zinc, he found that a variable mixture of inflammable air with this new air was obtained. He noticed that water impregnated with the new gas tasted very acid and dissolved iron very fast, yielding inflammable air. Finally, suspecting that the new air might come from the spirit of salt and not from the metal, he heated the spirit of salt alone, and found that "this air was immediately produced in as great plenty as before." He therefore rightly concluded that this "air is in fact nothing more than the vapour or fumes of spirit of salt," "and therefore may be very properly called an *acid air,* or more restrictively, the *marine acid air.*" Priestley, therefore, was the first to isolate hydrogen chloride, and to show that its solution in water was the well-known acid then called "spirits of salt" or "marine acid." This discovery occurred in 1772.

In 1773 it occurred to Priestley to apply the method he had used to obtain his "marine acid air" to see whether an alkaline air might be obtained from substances containing volatile alkali. He procured some "volatile spirit of sal ammoniac" (that is, ammonia water), placed it in a thin phial and heated it with a candle. A great quantity of vapor was discharged, which, collected over mercury, "continued in the form of a transparent and permanent air, not at all condensed by cold." Sal volatile (that is, ammonium carbonate) and other "salts obtained by the distillation of sal volatile with fixed alkalies," were tried but found to yield much fixed air also, so that he eventually used the mixture then customary for preparing the "volatile spirit of sal ammoniac," viz., one part of sal ammoniac with three parts of slaked lime, which furnished him a large and easily controlled supply of pure "alkaline air."

Having found that this new air was extremely soluble in water and that the solution was a very strong volatile spirit of sal ammoniac, Priestley next was curious to find out whether this alkaline air mixed with his marine-acid-air might not give a neutral air, "and perhaps this very same thing with common air." But, brought together, these two airs produced a "beautiful white cloud" which, when it had settled, he found to be common sal ammoniac (ammonium chloride). Priestley found the new gas, when mixed with fixed air, to yield oblong and slender crystals which "must be the same thing with the volatile alkalies which chemists get in a solid form by the distillation of sal ammoniac with fixed alkaline salts (that is, sal volatile)."

Priestley conducted many experiments with his alkaline air, as he had with his acid air, by means of which the more obvious physical and chemical properties were made known.

The isolation of the marine acid air suggested to Priest-

ley that airs might similarly be obtained from other
acids, and the "volatile vitriolic acid" was the first he
chose to investigate. He therefore wrote to his friend Mr.
Lane to send him a quantity of that substance, but, by a
misunderstanding, something else was sent, and the matter
went over till he met Mr. Lane, who told him that if he
would only heat "any oily or greasy matter" with oil of
vitriol, he would easily procure the "volatile or sulphureous
vitriolic acid." It was not, however, until the 26th of
November, 1774, that he was able to pursue this investigation. As, according to the theory of phlogiston, the volatile vitriolic acid was phlogisticated oil of vitriol, any solution rich in phlogiston heated with oil of vitriol should
give the volatile acid. He soon succeeded in producing the
gas from olive oil and oil of vitriol and later from oil of
vitriol heated with charcoal, mercury, and other substances. Collecting the gaseous product (sulphur dioxide) over
mercury was again his method for obtaining it in form
to study its properties.

Having in 1774 procured a lens of twelve inches diameter and twenty inches focal distance, Priestley "proceeded
with great alacrity to examine, by the help of it, what
kind of air a great variety of substances, natural and
factitious, would yield . . . on the 1st of August,
1774, I endeavored to extract air from *mercurius calcinatus
per se,* and I presently found that, by means of this lens,
air was expelled from it very readily." [26]

The substance he used was the red oxide of mercury obtained by heating mercury in air. He found that the air
so obtained was not imbibed by water.

"But what surprised me more than I can well express
was, that a candle burned in this air with a remarkably vigorous flame, very much like that enlarged flame with which
a candle burns in nitrous air exposed to iron or liver of
sulphur (that is, nitrous oxide reduced from nitric oxide,
his 'nitrous air'); but as I have got nothing like this re-

---

[26] Joseph Priestly, *Experiments and Observations on Different Kinds of Air,*
2d ed., London, 1776, II, Sec. III, p. 29 *ff.* "Of Dephlogisticated Air,
and of the Constitution of the Atmosphere."

markable appearance from any kind of air besides this particular modification of nitrous air, and I knew no nitrous air was used in the preparation of *mercurius calcinatus,* I was utterly at a loss how to account for it.''

At this time also, he tried his lens on ''red precipitate'' (that is, mercuric oxide made by dissolving mercury in nitric acid and igniting), and, obtaining similar results, he imagined something might have been communicated to it from the nitrous acid (our nitric acid), and that possibly also the *mercurius calcinatus* had collected something of nitre, in that state of heat, from the atmosphere. Priestley also found that *red lead* (minium) yielded the same gas but mixed with some fixed air, manifestly owing to impurities in his material. In October of the same year, his then patron, Lord Shelburne, took Priestley to the continent for a few weeks. While in Paris, he visited Lavoisier and other chemists, and in Lavoisier's laboratory he told this chemist, and several others present, of the strange air he had just obtained from *mercurious calcinatus* and from *red lead.* This announcement of Priestley's discovery while he had as yet but begun his investigation, and had as yet no name for his new gas, without doubt seemed much more sigificant to Lavoisier than it did to Priestley, for Lavoisier had himself already been occupied with the problems of the calcination of the metals, and with the general subject of pneumatic chemistry. On November first, 1772, Lavoisier had deposited a sealed note with the Secretary of the Academy of Sciences, in which he states that he has discovered that sulphur and phosphorus when burned gained weight.[27] ''This increase of weight is due to a great quantity of air which becomes fixed during the combustion and which combines with the vapours.'' He expresses his conviction that the same is true of all combustions and calcinations. In December of the following year (1773) he laid before the Academy a treatise in two parts, the first

---

[27] *Oeuvres de Lavoisier,* Paris, Imprimerie Impériale, Tome I, 1864, pp. 445–666. Marggraf had previously noticed the gain of weight in phosphorus on burning. (*See* ante, p. 440.)

being an historical review upon "Elastic Emanations" which are disengaged during combustion, fermentation and effervescence, from Van Helmont's time on, including very completely Priestley's experiments to that time. The second part of the work consists of an account of many experiments by Lavoisier himself upon changes taking place in calcination, the evolution or fixation of gases, etc., with careful data upon the changes of weight in these reactions. The trend of his thought may be gathered by the generalizations he draws in Chapter VI of this work, viz.:

1. That the calcination of metals when they are contained in a portion of air confined in a glass bell jar does not take place with quite the same facility as in free air.

2. That this calcination even has limits, that is to say when a certain portion of metal has been reduced to a calx in a given quantity of air, it is no longer possible to carry it beyond that calcination in the same air.

3. That in proportion to the calcination occurring there is a diminution of the volume of the air, and that this diminution is nearly proportional to the increase in weight of the metal.

4. That in comparing these facts with those reported in the preceding chapter, it would appear proven, that there combines with the metals during their calcination, an elastic fluid which becomes fixed, and it is to this fixation that is due their augmentation in weight.

5. That several circumstances would seem to tend to the belief that all of the air that we breathe is not fit to be fixed for entering into combination of metallic calxes, but that there exists in the atmosphere a particular elastic fluid which occurs mixed with the air, and that at the moment when the quantity of this fluid contained under the bell jar is exhausted, that the calcination can no longer take place, etc.[28]

It is manifest from Lavoisier's treatise that while skeptical as to the phlogistic theory, which he alludes to as

---

[28] *Oeuvres de Lavoisier*, Paris, Imprimerie Impériale, Tome I, 1864, p. 620.

the theory of the followers of Stahl, he was as yet not ready formally to advance a substitute.

We can imagine then with what interest and with how much greater realization of the importance of the new discovery, Lavoisier listened to Priestley's account of the new gas which supported combustion with such great energy. In the month of November 1774, the month following Priestley's visit, he began a verification of Priestley's experiment of heating *mercury precipitate per se* by the lens and in collecting and examining the properties of the air given off. The paper in which he announced the results of his experiment was reported in Rozier's Journal for May, 1775, and the memoir is on "The principle which combines with metals during their calcination, and which augments their weight." [29]

Lavoisier describes the well-known properties of this air, but makes no mention of Priestley's work on that subject, though in later writings he acknowledges his priority. He concludes his paper by expressing the belief that all metallic calxes, could we decompose them without reducing media such as charcoal, would also give this "purer part" of the air we breathe, and finally notes that as *mercurius precipitatus per se* heated with charcoal gives fixed air and mercury only, this fixed air "is the result of the combination of this eminently respirable portion of the air with the charcoal."

Priestley, after his return from the continent in November, 1774, did not take up the more extensive study of the new gas he had obtained from *mercurius calcinatus* until May 1, 1775. He then found that when tested for purity by his usual test, the nitrous air, that the new gas was much *purer* than common air, "even between five and six times as good as the best common air that I have ever met with." "Being now fully satisfied with respect to the nature of this new species of air, viz., that, being capable of taking more phlogiston from nitrous air, it therefore contains less

[29] *Oeuvres de Lavoisier*, Tome II, p. 122, *ff.*

of this principle: my next inquiry was, by what means it becomes to be so pure, or philosophically speaking, to be so much *dephlogisticated.*''

By *phlogisticated air* Priestley understood any air which had been rendered noxious, that is, a nonsupporter of combustion or respiration, this condition being generally recognized by chemists of the time to be produced by the phlogiston given off when substances were burned or when metals were calcined.[30]

His new gas supported combustion or respiration to a higher degree than common air, and therefore had a greater capacity for phlogiston than common air, and was, therefore, in relation to that, dephlogisticated. Priestley believed all gases to contain phlogiston, and the dephlogisticated air was, in his opinion, only relatively dephlogisticated. ''It is pleasing'' he says ''to observe how readily and perfectly dephlogisticated air mixes with phlogisticated air, so that the purity of the mixture may be accurately known from the quantity and the quality of the two kinds of air before their mixture.''

Priestley was far from any correct understanding of the nature of these gases. While he believed that his dephlogisticated air contained less of phlogiston than common air, and still less than phlogisticated air, yet phlogisticated air itself he conceived to consist of nitrous air and phlogiston, and common atmospheric air he considered to consist of ''the nitrous acid and earth, with so much phlogiston as is necessary to its elasticity and likewise so much more as is required to bring it from its state of perfect purity to the mean condition in which we find it.''[31] Priestley's ability in the realm of chemical philosophy was in no way commensurate with his enthusiasm and skill in experimentation or the acuteness of his power of observation. Phlogiston was to him a sort of mystical element which he used very ingeniously but not always consistently to solve his theoretical problems.

---

[30] Cf. Priestley, *Different Kinds of Air*, 2d ed., 1775, I, p. 178.
[31] Priestley, *Different Kinds of Air*, 2d ed., 1776, II, p. 55.

In 1777 appeared the notable work of Scheele on *Air and Fire* already referred to.[32]   It will be noted that this work contained many of the discoveries made by Priestley, and as Scheele's manuscript had been delayed for some two years in printing, the work of Scheele was independent of Priestley's publication and accomplished about the same time.   Scheele, however, interpreted his results as Priestley does in terms of the phlogistic hypothesis.[33]

It is evident that by 1777 Lavoisier was convinced that the phlogiston hypothesis was untrue to the facts as well as embarassing to the development of the science.   There were, however, certain unsolved problems which stood in the way of the general acceptance of the explanations from Lavoisier's point of view.   The principal one of these was connected with the nature of water.   The general opinion of water was that it was an element.   Any reaction which we should interpret as involving a decomposition of water had usually been explained by some combination of water with phlogiston or other material.   In 1783, however, Henry Cavendish proved that "inflammable air" combined with "dephlogisticated air" to form water and water only.   As Cavendish then considered inflammable air as phlogiston, this discovery Cavendish interpreted as proving that dephlogisticated air (that is, oxygen) was only water deprived of phlogiston.   In June of this year, Sir Charles Blagden, a mutual friend of Cavendish and Lavoisier, communicated to Lavoisier Cavendish's discovery and his interpretation.   That this announcement should have been, with his clearer viewpoint on oxidation phenomena very important and clarifying, may be easily understood.   He at once repeated this experiment of Cavendish and presented his results to the *Académie des Sciences* on November 12th, 1783.   An abstract was published in the December 1783 issue of Rozier's *Observations sur la Physique*. This was before Cavendish had formally made his an-

---

[32] *See* ante, p. 456.
[33] *See* ante, p. 450.

nouncement to the Royal Society in his *Experiments on Air*, January 15, 1784. Lavoisier, in this announcement, makes no reference to Cavendish as the first discoverer, though in the revised memorial printed in 1784, he says:

"This was on June 24, 1783, that we made this experiment, M. la Place and I, in the presence of MM. le Roi, de Vandermonde and other members of the Academy and M. Blagden, present Secretary of the Royal Society of London; the latter informs us that M. Cavendish had already tried burning inflammable air in closed vessels and that he had obtained a very sensible quantity of water."[34]

The question of the priority of the discovery of the composition of water gave rise to an extensive controversy between advocates of Cavendish, Lavoisier, Watt, and Priestley. The mass of evidence and argument cannot be summarized here. It must suffice to say that the final verdict is that, while Watt and Priestley had observed that the combustion of inflammable air in common air or in dephlogisticated air was accompanied by deposition of moisture, they had no realization of the significance of the phenomenon nor of the quantitative relation of the reaction. It is conceded that to Cavendish is due the credit of discovering that the two gases united completely to form water and water only, and that Lavoisier undoubtedly obtained his first knowledge of the reaction through Blagden from Cavendish. It is also true that Lavoisier was the only one of these men to comprehend the nature of the reaction, all the others being confused by their particular phlogistic hypotheses.[35]

---

[34] Lavoisier, *Oeuvres*, Tome II, p. 338.

[35] The evidence and arguments in the so-called "Water-Controversy" may be found in the following works:

James P. Muirhead; *Correspondence of the late James Watt on his Discovery of the Composition of Water*, etc., London, 1846.

George Wilson, M.D.; *The Life of the Honorable Henry Cavendish*, etc. London, 1851, pp. 265–445.

Hermann Kopp; *Beiträge zur Geschichte der Chemie*, Th. III, *Braunschweig*, 1875, pp. 235–310.

M. Berthelot; *La Révolution Chimique Lavoisier*, Paris, 1890, pp. 109–133.

G. W. A. Kahlbaum and August Hoffman; *Die Einführung der Lavoisier' schen Theorie in besonderen in Deutschland: Ueber den Anteil Lavoisier's an der Feststellung der das Wasser Zusammensetzenden Gase*, Leipzig, 1897, pp. 150–165.

The demonstration of the composition of water may be said to have removed the last obstacle to the substitution of Lavoisier's theory of oxidation for the phlogistic hypothesis. Cavendish was evidently impressed by Lavoisier's interpretation of the decomposition of water, for in his paper of January, 1784, he says:

"It seems, therefore, from what has been said, as if the phenomena of nature might be explained very well on this principle, without the help of phlogiston; and indeed, as adding dephlogisticated air to a body comes to the same thing as depriving it of its phlogiston and adding water to it, and as there are perhaps no bodies entirely destitute of water, and as I know no way by which phlogiston can be transferred from one body to another, without leaving it uncertain whether water is not at the same time transferred, it will be very difficult to determine by experiment which of these opinions is the truest, but as the commonly received principle of phlogiston explains all phenomena at least as well as Mr. Lavoisier's, I have adhered to that." [36]

It will be recalled that Scheele also, when informed in the year before his death, of the discovery of the composite nature of water was sufficiently interested to confirm the result of burning specially dried "inflammable air" and "fire-air," though he also preferred his complex assumption that "fire air" (or oxygen) was a composition of a saline principle, phlogiston, and water, rather than that it was simply an elementary constituent of water.

Another important discovery by Cavendish is based upon an observation of Priestley. Priestley had experimented by passing the electric spark through air confined over water colored with litmus, and found that the air was diminished in volume and that the litmus was reddened. As Priestley believed that electricity was another form of phlogiston, his results were puzzling to him. His curiosity excited by Priestley's observations, Cavendish also attacked the problem. This resulted in his proof that, by this means, practically all the phlogisticated air could by a sufficient

---

[36] *Scientific Papers of the Hon. Henry Cavendish,* II, pp. 180, 181.

excess of dephlogisticated air be converted into an acid. This acid Cavendish absorbed by an alkali (soap solution) and eventually recovered as niter.

"We may safely conclude, [he says] that in the present experiments the phlogisticated air was enabled by means of the electric spark to unite to form a chemical combination with the dephlogisticated air, and was thereby reduced to nitrous (that is, our "nitric") acid, which united to the soap-lees and formed a solution of nitre. . . . A furthur confirmation of it is, that, as far as I can perceive, no diminution of air is produced when the electric spark is passed either through pure dephlogisticated air, or through perfectly phlogisticated air, which indicates the necessity of a combination between these two airs to produce the acid."

In connection with this work, Cavendish used "a solution of liver of sulphur" to absorb the uncombined excess of oxygen—

"after which only a small bubble of air remained unabsorbed, which certainly was not more than 1/120 of the bulk of the phlogisticated air let up into the tube, so that if there is any part of the phlogisticated air of our atmosphere which differs from the rest, and cannot be reduced to nitrous acid, we may safely conclude that it is not more than 1/120 part of the whole." [37]

This small volume of air, ignored for a hundred years by later experimenters, was presumably argon and its related gases. That Cavendish's estimate of 1/120 of the volume of the nitrogen used, or .65 volume per cent of the atmosphere, is smaller than the actual content (about .93 volume percent as at present determined) is doubtless due to the fact of the solubility of argon in water.

Lavoisier now considered the phlogiston theory as virtually overthrown, and turned to the organization of his new philosophy, called for a time the antiphlogistic philosophy, and now recognized generally as the foundation of the modern theory of oxidation and reduction.

---

[37] *Scientific Papers of Hon. Henry Cavendish*, June, 1785, II, p. 193.

CHAPTER XIII

EARLY IDEAS OF CHEMICAL "AFFINITY"

Doubtless the earliest experimenters in chemistry recognized that chemical action, sometimes energetic and sometimes sluggish and incomplete, was due to peculiar forces or attractions which caused these differences. The earliest chemists were, however, not primarily interested in accounting for such facts by physical causes. They were satisfied with noticing the facts, considering the causes as manifestations of divine intention or of mysterious occult powers. In later periods of development, it seems to have been considered that the cause which stimulated chemical combination was that substances which combined, did so because they were in some respects alike; "like likes like," "similia similibus" are phrases which embody, in a manner, very ancient symbolism. The word "affinity"—*affinitas,* as employed by early writers—implies the idea of a resemblance or similarity in some respects between the reacting bodies. Albertus Magnus, in the thirteenth century, uses the word "affinitas" in this sense when he says that "sulphur destroys the metals because of its natural affinity to them." J. R. Glauber, in his *Novi Furni Philosophici* (1648), has the same notion when he says, "For sand and its like have a great community ("Gemeinschaft") with the salt of tartar (that is, potassium carbonate) and they love each other very much, so that neither of them willingly parts from the other."

It will be recalled that Boyle, in his *Sceptical Chymist*[1] (1680), protests against the prevalent accrediting to material substances of the ideas of antipathy and sympathy,

---

[1] *See* ante, pp. 403–404.

or enmity and amity, these qualities being attributes of
the human mind and not common to inanimate bodies.

Perhaps the earliest attempt to give to this force of
chemical affinity a more precise definition was by Isaac
Newton in his *Opticks*.[2] His consideration of the subject
is in No. 31, the last of a series of queries propounded to
the reader at the close of the last book of his *Opticks* "in
order," as he says, "to a farther search to be made by
others."

"Have not the small Particles of Bodies certain Powers,
Virtues or Forces, by which they act at a distance, not
only upon the Rays of Light for reflecting, refracting and
inflecting them, but also upon one another for producing
a great part of the Phenomena of Nature? For it's well
known that Bodies act one upon another by the attractions
of Gravity, Magnetism and Electricity; and instances shew
the Tenor and Course of Nature, and make it not improbable but that there may be more attractive Powers than
these. For Nature is very consonant and conformable to
herself. How these Attractions may be performed, I do
not here consider. What I call attraction may be performed
by impulse, or by some other means unknown to me. I
use that Word here to signify only in general any Force by
which Bodies tend towards one another, whatsoever be the
Cause. For we must learn from the Phaenomena of Nature
what Bodies attract one another, and what are the Laws
and Properties of the Attraction, before we enquire the
Cause by which the Attraction is performed. The Attractions of Gravity, Magnetism and Electricity, reach to very
sensible distances, and so have been observed by vulgar
Eyes, and there may be others which reach to so small distances as hitherto escape Observation; and perhaps electrical Attraction may reach to such small distances, even
without being excited by Friction.

"For when Salt of Tartar [that is, carbonate of potas-
sium] runs *per deliquium* [that is, deliquesces spon-
taneously] is not this done by an Attraction between the

---

[2] 1st ed., 1701, 2d ed., London, 1718. It is this second edition from which
the quotations are made, p. 350, *ff*.

Particles of the Salt of Tartar, and the Particles of the
Water which float in the Air in the form of Vapours? And
why does not Common Salt, or Saltpeter, or Vitriol, run
per deliquium, but for want of such an attraction? Or why
does not Salt of Tartar draw more Water out of the Air
than in a certain Proportion to its quantity, but for want
of an attractive Force after it is satiated with Water?
And whence is it but from this attractive Power that Water
which alone distils with a gentle lukewarm Heat, will not
distil from the Salt of Tartar without a great Heat? And
is it not from the like attractive Power between the Particles of Oil of Vitriol and the Particles of Water, that Oil
of Vitriol draws to it a good quantity of Water out of the
Air, and after it is satiated draws no more, and in Distillation lets go this Water very difficultly? And when Water
and Oil of Vitriol poured successively into the same Vessel
grow very hot in the mixing, does not this Heat argue
a great Motion in the parts of the Liquors? And
does not this Motion argue that the Parts of the two
Liquors in mixing coalesce with Violence and by consequence rush towards one another with an accellerated
Motion? . . . When Salt of Tartar *per deliquium,*
being poured into the solution of any Metal, precipitates the Metal and makes it fall down to the bottom of the Liquor in the form of Mud: does not
this argue that the acid particles are attracted more
strongly by the Salt of Tartar than by the Metal,
and by the Stronger Attraction go from the Metal to the
Salt of Tartar? . . . The parts of all homogeneal hard
Bodies which fully touch one another, stick together very
strongly. And for explaining how this may be, some have
invented hooked Atoms, which is begging the Question;
and others tell us that Bodies are glued together by rest,
that is by an occult Quality, or rather by nothing; and
others, that they stick together by conspiring Motions, that
is by relative rest among themselves. I had rather
infer from their Cohesion, that their Particles attract one
another by some Force, which in immediate Contact is exceeding strong, at small distances performs the chymical

Operations above mentioned, and reaches not far from the Particles with any sensible Effect.''

Newton adduces many different chemical reactions to illustrate his point of view and mentions other attractions, such as cohesion and capillary attraction, advancing numerous hypotheses, not all of which are at present justified. With respect to chemical attraction, however, he recognizes the varying degrees of attraction among similar actions, as when—

''a Solution of Copper dissolves Iron immersed in it and lets go the Copper, or a solution of Silver dissolves the Copper and lets go the Silver or a solution of Mercury in Aqua Fortis being poured upon Iron, Copper, Tin or Lead dissolves the Metal and lets go the Mercury, does not this argue that the Acid Particles of the Aqua Fortis are attracted more strongly . . . by Iron than by Copper, and more strongly by Copper than by Silver, and more strongly by Iron, Copper, Tin and Lead, than by Mercury?''

It may well be that when Newton speaks of explanations based on hooked atoms or on conspiring motions, he is referring to some speculations of Boyle, Lémery, and others of his predecessors, who sought to explain the mechanism of chemical action by the shapes of the ultimate particles and their interpenetrations or entanglements. Boyle and Lémery were believers in the corpuscular or atomic structure of matter, and both attributed to the physical structure and motions of these corpuscles many properties of substances otherwise unexplained.

This suggestion of Newton's of the existence of a special kind of attraction for chemical actions differing in its manifestation from the ordinary phenomena of gravitation, magnetism, or electricity, and subject to laws of its own, as yet unknown, made immediate impression on chemical thought. Its tendency was to cause chemists to think of chemical action in terms of mechanical forces, that is as an attraction producing motion of some kind among the minuter particles or atoms of bodies. In the version of Boerhaave's *Chemistry*, published in 1727, by Drs. Shaw and Chambers, the above article of Newton's is cited in a footnote,[3] quite extensively, and very appreciatively. In the text also, the function of chemistry is defined in the following manner:

"All the operations therefore which chemistry performs on bodies are mere changes in respect of Motion. Now a body may be changed in motion two ways: either when its whole bulk is removed from place to place, which does not come under the consideration of Chemistry, but of mechanics; or, when its parts are changed among themselves, that is when there is a transposition of its constituent parts."

These changes, however, do not go so far as to produce alterations in the elements themselves:

"Art goes no farther than to elements. . . . And hence Chemistry may be defined as the art of Changing bodies by solution or coagulation. In effect Chemistry in all its latitude is either the separating of parts before united, or uniting parts before separated, that is either the adding of bulk to bulk or separating of bulk from bulk."[4]

Boerhaave uses the term "affinitas" in his Latin treatise, but no longer in the sense of the ancients, implying a likeness of properties or contents of the reacting bodies, but it is applied to the tendency to react between bodies of opposite as well as of similar qualities, as with Newton. Writers after Boerhaave use apparently the term "affinity" as attraction, with Newton's significance for a specific attraction between reacting bodies. We find, also that the emphasis of attention is rather upon the limitations and laws of attraction than on its ultimate cause which indeed is little comprehended today.

Buffon, the celebrated French naturalist, about 1778 advanced the proposition that the phenomena of chemical affinity could be accounted for by the force of gravitation, the manifestations of its action being modified by the small distances between particles and by their varying shapes.

---

[3] *A New Method of Chemistry,* written by H. Boerhaave. Translated by P. Shaw, M.D. and E. Chambers, Gent. London, 1727, p. 170, *ff.*

[4] Boerhaave, *op. cit.,* p. 174. It will be recalled that Boerhaave, in his (Latin) *Elementa Chemiae* of 1732, declines responsibility for any previous version of his chemistry.

This theory was endorsed by Bergman, by de Morveau, and others.

The attention of many chemists from about this time was devoted to ascertaining the laws and generalizations of chemical attraction or affinity. The first serious attempt to systematize the relative affinities between substances was that of Étienne François Geoffroy (1672–1721), Professor of Chemistry at the Jardin du Roi from 1712 to 1731. He presented a memoir to the Academy of Sciences at Paris in 1718, entitled *Table of the different Connections ("rapports") observed in Chemistry between different Substances*. In this he lays down as his fundamental law: "Whenever two substances having some tendency to combine with each other are found combined and there enters a third which has more affinity with one of the two, it unites with that one, setting the other free."

On this basis he constructed his table showing the relative affinities of many substances as he had determined them. His table was printed in chemical symbols—or shorthand.[5] The principle of its arrangement may be illustrated by the following translation into the English language of the first four of the sixteen columns. The substances at the head of the columns are related to those below in the order of diminishing affinities.

RELATIVE AFFINITIES

Acids	Acid of Sea-salt	Nitrous Acid (our nitric acid)	Absorbent Earth
Fixed Alkali	Tin	Iron	Vitriolic Acid
Volatile Alkali	Regulus of	Copper	Nitrous Acid
Absorbent Earth	Antimony	Lead	Acid of Sea-salt
Metals	Copper	Mercury	
	Silver	Silver	
	Mercury		

[5] *See* Muir, *History of Chemical Theories and Laws*, for a facsimile of the original table, p. 382.

Geoffroy meant by this that any one of the substances in a certain column had a greater affinity for the substance at the head of the column than any lower substance, and would therefore displace such substances from their combinations with the substance at the head. The example of Geoffroy stimulated many chemists to improve or to extend his tables of affinities. Gilbert, de Limbourg, de Machy, de Fourcroy, Wenzel, Rouelle, de Morveau, and Bergman are among those who helped in developing the affinity relations in the eighteenth century.

All these tables assumed that there existed a certain constant value for affinity, but the data varied naturally according to the conditions under which they were determined.

Wenzel (1777) endeavored to determine the relative affinities of different metals for the same solvent by making cylinders of standard size, covering with a protecting varnish all but the surface of one end of the cylinder, and determining the relative affinities by the relative velocities of the solvent action. He did not succeed, however, in obtaining results that were accurate.

Two very able chemists of the latter part of the eighteenth century devoted much attention to determining the relative affinities of chemical substances. These were Torbern Bergman, who presented his paper on Affinity at the Upsala Academy in 1775, and Guyton de Morveau, of Dijon, who published in the *Élémens de Chymie, Théorique et Pratique* (1777), a discussion of the subject, and later wrote for the *Encyclopédie Méthodique*[6] a more elaborate discussion. Both these chemists believed in the existence of a constant value for these affinities, though both realized the difficulties in the way of obtaining their values, owing to disturbing factors. Both recognized the disturbing influence of excesses of a reacting body, and the variations resulting from determinations at different temperatures. The tables of affinity which they constructed were the

6 Article ''Chymie,'' I, 786.

expressions of their judgement, from experimental data of various kinds, as to the normal relative affinities at ordinary working temperatures. Bergman, indeed, constructed tables for both the wet way and the dry way, thus recognizing the influence of the wide range of temperatures. His elaborate tables of affinity consisted of fiftynine columns headed by as many substances, acids, alkalies, the calxes of the metals, etc., with all other substances known to combine with them arranged below in the diminishing order of their supposed affinities. From these tables it was assumed that chemists would be able to foresee the course of any action between the corresponding substances. He calls them tables of "Simple Elective Attractions." They may be illustrated by the following translation of columns one (1) and forty-eight (48).[7] (See p. 507).

The tables of affinities and particularly those of Bergman made a strong appeal to the chemists of the latter period of the eighteenth century. Lavoisier evidently was strongly impressed that in that direction lay the hope of developing chemistry to a true science, though he perhaps, more than any other appreciated the obstacles that lay in the way of that development. His latest discussion of the subject was in his comments upon Kirwan's book on Phlogiston, which it may be recalled was translated into French by Madame Lavoisier, with comments by Lavoisier, Monge, de Morveau, Laplace, Berthollet, and de Fourcroy. Kirwan had cited the table of affinities of oxygen from Lavoisier with several criticisms, and to these criticisms Lavoisier replies at length.[8] He begins:

"Mr. Kirwan, in the defects that he takes exception to in my table of the affinities of the oxygen principle with various substances, does not judge me more severely than I have judged myself, but he should be warned that all the objections he makes against this table I have made before he did, and perhaps in a stronger manner."

---

[7] Adapted from *Traité des Affinités Chymiques ou Attractions Électives:* traduit du Latin, sur la dernière édition de Bergman, Paris, 1788.

[8] *Essai sur la Phlogistique*, etc., traduit de l'Anglais de M. Kirwan, Paris, Metallic calxes

Column 1 Sulphuric Acid		Column 48 Calx of Mercury	
By the wet way	By the dry way	By the wet way	By the dry way
2 Baryta pure	Baryta pure	Acid sebacic	Gold
3 Potash "	Potash "	" hydrochloric	Silver
4 Soda "	Soda "	" oxalic	Platinum
5 Quick lime	Quick lime	" karabic	Lead
6 Ammonia pure	Magnesia pure	" arsenic	Tin
7 Magnesia "	Metallic calces	" phosphoric	Zinc
8 Alumina			
	Ammonia		Bismuth
9 Calx of zinc		" sulphuric	
10 " " iron	Alumina pure	" lactic	Copper
11 " " man-ganese		" tartaric	Antimony
12 " " cobalt		" citric	Arsenic
13 " " nickel		" formic	Iron
14 " " lead		" tungstic?	Saline liver of Sul-
15 " " tin		" malusic	phur?
16 " " copper		" nitric	
17 " " bismuth		" fluorhydric	
18 " " anti-mony		" acetic	
19 " " arsenic		" carbonic	
20 " " mercury			
21 " " silver			
22 " " gold			
23 " " plat-inum			
24 " " water			
25 " " alcohol			

Lavoisier then cites verbally from his *Mémoire* presented to the Académie des Sciences in 1782, in which that table of affinities was first printed. In this treatment, he begins by stating that he is not ignorant of the difficulties involved in making a table of affinities. And first, he says, all such tables represent only simple affinities while we recognize that there exist cases of double, triple, and much more complicated affinities. Next, the influence of temperature is considered, which complicates reactions by melting or vaporizing or otherwise affecting bodies in a way which alters their relative affinities. Mr. Bergman, he says, has sought to remedy this inconvenience by dividing his tables into two parts, one presenting the results of experiments in the wet way, and the other by the dry way, but to obtain tables rigorously in accord with experience, it would be necessary to make a table for each degree of the thermometer.

"A second fault of our tables of affinity is that they take no account of the influence of the attraction of water, and perhaps even of the decomposition of water in reactions by the wet way, because that acts as a real disturbance which ought to enter into account.

The third 'imperfection' of the affinity tables is in their inability to express changes which occur in the force of attraction, owing to the different degrees of saturation of substances. Thus sulphur and oxygen, in sulphuric acid, have a different attraction from that which these two substances have in sulphurous acid. Hydrochloric acid shows similar differences, and nitrogen, he says, is capable of combining with oxygen in a very great number of degrees of saturation.

"This which I have said against the tables of affinity in general naturally applies to the one I am presenting, but I think, nevertheless, that it may have some utility at least in so far as the more numerous experiences and the applications of calculation to chemistry place us in position to carry forward our views. Perhaps some day the precision of the data will lead to the point that the mathematician will be able to calculate in his study the phenomena of any chemical combination whatsoever, in the same manner, so to say, as he calculates the movement of the celestial bodies."

After the quotation from the memoir of 1782, Lavoisier states that in the four years since that presentation, he sees little to add to what he then said. He adds but two suggestions; first that we should avoid the mistake of supposing that one substance necessarily seizes on all of that substance for which it has the greatest affinity. As when sulphuric acid is boiled with mercury, copper, etc. In this case, only a part of the acid combines with the metal, and it is necessary to consider the oxygen as obeying two unequal forces, it is partly attracted by the metal, converting that to the oxide, partly by the sulphur, forming the sulphurous oxide. "In the second place, when I wrote in 1782, the decomposition of water was only a suspicion. The now proven decomposition of water obliges us to consider in a very different manner all affinities taking place in dilute water solution."

We possess only one later reference by Lavoisier to the affinity tables, viz., in the *Traité Élémentaire de Chymie* (1789). In the Preliminary Discourse, he says:

"This rigorous rule, from which I have not been able to deviate, of forming no conclusions beyond what experiments present, and of never supplying the absence of facts, has not permitted me to include in this work that part of chemistry the most susceptible, perhaps, of some day becoming an exact science; this is the part which treats of chemical affinities or elective attractions. Messrs. Geoffroy, Gilbert, Bergman, Scheele, de Morveau, Kirwan, and many others have collected a great number of particular facts, which only await the places which should be assigned to them; but the principal data are lacking, or at least those we have, are not sufficiently exact nor sufficiently certain to become the fundamental basis upon which can rest so important a part of chemistry. The science of affinities is moreover to ordinary chemistry as the transcendental geometry is to elementary geometry; I have not believed I ought to complicate by such great difficulties the simple and easy elements which will be, as I hope, in reach of a very great number of readers.

"Perhaps a sentiment of *amour propre* has given weight to these reflections, without my perceiving it. M. de Morveau is at the point of publishing the article *Affinité* of the *Encyclopédie Méthodique,* and I have good reasons to fear working in competition with him."

No better statement of the limitations of the affinity

problem was possible in the eighteenth century. Lavoisier's realization of the importance of the subject is justified by the results of the researches in the nineteenth and twentieth centuries of many of the ablest investigators, Berthollet, Berzelius, Davy, Faraday, Guldberg and Waage, Berthelot, Ostwald, Van't Hoff, Arrhenius, and many others. It is worthy of note that Lavoisier, in the treatise upon the new nomenclature in Part two of his *Traité Élémentaire,* in treating of the nomenclature of the salts of the various acids, arranges the bases under each of the acids "in the order of their affinities with this acid"; and this order is essentially the same as in Bergman's tables in the wet way.[9]

---

[9] Excellent articles on the development of the theories of chemical affinity are in: Raoul Jagnaux, *Histoire de la Chimie,* Affinité Chimique, 1891, I, pp. 300–360; Wilhelm Ostwald, *Lehrbuch der Allgemeinen Chemie,* 2te Auflage, 1896–1902, II, 2, *Verwandtschaftslehre,* pp. 18–198; M. M. Pattison Muir, *Chemical Theories and Laws,* 1907. Chap. XIV, pp. 379–430.

CHAPTER XIV

LAVOISIER AND THE CHEMICAL REVOLUTION

The history of the antiphlogistic theory would not be complete without giving credit to a Russian physicist and chemist, whose activity in chemistry was during the period of the most rapid development and spread of the theory of phlogiston (1741 to 1756). Michael W. Lomonossoff was born in 1711, the son of a peasant in the north of Russia. Against his father's wishes, he left his home at about twenty years of age (1731) to seek an education in Moscow. Here he studied, much burdened by poverty, for five years. A call came in 1735 from the Academy of Sciences in St. Petersburg for nomination of the best and most worthy students of the Moscow Academy to be sent abroad for study. Lomonossoff was among those chosen. He thus was enabled to study at Marburg and Freiberg for five or six years, devoting his attention largely to mathematics, physics, chemistry, and metallurgy. Returning to St. Petersburg in 1741, he was appointed an adjunct of the Academy and in 1745 was made professor of chemistry. Here he remained till his death in 1765.

Lomonossoff was a man of unusual versatility; his reputation as a poet was well recognized. He wrote a grammar and a rhetoric. He is credited with founding the art of mosaics in Russia, and wrote works on geography, astronomy, and metallurgy. Of his work in chemistry, strangely enough, only fragments have been preserved and apparently they made little or no impression upon the chemists of Europe of his time, and his work and his name seem to have been lost to chemical literature until the distinguished Russian chemist, Professor B. N. Menschutkin,

in 1904, collected and published his surviving notes.[1]
Though all his laboratory notes are missing and
his lecture notes are merely in the form of condensed digests, the chemical ideas are so sane and so
far in advance of such able contemporaries as Pott, Marggraf, Macquer, Cullen, etc., that, reading from the point
of view of present knowledge, it seems strange that they
could have been so neglected and forgotten. It is probable,
however, that they were so far out of sympathy with the
current ideas among the chemical thinkers of the very
popular phlogistic hypothesis, that they had no weight at
the time and perhaps, therefore, were never published in
generally accessible or popular form.

Lomonossoff approached chemistry from the point of
view of the physicist and mathematician. He believed that
the changes of matter should be capable of explanation on
the basis of mechanics, that they were due to motions of
the constituent particles. These particles consisted of
"elementa" or of corpuscles, elementa being portions
of a body which are composed of no smaller or different
kinds of parts (corresponding somewhat to the definition
of atom before radioactive phenomena were discovered),
corpuscles being the word used by Boyle, and used by
Lomonossoff as indicating the union of elementa to a
minute or inconsiderable mass (something like our molecule). These corpuscles are "homogeneous" when composed of the same kind of elementa (like our molecule of
an element), "heterogeneous" when composed of different
kinds of elementa (like our molecules of compounds) or,
when differently combined, or in different numbers. By
principium, Lomonossoff means any body which consists of
the same kind of corpuscles, that is any homogeneous substance.

---

[1] *M. W. Lomonossoff als Physiko-chemiker,* St. Petersberg, 1904. (In the
Russian language.)   Translated in great part by Dr. Max Speter and published as No. 178 of Ostwald's *Klassiker der Exakten Wissenschaften,* Leipzig,
1910.   See also the brief memoir by Alexander Smith, "An Early Physical
Chemist—W. M. Lomonossoff," *Journal of American Chemical Society,* 1912,
p. 109.

It was Lomonossoff's idea that it should be eventually possible by mathematics and mechanics to develop the science of changes in matter from the motions of these elementa and corpuscles on the assumption that heat was the cause of these motions. He adhered to the principles set forth by Gassendi and Descartes that heat is a mode of motion. This concept, obscured by the material concept of heat of the phlogiston hypothesis, but accepted by Lomonossoff, may be said to have protected Lomonossoff from many errors which confused his contemporaries. It will be remembered that experiments of Boyle, which satisfied him that the gain in weight of metals heated in contact with more or less air, was due to the absorption of some element of fire, had been quite generally accepted, although Mayow had a much clearer idea of the source of this added weight as coming from the "igneous particles" of the air.

Lomonossoff was prompted in 1756 to repeat Boyle's experiments, and he says:

"I have conducted experiments in air-tight sealed glass vessels, to ascertain whether the weight of the metals increases on account of the heat. These attempts showed that the opinion of the celebrated Robert Boyle is false, for without the admission of external air, the weight of the burned metal remains the same." [2]

These experiments of Lomonossoff, were some eighteen years previous to similar demonstrations by Lavoisier. It is in this proof and his rejection of the phlogiston hypothesis as an unnecessary hypothesis that Lomonossoff is a forerunner of Lavoisier. In his *Gedanken über die Ursachen der Wärme und Kälte* (1744 to 1747), Lomonossoff says:

"From all which we conclude that it is quite superfluous to attribute the heat of bodies to a subtle, specially devised matter. Heat on the contrary consists in an internal circular motion of the combined matter of the substance, etc." [3]

Antoine Laurent Lavoisier was born at Paris in 1743.

---

[2] Ostwald, *Klassiker*, No. 178, p. 51.
[3] Ostwald, *op. cit.*, No. 178, "Lomonossoff," p. 27.

His family was of the people rather than of the aristocracy. Antoine Lavoisier, who died in 1620, was a postrider or postillion, as was also his son of the same name, who became master of the post at Villers-Cotterets. His son, of the same name, was a bailiff; his son, Nicolas, a merchant; his son, Antoine, attorney or procurator of the bailliwick of Villers-Cotterets. His son, Jean Antoine, was procurator of the parliament at Paris, and married a Mlle. Punctis, daughter of a wealthy advocate. The great Lavoisier was the only son of this couple. A daughter died at the age of fifteen, leaving him the only child. His mother died also while he was a mere child, and his grandmother and an unmarried aunt, Mlle. Punctis, had the bringing up of the young Lavoisier, his father having come to live with them after the loss of his wife.

All three were devoted to the boy and there was fortune enough in the family so that no expense was spared in his education. He was educated at the Mazarin College, then distinguished for its courses in the sciences. Lavoisier distinguished himself in his studies. His first bent was toward literature; in 1760 he took the second prize in rhetoric. Soon, however, he developed a taste for mathematics and physical science, although pursuing legal studies as his main interest, eventually receiving the bachelor's degree in law, and obtaining an appointment as advocate or procurator to Parliament, the position previously occupied by his father.

His scientific studies were pursued with zeal, however, and in many lines. In mathematics and astronomy he was under the guidance of the eminent astronomer, Abbé de la Caille, in botany under Bernard de Jussieu, in mineralogy and geology under the the eminent Guettard, and in chemistry under Rouelle, an inspiring teacher and a distinguished chemist. Anatomy and physiology also claimed his attention to some degree. In his earlier years he devoted much attention to meteorology and to the construction of accurate barometers and other instruments.

In his twentieth year, he was already in correspondence with many of the most distinguished mathematicians, meteorologists, and astronomers of his time. In 1763, he accompanied Guettard on geological expeditions, paying attention also to botanical observations in the field.

At the age of twenty-two (1765), he presented his first paper before the Academy of Sciences, on the analysis of gypsum, in which he explains the action of the plaster of Paris in setting, as due to the reunion with expelled water of crystallization. In this paper also he determined the solubility of various specimens of gypsum (1 part to 426–476 water). The composition of gypsum, however, had been previously determined by Marggraf in Berlin in 1750, which Lavoisier acknowledges in an appended note, as having been brought to his notice since the reading of his paper.

In 1765 we find him presenting an essay in competition for a prize of the Academy, offered at the request of the king's ministry, for the best essay on the methods of lighting the streets of a large city at night. For this essay, he received a gold medal from the king. In the course of preparation of the essay, he made many experiments on lamps, reflectors, illuminating oils, with careful estimates of costs. It is related of him that, in order to make his eyes more sensitive for photometric purposes, he remained in a darkened chamber for six weeks.

In 1767, he accompanied Guettard on a royal commission to Alsace and Lorraine for the purpose of preparing a mineralogical atlas of France. In 1768, he was elected to the Academy of Sciences, though the appointment by the king was delayed, an older man receiving the honor. A few months later, however, Lavoisier, who in the meantime occupied a position created for him by the king's ministry, as adjunct chemist, received full standing in the Academy. He was then but twenty-five years old. From the time of his entrance he took a very active part both in the scientific and in the administrative work of the Academy, an activity which terminated only with the abolition of the Academy during the Reign of Terror.

In this same year—1768—Lavoisier, through his father's exertions and his own, became a member of the "Ferme Générale." This was a great corporation which, under charter of the king, had control of the leases of royal domains, the enforcing of laws pertaining to indirect taxes, customs, and revenues, the sale of salt and tobacco, and other large sources of revenue. For these privileges, the Ferme paid large annual royalties to the royal treasury. Lavoisier, at first, had a third of a share, but some years later increased his holding to a full share. His third of a share cost him about 520,000 francs. His income from his whole share in the ferme is said to have varied from 60,000 to 139,000 francs per annum. It is no matter of surprise that much odium should have been attached by the people to an organization with such power and such wealth. At times it was doubtless a power that was used unscrupulously, and under a corrupt court there was much corruption connected with its administration. The anecdotes, however, which illustrate this, usually—perhaps always—refer to a time previous to Lavoisier's connection with the Ferme. Lavoisier threw himself into the management of the organization with characteristic energy, and business sagacity. His influence seems always to have been used for better business methods and for honest administration. Nevertheless, the unpopularity of the company was the agency that finally brought the career of Lavoisier to its untimely end.

In 1771, Lavoisier married the fourteen year old daughter of M. Paulze, a wealthy member of the Ferme, though not so wealthy as was Lavoisier himself. His marriage seems to have been a happy one, and during all his later scientific work his wife was a zealous and able assistant in his laboratory and in his writing. After his death, also, she edited and published much of his scientific work.

The king was persuaded to separate the manufacture of

saltpeter, and gunpowder from the general management of the Ferme for the more perfect development of the industry, and Lavoisier was appointed one of the three commissioners to take charge of this department. He at once undertook the scientific development of this industry and soon brought the French powders to the position of the best in Europe. Several of his more laborious investigations were directed to the chemical questions involved in this work. From this time on, we find him connected with numerous important commissions and occupying the most varied posts of responsibility. He was at one time or another President of the Academy of Sciences, chief of the Bureau of Accounts, member of the commission of the National Treasury, member of the Orleans Assembly, member of the National Assembly, member of the commission for the revision of weights and measures, and of other commissions.

But the times were becoming stormy with the advance of the revolutionary movement, and Lavoisier, as a noble, a man of wealth, and one who had received many royal commissions, became more and more unacceptable to the radical element of the commune. The existence of the Academy of Sciences, as well as of all other institutions operating under royal charters, was threatened. Here Lavoisier proved his devotion to the Academy by his persistent efforts to maintain its integrity. He freely advanced money to sustain its scientific work, and endeavored to awaken a feeling of respect for its services. His efforts were futile; and in August, 1793, the Academy was abolished by a decree of the National Assembly.

Lavoisier began to feel his own insecurity; he was personally attacked in pamphlets. Gradually he withdrew from his public offices, giving his attention more completely to the work of the commission of weights and measures, then laboring with the determination of the standards of the metric system.

Finally came the blow which was to prove fatal to La-

voisier. The members of the Ferme Générale were arrested on the charge of having oppressed the people and
robbed the public treasury. An examination of their accounts was held which lasted several months and resulted
finally in the order for confiscation of the property of the
members of the Ferme and the handing over of their persons to the committee of public safety. The trial was
brief, as was usual with that body, preserving but a pretense of formality, and the death sentence was followed,
within twenty-four hours, by the execution on the guillotine
of thirty-two members of the Férme. M. Paulze, his
father-in-law, preceded Lavoisier to the block, and a moment later fell the head of France's greatest chemist.

The national repentance came soon. By a strange coincidence, the same man Dupin who had presented the de-
nunciation of the Fermiers Générales in the National Assembly, introduced, one year later, into the convention, a
resolution for the restitution to the widows and heirs of
the property of the "financiers unjustly condemned," and
this meager justice was accomplished.

In October, 1795, the Lycée des Arts, unveiled a bust of
Lavoisier with this inscription:

> Victime de la tyrannie,
> Ami des arts tant respecté,
> Il vit toujours par le génie
> Et sert encore l'humanité.

In August, 1796, the same society honored his memory
with a grand funeral ceremonial in the presence of three
thousand people, and a laurel-crowned bust of Lavoisier
was unveiled with impressive ceremony.

The last letter written by Lavoisier seems to have been a
letter to a cousin Augez de Villers, probably written after
the mock trial before the committee of public safety.

"I have achieved [he says], a passably long career, above
all very happy, and I believe that my memory will be accompanied with some regrets, perhaps with some glory.
What more could I desire? The emergencies in which I

find myself enveloped will probably avert from me the inconveniences of old age. I shall die complete, which is an advantage that I ought to estimate with the number of those I have enjoyed. If I experience some painful sentiments, it is that I have not done more for my family; to be deprived of everything and not to be able to give to them, to her, nor to you, some pledge of my attachment and of my gratitude. It is then true that the exercise of all the social virtues, important services rendered to my country, a career usefully employed for the progress of the arts and of human knowledge, do not suffice to preserve a man from a disastrous end and to prevent him from perishing like a guilty person.

"I write to you to-day because tomorrow it will perhaps not be permitted me to do it, and because it is a sweet consolation for me to occupy myself with you and with persons who are dear to me in these last moments. Do not forget those near who are interested in me, that this letter may be communicated to them. It is probably the last that I shall write you." [4]

LAVOISIER.

The work of Lavoisier covers a wide range of subjects. It has been collected and published in six large quarto volumes by the French government.[5] Many of these writings are reports written in his official capacity in the various bureaus and commissions of which he was a member. Such are, among many others, papers and reports upon:

Saltpeter production
Solid foods for use of sailors
The adulteration of cider
Report on projects for the removal of the abattoirs from the middle of Paris
Reports on the hospitals of Paris
Papers relating to the Bureau of weights and measures
Reports on agriculture, mines and mining.

---

[4] *Lavoisier, 1743–1794, d'après sa correspondence, ses manuscrits,* etc. Grimaux, Paris, 1888, pp. 296, 297.
[5] *Oeuvres de Lavoisier,* six volumes quarto, Paris, 1862–1893.

There are a number of papers upon physical subjects: essays upon the application of the heat of the sun's rays (by burning glasses); determining the specific heat of liquids and gases; the expansion of solids (with Laplace); the use of an ice-calorimeter for determining heat of combustion and specific heats (with Laplace); the weight of a cubic foot of water and the contents of the pint at Paris; observations on the great cold at Paris of 1776 (meteorological). His first chemical work was upon gypsum, previously alluded to, and published in 1765.

In 1770, was published the proof that water is not converted to some extent into earth by repeated distillation. Boyle had announced such to be the case; Boerhaave had re-investigated the question and found the contrary, but the error still had adherents; and Lavoisier proved by a carefully controlled investigation, that this "earth" was only the result of the corrosion of the glass vessel and was equal in weight to the loss in the weight of the retort.[6]

In 1774, he demonstrated that the gain of weight in the calcining of metals (tin) is at the expense of the air, and that the loss in weight of air equals the gain in weight of the tin. Boyle had experimented in a similar way but through oversight had come to wrong conclusions.[7] Lavoisier also conducted similar experiments on the burning of sulphur and phosphorus.

In 1775, Lavoisier published his paper on the composition of the air. In this historically interesting memoir, Lavoisier refers first to the action of heat upon a mixture of iron calx (oxide) and charcoal in giving "fixed air," and to the similar action of mercury precipitate and charcoal. He then describes an experiment, in which he subjected the mercury precipitate to strong heat by itself and collected the expelled gas over water. This gas on examination gave properties familiar to us as those of oxygen.

The testimony seems conclusive that Lavoisier had re-

---

[6] See Muir, *History of Chemical Theories and Laws*, New York and London, 1907, p. 49.

[7] *See ante*, p. 407.

ceived personally from Priestley cognizance of his experiments along a similar line and with similar results. It has been a shadow on Lavoisier's fame that in this paper no allusion to Priestley's work is made. Nevertheless, the memoir is almost epoch-making in chemical thought on account of the clear inferences made by Lavoisier, in contrast with Priestley's deductions as published shortly after Lavoisier's paper. Priestley, dominated by the phlogiston theory, supposed the resulting gas to be common air partly freed from the mystic phlogiston. Lavoisier, skeptical as to this theory, at once drew a logical scientific conclusion.

After noting the greater readiness of the air to support combustion, and applying to it the term "more pure" than common air, he says: "It appeared proved after that, that the principle which combines with metals during calcination and which augments the weight of them is no other than that "more pure" portion of the air that surrounds us, which we breathe and which passes in this process from a condition of expansibility into one of solidity; if then we obtain it in the condition of "fixed air" in all metallic reductions when we use charcoal, it is to the combination of this last with the pure portion of the air that this result is due, and it is very probable that all the metallic calxes would give, just as mercury does, this air eminently respirable if we could reduce them all without addition [of carbon] as we reduce the mercury precipitate (that is, oxide of mercury, as we now call it.)"

He proceeds to apply the same reasoning to the action of nitrates and carbon in giving fixed air, to show that nitrates must contain pure air. Finally he concludes:

"Since the carbon disappears completely in the revivication of the calx of mercury, and we obtain nothing but fixed air and mercury, we are forced to conclude that the principle to which up to now has been given the name fixed air is the result of the combination of the eminently respirable portion of the air with the carbon."

He thus ignores entirely phlogiston as a term in the chemical equation, and begins the antiphlogistic campaign.

We find Lavoisier now making many experiments with reference to the function in combination of the "air more pure." In 1777, he proves by heating mercury with oil of vitriol that sulphurous acid gas is produced and mercury precipitate left, which on heating can be made to give off the air "more pure," thus demonstrating the fact, to use our modern vocabulary, that sulphuric acid is a combination of oxygen and sulphurous acid. He also demonstrates in the same year that the conversion of pyrites into green vitriol is due to the union of the iron and the sulphur with the "air more pure." In the same year, also, he explains to the Academy his theory of combustion, which process he summarizes as consisting of four phenomena:

1. Heat or light is disengaged.
2. Substances burn only in "air pure."
3. The "air pure" which disappears is equal in weight to the augmentation in weight of the burned body.
4. The product of the combustion is an acid body.

He demonstrates that both bases and acids contain this "air pure."

On September 5, 1777, Lavoisier presented a paper to the Academy on "Considerations upon the Nature of Acids," in which occurs this noteworthy passage:

"I have already imparted to the Academy my first essays upon this subject. I have demonstrated to it in the preceding memoirs as far as it is possible to demonstrate in physics and chemistry, that the air more pure, that to which M. Priestley has given the name of dephlogisticated air, enters as a constituent part into the composition of several acids and notably of phosphoric, vitriolic, and nitrous (our nitric) acid. More numerous experiments place me to-day in the position of generalizing these conclusions and of advancing the proposition that the air more pure—the air eminently respirable, is the constituting principle of acidity, that this principle is common to all acids, and that there enter into the composition of each of them one or more other principles which differentiate them and which constitute them as one acid rather than

another. From these facts, which I regard already as very
solidly established, I will designate hereafter the 'dephlogisticated air' or 'air eminently respirable,' in the state of
combination and of fixity, by the name of 'acidifying principle' or if one likes better the same meaning under a
Greek word by that of 'oxygine principle.' This name will
save periphrasis, will introduce greater conciseness in my
manner of expressing myself, and will prevent misunder-
standings into which we shall be liable to fall if I employ
the word air.''

We know that this conclusion of Lavoisier regarding
the relation of oxygen to acids was too sweeping, as there
are acids which do not contain oxygen and oxides which
are base-forming and yet the generalization was at the
time extremely important, and the combustion theory itself a clear exposition of the general facts.

Under the stimulus of his new point of view, we find Lavoisier making many investigations, repeating numerous
experiments of other chemists, sometimes giving careful
and detailed credit to his predecessors, sometimes making
no references whatever to previous work, but reinterpreting
the results obtained in the light of his new point of view.
Thus he repeats and extends Priestley's investigations upon
respiration, and explains the function of oxygen in respiration.

It is to be noted that Lavoisier makes no direct or formal
attack in his earlier work upon the phlogiston theory, but
quietly leaves it out of account.

The phenomena of heat require explanation, however,
and he expresses himself in favor of the material theory
of heat—as an imponderable fluid pervading all space,
which condensing in the pores of a substance accounts for
the various phenomena of absorption or evolution of heat.
The physicists, in fact, were divided for a long time after
Lavoisier upon the nature of heat—whether it were a
mode of motion or an imponderable fluid. An English
writer, Metcalfe, in a two volume work on caloric, 1837,
presents the material theory about as strongly as possible.

It will be impossible to review in detail the particular investigations which Lavoisier carried on in extension of his theory.

The relation of Lavoisier to the discovery by Cavendish of the composition of water and his clearer concept of the nature of this process has already been discussed.[8]

In the year 1783, Lavoisier makes a formal attack upon the phlogiston theory in a memoir to the Academy, *Réflexions sur le Phlogistique,* a few lines of which it will be interesting to quote:

"In the course of the memoirs that I have communicated to the Academy, I have passed in review the principal phenomena of chemistry. I have insisted upon those which accompany combustion, the calcination of metals, and, in general, all the operations where there is absorption and fixation of air. I have deduced all the explanations from a simple principle. This is that the air pure, the vital air, is composed of a simple principle which is peculiar to it, which forms the base of it, and which I have called "principe oxygine"—combined with the material of fire or heat. This principle once admitted, the chief difficulties of chemistry have appeared to vanish and be dissipated, and all phenomena have been explained with astonishing simplicity.

"But if everything is explained in chemistry in a satisfactory manner without the aid of phlogiston, it is, by that only, infinitely probable that this principle does not exist, that it is a hypothetic entity, a gratuitous supposition, and surely it is according to logical principles not to multiply entities (êtres) unnecessarily. Perhaps I might have held to negative proofs, and contented myself with having proved that we can account for phenomena without phlogiston better than with phlogiston; but it is time that I explain myself in a manner more precise and formal upon an opinion that I regard as a sad error in chemistry—and which appears to me to have retarded progress by the bad method of philosophizing that it has introduced."

There follows a critical analysis of various of the doctrines of Stahl and their relation to observed facts; concluding with the expression that he does not expect his views to be accepted immediately.

"It is for time to confirm or to destroy the opinions I have presented. Meanwhile I see with great satisfaction that the young people who commence the study of chemistry without prejudices, and the geometers and physicists who have new heads for chemical facts, no longer believe in phlogiston in the sense in which Stahl presented it, and regard all this doctrine as a scaffolding more embarrassing than useful to continue the edifice of chemical science."[9]

Soon after the discovery of oxygen, Lavoisier had recognized that, as "fixed air" was obtained by the combustion of charcoal or the diamond, it was composed simply of carbon and oxygen. In his memoir with Laplace, *Sur Chaleur*, presented in 1782 before the Academy,[10] the quantitative composition of the oxide of carbon was approximately estimated. In a treatise presented in the same year (1783, though included in the volume for 1781) Lavoisier made a more accurate estimate from heating charcoal with minium, showing:

CARBON DIOXIDE

	Lavoisier	Actual Composition
Carbon.	72.125	72.727
Oxygen.	27.875	27.273

This was an important and original discovery.

In 1782, Lavoisier constructed a blast for oxygen for the purpose of producing high temperatures, and by its means first succeeded in melting platinum. The essay in which he announces this work is also interesting, because it con-

---

[9] *Oeuvres de Lavoisier*, Paris, 1862–1893, Tome II, p. 655.
[10] Published in the volume of memoirs for 1780 which was not printed until 1783. This habit of including in the memoirs for a particular year articles of importance of later origin is frequently confusing. As there was often a delay in printing of the memoirs of as much as three years, this was a possibility several times utilized by Lavoisier.

tains at the same time the first formal recognition of the priority of Priestley in the discovery of oxygen, and another quiet ignoring of Priestley's priority in the use of the oxygen blast. "This air," says Lavoisier, "which M. Priestley has discovered about the same time as I and I believe even before me." Of his present problem he says: "This idea has doubtless presented itself to many other persons before me, and I am even assured that M. Achard, a celebrated Berlin chemist, has made applications of it." About six years previously, Priestley, in his then celebrated work on different kinds of air,[11] had written:

"Nothing however would be easier than to augment the force of fire to a prodigious degree by blowing it with de-phlogisticated air, instead of common air. This I have tried, in the presence of my friend Mr. Magellan, by filling a bladder with it and puffing it through a small glass tube upon a piece of lighted wood, but it would be very easy to supply a pair of bellows with it from a large reservoir. Possibly much greater things might be effected by chymists in a variety of respects with the prodigious heat which this air may be the means of affording them. I had no sooner mentioned the discovery of this kind of air to my friend Mr. Mitchell than this use of it occurred to him. He observed that possibly platina might be melted by means of it."

Lavoisier makes no reference to this in his memoir.

In line with previous experiments of Bergman on the relative affinities of combination, are his experiments on the affinity of oxygen for different substances.[12] His conception underlying his problem is thus briefly suggested:

"To form precise ideas upon these phenomena, it is necessary to represent all bodies of nature as immersed in a fluid, elastic, very rarefied, very light, known as the igneous fluid, the principle of heat; this fluid which penetrates them all tends constantly to scatter their parts and

---

[11] *Different Kinds of Air,* 2d ed., 1776, II, p. 100.
[12] Read before the Academy of Sciences in 1783.   *Oeuvres de Lavoisier,* Tome II, p. 546 *ff.*

would accomplish it if they were not retained by the attraction that they exercise upon one another; it is this attraction that one is accustomed to call by the name of the 'affinity of aggregation.' "

The following is his table of affinities, interesting as being the list, as he says, of nearly all the substances with which oxygen combines. The arrangement is in the order of decreasing affinities toward oxygen.

PRINCIPE OXYGINE
Unknown principle of muriatic acid (or muriatic principle)
[this is chlorine]
Carbonaceous substance
Zinc
Iron
Inflammable principle of water   [hydrogen]
Regulus of Manganese
Cobalt
Nickel
Lead
Tin
Phosphorus of Kunckel
Copper
Bismuth
Regulus of Antimony
Mercury
Silver
Regulus of Arsenic
Sugar
Sulphur
Nitrous air
Principle of Heat
Gold
Fuming Muriatic acid of common Nitrous acid [nitric acid]
Calx of Manganese

The work of Lavoisier on substances of organic origin was epoch-making. His predecessors and his contemporaries had prepared and studied extensively organic substances, but they had only vague notions of their nature. The phlogistic hypothesis was the greatest obstacle in the way of clear ideas. Lavoisier, having broken away from that theory, was in a position to attack the problem of the real nature of organic substances; and so soon as he had realized that fixed air was only a compound of carbon and oxygen and that water was composed of hydrogen and oxygen, he was quick to draw the inference that organic substances, which gave mainly fixed air and water by their burning, must be composed largely of carbon, hydrogen, and oxygen. He made many analyses of organic bodies to show their elementary composition, that is the proportions of the simple bodies of elements which made up these substances. He was the first to devise methods for elementary analysis of organic bodies in so far as their carbon, hydrogen, and oxygen contents were concerned. His results, to be sure, were often inaccurate. One reason for this was that his estimate of the quantity of hydrogen and oxygen in water was inaccurate. He says, in his *Traité Élémentaire de Chimie*, 1789,[13] "It is by an experiment of this kind that we have recognized, M. Meusnier and I, that 85 parts by weight of oxygen, and 15 parts, similarly by weight, of hydrogen, are necessary to make 100 parts of water." As the true proportion by weight is 88.9 of oxygen to 11.1 of hydrogen, this discrepancy alone was enough to create serious errors of analysis, as the hydrogen was usually determined from the weight of water produced. His analysis of sugar (on page 142 of the same work) is given in the following table, together with the correct values:

	Lavoisier's	Instead of the Correct Composition
Hydrogen	8 parts	6.43 parts
Oxygen	64 "	51.46 "
Carbon	28 "	42.11 "
	100	100.00

That his determinations of the elementary composition of organic substances were not accurate by present standards is a matter of slight significance, in consideration of the fact that he was the first to recognize the common elementary constituents of organic bodies, and the first to devise a method for their determination. As stated by

13 1st ed., Paris, 1789, Tome I, p. 100.

August Kekulé, referring to the beginnings of organic chemistry:

"Lavoisier's investigations had broken the path; they had, namely, exploited a method of analysis which was soon improved by Gay Lussac and Thénard, by Saussure, and by Bergelius to attain finally under Liebig's hands such a degree of simplicity and precision that later decades could only retain the method in general, while adding, indeed, for special cases, some modifications."

Chemists of the latter part of the eighteenth century were seriously impressed with the necessity of a more systematic nomenclature of chemistry. Guyton de Morveau, professor of chemistry at Dijon from about 1782, undertook the task of devising a new system, in correspondence with Bergman of Sweden, and with other chemists. Lavoisier, with his more advanced insight into chemical theory, saw the necessity more keenly and realized its importance to his new antiphlogistic chemistry. He endeavored to gain the adherence of influential French chemists to this theory. De Morveau was doubtful, until he had a personal session with Lavoisier, whether the phlogistic hypothesis could be entirely dispensed with. Fourcroy was unconvinced until 1786. After many conferences, however, by 1787, they were united in a plan for the new nomenclature and in that year was published the result of these conferences, in a volume entitled *Méthode de Nomenclature Chimique,* proposed by MM. de Morveau, Lavoisier, Berthollet and de Fourcroy, to which is joined a new system of characters adapted to this nomenclature by MM. Hassenfratz and Adet.[14]

This work was of great importance, appearing as the phlogiston theory was tottering. It consists of several distinct articles, first, a memoir by Lavoisier on the necessity of reforming chemical nomenclature, read at the Academy of Sciences, April 18, 1787, followed by a memoir upon the development of the principles of a methodic nomenclature, read on May 2, before the Academy by de

---

[14] Paris, 1787, under the privilege of the *Academy of Sciences.*

Morveau; appendices. containing the nomenclature of some compound substances, which combine sometimes like simple bodies; a memoir by de Fourcroy, explaining the tables of nomenclature (thirty-seven octavo pages); a directory of the new nomenclature in ninety-four pages, and the symbols prepared by Hassenfratz and Adet, a chemical shorthand by which the names of elements and compounds could be replaced by symbols. This system never came into general use, and symbols, in so far as they were used by chemists, were of the already developed systems, until Dalton's concept of the atomic weights and symbols had been simplified by Berzelius (in 1815) into the system still in use.

The new nomenclature consisted essentially in the substitution for the medley of empirical names of substances of names intended to express the composition, for instance, "ferrous sulphate" for "green vitriol," "alkaline sulphide" instead of "liver of sulphur," etc. Objection to these changes, made at the time of the discussion by some participants, was that this new nomenclature depended on suppositions that might be wrong. Indeed, many of the names then suggested convey the mistaken ideas of the time, as when what we know as chlorine gas was called "gas acide muriatique oxyginé," on the supposition which prevailed before Davy and Faraday that chlorine contained oxygen combined with an as yet unknown element. The opponents of the system contended that an empirical name was at least not liable to confuse the users by being complicated with theories which might be mistaken. The system of nomenclature in use in mineralogy is a good illustration of the older system of nomenclature in chemistry, the names making no pretense to define the constitution of the mineral. The system of nomenclature suggested by the French chemists was soon translated into the other modern languages and into Latin, and, with such modifications as increasing knowledge necessitated, it is the system at present used.

Just how much of this work on nomenclature is due to

Lavoisier and how much to his collaborators is not easy to say, but from our knowledge of his general views and character it is safe to assert that his mind was extremely influential, if not indeed controlling.

Two years later (1789), Lavoisier published his celebrated *Traité Élémentaire de Chimie.*[15] In the preliminary discourse, he states:

"When I undertook this work, I had the purpose only of giving further development to the memoir that I read at the public session of the Academy of Sciences, in the month of April, 1787, upon the necessity of reforming and perfecting the nomenclature of chemistry. . . . And indeed while I believed that I had only the purpose of perfecting the language of chemistry, my work was insensibly transformed under my hands, without my being able to prevent it, into an *Elementary Treatise of Chemistry.* . . . . . It may be permitted me to add that he who enters upon a scientific career is in a situation less advantageous than the child even, who is acquiring his first ideas—if the child is deceived in the salutary or injurious effects of the objects which surround him, nature gives him numerous means of correcting his ideas. At each instant the judgment which he has made is corrected by experience. Privation or grief come as consequences of a false judgment, enjoyment or pleasure as the consequence of a correct judgment. We do not delay, under such masters, in becoming consistent, and we soon reason rightly when we can reason otherwise only at the cost of privation or suffering.

"It is not the same in the study and practice of the sciences—the false judgments we make do not concern either our existence or our welfare. No physical interest compels us to correct ourselves. Imagination, on the contrary, which constantly tends to carry us away from the truth; self-esteem, and that self-confidence with which self-esteem so easily inspires us, solicit us to draw conclusions which do not follow directly from the facts, so that we are in a fashion interested in deceiving ourselves. It is then

---

[15] Lavoisier, *Traité Élémentaire de Chimie*, présenté dans un ordre nouveau et d'après les découvertes modernes, avec figures, Paris, 1789.

not surprising that, in the physical sciences in general, men have often *assumed* instead of *concluded*,—that assumptions transmitted from age to age have become more and more imposing by the weight of the authority they have acquired, and that they have finally been adopted and regarded as fundamental truths even by very good minds.

"The only mode of preventing these errors consists in suppressing, or at least in simplifying as much as possible, the reasoning which is from ourselves, and which can only mislead us—to subject it constantly to the test of experience, to preserve only the facts which are the data of nature and which cannot mislead us, to seek only the truth in the natural series of experiments and observations in the same manner as the mathematicians arrive at the solution of a problem, by the simple arrangement of the data, and by reducing the reasoning to operations so simple, to reasonings so short, that they never lose sight of the evidence which serves as their guide.

"Convinced of these truths, I have imposed upon myself the role of never proceeding except from the known to the unknown, of deducing no consequence which is not derived directly from experience and observation, and of arranging the facts and chemical truths in the order most appropriate to facilitate the understanding of them by beginners. It was impossible in accommodating myself to this plan not to depart from the usual paths. It is indeed a common fault of all the courses and treatises on chemistry to assume in the first steps, knowledge which the student or reader can only acquire in subsequent lessons. Nearly all begin with a treatment of the principles of bodies, without explaining the table of affinities, without noticing that we are obliged to pass in review from this first day the principal phenomena of chemistry, to use expressions which have not been defined, and to assume that knowledge as already acquired by those to whom we propose to impart it. It is also to be recognized that we learn only a little in a first course of chemistry; that one year scarcely suffices to familiarize the ear with the language, the eye with the apparatus, and that it is nearly impossible to make a chemist in less than three or four years."

We might expect a good treatise, based on such sound philosophy. It is divided formally into three parts. The first part deals with the various gases, their formation and properties, and with the combinations of caloric with bodies, for Lavoisier still held it necessary to consider heat as a fluid substance. He says, in the *Traité*:

"We have consequently designated the cause of heat, the highly elastic fluid which produces it, by the name of caloric. Independently of the fact that this expression fulfills our object in the system we have adopted, it has still another advantage; this is its power of being adapted to all kinds of opinions, since, rigorously speaking, we are not even obliged to suppose that caloric is a real substance; it suffices, as we shall see better on reading that which is to follow, that this may be any cause whatever which separa-ates the molecules of matter, and we can thus consider its effects in an abstract and mathematical way."

In this first part, also, Lavoisier considers the subjects of oxidation, fermentation, putrefaction, the composition of air and water and of acids, bases, and salts in general.

His second part deals with the combination of acids with salt-forming bases and the formation of neutral salts. He begins this section with a table of "simple substances" or at least those that "the present state of our knowledge obliges us to consider as such." This table is largely the same as the one which was presented in the *Nomenclature Chimique* by de Morveau, Lavoisier, Berthollet and de Fourcroy, under the title of *Substances not Decomposed*. The main difference is in the omission by Lavoisier of the radicals of many organic acids included in the previous table. By radical was here meant that portion of the acid other than the oxygen, which was supposed to be the acidi-fying principle. The omission of the alkalies, potash and soda from the list is not significant, for these bases had not yet been decomposed, and Lavoisier frequently includes them among the simple bodies, in subsequent tables of the salts and other compounds of simple bodies. Lavoisier's table, translated into English, is as follows:

TABLE OF SIMPLE SUBSTANCES[16]

	New Names	Corresponding Old Names
Simple substances which belong to the three kingdoms and which may be considered as the elements of bodies.	Light	Light
	Caloric	Heat Principle of Heat Igneous Fluid Fire Matter of Fire and of Heat
	Oxygen	Dephlogisticated Air Empyreal air Vital air Base of Vital air
	Nitrogen ("Azote")	Phlogisticated air or gas Mephites Base of Mephites
	Hydrogen	Inflammable gas Base of inflammable gas

Simple non-metallic oxidable and acidifiable.	Sulphur
	Phosphorus
	Carbon
	Muriatic radical
	Fluoric "
	Boracic "
		Sulphur
		Phosphorus
		Pure charcoal
		Unknown
		    "
		    "

Simple metallic substances oxidable and acidifiable	Antimony Silver (Argent) Arsenic Bismuth Cobalt Tin Iron Manganese Mercury Molybdenum Nickel Gold Platina Lead Tungsten Zinc	Antimony Silver Arsenic Bismuth Cobalt Tin Iron Manganese Mercury Molybdenum Nickel Gold Platina Lead Tungsten Zinc
Simple earthy substances salifiable	Lime Magnesia Baryta Alumina Silica	Calcareous earth, lime Magnesia, base of Epsom salts Barytes, heavy earth Clay, earth of Alum, base of Alum Silicious earth, vitrifiable earth

[16] (See footnote p. 535).

The second part of the *Traité* consists mainly of tables
of nomenclature of the compounds of these simple substances with oxygen, hydrogen, sulphur, and of their salts
which are formed with all the known acids, inorganic or
organic, together with such observations and comment upon
the tables as is needed render them clear to the reader.

It is interesting to see that in these tables all elements
in the gaseous state are listed as combinations of caloric
with the element under consideration. Thus, caloric combined with oxygen gives oxygen gas; caloric combined with
hydrogen gives hydrogen gas, caloric combined with sulphur gives sulphur vapor. Caloric, he considers the material fluid which separates their particles to form the gaseous conditions.[18]

A portion of the table on combinations of bases with sulphuric acid will illustrate the character of the many tables
in this second part of the treatise, (See p. 536.)

The third part of the work treats of the apparatus and
methods of chemical experimentation. It is interesting to
note that he describes in detail the pneumatic trough for
manipulation of gases over water and over mercury, crediting the invention to Priestley. The many engraved plates
illustrating apparatus of all kinds bear the signature of
Paulze-Lavoisier, indicating the coöperation of the brilliant
Mme. Lavoisier in the labor of this part of the work. A
final section of the work is devoted to tables for the use
of chemists; weights and measures, specific weights, and
density of gases, liquids, minerals and rocks.

The Treatise on Elementary Chemistry, as published in
1789, was never changed by Lavoisier. Robert Kerr, the
English translator of the work says in the preface to the
third English edition:

"A new edition of the original having appeared in Paris

[17] Lavoisier, *Traité de Chimie*, 1st ed., 1789, Tome I, p. 192; or *Oeuvres de Lavoisier*, 1864, Tome I, p. 135.
[18] *See* ante, p. 533.

TABLE OF COMBINATIONS OF SULPHURIC ACID OR OXYGENATED SULPHUR WITH SALTFORMING BASES IN THE ORDER OF THEIR AFFINITY WITH THIS ACID, BY THE WET WAY.

New Nomenclature Combinations of Sulphuric Acid with:			Old Nomenclature Combinations of Vitriolic Acid with:		
Nos.	Names of Bases	Neutral salts resulting	Nos.	Names of Bases	Neutral salts resulting
1	Baryta	Sulphate of Baryta	1	Heavy Earth	Vitriol of Heavy earth / Heavy spar
2	Potash	Sulphate of Potash	2	Fixed vegetable alkali	Vitriolated tartar / Sel de duobus / Areanum duplicatum
3	Soda	Sulphate of Soda	3	Fixed mineral alkali	Glauber's salt
4	Lime	Sulphate of Lime	4	Calcareous earth	Selenite, gypsum / Calcareous vitriol
5	Magnesia	Sulphate of Magnesia	5	Magnesia	Vitriol of Magnesia / Epsom Salt, Sedlitz Salt
6	Ammoniac	Sulphate of Ammoniac	6	Volatile Alkali	Secret ammoniacal Salt of Glauber
7	Alumina etc.	Sulphate of Alumina etc.	7	Earth of Alum etc.	Alum etc.

in the winter of 1792–3, expectations were formed that the author might have made considerable improvements: but from a correspondence with Mr. Lavoisier, the translator is enabled to say that the new edition, having been printed without his knowledge, is entirely a transcript from the former.''

And in a postscript to the fourth English edition, Mr. Kerr says:

"Had Lavoisier lived, as expressed in a letter received from him by the translator a short while before his massacre, it was his intention to have republished these *Elements* in an entirely new form, composing a Complete System of Chemical Philosophy, and as a mark of his satisfaction with the fidelity of this translation he proposed to have conveyed to the translator, sheet by sheet, as it should come from the press, that new and invaluable work, alas! now for ever lost.'' [19]

The success of Lavoisier's *Traité* was enormous, as says Professor Grimaux, the capable biographer of Lavoisier. It was at once translated into foreign languages. From all sides felicitations came to the author who could finally enjoy a complete victory over the old theory of phlogiston.

"The *Traité Élémentaire de Chimie* [says Grimaux] marks the definite separation between the chemistry of Stahl and the real chemistry. Written less than twenty years after the work of Baumé, it differs so much in the ideas and language of chemistry, that it seems as if a century might have intervened between the two. Scarcely can we read the first, it is strange to us by its superannuated theories, its method of reasoning, its nomenclature, and classification, while the treatise of Lavoisier seems to us as if written yesterday, it is our contemporary. With the exception of some obstinate resistance from a genius like Priestley or from mediocre men like Baumé, from this moment, the pneumatic theory conquered the world of scholars. One of the most illustrious chemists of Europe, Black, honored his old age by rallying to the new doctrine.

---

[19] From the fourth edition of Kerr's translation "with considerable additions,'' Edinburgh, 1799. See advertisement, p. vii and xi.

'I seek to make my pupils understand' he writes, 'the principles and explanations of the new system that you have so happily devised, and I am beginning to recommend it to them as simpler, easier, and better sustained by facts than the old system.' "

In 1790, Chaptal, Professor of Chemistry at Montpelier, wrote his *Éléments de Chymie* based on the new system and Lavoisier wrote to him as follows (1791):

"To see you adopt the principles which I first announced is to me a real joy. The conquest of yourself, M. de Morveau, and of a small number of chemists scattered through Europe is all that I had the ambition of accomplishing, and the success surpasses my hopes, for I receive from all sides letters which announce new proselytes, and I see now that only aged persons who have no longer the courage to begin again their studies, or who can no longer turn their imagination to a new order of things, still hold to the doctrine of phlogiston. All young people adopt the new doctrine, and from this I conclude that the revolution in chemistry is accomplished." [20]

It should be recalled that of the most distinguished among the upholders of phlogiston, Macquer, Marggraf, Bergman, Scheele were dead when Lavoisiers *Traité* appeared in 1789. Black and Kirwan adopted the new chemistry, though they were both advanced in years. Cavendish made no acknowledgment of conviction, though he made no later contribution to the discussion. Priestley alone, among the men of recognized eminence, continued to endeavor to uphold the ancient system.

It is tempting to speculate, vain though it be, on what might have been the influence of Lavoisier on the development of chemistry in the next twenty years, had he lived to attain the Biblical limit of useful years, for he was but fifty years of age at his untimely ending. From the general acceptance of the chemical philosophy presented by Lavoisier, a new zest entered into chemical research. Phlogiston, with its obscuring influence upon chemical reactions,

[20] *Lavoisier, 1743–1794, etc.*, par Edouard Grimaux, Paris, 1888, pp. 125, 126.

being eliminated, quantitative determinations could be criticized upon the basis of a confidence in the conservation of mass. The work of J. B. Richter between 1792 and 1802 established the important doctrine of equivalent weights of bases and acids. Berthollet's *Essai d'une Statique Chimique* (1802), challenging the idea of constant affinities, and of constant composition of chemical compounds, excited great interest and his controversy with Proust was keenly followed by the chemical world. Dalton's concept that the elements were composed of homogeneous atoms of constant weight and that compounds were formed by the union of these atoms in simple numerical proportions, gave a new interest to the "atoms," "corpuscles," or minute "particles" which were the basis of speculation of earlier chemists, and founded our atomic and molecular theory. The extension of the application of electricity to chemical experimentation and theory by Davy, Faraday, and Berzelius, and others early in the nineteenth century also opened a vast field of inquiry and research. These influences so broadened and transformed the domain of chemical study as to make it evident that the logical separation of early from modern chemistry is most clearly marked by the acceptance of the Antiphlogistic Philosophy at the close of the eighteenth century.

BIBLIOGRAPHY

Works used by the author, not including serial publications nor proceedings of learned societies:

AGRICOLA, GEORGIUS.
    De mensuris et ponderibus Romanorum atque Graecorum.
    De externis mensuris et ponderibus.
    Ad ea quae Andreas Alcistus denuo disputavit, etc.
    De mensuris quibus intervalla metimur.
    De restituendis ponderibus atque mensuris.
    De precio metallorum et monetis.
               Basileae, 1550.

AGRICOLA, GEORGIUS.
    De ortu et causis subterraneorum.
    De natura eorum quae effluunt terra.
    De natura fossilium.
    De veteribus et novis metallis.
    Bermannus sive de re metallica dialogus.
               Basileae, 1558.
    De re metallica.   De animantibus subterraneis.
               Basileae, 1561.

    De re metallica, tr. from the 1st Latin ed. of 1556 by **H. C.** Hoover and L. H. Hoover, London, 1912.

AGRIPPA, HENRICUS CORNELIUS, AB NETTESHEYM.
    De incertitudine & vanitate omnium scientiarum & artium liber, Francofurti & Lipsiae, J. A. Plener, 1593.

ALBERTUS MAGNUS.
    Opera omnia, Paris, 1890–1899, 38 vols.

ALLBUTT, THOMAS CLIFFORD.
    Science and medieval thought, London, 1901.

ARISTOTELES (pseudo—).
    Das steinbuch des Aristoteles.   Mit literargeschichtlichen untersuchungen von Julius Ruska, Heidelberg, 1912.

ARNALDUS DE VILLANOVA.
    Opera omnia, Basileae, C. Vvaldkirch, 1585.

ASHMOLE, ELIAS.
Theatrum Chemicum Britannicum, London, 1652.
AUDIAT, LOUIS.
Bernard Palissy; Étude sur sa vie et ses travaux, Paris, 1868.
BACKER, H. J.
Oude chemische werktuigen en laboratoria van Zosimos tot Boerhaave, Groningen, 1918.
BACON, FRANCIS.
Works, new ed. in ten vols., London, 1826.
BACON, ROGER.
"De alchemia libellus." (In *Theatrum Chemicum*, Urselli, Zetzner, 1602, vol. ii, p. 433–441.)
Mirror of Alchimy, London, 1597.
Opera hactenus inedita, Robert Steele ed., Oxonii, 1909–1913, fasc. 2–4.
Opera quaedam hactenus inedita, J. S. Brewer ed., London, 1859. Vol. 1.
Opus majus, J. H. Bridges ed., Oxford, 1897–1900, 3 vols.
Part of the Opus tertium, A. G. Little ed., Aberdeen, 1912.
BAEUMKER, CLEMENS.
Der problem der materie in der griechischen philosophie, Münster, 1890.
BARBA, ALVARO ALONSO.
Traité de l'art métalique, Paris, 1730.
BARTHOLOMAEUS ANGLICUS.
Liber de proprietatibus rerum, Argentine, 1505.
Mediaeval lore from Bartholomaeus Anglicus, By Robert Steele, London, 1907.
BASILIUS VALENTINUS.
Chymische schriften: zum dritten mahl zusammen gedruckt, Hamburg. 1700, 2 pts.
The Triumphal Chariot of Antimony, A. E. Waite ed., London, 1893.
BAUER, ALEXANDER.
Chemie und alchymie in Österreich bis zum beginnenden XIX. jahrhundert, Wien, 1883.
BECKMAN, JOHN.
A history of inventions and discoveries, 3d ed., London, 1817, 4 vols.
BEITRAGE *aus der geschichte der chemie dem gedächtnis von Georg W. A. Kahlbaum*, hrsg. von Paul Diegart, Leipzig und Wien, 1909.

BERGMAN, SIR TORBERN OLOF.
Analyse du fer, tr. par M. Grignon, Paris, 1783.
Manuel du minéralogiste; ou, Sciagraphie du règne minéral, distribuée d'après l'analyse chimique, tr. par Mongez, Nouvelle éd., Paris, 1792, 2 vols.
Opuscules chymiques et physiques, tr. par M. de Morveau, Dijon, 1780–1785, 2 vols. in 1.
Physical and Chemical Essays, London, 1784–1791, 3 vols.
Traité des affinités chymiques, Paris, 1788.
BERTHELOT, MARCELLIN.
Archéologie et histoire des sciences, Paris, 1906.
Collection des anciens alchimistes grecs, Paris, 1887–1888, 4 pts.
Die chemie im altertum und im mittelalter, aus dem französischen übertragen von Emma Kalliwoda, Leipzig und Wien, 1909.
Introduction a l'étude de la chimie des anciens et du moyen age, Paris, 1889.
La chimie au moyen age, Paris, 1893, 3 vols.
La révolution chimique, Lavoisier, Paris, 1890.
Les origines de l'alchimie, Paris, 1885.
BERTRAND.
Reflexions nouvelles sur l'acide et sur l'alcali, Lyon, 1683.
BIRINGUCCIO, VANOCCIO.
De la pirotechnia, Libri. X, Venetia, 1540.
La pyrotechnie ou art du feu, contenant dix livres, tr. d'Italien en françois par Jaques Vincent, Paris, G. Iullian, 1572.
BOERHAAVE, HERMANN.
A new method of chemistry. tr. from the printed ed., collated with the best manuscript copies, by P. Shaw and E. Chambers, London, 1727.
Elementa chemiae, Lugduni Batavorum, 1732, 2 vols.
BOYLE, ROBERT.
Certain physiological essays and other tracts, 2d ed., Increased by the addition of a discourse about the absolute rest in bodies, London, 1669.
Essays of the strange subtilty, great efficacy, determinate nature of effluviums, London, 1673.
The sceptical chymist: or Chymico-physical doubts & paradoxes, Oxford, 1680.
Works, London, 1744, 5 vols.
BRAJENDRANATH SEAL.
The positive sciences of the ancient Hindus, London, 1915.
BREASTED, JAMES HENRY.
Ancient times, a history of the early world, Boston, 1916.

BROWN, JAMES CAMPBELL.
> *A history of chemistry from the earliest times till the present day,* Philadelphia, 1913.

BRUNSCHWIG, HIERONYMUS.
> *Liber de arte distillandi de simplicibus,* das buch der rechten kunst zu distilieren die eintzige ding, Strassburg, 1500.

BURNAM, JOHN M.
> *A classical technology,* ed. *from Codex Lucensis,* 490, Boston, 1920.
>
> *Recipes from Codex Matritensis* A 16 (ahora 19) Cincinnati, 1912. (University of Cincinnati studies, Ser. II, Vol. VIII.)

CAP, PAUL ANTOINE.
> *Études biographiques pour servir à l'histoire des sciences,* 1.–2. ser., Paris, 1857–1864.

CAVENDISH, HENRY.
> *Scientific papers,* Sir Edward Thorpe ed., Cambridge, 1921, Vol. II.

CHARLES, EMILE AUGUSTE.
> *Roger Bacon, sa vie, ses ouvrages, ses doctrines, d'après des textes inédits,* Bordeaux, 1861.

CHRIST, WILHELM VON.
> *Geschichte der griechischen litteratur,* 5. aufl., München, 1908–1913, 3 vols.

CROLL, OSWALD.
> *Basilica chymica,* Venetiis, 1643. (Contains the *Praefatio admonitoria* and the *Tractatus de signaturis.*)

DANNEMANN, FRIEDRICH.
> *Aus der werkstatt grosser forscher,* 3. aufl., des 1. bd. des "Grundriss einer geschichte der naturwissenschaften," Leipzig, 1908.
>
> *Die naturwissenschaften in ihrer entwicklung und in ihrem zusammenhange.* Leipzig, 1910–13, 4 vols.

DARIOT, CLAUDE.
> *Der guldin arch, schatz und kunstkammer.* Basel, 1614. Der ander theil. (Contains the German translation of the Experimenta of Raymundus Lullus.)

DARMSTAEDTER, ERNST.
> *Die Alchemie des Geber,* übersetzt und erklärt, Berlin, 1922.

DARMSTAEDTER, LUDWIG.
> *Handbuch zur geschichte der naturwissenschaften und der technik,* 2. aufl., Berlin, 1908.

DEUSSEN, P.
> *Allgemeine geschichte der philosophie,* Leipzig, 1906–1920, 2 vols. in 6.

DIETERICI, FRIEDRICH.
   Die philosophie der Araber im IX. und X. jahrhundert n. Chr.,
   Leipzig, 1879, buch. 2.
DIODORUS SICULUS.
   Historical library. tr. by G. Booth, London, 1814, 2 vols.
DIOSCORIDES, PEDANIUS.
   De materia medica. C. Sprengel ed., Leipzig, 1829–1830, 2 vols.
   Des Pedanios Dioskorides aus Anazarbos Arzneimittellehre in
   fünf Büchern, Übersetzt und mit Erklärungen versehen von
   Prof. Dr. J. Berendes, Stuttgart, 1902.
DORN, GERARDUS.
   Clavis totius philosophiae chymisticae, Lugduni, Haeredes
   Iacobi, 1567.
DUHEM, PIERRE.
   Études sur Leonard de Vinci, Paris, 1906–13, 3 vols.
EHRENFELD, RICHARD.
   Grundriss einer entwicklungsgeschichte der chemischen atomis-
   tik, Heidelberg, 1906.
EHRMANN, F. L.
   Déscription et usage de quelques lampes à air inflammable,
   Strassbourg, 1780.
ERASTUS, THOMAS.
   Explicatio quaestionis famosae, cum gratia & privilegio
   Caesareo, 1572. (Contains his *Epistola de natura*), Basileae,
   1572.
FERGUSON, JOHN.
   Bibliotheca chemica, Glasgow, 1906, 2 vols.
FIGUIER, LOUIS.
   L' alchimie et les alchimistes, 3d. ed., Paris, 1860.
FOURCROY, ANTOINE FRANCOIS DE.
   Mémoires et observations de chimie, Paris, 1784.
GARBE, RICHARD.
   The philosophy of ancient India, Chicago, 1897.
GESSMANN, G. W.
   Die geheimsymbole der chemie und medicin des mittelalters,
   Graz, 1899.
GLASER, CHRISTOPHLE.
   Traité de la chymie, 2d. ed., Paris, 1668.
   Traité de la chymie, 4th. ed., Bruxelles, 1676.
GLAUBER, JOHANN RUDOLPH.
   Furni novi philosophici, sive Descriptio artis destillatoriae
   novae, Amsterodami, 1651.
   Opera chymica, Franckfurt am Mayn, 1658–59, 2 vols.

GMELIN, JOHANN FRIEDRICH.
  Geschichte der chemie, Göttingen, 1797–99, 3 vols.
GRIMAUX, ÉDOUARD.
  Lavoisier, 1743–1794, Paris, 1888.
GROTHE, HERMANN.
  Leonardo da Vinci als ingenieur und philosoph, Berlin, 1874.
HAESER, HEINRICH.
  Lehrbuch der geschichte der medicin, Jena, 1875, Vol. I.
HALES, STEPHEN.
  La statique des végétaux, et l'analyse de l'air, tr. de l'Anglois,
    par M. de Buffon, Paris, 1735.
  *Vegetable staticks or, An account of some statical experiments
    on the sap in vegetables,* London, 1727–33, 2 vols. Vol. II
    has title, *Statical essays.*
HARTMANN, R. JULIUS.
  Theophrast von Hohenheim. Stuttgart und Berlin, 1904.
HAUREAU, B.
  Arnald of Villanova. (In *Histoire littéraire de la France,*
    1881, Vol. 27.)
  Raimundus Lullus. (In *Histoire littéraire de la France, Vol.*
    29.)
HELMONT, JEAN BAPTISTE VAN.
  Opera omnia, Francofurti, 1682. Contains his *Opuscula
    medica inaudita,* Francofurti, 1682.
HOEFER, FERDINAND.
  Histoire de la chimie, Paris, 1842–43, 2 vols.
  Histoire de la chimie, 2d. ed., rev. et augm., Paris, 1869, 2 vols.
  La chimie enseignée par la biographie de ses fondateurs, Paris,
    1865.
HOLLANDUS, JOHANN ISAAC.
  Die hand der philosophen, Franckfurt, MDCLXIVI, 1667. (?)
  Sammlung unterschiedlicher bewahrter chymischer schriften,
    Wien, 1773. (Manifestly the same as the 1746 ed., but with
    a new t.-p. Cf. Ferguson, *Bibliotheca chemica,* Vol. I.)
HOOKE, ROBERT.
  Philosophical experiments and observations, pub. by W.
    Derham, London, 1726.
  *Posthumous works, containing his Cutlerian lectures, and
    other discourses,* London, 1705.
ISIDORUS.
  Etymologiarum sive originum libri XX. Recognovit W. M.
    Lindsay, Oxonii, 1911, 2 vols.

JABIR IBN HAIYAN, AL-TARUSUSI.
Alchemiae Gebri arabis philosophi solertissimi libri, Bernae,
1545. (Contains his *De investigatione perfectione metal-
lorum; Summae perfectionis metallorum; De inventione
veritatis; De fornacibus construendis; Speculum alchemiae,
R. Bachonis; Correctorium alchemiae, Richardi Anglici;
Rosarius minor, de alchemia, incerti authoris; Liber secretorum alchemiae Calidis filii Jazichi Judaei; Tabula smaragdina de alchemia, Hermetis commentarius.*)
Works, Englished by Richard Russell, London, 1678.
JAGNAUX, RAOUL.
Histoire de la chimie, Paris, 1891, 2 vols.
JOLY, GABRIEL.
*Trois anciens traictez de la philosophie naturelle; 1. Les
sept chapites dorez . . . & La table d'esmeraude
d'Hermes Trismegiste; 2. La response de Messire Bernard
conte de la march Treuisane, à Thomas de Boulongne . . . ;
3. La chrysopée de Iean Aurelle Augurel, qui enseigne
l'art de faire l'or . . .* Paris, C. Hulpeau, 1626.
JUNCKER, JOHANN.
*Conspectus chemiae theoretico-practicae in forma tabularum,
e dogmatibus Becheri et Stahlii potissimum explicantur,*
Halae Magd., 1730.
KAHLBAUM, GEORG W. A.
*Die einführung der Lavoisier'schen theorie im besonderen in
Deutschland,* Leipzig, 1897.
Beiträge aus der Geschichte du Chemie dem Gedrächtnis von,
hrsg von Paul Diergart, Leipzig und Wien, 1909.
KIRWAN, RICHARD.
Essai sur le phlogistique, et sur la constitution des acides, tr.
de l'Anglois avec des notes de MM. de Morveau, Lavoi-
sier. . . , Paris, 1788.
KOPP, HERMANN.
Beiträge zur geschichte der chemie, Braunschweig, 1869–75,
3 pts.
Die alchemie in älterer und neuerer zeit, Heidelberg, 1886,
2 pts.
Die entwickelung der chemie in der neuren zeit, München,
1873.
Geschichte der chemie, Braunschweig, 1843–47, 4 vols.
LADENBURG, A.
Handwörterbuch der chemie, Breslau, 1884, Bd. 2.

LA FONTAINE, JEAN DE.
    La fontaine des amoureux de science. Pub. par Ach. Genty.
        Paris, 1861.
LAGERCRANTZ, OTTO.
    Papyrus Graecus Holmiensis, Upsala, 1913.
LA METHERIE, JEAN CLAUDE DE.
    Essai analytique sur l'air pur. Paris, 1785.
LAMINNE, JACQUES.
    Les quatre éléments; le feu, l'air, l'eau, la terre, Bruxelles,
        1904.
LANGLOIS, CH. V.
    La connaissance de la nature et du monde au moyen age,
        Paris, 1911.
LASSWITZ, KURD.
    Geschichte der atomistik vom mittelalter bis Newton, Hamburg,
        und Leipzig, 1890, 2 vols.
LATZ, GOTTLIEB.
    Die alchemie, Bonn, 1869.
LAVOISIER, A. L.
    Oeuvres de Lavoisier, Pub. par les soins de Son Excellence le
        ministre de l'instruction publique et des cultes, Paris,
        1864–93, 6 vols.
    Traité Élémentaire. 2 vols. Paris, 1789.
    Various papers of the Fermiers-généraux, concerning the trial
        and condemnation of Lavoisier.
LAYARD, SIR AUSTEN H.
    Discoveries among the ruins of Nineveh and Babylon, New
        York, 1859.
LEFEBURE, NICOLAS.
    Traicté de la chymie, 2d. ed., Paris, 1669.
LEMERY, NICHOLAS.
    Cours de chymie, 9th. ed., Paris, 1701.
LEMNIUS, LEVINUS.
    Les occultes merveilles et secretz de nature, tr. de Latin en
        Francois par I. G. P., Paris, G. du Pré, 1574.
LENGLET DUFRESNOY, NICOLAS.
    Histoire de la philosophie hermétique, Paris, 1742, 3 vols.
LEONARDO DA VINCI.
    Leonardo da Vinci, der denker, forscher und poet, Ubersetz-
        ung von Marie Herzfeld, 3. umgearb. aufl., Jena, 1911.
LESLIE, P. DUGUD.
    A philosophical inquiry into the cause of animal heat, Lon-
        don, 1778.

LESSING, GOTTHOLD EPHRAIM.
Sämtliche schriften, hrsg. von Karl Lachmann, 3. aufl. be-
sorgt durch Franz Muncker, 14. bd., Leipzig, 1898.

LIEBAUT, JEAN.
Quatre livres des secrets de medecine, et de la philosophie
chymique, dernière ed. Roven, T. Reinsart, M. VI. C. [pref.
1573.]

LIPPMANN, EDMUND O. VON.
Abhandlungen und vorträge zur geschichte der naturwissen-
schaften, Leipzig, 1906-13, 2 vols.
Entstehung und ausbreitung der alchemie, Berlin, 1919.
Geschichte des zuckers, Leipzig, 1890.

LITTLE, ANDREW GEORGE, ED.
Roger Bacon essays, Oxford, 1914.

LÖWIG, CARL.
Jeremias Benjamin Richter, der entdecker der chemischen pro-
portionen, Breslau, 1874.

LUCRETIUS.
On the nature of things, tr. by Cyril Bailey, Oxford, 1910.

LULLUS, RAYMUNDUS.
Testamentum. . . Item eiusdem compendium animae trans-
mutationis artis metallorum.   Secunda ed., Coloniae Agrip-
pinae, I. Birkmann, 1573.

MABILLEAU, LEOPOLD.
Histoire de la philosophie atomistique, Paris, 1895.

MACBRIDE, DAVID.
Essais d'expériences, tr. de l'Anglois par M. Abbadie, Paris,
1766.

MACQUER, PIERRE JOSEPH.
Dictionnaire de chymie, 2d. éd., Paris, 1777, 3 vols.
Élémens de chimie-pratique, 2d. éd., Paris, 1754-1756.
Élémens de chymie-théorique, Paris, 1749.
Élémens de chymie-théorique, Nouvélle éd., Paris, 1756.
Plan d'un cours de chymie expérimentale et raisonnée, avec un
discours historique sur la chymie, par. M. Macquer & M.
Baume, Paris, 1757.

MAGNUS, HUGO.
Paracelsus, der überarzt, Breslau, 1906.

MANGETUS, JO. JACOBUS.
Bibliotheca chemica curiosa, Genevae, 1702, 2 vols.

MARCUS GRAECUS.
Le livre des feux, tr. par A. Poisson, Paris, 1891.

MARGGRAF, ANDREUS SIEGMUND.
Chymische schriften, Berlin, 1761-67, 2 vols.

Masson, John.
  The atomic theory of Lucretius contrasted with modern doctrines of atoms and evolution, London, 1884.
Mayow, John.
  Medico-physical works, Chicago, 1908. *Alembic club reprints*, No. 17.
  Untersuchungen über den salpeter und den salpetrigen luftgeist, das brennen und das athmen, hrsg. von F. G. Donnan, Leipzig, 1901.
Methode de Nomenclature de Chimique, *proposé par MM. de Morveau, Lavoisier, Berthollet, & de Fourcroy*, Paris, 1787.
Meyer, Ernst von.
  Geschichte der chemie von den ältesten zeiten bis zur gegenwart, 3. verb. und verm. aufl., Leipzig, 1905.
  A history of chemistry from earliest times to the present day, 3d. English ed., London, 1906.
Mook, Friedrich.
  Theophrastus Paracelsus, eine kritische studie, Würzburg, 1876.
Moore, F. J.
  A history of chemistry, 1st ed., New York, 1918.
Morley, Henry.
  The life of Henry Cornelius Agrippa von Nettesheim, London, 1856, 2 vols.
Morveau, Louis Bernard Guyton de.
  Digressions académiques ou essais sur quelques sujets de phisique de chymie & d'histoire naturelle, Dijon, 1762.
  Élémens de chymie, théorique et pratique, Dijon, 1777–78, 3 vols.
Mosso, Angelo.
  The dawn of Mediterranean civilisation, tr. by Marian C. Harrison, London, 1910.
Muir, M. M. Pattison.
  A history of chemical theories and laws, 1st ed., New York, 1907.
  Heroes of science, Chemists, London, 1883.
Muller, Max.
  The six systems of Indian philosophy, New York, 1899.
Muspratt, James Sheridan.
  Muspratt's theoretische, praktische und analytische chemie, 4. aufl., Braunschweig, 1888–1905, 8 vols.
Neri, Antonio.
  Art de la verrerie de Neri, Merret et Kunckel, tr. de l'Allemand, par M. D., Paris 1752.

NETZHAMMER, RAYMUND.
  Theophrastus Paracelsus, Einsiedeln, 1901.
NEUBURGER, ALBERT.
  Die technik des altertums, 2. verb. aufl., Leipzig, 1921.
NEUMANN, CASPAR.
  Chemical works, abridged and methodized, by William Lewis,
    London, 1759.
OSTWALD, WILHELM.
  Der werdegang einer wissenschaft, 2 verm. und verb. aufl. der
    "Leitlinien der chemie," Leipzig, 1908.
  Lehrbuch der Allgemeinen Chemie, 2te Auflage, Bd. I, Leip-
    zig, 1891.
PALISSY, BERNARD.
  Les oeuvres, pub. par A. France, Paris, 1880.
  Papyrus graecus holmiensis (P. holm.); recepte für silber,
    steine und purpur, bearb. von Otto Lagercrantz, Uppsala,
    1913.
PARACELSUS.
  Hermetic and alchemical writings, Arthur Edward Waite, ed.,
    London, 1894, 2 vols.
  Opera, bücher und schrifften . . . Durch Joannem Huserum
    Brisgoium in zehen underschiedliche theil, in truck gegeben.
    Strassburg, 1616, 2 vols. (Contains his *Chirurgische bücher
    und schrifften,* Strassburg, 1618.)
PHILLIPPS, THOMAS.
  The Mappae clavicula, communicated by T. Phillipps and pre-
    sented by Albert Way, London, 1847. (In Archaeologia.
    Vol. 32.)
  *Philosophie naturelle de trois anciens philosophes renommez
    Artephius, Flamel, & Synesius traitant de l'art occulte, &
    de la transmutation metallique,* dernière éd., Paris, 1682.
PLATO.
  Phaedo, tr. by Henry Cary, London and New York. (Every-
    man's Library, No. 456.)
  Timaeus, R. D. Archer-Hinde ed., London, 1888.
PLINIUS SECUNDUS, C.
  The natural history of Pliny, tr. by John Bostock and H. T.
    Riley, London, 1856–93, 6 vols. Bohn's classical library.
  Naturalis Historiae, Libri xxxvii, London, 1826. Vol. XII.
PORTA, JOHAN BAPTISTA.
  De distillatione, lib. IX, Romae, 1608.
  Magiae naturalis, libri viginti, Amstelodami, 1664.

PRIESTLEY, JOSEPH.

Experiments and observations on different kinds of air, London. 1775–77, 3 vols. (vols. 1 & 2, 2d ed.)

Réflexions sur la doctrine due phlogistique et la décomposition de l'eau, tr. par P. A. Adet, Paris, 1798.

Scientific correspondence, Henry Carrington Bolton ed., New York, 1892.

RAMSAY, WILLIAM.

The gases of the atmosphere, the history of their discovery, London and New York, 1896.

RECUEIL *des mémoires les plus intéressants de chymie, et d'histoire naturelle,* contenus dans les Actes de l'Académie d'Upsal et dans les Mémoires de l'Académie royale des sciences de Stockholm; publiés depuis 1720 jusqu'en 1760, Paris, 1764, 2 vols.

REDGROVE, H. STANLEY.

Alchemy, ancient and modern, London, 1911.

REY, JEAN.

Essais, réimpression de l'éd. de 1630, Paris, 1896.

Essays of Jean Rey. On an enquiry into the cause wherefore tin and lead increase in weight on calcination, (1630) Edinburgh, 1895, *Alembic Club Reprints,* No. 11.

ROBINSON, VICTOR.

Pathfinders in medicine, New York, 1912.

ROSE, T. KIRKE.

The metallurgy of gold, 5th ed., London, 1906.

ROSE, V.

Aristoteles de lapidibus und Arnoldus Saxo. (In Zeitschrift für deutsche alterthum, Vol. 18, 1875.)

ROULAND.

Tableau historique des propriétés et des phénomènes de l'air, Paris, 1784.

RULAND, MARTIN.

Lexicon alchemiae, sive; Dictionarium alchemisticum, Francofurtensium Repub., 1612.

SCHEELE, CARL WILHELM.

Sämmtliche physische und chemische werke, Berlin, 1793. 2. unveränderte aufl., Berlin, 1891, 2 vols. in 1.

Traité chimique de l'air et du feu, tr. par le baron de Dietrich, 2. éd., Paris, 1787, and supplement, Paris, 1785.

SCHELENZ, HERMANN.

Geschichte der pharmazie, Berlin, 1904.

SCHLEGEL, E.
Paracelsus in seiner bedeutung für unsere zeit, München, 1907.
SCHMIEDER, KARL CHRISTOPH.
Geschichte der alchemie, Halle, 1832.
SCHUBERT, EDUARD.
*Paracelsus-forschungen, von Eduard Schubert und Karl Sud-
hoff,* Frankfurt a. M., 1887–1889, 1.–2. hft.
SENEBIER, JEAN.
Recherches analytiques sur la nature de l'air inflammable,
Geneve, 1784.
STAHL, GEORG ERNEST.
Fundamenta chymiae dogmaticae et experimentalis, ed. se-
cunda, emendatior et auctior, Norimbergae, 1746, 2 pts. in
1 vol.
Traité des sels, Paris, 1771.
Traité du soufre, Paris, 1766.
STILLMAN, JOHN MAXSON.
Theophrastus Bombastus von Hohenheim called Paracelsus,
Chicago, 1920.
STODDART, ANNA M.
The life of Paracelsus, London, 1911.
STRABO.
The geography of Strabo, with an English translation by H. L.
Jones, Vol. I, London, 1917, (Loeb classical library).
STRUNZ, FRANZ.
Geschichte der naturwissenschaften im mittelalter, Stuttgart,
1910.
Johann Baptist van Helmont (1577–1644), Ein beitrag zur
geschichte der naturwissenschaften, Leipzig und Wien, 1907.
Naturbetrachtung und naturkenntniss im altertum, Hamburg
und Leipzig, 1904.
Über die vorgeschichte und die anfänge der chemie, Leipzig
und Wien, 1906.
SUDHOFF, KARL.
Hohenheims literarische hinterlassenschaft, Roma, 1904.
Versuch einer kritik der echtheit der Paracelsischen schriften,
Berlin, 1894–1899. 2 vols.
SYLVIUS, FRANCISCUS DE LE BOE.
Opera medica, Venetiis, Hertz, 1696.
TACHENIUS, OTTO.
Hippocrates chimicus, ed. tertia, Lvgd. Batavor, 1671. (Con-
tains his *Antiquissima Hippocraticae medicinae clavis,* ed.
tertia, Lugduni Batavor, 1671.)
THEATRUM CHEMICUM, ed. by Zetzner, Vol. III, Ursellis, 1602.

THEOPHRASTUS.
> *Enquiry into plants and minor works on odours and weather signs*, tr. by Sir Arthur Hort, London, 1916, 2 vols.
> *History of stones*, Greek text, with an English version . . . by John Hill, London, 1746.

THOMSON, THOMAS.
> *The history of chemistry*, London, 1830–1831, 2 vols.

THORNDIKE, LYNN.
> *A history of magic and experimental science during the first thirteen centuries of our era*, New York, 1923.

THORPE, SIR THOMAS EDWARD.
> *A dictionary of applied chemistry*, rev. and enl. ed., London, 1912–1913, 5 vols.
> *Joseph Priestley*, London and New York, 1906.

TILDEN, SIR WILLIAM A.
> *Famous chemists, the men and their work*, London and New York, 1921.

VINCENT DE BEAUVAIS.
> *Speculum naturale*, Nuremberg, Koberger, 1485, 2 vols.

VITRUVIUS POLLIO.
> *De architectura libri decem*, V. Rose ed., Leipzig, 1899.
> *De architectura libri decem*, H. Müller-Strübing ed., Lipsiae, 1867.
> *The ten books on architecture*, tr. by Morris Hicky Morgan, Cambridge, 1914.

WAITE, ARTHUR EDWARD.
> *Lives of alchemystical philosophers*, London, 1888.

WALSH, JAMES JOSEPH.
> *Catholic churchmen in science*, Philadelphia, 1917, 3d ser.
> *The popes and science*, London, 1912.
> *The thirteenth, greatest of centuries*, New York, 1911.

WATT, JAMES.
> *Correspondence . . . on his discovery of the theory of the composition of water*, James Patrick Muirhead ed., London, 1846.

WILKINSON, SIR JOHN GARDNER.
> *Manners and customs of the ancient Egyptians*, 3d. ed., London, 1847, 5 vols.

WILSON, GEORGE.
> *The life of the Hon. Henry Cavendish*, London, 1851.

ZWEMER, SAMUEL M.
> *Raymund Lull, first missionary to the moslems*, New York and London, 1902.

(1)

INDEX

Carbon dioxide, 62, 477, 478, 525.
See also Fixed air.
Cardanus, 405.
Cavendish, Henry, phlogistonist,
442, 461, 538; life and work,
469-476; on nitrogen, 476; ni-
trous air, 487; eudiometer, 487,
488; hydrochloric acid and cop-
per, 488; composition of water,
495-497; on Lavoisier's theory,
497; electric current on air, 497,
498.
Cellini, Benvenuto, 331, 332.
Cements, 28.
Centumpondium, 304.
Cerussa, 19, 68. *See also* White
lead.
Chalchanthon, 42, 43, 186.
Chalcos, 65. *See also* Aes.
Chaos, 322.
Chaptal, 538.
Charcoal burning, 22, 23.
Charles, M., 202.
Charles, V., 274.
Chaucer, 275.
Chemeia, origin of word, 135-136.
Chemical affinity, 499-510; Sir
Isaac Newton on, 500-502; Boer-
haave on, 502, 503; Buffon on,
503; of oxygen, 527; constancy
of values, 539.
Chemical arts, ancient, 1.
Chemical attraction. *See* Chemical
affinity.
Chemistry, origin of word, 136;
of Middle Ages, 184-229; steril-
ity in fourteenth and fifteenth
centuries, 275; teaching of in
eighteenth century, 425.
China, ancient arts, 98.
Chlorine, 460.
Christian church, rise and influence,
138-141.
Chrysocolla, 18, 33, 42, 83, 84.
Cinnabar, 18, 30-32, 186, 204, 222,
250. *See also* Minium.
Clavis Philosophorum, 284.
Clement IV, 258.
Cleopatra, 151.
Cleves, Gaston, 368.
Climia, 255.
Cobalt, 313.
Coccus, 71.

Coerulium, 34.
Columella, 153.
Combustion, theories, 246, 404-416;
Stahl on, 428; Scheele on, 458,
459.
Communium Naturalium, of Bacon,
260.
Compendium Studii Theologiae, of
Bacon, 259.
Compendium Studii Philosophiae,
of Bacon, 258.
Compositio sisami, 194.
Compositiones ad Tingenda, 185.
Constantinople, capture by Turks,
301.
Copper, 42, 65, 66, 283, 284; early
use, 2, 4; oxide as pigment, 37;
ores of, 67; tinning of, 68; pre-
cipitation of by iron not due to
transmutation, Sala, 380.
Corpuscular theory, 416, 417, 512.
Correctorium Alchemiae, 213.
Costa, Enrico Mendez da, 439.
Cours de Chymie, of Lémery, 398.
Cremer, John, 297.
Crete, 52.
Crollius, Oswald, 354.
Crusades, 230.
Crystal, 76, 89, 90.
Cullen, William, 463-464.
Cupellation, 224, 304, 305.
Cyanos, 19, 44.

Dalton, 539.
Dante, 275.
Darmstaedter, Ernst, 279.
Daumon, 238.
Davy, Sir Humphry, 15, 460, 510,
530, 539.
De Aluminibus et Salibus, 238, 239,
242.
De Anima in Arte Alchemiae, 217.
De Artibus Romanorum, 219, 220.
De Natura Fossilium, 337.
De Natura Rerum, 234.
De Re Metallica, 341-345.
De Rebus Metallicis et Mineralibus,
249.
Dee, John, 368.
Delisle, Leopold, 234.
Democritus of Abdera, 16, 25, 26,
118-120.

(1)

CATALOGUE OF DOVER BOOKS

CHEMISTRY AND PHYSICAL CHEMISTRY

ORGANIC CHEMISTRY, F. C. Whitmore. The entire subject of organic chemistry for the practicing chemist and the advanced student. Storehouse of facts, theories, processes found elsewhere only in specialized journals. Covers aliphatic compounds (500 pages on the properties and synthetic preparation of hydrocarbons, halides, proteins, ketones, etc.), alicyclic compounds, aromatic compounds, heterocyclic compounds, organophosphorus and organometallic compounds. Methods of synthetic preparation analyzed critically throughout. Includes much of biochemical interest. "The scope of this volume is astonishing," INDUSTRIAL AND ENGINEERING CHEMISTRY. 12,000-reference index. 2387-item bibliography. Total of x + 1005pp. 5⅜ x 8. Two volume set.

S700 Vol I Paperbound **$2.25**
S701 Vol II Paperbound **$2.25**
The set **$4.50**

THE MODERN THEORY OF MOLECULAR STRUCTURE, Bernard Pullman. A reasonably popular account of recent developments in atomic and molecular theory. Contents: The Wave Function and Wave Equations (history and bases of present theories of molecular structure); The Electronic Structure of Atoms (Description and classification of atomic wave functions, etc.); Diatomic Molecules; Non-Conjugated Polyatomic Molecules; Conjugated Polyatomic Molecules; The Structure of Complexes. Minimum of mathematical background needed. New translation by David Antin of "La Structure Moleculaire." Index. Bibliography. vii + 87pp. 5⅜ x 8½.

S987 Paperbound **$1.00**

CATALYSIS AND CATALYSTS, Marcel Prettre, Director, Research Institute on Catalysis. This brief book, translated into English for the first time, is the finest summary of the principal modern concepts, methods, and results of catalysis. Ideal introduction for beginning chemistry and physics students. Chapters: Basic Definitions of Catalysis (true catalysis and generalization of the concept of catalysis); The Scientific Bases of Catalysis (Catalysis and chemical thermodynamics, catalysis and chemical kinetics); Homogeneous Catalysis (acid-base catalysis, etc.); Chain Reactions; Contact Masses; Heterogeneous Catalysis (Mechanisms of contact catalyses, etc.); and Industrial Applications (acids and fertilizers, petroleum and petroleum chemistry, rubber, plastics, synthetic resins, and fibers). Translated by David Antin. Index. vi + 88pp. 5⅜ x 8½.

S998 Paperbound **$1.00**

POLAR MOLECULES, Pieter Debye. This work by Nobel laureate Debye offers a complete guide to fundamental electrostatic field relations, polarizability, molecular structure. Partial contents: electric intensity, displacement and force, polarization by orientation, molar polarization and molar refraction, halogen-hydrides, polar liquids, ionic saturation, dielectric constant, etc. Special chapter considers quantum theory. Indexed. 172pp. 5⅜ x 8.

S64 Paperbound **$1.65**

THE ELECTRONIC THEORY OF ACIDS AND BASES, W. F. Luder and Saverio Zuffanti. The first full systematic presentation of the electronic theory of acids and bases—treating the theory and its ramifications in an uncomplicated manner. Chapters: Historical Background; Atomic Orbitals and Valence; The Electronic Theory of Acids and Bases; Electrophilic and Electrodotic Reagents; Acidic and Basic Radicals; Neutralization; Titrations with Indicators; Displacement; Catalysis; Acid Catalysis; Base Catalysis; Alkoxides and Catalysts; Conclusion. Required reading for all chemists. Second revised (1961) eidtion, with additional examples and references. 3 figures. 9 tables. Index. Bibliography xii + 165pp. 5⅜ x 8.

S201 Paperbound **$1.50**

KINETIC THEORY OF LIQUIDS, J. Frenkel. Regarding the kinetic theory of liquids as a generalization and extension of the theory of solid bodies, this volume covers all types of arrangements of solids, thermal displacements of atoms, interstitial atoms and ions, orientational and rotational motion of molecules, and transition between states of matter. Mathematical theory is developed close to the physical subject matter. 216 bibliographical footnotes. 55 figures. xi + 485pp. 5⅜ x 8.

S95 Paperbound **$2.55**

THE PRINCIPLES OF ELECTROCHEMISTRY, D. A. MacInnes. Basic equations for almost every subfield of electrochemistry from first principles, referring at all times to the soundest and most recent theories and results; unusually useful as text or as reference. Covers coulometers and Faraday's Law, electrolytic conductance, the Debye-Hueckel method for the theoretical calculation of activity coefficients, concentration cells, standard electrode potentials, thermodynamic ionization constants, pH, potentiometric titrations, irreversible phenomena, Planck's equation, and much more. "Excellent treatise," AMERICAN CHEMICAL SOCIETY JOURNAL. "Highly recommended," CHEMICAL AND METALLURGICAL ENGINEERING. 2 Indices. Appendix. 585-item bibliography. 137 figures. 94 tables. ii + 478pp. 5⅝ x 8⅜.

S52 Paperbound **$2.45**

THE PHASE RULE AND ITS APPLICATION, Alexander Findlay. Covering chemical phenomena of 1, 2, 3, 4, and multiple component systems, this "standard work on the subject" (NATURE, London), has been completely revised and brought up to date by A. N. Campbell and N. O. Smith. Brand new material has been added on such matters as binary, tertiary liquid equilibria, solid solutions in ternary systems, quinary systems of salts and water. Completely revised to triangular coordinates in ternary systems, clarified graphic representation, solid models, etc. 9th revised edition. Author, subject indexes. 236 figures. 505 footnotes, mostly bibliographic. xii + 494pp. 5⅜ x 8.

S91 Paperbound **$2.50**

THE SOLUBILITY OF NONELECTROLYTES, Joel H. Hildebrand and Robert L. Scott. The standard work on the subject; still indispensable as a reference source and for classroom work. Partial contents: The Ideal Solution (including Raoult's Law and Henry's Law, etc.); Nonideal Solutions; Intermolecular Forces; The Liquid State; Entropy of Athermal Mixing; Heat of Mixing; Polarity; Hydrogen Bonding; Specific Interactions; "Solvation" and "Association"; Systems of Three or More Components; Vapor Pressure of Binary Liquid Solutions; Mixtures of Gases; Solubility of Gases in Liquids; of Liquids in Liquids; of Solids in Liquids; Evaluation of Solubility Parameters; and other topics. Corrected republication of third (revised) edition. Appendices. Indexes. 138 figures. 111 tables. 1 photograph. iv + 488pp. 5⅜ x 8½.
S1125 Paperbound **$2.50**

TERNARY SYSTEMS: INTRODUCTION TO THE THEORY OF THREE COMPONENT SYSTEMS, G. Masing. Furnishes detailed discussion of representative types of 3-components systems, both in solid models (particularly metallic alloys) and isothermal models. Discusses mechanical mixture without compounds and without solid solutions; unbroken solid solution series; solid solutions with solubility breaks in two binary systems; iron-silicon-aluminum alloys; allotropic forms of iron in ternary system; other topics. Bibliography. Index. 166 illustrations. 178pp. 5⅝ x 8⅜.
S631 Paperbound **$1.50**

THE KINETIC THEORY OF GASES, Leonard B. Loeb, University of California. Comprehensive text and reference book which presents full coverage of basic theory and the important experiments and developments in the field for the student and investigator. Partial contents: The Mechanical Picture of a Perfect Gas, The Mean Free Path—Clausius' Deductions, Distribution of Molecular Velocities, discussions of theory of the problem of specific heats, the contributions of kinetic theory to our knowledge of electrical and magnetic properties of molecules and its application to the conduction of electricity in gases. New 14-page preface to Dover edition by the author. Name, subject indexes. Six appendices. 570-item bibliography. xxxvi + 687pp. 5⅜ x 8½.
S942 Paperbound **$3.50**

IONS IN SOLUTION, Ronald W. Gurney. A thorough and readable introduction covering all the fundamental principles and experiments in the field, by an internationally-known authority. Contains discussions of solvation energy, atomic and molecular ions, lattice energy, transferral of ions, interionic forces, cells and half-cells, transference of electrons, exchange forces, hydrogen ions, the electro-chemical series, and many other related topics. Indispensable to advanced undergraduates and graduate students in electrochemistry. Index. 45 illustrations. 15 tables. vii + 206pp. 5⅜ x 8½.
S124 Paperbound **$1.!5**

IONIC PROCESSES IN SOLUTION, Ronald W. Gurney. Lucid, comprehensive examination which brings together the approaches of electrochemistry, thermodynamics, statistical mechanics, electroacoustics, molecular physics, and quantum theory in the interpretation of the behavior of ionic solutions—the most important single work on the subject. More extensive and technical than the author's earlier work (IONS IN SOLUTION), it is a middle-level text for graduate students and researchers in electrochemistry. Covers such matters as Brownian motion in liquids, molecular ions in solution, heat of precipitation, entropy of solution, proton transfers, dissociation constant of nitric acid, viscosity of ionic solutions, etc. 78 illustrations. 47 tables. Name and subject index. ix + 275pp. 5⅜ x 8½.
S134 Paperbound **$1.85**

CRYSTALLOGRAPHIC DATA ON METAL AND ALLOY STRUCTURES, Compiled by A. Taylor and B. J. Kagle, Westinghouse Research Laboratories. Unique collection of the latest crystallographic data on alloys, compounds, and the elements, with lattice spacings expressed uniformly in absolute Angstrom units. Gathers together previously widely-scattered data from the Power Data File of the ATSM, structure reports, and the Landolt-Bornstein Tables, as well as from other original literature. 2300 different compounds listed in the first table. Alloys and Intermetallic Compounds, with much vital information on each. Also listings for nearly 700 Borides, Carbides, Hydrides, Oxides, Nitrides. Also all the necessary data on the crystal structure of 77 elements. vii + 263pp. 5⅜ x 8.
S1013 Paperbound **$2.25**

MATHEMATICAL CRYSTALLOGRAPHY AND THE THEORY OF GROUPS OF MOVEMENTS, Harold Hilton. Classic account of the mathematical theory of crystallography, particularly the geometrical theory of crystal-structure based on the work of Bravais, Jordan, Sohncke, Federow, Schoenflies, and Barlow. Partial contents: The Stereographic Projection, Properties Common to Symmetrical and Asymmetrical Crystals, The Theory of Groups, Coordinates of Equivalent Points, Crystallographic Axes and Axial Ratios, The Forms and Growth of Crystals, Lattices and Translations, The Structure-Theory, Infinite Groups of Movements, Triclinic and Monoclinic Groups, Orthorhombic Groups, etc. Index. 188 figures. xii + 262pp. 5⅜ x 8½.
S1058 Paperbound **$2.00**

CLASSICS IN THE THEORY OF CHEMICAL COMBINATIONS. Edited by O. T. Benfey. Vol. I of the Classics of Science Series, G. Holton, Harvard University, General Editor. This book is a collection of papers representing the major chapters in the development of the valence concept in chemistry. Includes essays by Wöhler and Liebig, Laurent, Williamson, Frankland, Kekulé and Couper, and two by van't Hoff and le Bel, which mark the first extension of the valence concept beyond its purely numerical character. Introduction and epilogue by Prof. Benfey. Index. 9 illustrations. New translation of Kekulé paper by Benfey. xiv + 191pp. 5⅜ x 8½.
S1066 Paperbound **$1.85**

ENGINEERING AND TECHNOLOGY

General and mathematical

ENGINEERING MATHEMATICS, Kenneth S. Miller. A text for graduate students of engineering to strengthen their mathematical background in differential equations, etc. Mathematical steps very explicitly indicated. Contents: Determinants and Matrices, Integrals, Linear Differential Equations, Fourier Series and Integrals, Laplace Transform, Network Theory, Random Function . . . all vital requisites for advanced modern engineering studies. Unabridged republication. Appendices: Borel Sets; Riemann-Stieltjes Integral; Fourier Series and Integrals. Index. References at Chapter Ends. xii + 417pp. 6 x 8½.　　　　　　S1121 Paperbound **$2.00**

MATHEMATICAL ENGINEERING ANALYSIS, Rufus Oldenburger. A book designed to assist the research engineer and scientist in making the transition from physical engineering situations to the corresponding mathematics. Scores of common practical situations found in all major fields of physics are supplied with their correct mathematical formulations—applications to automobile springs and shock absorbers, clocks, throttle torque of diesel engines, resistance networks, capacitors, transmission lines, microphones, neon tubes, gasoline engines, refrigeration cycles, etc. Each section reviews basic principles of underlying various fields: mechanics of rigid bodies, electricity and magnetism, heat, elasticity, fluid mechanics, and aerodynamics. Comprehensive and eminently useful. Index. 169 problems, answers. 200 photos and diagrams. xiv + 426pp. 5⅜ x 8½.　　　　　　S919 Paperbound **$2.50**

MATHEMATICS OF MODERN ENGINEERING, E. G. Keller and R. E. Doherty. Written for the Advanced Course in Engineering of the General Electric Corporation, deals with the engineering use of determinants, tensors, the Heaviside operational calculus, dyadics, the calculus of variations, etc. Presents underlying principles fully, but purpose is to teach engineers to deal with modern engineering problems, and emphasis is on the perennial engineering attack of set-up and solve. Indexes. Over 185 figures and tables. Hundreds of exercises, problems, and worked-out examples. References. Two volume set. Total of xxxiii + 623pp. 5⅜ x 8.
S734 Vol I Paperbound **$1.85**
S735 Vol II Paperbound **$1.85**
The set **$3.70**

MATHEMATICAL METHODS FOR SCIENTISTS AND ENGINEERS, L. P. Smith. For scientists and engineers, as well as advanced math students. Full investigation of methods and practical description of conditions under which each should be used. Elements of real functions, differential and integral calculus, space geometry, theory of residues, vector and tensor analysis, series of Bessel functions, etc. Each method illustrated by completely-worked-out examples, mostly from scientific literature. 368 graded unsolved problems. 100 diagrams. x + 453pp. 5⅝ x 8⅜.　　　　　　S220 Paperbound **$2.00**

THEORY OF FUNCTIONS AS APPLIED TO ENGINEERING PROBLEMS, edited by R. Rothe, F. Ollendorff, and K. Pohlhausen. A series of lectures given at the Berlin Institute of Technology that shows the specific applications of function theory in electrical and allied fields of engineering. Six lectures provide the elements of function theory in a simple and practical form, covering complex quantities and variables, integration in the complex plane, residue theorems, etc. Then 5 lectures show the exact uses of this powerful mathematical tool, with full discussions of problem methods. Index. Bibliography. 108 figures. x + 189pp. 5⅜ x 8.
S733 Paperbound **$1.35**

Aerodynamics and hydrodynamics

AIRPLANE STRUCTURAL ANALYSIS AND DESIGN, E. E. Sechler and L. G. Dunn. Systematic authoritative book which summarizes a large amount of theoretical and experimental work on structural analysis and design. Strong on classical subsonic material still basic to much aeronautic design . . . remains a highly useful source of information. Covers such areas as layout of the airplane, applied and design loads, stress-strain relationships for stable structures, truss and frame analysis, the problem of instability, the ultimate strength of stiffened flat sheet, analysis of cylindrical structures, wings and control surfaces, fuselage analysis, engine mounts, landing gears, etc. Originally published as part of the CALCIT Aeronautical Series. 256 Illustrations. 47 study problems. Indexes. xi + 420pp. 5⅜ x 8½.
S1043 Paperbound **$2.25**

FUNDAMENTALS OF HYDRO- AND AEROMECHANICS, L. Prandtl and O. G. Tietjens. The well-known standard work based upon Prandtl's lectures at Goettingen. Wherever possible hydrodynamics theory is referred to practical considerations in hydraulics, with the view of unifying theory and experience. Presentation is extremely clear and though primarily physical, mathematical proofs are rigorous and use vector analysis to a considerable extent. An Engineering Society Monograph, 1934. 186 figures. Index. xvi + 270pp. 5⅜ x 8.
S374 Paperbound **$1.85**

FLUID MECHANICS FOR HYDRAULIC ENGINEERS, H. Rouse. Standard work that gives a coherent picture of fluid mechanics from the point of view of the hydraulic engineer. Based on courses given to civil and mechanical engineering students at Columbia and the California Institute of Technology, this work covers every basic principle, method, equation, or theory of interest to the hydraulic engineer. Much of the material, diagrams, charts, etc., in this self-contained text are not duplicated elsewhere. Covers irrotational motion, conformal mapping, problems in laminar motion, fluid turbulence, flow around immersed bodies, transportation of sediment, general charcteristics of wave phenomena, gravity waves in open channels, etc. Index. Appendix of physical properties of common fluids. Frontispiece + 245 figures and photographs. xvi + 422pp. 5⅜ x 8. **S729 Paperbound $2.25**

WATERHAMMER ANALYSIS, John Parmakian. Valuable exposition of the graphical method of solving waterhammer problems by Assistant Chief Designing Engineer, U.S. Bureau of Reclamation. Discussions of rigid and elastic water column theory, velocity of waterhammer waves, theory of graphical waterhammer analysis for gate operation, closings, openings, rapid and slow movements, etc., waterhammer in pump discharge caused by power failure, waterhammer analysis for compound pipes, and numerous related problems. "With a concise and lucid style, clear printing, adequate bibliography and graphs for approximate solutions at the project stage, it fills a vacant place in waterhammer literature," WATER POWER. 43 problems. Bibliography. Index. 113 illustrations. xiv + 161pp. 5⅜ x 8½. **S1061 Paperbound $1.65**

AERODYNAMIC THEORY: A GENERAL REVIEW OF PROGRESS, William F. Durand, editor-in-chief. A monumental joint effort by the world's leading authorities prepared under a grant of the Guggenheim Fund for the Promotion of Aeronautics. Intended to provide the student and aeronautic designer with the theoretical and experimental background of aeronautics. Never equalled for breadth, depth, reliability. Contains discussions of special mathematical topics not usually taught in the engineering or technical courses. Also: an extended two-part treatise on Fluid Mechanics, discussions of aerodynamics of perfect fluids, analyses of experiments with wind tunnels, applied airfoil theory, the non-lifting system of the airplane, the air propeller, hydrodynamics of boats and floats, the aerodynamics of cooling, etc. Contributing experts include Munk, Giacomelli, Prandtl, Toussaint, Von Karman, Klemperer, among others. Unabridged republication. 6 volumes bound as 3. Total of 1,012 figures, 12 plates. Total of 2,186pp. Bibliographies. Notes. Indices. 5⅜ x 8. **S328-S330 Paperbound The Set $13.50**

APPLIED HYDRO- AND AEROMECHANICS, L. Prandtl and O. G. Tietjens. Presents, for the most part, methods which will be valuable to engineers. Covers flow in pipes, boundary layers, airfoil theory, entry conditions, turbulent flow in pipes, and the boundary layer, determining drag from measurements of pressure and velocity, etc. "Will be welcomed by all students of aerodynamics," NATURE. Unabridged, unaltered. An Engineering Society Monograph, 1934. Index. 226 figures, 28 photographic plates illustrating flow patterns. xvi + 311pp. 5⅜ x 8. **S375 Paperbound $2.00**

SUPERSONIC AERODYNAMICS, E. R. C. Miles. Valuable theoretical introduction to the supersonic domain, with emphasis on mathematical tools and principles, for practicing aerodynamicists and advanced students in aeronautical engineering. Covers fundamental theory, divergence theorem and principles of circulation, compressible flow and Helmholtz laws, the Prandtl-Busemann graphic method for 2-dimensional flow, oblique shock waves, the Taylor-Maccoll method for cones in supersonic flow, the Chaplygin method for 2-dimensional flow, etc. Problems range from practical engineering problems to development of theoretical results. "Rendered outstanding by the unprecedented scope of its contents . . . has undoubtedly filled a vital gap," AERONAUTICAL ENGINEERING REVIEW. Index. 173 problems, answers. 106 diagrams. 7 tables. xii + 255pp. 5⅜ x 8. **S214 Paperbound $1.45**

HYDRAULIC TRANSIENTS, G. R. Rich. The best text in hydraulics ever printed in English . . . by one of America's foremost engineers (former Chief Design Engineer for T.V.A.). Provides a transition from the basic differential equations of hydraulic transient theory to the arithmetic intergration computation required by practicing engineers. Sections cover Water Hammer, Turbine Speed Regulation, Stability of Governing, Water-Hammer Pressures in Pump Discharge Lines, The Differential and Restricted Orifice Surge Tanks, The Normalized Surge Tank Charts of Calame and Gaden, Navigation Locks, Surges in Power Canals—Tidal Harmonics, etc. Revised and enlarged. Author's prefaces. Index. xiv + 409pp. 5⅜ x 8½. **S116 Paperbound $2.50**

HYDRAULICS AND ITS APPLICATIONS, A. H. Gibson. Excellent comprehensive textbook for the student and thorough practical manual for the professional worker, a work of great stature in its area. Half the book is devoted to theory and half to applications and practical problems met in the field. Covers modes of motion of a fluid, critical velocity, viscous flow, eddy formation, Bernoulli's theorem, flow in converging passages, vortex motion, form of effluent streams, notches and weirs, skin friction, losses at valves and elbows, siphons, erosion of channels, jet propulsion, waves of oscillation, and over 100 similar topics. Final chapters (nearly 400 pages) cover more than 100 kinds of hydraulic machinery: Pelton wheel, speed regulators, the hydraulic ram, surge tanks, the scoop wheel, the Venturi meter, etc. A special chapter treats methods of testing theoretical hypotheses: scale models of rivers, tidal estuaries, siphon spillways, etc. 5th revised and enlarged (1952) edition. Index. Appendix. 427 photographs and diagrams. 95 examples, answers. xv + 813pp. 6 x 9. **S791 Clothbound $8.00**

FLUID MECHANICS THROUGH WORKED EXAMPLES, D. R. L. Smith and J. Houghton. Advanced text covering principles and applications to practical situations. Each chapter begins with concise summaries of fundamental ideas. 163 fully worked out examples applying principles outlined in the text. 275 other problems, with answers. Contents: The Pressure of Liquids on Surfaces; Floating Bodies; Flow Under Constant Head in Pipes; Circulation; Vorticity; The Potential Function; Laminar Flow and Lubrication; Impact of Jets; Hydraulic Turbines; Centrifugal and Reciprocating Pumps; Compressible Fluids; and many other items. Total of 438 examples. 250 line illustrations. 340pp. Index. 6 x 8⅞.   S981 Clothbound **$6.00**

THEORY OF SHIP MOTIONS, S. N. Blagoveshchensky. The only detailed text in English in a rapidly developing branch of engineering and physics, it is the work of one of the world's foremost authorities—Blagoveshchensky of Leningrad Shipbuilding Institute. A senior-level treatment written primarily for engineering students, but also of great importance to naval architects, designers, contractors, researchers in hydrodynamics, and other students. No mathematics beyond ordinary differential equations is required for understanding the text. Translated by T. & L. Strelkoff, under editorship of Louis Landweber, Iowa Institute of Hydraulic Research, under auspices of Office of Naval Research. Bibliography. Index. 231 diagrams and illustrations. Total of 649pp. 5⅜ x 8½.   Vol. I: S234 Paperbound **$2.00**
Vol. II: S235 Paperbound **$2.00**

THEORY OF FLIGHT, Richard von Mises. Remains almost unsurpassed as balanced, well-written account of fundamental fluid dynamics, and situations in which air compressibility effects are unimportant. Stressing equally theory and practice, avoiding formidable mathematical structure, it conveys a full understanding of physical phenomena and mathematical concepts. Contains perhaps the best introduction to general theory of stability. "Outstanding," Scientific, Medical, and Technical Books. New introduction by K. H. Hohenemser. Bibliographical, historical notes. Index. 408 illustrations. xvi + 620pp. 5⅜ x 8⅜.   S541 Paperbound **$3.50**

THEORY OF WING SECTIONS, I. H. Abbott, A. E. von Doenhoff. Concise compilation of subsonic aerodynamic characteristics of modern NASA wing sections, with description of their geometry, associated theory. Primarily reference work for engineers, students, it gives methods, data for using wing-section data to predict characteristics. Particularly valuable: chapters on thin wings, airfoils; complete summary of NACA's experimental observations, system of construction families of airfoils. 350pp. of tables on Basic Thickness Forms, Mean Lines, Airfoil Ordinates, Aerodynamic Characteristics of Wing Sections. Index. Bibliography. 191 illustrations. Appendix. 705pp. 5⅜ x 8.   S558 Paperbound **$3.25**

WEIGHT-STRENGTH ANALYSIS OF AIRCRAFT STRUCTURES, F. R. Shanley. Scientifically sound methods of analyzing and predicting the structural weight of aircraft and missiles. Deals directly with forces and the distances over which they must be transmitted, making it possible to develop methods by which the minimum structural weight can be determined for any material and conditions of loading. Weight equations for wing and fuselage structures. Includes author's original papers on inelastic buckling and creep buckling. "Particularly successful in presenting his analytical methods for investigating various optimum design principles," AERONAUTICAL ENGINEERING REVIEW. Enlarged bibliography. Index. 199 figures. xiv + 404pp. 5⅝ x 8⅜.   S660 Paperbound **$2.50**

Electricity

TWO-DIMENSIONAL FIELDS IN ELECTRICAL ENGINEERING, L. V. Bewley. A useful selection of typical engineering problems of interest to practicing electrical engineers. Introduces senior students to the methods and procedures of mathematical physics. Discusses theory of functions of a complex variable, two-dimensional fields of flow, general theorems of mathematical physics and their applications, conformal mapping or transformation, method of images, freehand flux plotting, etc. New preface by the author. Appendix by W. F. Kiltner. Index. Bibliography at chapter ends. xiv + 204pp. 5⅜ x 8½.   S1118 Paperbound **$1.50**

FLUX LINKAGES AND ELECTROMAGNETIC INDUCTION, L. V. Bewley. A brief, clear book which shows proper uses and corrects misconceptions of Faraday's law of electromagnetic induction in specific problems. Contents: Circuits, Turns, and Flux Linkages; Substitution of Circuits; Electromagnetic Induction; General Criteria for Electromagnetic Induction; Applications and Paradoxes; Theorem of Constant Flux Linkages. New Section: Rectangular Coil in a Varying Uniform Medium. Valuable supplement to class texts for engineering students. Corrected, enlarged edition. New preface. Bibliography in notes. 49 figures. xi + 106pp. 5⅜ x 8.   S1103 Paperbound **$1.25**

INDUCTANCE CALCULATIONS: WORKING FORMULAS AND TABLES, Frederick W. Grover. An invaluable book to everyone in electrical engineering. Provides simple single formulas to cover all the more important cases of inductance. The approach involves only those parameters that naturally enter into each situation, while extensive tables are given to permit easy interpolations. Will save the engineer and student countless hours and enable them to obtain accurate answers with minimal effort. Corrected republication of 1946 edition. 58 tables. 97 completely worked out examples. 66 figures. xiv + 286pp. 5⅜ x 8½.   S974 Paperbound **$1.85**

GASEOUS CONDUCTORS: THEORY AND ENGINEERING APPLICATIONS, J. D. Cobine. An indispensable text and reference to gaseous conduction phenomena, with the engineering viewpoint prevailing throughout. Studies the kinetic theory of gases, ionization, emission phenomena; gas breakdown, spark characteristics, glow, and discharges; engineering applications in circuit interrupters, rectifiers, light sources, etc. Separate detailed treatment of high pressure arcs (Suits); low pressure arcs (Langmuir and Tonks). Much more. "Well organized, clear, straightforward," Tonks, Review of Scientific Instruments. Index. Bibliography. 83 practice problems. 7 appendices. Over 600 figures. 58 tables. xx + 606pp. 5⅜ x 8.   S442 Paperbound **$3.25**

INTRODUCTION TO THE STATISTICAL DYNAMICS OF AUTOMATIC CONTROL SYSTEMS, V. V. Solodovnikov. First English publication of text-reference covering important branch of automatic control systems—random signals; in its original edition, this was the first comprehensive treatment. Examines frequency characteristics, transfer functions, stationary random processes, determination of minimum mean-squared error, of transfer function for a finite period of observation, much more. Translation edited by J. B. Thomas, L. A. Zadeh. Index. Bibliography. Appendix. xxii + 308pp. 5⅜ x 8.   S420 Paperbound **$2.35**

TENSORS FOR CIRCUITS, Gabriel Kron. A boldly original method of analyzing engineering problems, at center of sharp discussion since first introduced, now definitely proved useful in such areas as electrical and structural networks on automatic computers. Encompasses a great variety of specific problems by means of a relatively few symbolic equations. "Power and flexibility . . . becoming more widely recognized," Nature. Formerly "A Short Course in Tensor Analysis." New introduction by B. Hoffmann. Index. Over 800 diagrams. xix + 250pp. 5⅜ x 8.   S534 Paperbound **$2.00**

SELECTED PAPERS ON SEMICONDUCTOR MICROWAVE ELECTRONICS, edited by Sumner N. Levine and Richard R. Kurzrok. An invaluable collection of important papers dealing with one of the most remarkable devolopments in solid-state electronics—the use of the **p-n** junction to achieve amplification and frequency conversion of microwave frequencies. Contents: General Survey (3 introductory papers by W. E. Danielson, R. N. Hall, and M. Tenzer); General Theory of Nonlinear Elements (3 articles by A. van der Ziel, H. E. Rowe, and Manley and Rowe); Device Fabrication and Characterization (3 pieces by Bakanowski, Cranna, and Uhlir, by McCotter, Walker and Fortini, and by S. T. Eng); Parametric Amplifiers and Frequency Multipliers (13 articles by Uhlir, Heffner and Wade, Matthaei, P. K. Tien, van der Ziel, Engelbrecht, Currie and Gould, Uenohara, Leeson and Weinreb, and others); and Tunnel Diodes (4 papers by L. Esaki, H. S. Sommers, Jr., M. E. Hines, and Yariv and Cook). Introduction. 295 Figures. xiii + 286pp. 6½ x 9¼.   S1126 Paperbound **$2.50**

THE PRINCIPLES OF ELECTROMAGNETISM APPLIED TO ELECTRICAL MACHINES, B. Hague. A concise, but complete, summary of the basic principles of the magnetic field and its applications, with particular reference to the kind of phenomena which occur in electrical machines. Part I: General Theory—magnetic field of a current, electromagnetic field passing from air to iron, mechanical forces on linear conductors, etc. Part II: Application of theory to the solution of electromechanical problems—the magnetic field and mechanical forces in non-salient pole machinery, the field within slots and between salient poles, and the work of Rogowski, Roth, and Strutt. Formery titled "Electromagnetic Problems in Electrical Engineering." 2 appendices. Index. Bibliography in notes. 115 figures. xiv + 359pp. 5⅜ x 8½.   S246 Paperbound **$2.25**

Mechanical engineering

DESIGN AND USE OF INSTRUMENTS AND ACCURATE MECHANISM, T. N. Whitehead. For the instrument designer, engineer; how to combine necessary mathematical abstractions with independent observation of actual facts. Partial contents: instruments & their parts, theory of errors, systematic errors, probability, short period errors, erratic errors, design precision, kinematic, semikinematic design, stiffness, planning of an instrument, human factor, etc. Index. 85 photos, diagrams. xii + 288pp. 5⅜ x 8.   S270 Paperbound **$2.00**

A TREATISE ON GYROSTATICS AND ROTATIONAL MOTION: THEORY AND APPLICATIONS, Andrew Gray. Most detailed, thorough book in English, generally considered definitive study. Many problems of all sorts in full detail, or step-by-step summary. Classical problems of Bour, Lottner, etc.; later ones of great physical interest. Vibrating systems of gyrostats, earth as a top, calculation of path of axis of a top by elliptic integrals, motion of unsymmetrical top, much more. Index. 160 illus. 550pp. 5⅜ x 8.   S589 Paperbound **$2.75**

MECHANICS OF THE GYROSCOPE, THE DYNAMICS OF ROTATION, R. F. Deimel, Professor of Mechanical Engineering at Stevens Institute of Technology. Elementary general treatment of dynamics of rotation, with special application of gyroscopic phenomena. No knowledge of vectors needed. Velocity of a moving curve, acceleration to a point, general equations of motion, gyroscopic horizon, free gyro, motion of discs, the damped gyro, 103 similar topics. Exercises. 75 figures. 208pp. 5⅜ x 8.   S66 Paperbound **$1.75**

STRENGTH OF MATERIALS, J. P. Den Hartog. Distinguished text prepared for M.I.T. course, ideal as introduction, refresher, reference, or self-study text. Full clear treatment of elementary material (tension, torsion, bending, compound stresses, deflection of beams, etc.), plus much advanced material on engineering methods of great practical value: full treatment of the Mohr circle, lucid elementary discussions of the theory of the center of shear and the "Myosotis" method of calculating beam deflections, reinforced concrete, plastic deformations, photoelasticity, etc. In all sections, both general principles and concrete applications are given. Index. 186 figures (160 others in problem section). 350 problems, all with answers. List of formulas. viii + 323pp. 5⅜ x 8.   S755 Paperbound **$2.00**

PHOTOELASTICITY: PRINCIPLES AND METHODS, H. T. Jessop, F. C. Harris. For the engineer, for specific problems of stress analysis. Latest time-saving methods of checking calculations in 2-dimensional design problems, new techniques for stresses in 3 dimensions, and lucid description of optical systems used in practical photoelasticity. Useful suggestions and hints based on on-the-job experience included. Partial contents: strained and stress-strain relations, circular disc under thrust along diameter, rectangular block with square hole under vertical thrust, simply supported rectangular beam under central concentrated load, etc. Theory held to minimum, no advanced mathematical training needed. Index. 164 illustrations. viii + 184pp. 6⅛ x 9¼.   S720 Paperbound **$2.00**

APPLIED ELASTICITY, J. Prescott. Provides the engineer with the theory of elasticity usually lacking in books on strength of materials, yet concentrates on those portions useful for immediate application. Develops every important type of elasticity problem from theoretical principles. Covers analysis of stress, relations between stress and strain, the empirical basis of elasticity, thin rods under tension or thrust, Saint Venant's theory, transverse oscillations of thin rods, stability of thin plates, cylinders with thin walls, vibrations of rotating disks, elastic bodies in contact, etc. "Excellent and important contribution to the subject, not merely in the old matter which he has presented in new and refreshing form, but also in the many original investigations here published for the first time," NATURE. Index. 3 Appendixes. vi + 672pp. 5⅜ x 8.   S726 Paperbound **$3.25**

APPLIED MECHANICS FOR ENGINEERS, Sir Charles Inglis, F.R.S. A representative survey of the many and varied engineering questions which can be answered by statics and dynamics. The author, one of first and foremost adherents of "structural dynamics," presents distinctive illustrative examples and clear, concise statement of principles—directing the discussion at methodology and specific problems. Covers fundamental principles of rigid-body statics, graphic solutions of static problems, theory of taut wires, stresses in frameworks, particle dynamics, kinematics, simple harmonic motion and harmonic analysis, two-dimensional rigid dynamics, etc. 437 illustrations. xii + 404pp. 5⅜ x 8½. S1119 Paperbound **$2.50**

THEORY OF MACHINES THROUGH WORKED EXAMPLES, G. H. Ryder. Practical mechanical engineering textbook for graduates and advanced undergraduates, as well as a good reference work for practicing engineers. Partial contents: Mechanisms, Velocity and Acceleration (including discussion of Klein's Construction for Piston Acceleration), Cams, Geometry of Gears, Clutches and Bearings, Belt and Rope Drives, Brakes, Inertia Forces and Couples, General Dynamical Problems, Gyroscopes, Linear and Angular Vibrations, Torsional Vibrations, Transverse Vibrations and Whirling Speeds (Chapters on vibrations considerably enlarged from previous editions). Over 300 problems, many fully worked out. Index. 195 line illustrations. Revised and enlarged edition. viii + 280pp. 5⅝ x 8¾.   S980 Clothbound **$5.00**

THE KINEMATICS OF MACHINERY: OUTLINES OF A THEORY OF MACHINES, Franz Reuleaux. The classic work in the kinematics of machinery. The present thinking about the subject has all been shaped in great measure by the fundamental principles stated here by Reuleaux almost 90 years ago. While some details have naturally been superseded, his basic viewpoint has endured; hence, the book is still an excellent text for basic courses in kinematics and a standard reference work for active workers in the field. Covers such topics as: the nature of the machine problem, phoronomic propositions, pairs of elements, incomplete kinematic chains, kinematic notation and analysis, analyses of chamber-crank trains, chamber-wheel trains, constructive elements of machinery, complete machines, etc., with main focus on controlled movement in mechanisms. Unabridged republication of original edition, translated by Alexander B. Kennedy. New introduction for this edition by E. S. Ferguson. Index. 451 illustrations. xxiv + 622pp. 5⅜ x 8½.   S1124 Paperbound **$3.00**

ANALYTICAL MECHANICS OF GEARS, Earle Buckingham. Provides a solid foundation upon which logical design practices and design data can be constructed. Originally arising out of investigations of the ASME Special Research Committee on Worm Gears and the Strength of Gears, the book covers conjugate gear-tooth action, the nature of the contact, and resulting gear-tooth profiles of: spur, internal, helical, spiral, worm, bevel, and hypoid or skew bevel gears. Also: frictional heat of operation and its dissipation, friction losses, etc., dynamic loads in operation, and related matters. Familiarity with this book is still regarded as a necessary prerequisite to work in modern gear manufacturing. 263 figures. 103 tables. Index. x + 546pp. 5⅜ x 8½.   S1073 Paperbound **$2.75**

Optical design, lighting

THE SCIENTIFIC BASIS OF ILLUMINATING ENGINEERING, Parry Moon, Professor of Electrical Engineering, M.I.T. Basic, comprehensive study. Complete coverage of the fundamental theoretical principles together with the elements of design, vision, and color with which the lighting engineer must be familiar. Valuable as a text as well as a reference source to the practicing engineer. Partial contents: Spectroradiometric Curve, Luminous Flux, Radiation from Gaseous-Conduction Sources, Radiation from Incandescent Sources, Incandescent Lamps, Measurement of Light, Illumination from Point Sources and Surface Sources, Elements of Lighting Design. 7 Appendices. Unabridged and corrected republication, with additions. New preface containing conversion tables of radiometric and photometric concepts. Index. 707-item bibliography. 92-item bibliography of author's articles. 183 problems. xxiii + 608pp. 5⅜ x 8½. **S242 Paperbound $3.25**

OPTICS AND OPTICAL INSTRUMENTS: AN INTRODUCTION WITH SPECIAL REFERENCE TO PRACTICAL APPLICATIONS, B. K. Johnson. An invaluable guide to basic practical applications of optical principles, which shows how to set up inexpensive working models of each of the four main types of optical instruments—telescopes, microscopes, photographic lenses, optical projecting systems. Explains in detail the most important experiments for determining their accuracy, resolving power, angular field of view, amounts of aberration, all other necessary facts about the instruments. Formerly "Practical Optics." Index. 234 diagrams. Appendix. 224pp. 5⅜ x 8. **S642 Paperbound $1.75**

APPLIED OPTICS AND OPTICAL DESIGN, A. E. Conrady. With publication of vol. 2, standard work for designers in optics is now complete for first time. Only work of its kind in English; only detailed work for practical designer and self-taught. Requires, for bulk of work, no math above trig. Step-by-step exposition, from fundamental concepts of geometrical, physical optics, to systematic study, design, of almost all types of optical systems. Vol. 1: all ordinary ray-tracing methods; primary aberrations; necessary higher aberration for design of telescopes, low-power microscopes, photographic equipment. Vol. 2: (Completed from author's notes by R. Kingslake, Dir. Optical Design, Eastman Kodak.) Special attention to high-power microscope, anastigmatic photographic objectives. "An indispensable work," J., Optical Soc. of Amer. "As a practical guide this book has no rival," Transactions, Optical Soc. Index. Bibliography. 193 diagrams. 852pp. 6⅛ x 9¼. Vol. 1 S366 Paperbound **$3.50** Vol. 2 S612 Paperbound **$2.95**

Miscellaneous

THE MEASUREMENT OF POWER SPECTRA FROM THE POINT OF VIEW OF COMMUNICATIONS ENGINEERING, R. B. Blackman, J. W. Tukey. This pathfinding work, reprinted from the "Bell System Technical Journal," explains various ways of getting practically useful answers in the measurement of power spectra, using results from both transmission theory and the theory of statistical estimation. Treats: Autocovariance Functions and Power Spectra; Direct Analog Computation; Distortion, Noise, Heterodyne Filtering and Pre-whitening; Aliasing; Rejection Filtering and Separation; Smoothing and Decimation Procedures; Very Low Frequencies; Transversal Filtering; much more. An appendix reviews fundamental Fourier techniques. Index of notation. Glossary of terms. 24 figures. XII tables. Bibliography. General index. 192pp. 5⅜ x 8. **S507 Paperbound $1.85**

CALCULUS REFRESHER FOR TECHNICAL MEN, A. Albert Klaf. This book is unique in English as a refresher for engineers, technicians, students who either wish to brush up their calculus or to clear up uncertainties. It is not an ordinary text, but an examination of most important aspects of integral and differential calculus in terms of the 756 questions most likely to occur to the technical reader. The first part of this book covers simple differential calculus, with constants, variables, functions, increments, derivatives, differentiation, logarithms, curvature of curves, and similar topics. The second part covers fundamental ideas of integration, inspection, substitution, transformation, reduction, areas and volumes, mean value, successive and partial integration, double and triple integration. Practical aspects are stressed rather than theoretical. A 50-page section illustrates the application of calculus to specific problems of civil and nautical engineering, electricity, stress and strain, elasticity, industrial engineering, and similar fields.—756 questions answered. 566 problems, mostly answered. 36 pages of useful constants, formulae for ready reference. Index. v + 431pp. 5⅜ x 8. **T370 Paperbound $2.00**

METHODS IN EXTERIOR BALLISTICS, Forest Ray Moulton. Probably the best introduction to the mathematics of projectile motion. The ballistics theories propounded were coordinated with extensive proving ground and wind tunnel experiments conducted by the author and others for the U.S. Army. Broad in scope and clear in exposition, it gives the beginnings of the theory used for modern-day projectile, long-range missile, and satellite motion. Six main divisions: Differential Equations of Translatory Motion of a projectile; Gravity and the Resistance Function; Numerical Solution of Differential Equations; Theory of Differential Variations; Validity of Method of Numerical Integration; and Motion of a Rotating Projectile. Formerly titled: "New Methods in Exterior Ballistics." Index. 38 diagrams. viii + 259pp. 5⅜ x 8½. **S232 Paperbound $1.75**

LOUD SPEAKERS: THEORY, PERFORMANCE, TESTING AND DESIGN, N. W. McLachlan. Most comprehensive coverage of theory, practice of loud speaker design, testing; classic reference, study manual in field. First 12 chapters deal with theory, for readers mainly concerned with math. aspects; last 7 chapters will interest reader concerned with testing, design. Partial contents: principles of sound propagation, fluid pressure on vibrators, theory of moving-coil principle, transients, driving mechanisms, response curves, design of horn type moving coil speakers, electrostatic speakers, much more. Appendix. Bibliography. Index. 165 illustrations, charts. 411pp. 5⅜ x 8.   S588 Paperbound **$2.25**

MICROWAVE TRANSMISSION, J. C. Slater. First text dealing exclusively with microwaves, brings together points of view of field, circuit theory, for graduate student in physics, electrical engineering, microwave technician. Offers valuable point of view not in most later studies. Uses Maxwell's equations to study electromagnetic field, important in this area. Partial contents: infinite line with distributed parameters, impedance of terminated line, plane waves, reflections, wave guides, coaxial line, composite transmission lines, impedance matching, etc. Introduction. Index. 76 illus. 319pp. 5⅜ x 8.
S564 Paperbound **$1.50**

MICROWAVE TRANSMISSION DESIGN DATA, T. Moreno. Originally classified, now rewritten and enlarged (14 new chapters) for public release under auspices of Sperry Corp. Material of immediate value or reference use to radio engineers, systems designers, applied physicists, etc. Ordinary transmission line theory; attenuation; capacity; parameters of coaxial lines; higher modes; flexible cables; obstacles, discontinuities, and injunctions; tunable wave guide impedance transformers; effects of temperature and humidity; much more. "Enough theoretical discussion is included to allow use of data without previous background," Electronics. 324 circuit diagrams, figures, etc. Tables of dielectrics, flexible cable, etc., data. Index. ix + 248pp. 5⅜ x 8.   S459 Paperbound **$1.65**

RAYLEIGH'S PRINCIPLE AND ITS APPLICATIONS TO ENGINEERING, G. Temple & W. Bickley. Rayleigh's principle developed to provide upper and lower estimates of true value of fundamental period of a vibrating system, or condition of stability of elastic systems. Illustrative examples; rigorous proofs in special chapters. Partial contents: Energy method of discussing vibrations, stability. Perturbation theory, whirling of uniform shafts. Criteria of elastic stability. Application of energy method. Vibrating systems. Proof, accuracy, successive approximations, application of Rayleigh's principle. Synthetic theorems. Numerical, graphical methods. Equilibrium configurations, Ritz's method. Bibliography. Index. 22 figures. ix + 156pp. 5⅜ x 8.
S307 Paperbound **'$1.85**

ELASTICITY, PLASTICITY AND STRUCTURE OF MATTER, R. Houwink. Standard treatise on rheological aspects of different technically important solids such as crystals, resins, textiles, rubber, clay, many others. Investigates general laws for deformations; determines divergences from these laws for certain substances. Covers general physical and mathematical aspects of plasticity, elasticity, viscosity. Detailed examination of deformations, internal structure of matter in relation to elastic and plastic behavior, formation of solid matter from a fluid, conditions for elastic and plastic behavior of matter. Treats glass, asphalt, gutta percha, balata, proteins, baker's dough, lacquers, sulphur, others. 2nd revised, enlarged edition. Extensive revised bibliography in over 500 footnotes. Index. Table of symbols. 214 figures. xviii + 368pp. 6 x 9¼.   S385 Paperbound **$2.45**

THE SCHWARZ-CHRISTOFFEL TRANSFORMATION AND ITS APPLICATIONS: A SIMPLE EXPOSITION, Miles Walker. An important book for engineers showing how this valuable tool can be employed in practical situations. Very careful, clear presentation covering numerous concrete engineering problems. Includes a thorough account of conjugate functions for engineers—useful for the beginner and for review. Applications to such problems as: Stream-lines round a corner, electric conductor in air-gap, dynamo slots, magnetized poles, much more. Formerly "Conjugate Functions for Engineers." Preface. 92 figures, several tables. Index. ix + 116pp. 5⅜ x 8½.   S1149 Paperbound **$1.25**

THE LAWS OF THOUGHT, George Boole. This book founded symbolic logic some hundred years ago. It is the 1st significant attempt to apply logic to all aspects of human endeavour. Partial contents: derivation of laws, signs & laws, interpretations, eliminations, conditions of a perfect method, analysis, Aristotelian logic, probability, and similar topics. xviii + 424pp. 5⅜ x 8.   S28 Paperbound **$2.00**

SCIENCE AND METHOD, Henri Poincaré. Procedure of scientific discovery, methodology, experiment, idea-germination—the intellectual processes by which discoveries come into being. Most significant and most interesting aspects of development, application of ideas. Chapters cover selection of facts, chance, mathematical reasoning, mathematics, and logic; Whitehead, Russell, Cantor; the new mechanics, etc. 288pp. 5⅜ x 8.   S222 Paperbound **$1.35**

FAMOUS BRIDGES OF THE WORLD, D. B. Steinman. An up-to-the-minute revised edition of a book that explains the fascinating drama of how the world's great bridges came to be built. The author, designer of the famed Mackinac bridge, discusses bridges from all periods and all parts of the world, explaining their various types of construction, and describing the problems their builders faced. Although primarily for youngsters, this cannot fail to interest readers of all ages. 48 illustrations in the text. 23 photographs. 99pp. 6⅛ x 9¼.
T161 Paperbound **$1.00**

PHYSICS

General physics

FOUNDATIONS OF PHYSICS, R. B. Lindsay & H. Margenau. Excellent bridge between semi-popular works & technical treatises. A discussion on methods of physical description, construction of theory; valuable for physicist with elementary calculus who is interested in ideas that give meaning to data, tools of modern physics. Contents include symbolism, mathematical equations; space & time foundations of mechanics; probability; physics & continua; electron theory; special & general relativity; quantum mechanics; causality. "Thorough and yet not overdetailed. Unreservedly recommended," NATURE (London). Unabridged, corrected edition. List of recommended readings. 35 illustrations. xi + 537pp. 5⅜ x 8.
S377 Paperbound **$3.00**

FUNDAMENTAL FORMULAS OF PHYSICS, ed. by D. H. Menzel. Highly useful, fully inexpensive reference and study text, ranging from simple to highly sophisticated operations. Mathematics integrated into text—each chapter stands as short textbook of field represented. Vol. 1: Statistics, Physical Constants, Special Theory of Relativity, Hydrodynamics, Aerodynamics, Boundary Value Problems in Math. Physics; Viscosity, Electromagnetic Theory, etc. Vol. 2: Sound, Acoustics, Geometrical Optics, Electron Optics, High-Energy Phenomena, Magnetism, Biophysics, much more. Index. Total of 800pp. 5⅜ x 8.   Vol. 1 S595 Paperbound **$2.25**
Vol. 2 S596 Paperbound **$2.25**

MATHEMATICAL PHYSICS, D. H. Menzel. Thorough one-volume treatment of the mathematical techniques vital for classic mechanics, electromagnetic theory, quantum theory, and relativity. Written by the Harvard Professor of Astrophysics for junior, senior, and graduate courses, it gives clear explanations of all those aspects of function theory, vectors, matrices, dyadics, tensors, partial differential equations, etc., necessary for the understanding of the various physical theories. Electron theory, relativity, and other topics seldom presented appear here in considerable detail. Scores of definitions, conversion factors, dimensional constants, etc. "More detailed than normal for an advanced text . . . excellent set of sections on Dyadics, Matrices, and Tensors," JOURNAL OF THE FRANKLIN INSTITUTE. Index. 193 problems, with answers. x + 412pp. 5⅜ x 8.   S56 Paperbound **$2.00**

THE SCIENTIFIC PAPERS OF J. WILLARD GIBBS. All the published papers of America's outstanding theoretical scientist (except for "Statistical Mechanics" and "Vector Analysis"). Vol I (thermodynamics) contains one of the most brilliant of all 19th-century scientific papers—the 300-page "On the Equilibrium of Heterogeneous Substances," which founded the science of physical chemistry, and clearly stated a number of highly important natural laws for the first time; 8 other papers complete the first volume. Vol II includes 2 papers on dynamics, 8 on vector analysis and multiple algebra, 5 on the electromagnetic theory of light, and 6 miscellaneous papers. Biographical sketch by H. A. Bumstead. Total of xxxvi + 718pp. 5⅜ x 8⅜.
S721 Vol I Paperbound **$2.50**
S722 Vol II Paperbound **$2.00**
The set **$4.50**

BASIC THEORIES OF PHYSICS, Peter Gabriel Bergmann. Two-volume set which presents a critical examination of important topics in the major subdivisions of classical and modern physics. The first volume is concerned with classical mechanics and electrodynamics: mechanics of mass points, analytical mechanics, matter in bulk, electrostatics and magnetostatics, electromagnetic interaction, the field waves, special relativity, and waves. The second volume (Heat and Quanta) contains discussions of the kinetic hypothesis, physics and statistics, stationary ensembles, laws of thermodynamics, early quantum theories, atomic spectra, probability waves, quantization in wave mechanics, approximation methods, and abstract quantum theory. A valuable supplement to any thorough course or text.
Heat and Quanta: Index. 8 figures. x + 300pp. 5⅜ x 8½.   S968 Paperbound **$2.00**
Mechanics and Electrodynamics: Index. 14 figures. vii + 280pp. 5⅜ x 8½.
S969 Paperbound **$1.85**

THEORETICAL PHYSICS, A. S. Kompaneyets. One of the very few thorough studies of the subject in this price range. Provides advanced students with a comprehensive theoretical background. Especially strong on recent experimentation and developments in quantum theory. Contents: Mechanics (Generalized Coordinates, Lagrange's Equation, Collision of Particles, etc.), Electrodynamics (Vector Analysis, Maxwell's equations, Transmission of Signals, Theory of Relativity, etc.), Quantum Mechanics (the Inadequacy of Classical Mechanics, the Wave Equation, Motion in a Central Field, Quantum Theory of Radiation, Quantum Theories of Dispersion and Scattering, etc.), and Statistical Physics (Equilibrium Distribution of Molecules in an Ideal Gas, Boltzmann statistics, Bose and Fermi Distribution, Thermodynamic Quantities, etc.). Revised to 1961. Translated by George Yankovsky, authorized by Kompaneyets. 137 exercises. 56 figures. 529pp. 5⅜ x 8½. S972 Paperbound **$2.50**

ANALYTICAL AND CANONICAL FORMALISM IN PHYSICS, André Mercier. A survey, in one volume, of the variational principles (the key principles—in mathematical form—from which the basic laws of any one branch of physics can be derived) of the several branches of physical theory, together with an examination of the relationships among them. Contents: the Lagrangian Formalism, Lagrangian Densities, Canonical Formalism, Canonical Form of Electrodynamics, Hamiltonian Densities, Transformations, and Canonical Form with Vanishing Jacobian Determinant. Numerous examples and exercises. For advanced students, teachers, etc. 6 figures. Index. viii + 222pp. 5⅜ x 8½.   S1077 Paperbound **$1.75**

Acoustics, optics, electricity and magnetism, electromagnetics, magneto-hydrodynamics

THE THEORY OF SOUND, Lord Rayleigh. Most vibrating systems likely to be encountered in practice can be tackled successfully by the methods set forth by the great Nobel laureate, Lord Rayleigh. Complete coverage of experimental, mathematical aspects of sound theory. Partial contents: Harmonic motions, vibrating systems in general, lateral vibrations of bars, curved plates or shells, applications of Laplace's functions to acoustical problems, fluid friction, plane vortex-sheet, vibrations of solid bodies, etc. This is the first inexpensive edition of this great reference and study work. Bibliography. Historical introduction by R. B. Lindsay. Total of 1040pp. 97 figures. 5⅜ x 8.
S292, S293, Two volume set, paperbound, **$4.70**

THE DYNAMICAL THEORY OF SOUND, H. Lamb. Comprehensive mathematical treatment of the physical aspects of sound, covering the theory of vibrations, the general theory of sound, and the equations of motion of strings, bars, membranes, pipes, and resonators. Includes chapters on plane, spherical, and simple harmonic waves, and the Helmholtz Theory of Audition. Complete and self-contained development for student and specialist; all fundamental differential equations solved completely. Specific mathematical details for such important phenomena as harmonics, normal modes, forced vibrations of strings, theory of reed pipes, etc. Index. Bibliography. 86 diagrams. viii + 307pp. 5⅜ x 8.
S655 Paperbound **$2.00**

WAVE PROPAGATION IN PERIODIC STRUCTURES, L. Brillouin. A general method and application to different problems: pure physics, such as scattering of X-rays of crystals, thermal vibration in crystal lattices, electronic motion in metals; and also problems of electrical engineering. Partial contents: elastic waves in 1-dimensional lattices of point masses. Propagation of waves along 1-dimensional lattices. Energy flow. 2 dimensional, 3 dimensional lattices. Mathieu's equation. Matrices and propagation of waves along an electric line. Continuous electric lines. 131 illustrations. Bibliography. Index. xii + 253pp. 5⅜ x 8.
S34 Paperbound **$2.00**

THEORY OF VIBRATIONS, N. W. McLachlan. Based on an exceptionally successful graduate course given at Brown University, this discusses linear systems having 1 degree of freedom, forced vibrations of simple linear systems, vibration of flexible strings, transverse vibrations of bars and tubes, transverse vibration of circular plate, sound waves of finite amplitude, etc. Index. 99 diagrams. 160pp. 5⅜ x 8.
S190 Paperbound **$1.50**

LIGHT: PRINCIPLES AND EXPERIMENTS, George S. Monk. Covers theory, experimentation, and research. Intended for students with some background in general physics and elementary calculus. Three main divisions: 1) Eight chapters on geometrical optics—fundamental concepts (the ray and its optical length, Fermat's principle, etc.), laws of image formation, apertures in optical systems, photometry, optical instruments etc.; 2) 9 chapters on physical optics—interference, diffraction, polarization, spectra, the Rayleigh refractometer, the wave theory of light, etc.; 3) 23 instructive experiments based directly on the theoretical text. "Probably the best intermediate textbook on light in the English language. Certainly, it is the best book which includes both geometrical and physical optics," J. Rud Nielson, PHYSICS FORUM. Revised edition. 102 problems and answers. 12 appendices. 6 tables. Index. 270 illustrations. xi +489pp. 5⅜ x 8½.
S341 Paperbound **$2.50**

PHOTOMETRY, John W. T. Walsh. The best treatment of both "bench" and "illumination" photometry in English by one of Britain's foremost experts in the field (President of the International Commission on Illumination). Limited to those matters, theoretical and practical, which affect the measurement of light flux, candlepower, illumination, etc., and excludes treatment of the use to which such measurements may be put after they have been made. Chapters on Radiation, The Eye and Vision, Photo-Electric Cells, The Principles of Photometry, The Measurement of Luminous Intensity, Colorimetry, Spectrophotometry, Stellar Photometry, The Photometric Laboratory, etc. Third revised (1958) edition. 281 illustrations. 10 appendices. xxiv + 544pp. 5½ x 9¼.
S319 Paperbound **$3.00**

EXPERIMENTAL SPECTROSCOPY, R. A. Sawyer. Clear discussion of prism and grating spectrographs and the techniques of their use in research, with emphasis on those principles and techniques that are fundamental to practically all uses of spectroscopic equipment. Beginning with a brief history of spectroscopy, the author covers such topics as light sources, spectroscopic apparatus, prism spectroscopes and graphs, diffraction grating, the photographic process, determination of wave length, spectral intensity, infrared spectroscopy, spectrochemical analysis, etc. This revised edition contains new material on the production of replica gratings, solar spectroscopy from rockets, new standard of wave length, etc. Index. Bibliography. 111 illustrations. x + 358pp. 5⅜ x 8½. S1045 Paperbound **$2.25**

FUNDAMENTALS OF ELECTRICITY AND MAGNETISM, L. B. Loeb. For students of physics, chemistry, or engineering who want an introduction to electricity and magnetism on a higher level and in more detail than general elementary physics texts provide. Only elementary differential and integral calculus is assumed. Physical laws developed logically, from magnetism to electric currents, Ohm's law, electrolysis, and on to static electricity, induction, etc. Covers an unusual amount of material; one third of book on modern material: solution of wave equation, photoelectric and thermionic effects, etc. Complete statement of the various electrical systems of units and interrelations. 2 Indexes. 75 pages of problems with answers stated. Over 300 figures and diagrams. xix +669pp. 5⅜ x 8. S745 Paperbound **$3.50**

MATHEMATICAL ANALYSIS OF ELECTRICAL AND OPTICAL WAVE-MOTION, Harry Bateman. Written by one of this century's most distinguished mathematical physicists, this is a practical introduction to those developments of Maxwell's electromagnetic theory which are directly connected with the solution of the partial differential equation of wave motion. Methods of solving wave-equation, polar-cylindrical coordinates, diffraction, transformation of coordinates, homogeneous solutions, electromagnetic fields with moving singularities, etc. Index. 168pp. 5⅜ x 8.   S14 Paperbound **$1.75**

PRINCIPLES OF PHYSICAL OPTICS, Ernst Mach. This classical examination of the propagation of light, color, polarization, etc. offers an historical and philosophical treatment that has never been surpassed for breadth and easy readability. Contents: Rectilinear propagation of light. Reflection, refraction. Early knowledge of vision. Dioptrics. Composition of light. Theory of color and dispersion. Periodicity. Theory of interference. Polarization. Mathematical representation of properties of light. Propagation of waves, etc. 279 illustrations, 10 portraits. Appendix. Indexes. 324pp. 5⅜ x 8.   S178 Paperbound **$2.00**

THE THEORY OF OPTICS, Paul Drude. One of finest fundamental texts in physical optics, classic offers thorough coverage, complete mathematical treatment of basic ideas. Includes fullest treatment of application of thermodynamics to optics; sine law in formation of images, transparent crystals, magnetically active substances, velocity of light, apertures, effects depending upon them, polarization, optical instruments, etc. Introduction by A. A. Michelson. Index. 110 illus. 567pp. 5⅜ x 8.   S532 Paperbound **$3.00**

ELECTRICAL THEORY ON THE GIORGI SYSTEM, P. Cornelius. A new clarification of the fundamental concepts of electricity and magnetism, advocating the convenient m.k.s. system of units that is steadily gaining followers in the sciences. Illustrating the use and effectiveness of his terminology with numerous applications to concrete technical problems, the author here expounds the famous Giorgi system of electrical physics. His lucid presentation and well-reasoned, cogent argument for the universal adoption of this system form one of the finest pieces of scientific exposition in recent years. 28 figures. Index. Conversion tables for translating earlier data into modern units. Translated from 3rd Dutch edition by L. J. Jolley. x + 187pp. 5½ x 8¾.   S909 Clothbound **$6.00**

ELECTRIC WAVES: BEING RESEARCHES ON THE PROPAGATION OF ELECTRIC ACTION WITH FINITE VELOCITY THROUGH SPACE, Heinrich Hertz. This classic work brings together the original papers in which Hertz—Helmholtz's protegé and one of the most brilliant figures in 19th-century research—probed the existence of electromagnetic waves and showed experimentally that their velocity equalled that of light, research that helped lay the groundwork for the development of radio, television, telephone, telegraph, and other modern technological marvels. Unabridged republication of original edition. Authorized translation by D. E. Jones. Preface by Lord Kelvin. Index of names. 40 illustrations. xvii + 278pp. 5⅜ x 8½.
S57 Paperbound **$1.75**

PIEZOELECTRICITY: AN INTRODUCTION TO THE THEORY AND APPLICATIONS OF ELECTRO-MECHANICAL PHENOMENA IN CRYSTALS, Walter G. Cady. This is the most complete and systematic coverage of this important field in print—now regarded as something of scientific classic. This republication, revised and corrected by Prof. Cady—one of the foremost contributors in this area—contains a sketch of recent progress and new material on Ferroelectrics. Time Standards, etc. The first 7 chapters deal with fundamental theory of crystal electricity. 5 important chapters cover basic concepts of piezoelectricity, including comparisons of various competing theories in the field. Also discussed: piezoelectric resonators (theory, methods of manufacture, influences of air-gaps, etc.); the piezo oscillator; the properties, history, and observations relating to Rochelle salt; ferroelectric crystals; miscellaneous applications of piezoelectricity; pyroelectricity; etc. "A great work," W. A. Wooster, NATURE. Revised (1963) and corrected edition. New preface by Prof. Cady. 2 Appendices. Indices. Illustrations. 62 tables. Bibliography. Problems. Total of 1 + 822pp. 5⅜ x 8½.
S1094 Vol. I Paperbound **$2.50**
S1095 Vol. II Paperbound **$2.50**
Two volume set Paperbound **$5.00**

MAGNETISM AND VERY LOW TEMPERATURES, H. B. G. Casimir. A basic work in the literature of low temperature physics. Presents a concise survey of fundamental theoretical principles, and also points out promising lines of investigation. Contents: Classical Theory and Experimental Methods, Quantum Theory of Paramagnetism, Experiments on Adiabatic Demagnetization. Theoretical Discussion of Paramagnetism at Very Low Temperatures, Some Experimental Results, Relaxation Phenomena. Index. 89-item bibliography. ix + 95pp. 5⅜ x 8.
S943 Paperbound **$1.35**

SELECTED PAPERS ON NEW TECHNIQUES FOR ENERGY CONVERSION: THERMOELECTRIC METHODS; THERMIONIC; PHOTOVOLTAIC AND ELECTRICAL EFFECTS; FUSION, Edited by Sumner N. Levine. Brings together in one volume the most important papers (1954-1961) in modern energy technology. Included among the 37 papers are general and qualitative descriptions of the field as a whole, indicating promising lines of research. Also: 15 papers on thermoelectric methods, 7 on thermionic, 5 on photovoltaic, 4 on electrochemical effect, and 2 on controlled fusion research. Among the contributors are: Joffe, Maria Telkes, Herold, Herring, Douglas, Jaumot, Post, Austin, Wilson, Pfann, Rappaport, Morehouse, Domenicali, Moss, Bowers, Harman, Von Doenhoef. Preface and introduction by the editor. Bibliographies. xxviii + 451pp. 6⅛ x 9¼.   S37 Paperbound **$3.00**

SUPERFLUIDS: MACROSCOPIC THEORY OF SUPERCONDUCTIVITY, Vol. I, Fritz London. The major work by one of the founders and great theoreticians of modern quantum physics. Consolidates the researches that led to the present understanding of the nature of superconductivity. Prof. London here reveals that quantum mechanics is operative on the macroscopic plane as well as the submolecular level. Contents: Properties of Superconductors and Their Thermodynamical Correlation; Electrodynamics of the Pure Superconducting State; Relation between Current and Field; Measurements of the Penetration Depth; Non-Viscous Flow vs. Superconductivity; Micro-waves in Superconductors; Reality of the Domain Structure; and many other related topics. A new epilogue by M. J. Buckingham discusses developments in the field up to 1960. Corrected and expanded edition. An appreciation of the author's life and work by L. W. Nordheim. Biography by Edith London. Bibliography of his publications. 45 figures. 2 Indices. xviii + 173pp. 5⅜ x 8⅜. S44 Paperbound **$1.75**

SELECTED PAPERS ON PHYSICAL PROCESSES IN IONIZED PLASMAS, Edited by Donald H. Menzel, Director, Harvard College Observatory. 30 important papers relating to the study of highly ionized gases or plasmas selected by a foremost contributor in the field, with the assistance of Dr. L. H. Aller. The essays include 18 on the physical processes in gaseous nebulae, covering problems of radiation and radiative transfer, the Balmer decrement, electron temperatives, spectrophotometry, etc. 10 papers deal with the interpretation of nebular spectra, by Bohm, Van Vleck, Aller, Minkowski, etc. There is also a discussion of the intensities of "forbidden" spectral lines by George Shortley and a paper concerning the theory of hydrogenic spectra by Menzel and Pekeris. Other contributors: Goldberg, Hebb, Baker, Bowen, Ufford, Liller, etc. viii + 374pp. 6⅛ x 9¼. S60 Paperbound **$2.95**

THE ELECTROMAGNETIC FIELD, Max Mason & Warren Weaver. Used constantly by graduate engineers. Vector methods exclusively: detailed treatment of electrostatics, expansion methods, with tables converting any quantity into absolute electromagnetic, absolute electrostatic, practical units. Discrete charges, ponderable bodies, Maxwell field equations, etc. Introduction. Indexes. 416pp. 5⅜ x 8. S185 Paperbound **$2.25**

THEORY OF ELECTRONS AND ITS APPLICATION TO THE PHENOMENA OF LIGHT AND RADIANT HEAT, H. Lorentz. Lectures delivered at Columbia University by Nobel laureate Lorentz. Unabridged, they form a historical coverage of the theory of free electrons, motion, absorption of heat, Zeeman effect, propagation of light in molecular bodies, inverse Zeeman effect, optical phenomena in moving bodies, etc. 109 pages of notes explain the more advanced sections. Index. 9 figures. 352pp. 5⅜ x 8. S173 Paperbound **$2.00**

FUNDAMENTAL ELECTROMAGNETIC THEORY, Ronold P. King, Professor Applied Physics, Harvard University. Original and valuable introduction to electromagnetic theory and to circuit theory from the standpoint of electromagnetic theory. Contents: Mathematical Description of Matter—stationary and nonstationary states; Mathematical Description of Space and of Simple Media—Field Equations, Integral Forms of Field Equations, Electromagnetic Force, etc.; Transformation of Field and Force Equations; Electromagnetic Waves in Unbounded Regions; Skin Effect and Internal Impedance—in a solid cylindrical conductor, etc.; and Electrical Circuits—Analytical Foundations, Near-zone and quasi-near zone circuits, Balanced two-wire and four-wire transmission lines. Revised and enlarged version. New preface by the author. 5 appendices (Differential operators: Vector Formulas and Identities, etc.). Problems. Indexes. Bibliography. xvi + 580pp. 5⅜ x 8½. S1023 Paperbound **$3.00**

Hydrodynamics

A TREATISE ON HYDRODYNAMICS, A. B. Basset. Favorite text on hydrodynamics for 2 generations of physicists, hydrodynamical engineers, oceanographers, ship designers, etc. Clear enough for the beginning student, and thorough source for graduate students and engineers on the work of d'Alembert, Euler, Laplace, Lagrange, Poisson, Green, Clebsch, Stokes, Cauchy, Helmholtz, J. J. Thomson, Love, Hicks, Greenhill, Besant, Lamb, etc. Great amount of documentation on entire theory of classical hydrodynamics. Vol I: theory of motion of frictionless liquids, vortex, and cyclic irrotational motion, etc. 132 exercises. Bibliography. 3 Appendixes. xii + 264pp. Vol II: motion in viscous liquids, harmonic analysis, theory of tides, etc. 112 exercises, Bibliography. 4 Appendixes. xv + 328pp. Two volume set. 5⅜ x 8.
S724 Vol I Paperbound **$1.75**
S725 Vol II Paperbound **$1.75**
The set **$3.50**

HYDRODYNAMICS, Horace Lamb. Internationally famous complete coverage of standard reference work on dynamics of liquids & gases. Fundamental theorems, equations, methods, solutions, background, for classical hydrodynamics. Chapters include Equations of Motion, Integration of Equations in Special Gases, Irrotational Motion, Motion of Liquid in 2 Dimensions, Motion of Solids through Liquid-Dynamical Theory, Vortex Motion, Tidal Waves, Surface Waves, Waves of Expansion, Viscosity, Rotating Masses of liquids. Excellently planned, arranged; clear, lucid presentation. 6th enlarged, revised edition. Index. Over 900 footnotes, mostly bibliographical. 119 figures. xv + 738pp. 6⅛ x 9¼. S256 Paperbound **$3.75**

HYDRODYNAMICS, H. Dryden, F. Murnaghan, Harry Bateman. Published by the National Research Council in 1932 this enormous volume offers a complete coverage of classical hydrodynamics. Encyclopedic in quality. Partial contents: physics of fluids, motion, turbulent flow, compressible fluids, motion in 1, 2, 3 dimensions; viscous fluids rotating, laminar motion, resistance of motion through viscous fluid, eddy viscosity, hydraulic flow in channels of various shapes, discharge of gases, flow past obstacles, etc. Bibliography of over 2,900 items. Indexes. 23 figures. 634pp. 5⅜ x 8. S303 Paperbound **$2.75**

Mechanics, dynamics, thermodynamics, elasticity

MECHANICS, J. P. Den Hartog. Already a classic among introductory texts, the M.I.T. professor's lively and discursive presentation is equally valuable as a beginner's text, an engineering student's refresher, or a practicing engineer's reference. Emphasis in this highly readable text is on illuminating fundamental principles and showing how they are embodied in a great number of real engineering and design problems: trusses, loaded cables, beams, jacks, hoists, etc. Provides advanced material on relative motion and gyroscopes not usual in introductory texts. "Very thoroughly recommended to all those anxious to improve their real understanding of the principles of mechanics." MECHANICAL WORLD. Index. List of equations. 334 problems, all with answers. Over 550 diagrams and drawings. ix + 462pp. 5⅜ x 8. S754 Paperbound **$2.00**

THEORETICAL MECHANICS: AN INTRODUCTION TO MATHEMATICAL PHYSICS, J. S. Ames, F. D. Murnaghan. A mathematically rigorous development of theoretical mechanics for the advanced student, with constant practical applications. Used in hundreds of advanced courses. An unusually thorough coverage of gyroscopic and baryscopic material, detailed analyses of the Coriolis acceleration, applications of Lagrange's equations, motion of the double pendulum, Hamilton-Jacobi partial differential equations, group velocity and dispersion, etc. Special relativity is also included. 159 problems. 44 figures. ix + 462pp. 5⅜ x 8. S461 Paperbound **$2.25**

THEORETICAL MECHANICS: STATICS AND THE DYNAMICS OF A PARTICLE, W. D. MacMillan. Used for over 3 decades as a self-contained and extremely comprehensive advanced undergraduate text in mathematical physics, physics, astronomy, and deeper foundations of engineering. Early sections require only a knowledge of geometry; later, a working knowledge of calculus. Hundreds of basic problems, including projectiles to the moon, escape velocity, harmonic motion, ballistics, falling bodies, transmission of power, stress and strain, elasticity, astronomical problems. 340 practice problems plus many fully worked out examples make it possible to test and extend principles developed in the text. 200 figures. xvii + 430pp. 5⅜ x 8. S467 Paperbound **$2.25**

THEORETICAL MECHANICS: THE THEORY OF THE POTENTIAL, W. D. MacMillan. A comprehensive, well balanced presentation of potential theory, serving both as an introduction and a reference work with regard to specific problems, for physicists and mathematicians. No prior knowledge of integral relations is assumed, and all mathematical material is developed as it becomes necessary. Includes: Attraction of Finite Bodies; Newtonian Potential Function; Vector Fields, Green and Gauss Theorems; Attractions of Surfaces and Lines; Surface Distribution of Matter; Two-Layer Surfaces; Spherical Harmonics; Ellipsoidal Harmonics; etc. "The great number of particular cases . . . should make the book valuable to geophysicists and others actively engaged in practical applications of the potential theory," Review of Scientific Instruments. Index. Bibliography. xiii + 469pp. 5⅜ x 8. S486 Paperbound **$2.50**

THEORETICAL MECHANICS: DYNAMICS OF RIGID BODIES, W. D. MacMillan. Theory of dynamics of a rigid body is developed, using both the geometrical and analytical methods of instruction. Begins with exposition of algebra of vectors, it goes through momentum principles, motion in space, use of differential equations and infinite series to solve more sophisticated dynamics problems. Partial contents: moments of inertia, systems of free particles, motion parallel to a fixed plane, rolling motion, method of periodic solutions, much more. 82 figs. 199 problems. Bibliography. Indexes. xii + 476pp. 5⅜ x 8. S641 Paperbound **$2.50**

MATHEMATICAL FOUNDATIONS OF STATISTICAL MECHANICS, A. I. Khinchin. Offering a precise and rigorous formulation of problems, this book supplies a thorough and up-to-date exposition. It provides analytical tools needed to replace cumbersome concepts, and furnishes for the first time a logical step-by-step introduction to the subject. Partial contents: geometry & kinematics of the phase space, ergodic problem, reduction to theory of probability, application of central limit problem, ideal monatomic gas, foundation of thermo-dynamics, dispersion and distribution of sum functions. Key to notations. Index. viii + 179pp. 5⅜ x 8. S147 Paperbound **$1.50**

ELEMENTARY PRINCIPLES IN STATISTICAL MECHANICS, J. W. Gibbs. Last work of the great Yale mathematical physicist, still one of the most fundamental treatments available for advanced students and workers in the field. Covers the basic principle of conservation of probability of phase, theory of errors in the calculated phases of a system, the contributions of Clausius, Maxwell, Boltzmann, and Gibbs himself, and much more. Includes valuable comparison of statistical mechanics with thermodynamics: Carnot's cycle, mechanical definitions of entropy, etc. xvi + 208pp. 5⅜ x 8. S707 Paperbound **$1.45**

PRINCIPLES OF MECHANICS AND DYNAMICS, Sir William Thomson (Lord Kelvin) and Peter Guthrie Tait. The principles and theories of fundamental branches of classical physics explained by two of the greatest physicists of all time. A broad survey of mechanics, with material on hydrodynamics, elasticity, potential theory, and what is now standard mechanics. Thorough and detailed coverage, with many examples, derivations, and topics not included in more recent studies. Only a knowledge of calculus is needed to work through this book. Vol. I (Preliminary): Kinematics; Dynamical Laws and Principles; Experience (observation, experimentation, formation of hypotheses, scientific method); Measures and Instruments; Continuous Calculating Machines. Vol. II (Abstract Dynamics): Statics of a Particle—Attraction; Statics of Solids and Fluids. Formerly Titled "Treatise on Natural Philosophy." Unabridged reprint of revised edition. Index. 168 diagrams. Total of xlii + 1035pp. 5⅜ x 8½.
Vol. I: S966 Paperbound **$2.35**
Vol. II: S967 Paperbound **$2.35**
Two volume Set Paperbound **$4.70**

INVESTIGATIONS ON THE THEORY OF THE BROWNIAN MOVEMENT, Albert Einstein. Reprints from rare European journals. 5 basic papers, including the Elementary Theory of the Brownian Movement, written at the request of Lorentz to provide a simple explanation. Translated by A. D. Cowper. Annotated, edited by R. Fürth. 33pp. of notes elucidate, give history of previous investigations. Author, subject indexes. 62 footnotes. 124pp. 5⅜ x 8.
S304 Paperbound **$1.25**

MECHANICS VIA THE CALCULUS, P. W. Norris, W. S. Legge. Covers almost everything, from linear motion to vector analysis: equations determining motion, linear methods, compounding of simple harmonic motions, Newton's laws of motion, Hooke's law, the simple pendulum, motion of a particle in 1 plane, centers of gravity, virtual work, friction, kinetic energy of rotating bodies, equilibrium of strings, hydrostatics, sheering stresses, elasticity, etc. 550 problems. 3rd revised edition. xii + 367pp. 6 x 9.
S207 Clothbound **$4.95**

THE DYNAMICS OF PARTICLES AND OF RIGID, ELASTIC, AND FLUID BODIES; BEING LECTURES ON MATHEMATICAL PHYSICS, A. G. Webster. The reissuing of this classic fills the need for a comprehensive work on dynamics. A wide range of topics is covered in unusually great depth, applying ordinary and partial differential equations. Part I considers laws of motion and methods applicable to systems of all sorts; oscillation, resonance, cyclic systems, etc. Part 2 is a detailed study of the dynamics of rigid bodies. Part 3 introduces the theory of potential; stress and strain, Newtonian potential functions, gyrostatics, wave and vortex motion, etc. Further contents: Kinematics of a point; Lagrange's equations; Hamilton's principle; Systems of vectors; Statics and dynamics of deformable bodies; much more, not easily found together in one volume. Unabridged reprinting of 2nd edition. 20 pages of notes on differential equations and the higher analysis. 203 illustrations. Selected bibliography. Index. xi + 588pp. 5⅜ x 8.
S522 Paperbound **$2.45**

A TREATISE ON DYNAMICS OF A PARTICLE, E. J. Routh. Elementary text on dynamics for beginning mathematics or physics student. Unusually detailed treatment from elementary definitions to motion in 3 dimensions, emphasizing concrete aspects. Much unique material important in recent applications. Covers impulsive forces, rectilinear and constrained motion in 2 dimensions, harmonic and parabolic motion, degrees of freedom, closed orbits, the conical pendulum, the principle of least action, Jacobi's method, and much more. Index. 559 problems, many fully worked out, incorporated into text. xiii + 418pp. 5⅜ x 8.
S696 Paperbound **$2.25**

DYNAMICS OF A SYSTEM OF RIGID BODIES (Elementary Section), E. J. Routh. Revised 7th edition of this standard reference. This volume covers the dynamical principles of the subject, and its more elementary applications: finding moments of inertia by integration, foci of inertia, d'Alembert's principle, impulsive forces, motion in 2 and 3 dimensions, Lagrange's equations, relative indicatrix, Euler's theorem, large tautochronous motions, etc. Index. 55 figures. Scores of problems. xv + 443pp. 5⅜ x 8.
S664 Paperbound **$2.50**

DYNAMICS OF A SYSTEM OF RIGID BODIES (Advanced Section), E. J. Routh. Revised 6th edition of a classic reference aid. Much of its material remains unique. Partial contents: moving axes, relative motion, oscillations about equilibrium, motion. Motion of a body under no forces, any forces. Nature of motion given by linear equations and conditions of stability. Free, forced vibrations, constants of integration, calculus of finite differences, variations, precession and nutation, motion of the moon, motion of string, chain, membranes. 64 figures. 498pp. 5⅜ x 8.
S229 Paperbound **$2.45**

DYNAMICAL THEORY OF GASES, James Jeans. Divided into mathematical and physical chapters for the convenience of those not expert in mathematics, this volume discusses the mathematical theory of gas in a steady state, thermodynamics, Boltzmann and Maxwell, kinetic theory, quantum theory, exponentials, etc. 4th enlarged edition, with new material on quantum theory, quantum dynamics, etc. Indexes. 28 figures. 444pp. 6⅛ x 9¼.
S136 Paperbound **$2.75**

THE THEORY OF HEAT RADIATION, Max Planck. A pioneering work in thermodynamics, providing basis for most later work, Nobel laureate Planck writes on Deductions from Electrodynamics and Thermodynamics, Entropy and Probability, Irreversible Radiation Processes, etc. Starts with simple experimental laws of optics, advances to problems of spectral distribution of energy and irreversibility. Bibliography. 7 illustrations. xiv + 224pp. 5⅜ x 8.
S546 Paperbound **$1.75**

FOUNDATIONS OF POTENTIAL THEORY, O. D. Kellogg. Based on courses given at Harvard this is suitable for both advanced and beginning mathematicians. Proofs are rigorous, and much material not generally avaliable elsewhere is included. Partial contents: forces of gravity, fields of force, divergence theorem, properties of Newtonian potentials at points of free space, potentials as solutions of Laplace's equations, harmonic functions, electrostatics, electric images, logarithmic potential, etc. One of Grundlehren Series. ix + 384pp. 5⅜ x 8.
S144 Paperbound **$2.00**

THERMODYNAMICS, Enrico Fermi. Unabridged reproduction of 1937 edition. Elementary in treatment; remarkable for clarity, organization. Requires no knowledge of advanced math beyond calculus, only familiarity with fundamentals of thermometry, calorimetry. Partial Contents: Thermodynamic systems; First & Second laws of thermodynamics; Entropy; Thermodynamic potentials: phase rule, reversible electric cell; Gaseous reactions: van't Hoff reaction box, principle of LeChatelier; Thermodynamics of dilute solutions: osmotic & vapor pressures, boiling & freezing points; Entropy constant. Index. 25 problems. 24 illustrations. x + 160pp. 5⅜ x 8.
S361 Paperbound **$1.75**

THE THERMODYNAMICS OF ELECTRICAL PHENOMENA IN METALS and A CONDENSED COLLECTION OF THERMODYNAMIC FORMULAS, P. W. Bridgman. Major work by the Nobel Prizewinner: stimulating conceptual introduction to aspects of the electron theory of metals, giving an intuitive understanding of fundamental relationships concealed by the formal systems of Onsager and others. Elementary mathematical formulations show clearly the fundamental thermodynamical relationships of the electric field, and a complete phenomenological theory of metals is created. This is the work in which Bridgman announced his famous "thermomotive force" and his distinction between "driving" and "working" electromotive force. We have added in this Dover edition the author's long unavailable tables of thermodynamic formulas, extremely valuable for the speed of reference they allow. Two works bound as one. Index. 33 figures. Bibliography. xviii + 256pp. 5⅜ x 8. S723 Paperbound **$1.75**

TREATISE ON THERMODYNAMICS, Max Planck. Based on Planck's original papers this offers a uniform point of view for the entire field and has been used as an introduction for students who have studied elementary chemistry, physics, and calculus. Rejecting the earlier approaches of Helmholtz and Maxwell, the author makes no assumptions regarding the nature of heat, but begins with a few empirical facts, and from these deduces new physical and chemical laws. 3rd English edition of this standard text by a Nobel laureate. xvi + 297pp. 5⅜ x 8.
S219 Paperbound **$1.85**

THE MATHEMATICAL THEORY OF ELASTICITY, A. E. H. Love. A wealth of practical illustration combined with thorough discussion of fundamentals—theory, application, special problems and solutions. Partial Contents: Analysis of Strain & Stress, Elasticity of Solid Bodies, Elasticity of Crystals, Vibration of Spheres, Cylinders, Propagation of Waves in Elastic Solid Media, Torsion, Theory of Continuous Beams, Plates. Rigorous treatment of Volterra's theory of dislocations, 2-dimensional elastic systems, other topics of modern interest. "For years the standard treatise on elasticity," AMERICAN MATHEMATICAL MONTHLY. 4th revised edition. Index. 76 figures. xviii + 643pp. 6⅛ x 9¼.
S174 Paperbound **$3.25**

STRESS WAVES IN SOLIDS, H. Kolsky, Professor of Applied Physics, Brown University. The most readable survey of the theoretical core of current knowledge about the propagation of waves in solids, fully correlated with experimental research. Contents: Part I—Elastic Waves: propagation in an extended plastic medium, propagation in bounded elastic media, experimental investigations with elastic materials. Part II—Stress Waves in Imperfectly Elastic Media: internal friction, experimental investigations of dynamic elastic properties, plastic waves and shock waves, fractures produced by stress waves. List of symbols. Appendix. Supplemented bibliography. 3 full-page plates. 46 figures. x + 213pp. 5⅜ x 8½.
S1098 Paperbound **$1.75**

Relativity, quantum theory, atomic and nuclear physics

SPACE TIME MATTER, Hermann Weyl. "The standard treatise on the general theory of relativity" (Nature), written by a world-renowned scientist, provides a deep clear discussion of the logical coherence of the general theory, with introduction to all the mathematical tools needed: Maxwell, analytical geometry, non-Euclidean geometry, tensor calculus, etc. Basis is classical space-time, before absorption of relativity. Partial contents: Euclidean space, mathematical form, metrical continuum, relativity of time and space, general theory. 15 diagrams. Bibliography. New preface for this edition. xviii + 330pp. 5⅜ x 8.
S267 Paperbound **$2.25**

ATOMIC SPECTRA AND ATOMIC STRUCTURE, G. Herzberg. Excellent general survey for chemists, physicists specializing in other fields. Partial contents: simplest line spectra and elements of atomic theory, building-up principle and periodic system of elements, hyperfine structure of spectral lines, some experiments and applications. Bibiliography. 80 figures. Index. xii + 257pp. 5⅜ x 8.
S115 Paperbound **$2.00**

THE PRINCIPLE OF RELATIVITY, A. Einstein, H. Lorentz, H. Minkowski, H. Weyl. These are the 11 basic papers that founded the general and special theories of relativity, all translated into English. Two papers by Lorentz on the Michelson experiment, electromagnetic phenomena. Minkowski's SPACE & TIME, and Weyl's GRAVITATION & ELECTRICITY. 7 epoch-making papers by Einstein: ELECTROMAGNETICS OF MOVING BODIES, INFLUENCE OF GRAVITATION IN PROPAGATION OF LIGHT, COSMOLOGICAL CONSIDERATIONS, GENERAL THEORY, and 3 others. 7 diagrams. Special notes by A. Sommerfeld. 224pp. 5⅜ x 8.
S81 Paperbound **$2.00**

EINSTEIN'S THEORY OF RELATIVITY, Max Born. Revised edition prepared with the collaboration of Gunther Leibfried and Walter Biem. Steering a middle course between superficial popularizations and complex analyses, a Nobel laureate explains Einstein's theories clearly and with special insight. Easily followed by the layman with a knowledge of high school mathematics, the book has been thoroughly revised and extended to modernize those sections of the well-known original edition which are now out of date. After a comprehensive review of classical physics, Born's discussion of special and general theories of relativity covers such topics as simultaneity, kinematics, Einstein's mechanics and dynamics, relativity of arbitrary motions, the geometry of curved surfaces, the space-time continuum, and many others. Index. Illustrations, vii + 376pp. 5⅜ x 8.
S769 Paperbound **$2.00**

ATOMS, MOLECULES AND QUANTA, Arthur E. Ruark and Harold C. Urey. Revised (1963) and corrected edition of a work that has been a favorite with physics students and teachers for more than 30 years. No other work offers the same combination of atomic structure and molecular physics and of experiment and theory. The first 14 chapters deal with the origins and major experimental data of quantum theory and with the development of conceptions of atomic and molecular structure prior to the new mechanics. These sections provide a thorough introduction to atomic and molecular theory, and are presented lucidly and as simply as possible. The six subsequent chapters are devoted to the laws and basic ideas of quantum mechanics: Wave Mechanics, Hydrogenic Atoms in Wave Mechanics, Matrix Mechanics, General Theory of Quantum Dynamics, etc. For advanced college and graduate students in physics. Revised, corrected republication of original edition, with supplementary notes by the authors. New preface by the authors. 9 appendices. General reference list. Indices. 228 figures. 71 tables. Bibliographical material in notes, etc. Total of xxiii + 810pp. 5⅜ x 8⅜.
S1106 Vol. I Paperbound **$2.50**
S1107 Vol. II Paperbound **$2.50**
Two volume set Paperbound **$5.00**

WAVE MECHANICS AND ITS APPLICATIONS, N. F. Mott and I. N. Sneddon. A comprehensive introduction to the theory of quantum mechanics; not a rigorous mathematical exposition it progresses, instead, in accordance with the physical problems considered. Many topics difficult to find at the elementary level are discussed in this book. Includes such matters as: the wave nature of matter, the wave equation of Schrödinger, the concept of stationary states, properties of the wave functions, effect of a magnetic field on the energy levels of atoms, electronic spin, two-body problem, theory of solids, cohesive forces in ionic crystals, collision problems, interaction of radiation with matter, relativistic quantum mechanics, etc. All are treated both physically and mathematically. 68 illustrations. 11 tables. Indexes. xii + 393pp. 5⅜ x 8½.
S1070 Paperbound **$2.35**

BASIC METHODS IN TRANSFER PROBLEMS, V. Kourganoff, Professor of Astrophysics, U. of Paris. A coherent digest of all the known methods which can be used for approximate or exact solutions of transfer problems. All methods demonstrated on one particular problem—Milne's problem for a plane parallel medium. Three main sections: fundamental concepts (the radiation field and its interaction with matter, the absorption and emission coefficients, etc.); different methods by which transfer problems can be attacked; and a more general problem—the non-grey case of Milne's problem. Much new material, drawing upon declassified atomic energy reports and data from the USSR. Entirely understandable to the student with a reasonable knowledge of analysis. Unabridged, revised reprinting. New preface by the author. Index. Bibliography. 2 appendices. xv + 281pp. 5⅜ x 8½.
S1074 Paperbound **$2.00**

PRINCIPLES OF QUANTUM MECHANICS, W. V. Houston. Enables student with working knowledge of elementary mathematical physics to develop facility in use of quantum mechanics, understand published work in field. Formulates quantum mechanics in terms of Schroedinger's wave mechanics. Studies evidence for quantum theory, for inadequacy of classical mechanics, 2 postulates of quantum mechanics; numerous important, fruitful applications of quantum mechanics in spectroscopy, collision problems, electrons in solids; other topics. "One of the most rewarding features . . . is the interlacing of problems with text," Amer. J. of Physics. Corrected edition. 21 illus. Index. 296pp. 5⅜ x 8. S524 Paperbound **$2.00**

PHYSICAL PRINCIPLES OF THE QUANTUM THEORY, Werner Heisenberg. A Nobel laureate discusses quantum theory; Heisenberg's own work, Compton, Schroedinger, Wilson, Einstein, many others. Written for physicists, chemists who are not specialists in quantum theory, only elementary formulae are considered in the text; there is a mathematical appendix for specialists. Profound without sacrifice of clarity. Translated by C. Eckart, F. Hoyt. 18 figures. 192pp. 5⅜ x 8.
S113 Paperbound **$1.35**

GEOLOGY, GEOGRAPHY, METEOROLOGY

PRINCIPLES OF STRATIGRAPHY, A. W. Grabau. Classic of 20th century geology, unmatched in scope and comprehensiveness. Nearly 600 pages cover the structure and origins of every kind of sedimentary, hydrogenic, oceanic, pyroclastic, atmoclastic, hydroclastic, marine hydroclastic, and bioclastic rock; metamorphism; erosion; etc. Includes also the constitution of the atmosphere; morphology of oceans, rivers, glaciers; volcanic activities; faults and earthquakes; and fundamental principles of paleontology (nearly 200 pages). New introduction by Prof. M. Kay, Columbia U. 1277 bibliographical entries. 264 diagrams. Tables, maps, etc. Two volume set. Total of xxxii + 1185pp. 5⅜ x 8. S686 Vol I Paperbound **$2.50**
S687 Vol II Paperbound **$2.50**
The set **$5.00**

TREATISE ON SEDIMENTATION, William H. Twenhofel. A milestone in the history of geology, this two-volume work, prepared under the auspices of the United States Research Council, contains practically everything known about sedimentation up to 1932. Brings together all the findings of leading American and foreign geologists and geographers and has never been surpassed for completeness, thoroughness of description, or accuracy of detail. Vol. 1 discusses the sources and production of sediments, their transportation, deposition, diagenesis, and lithification. Also modification of sediments by organisms and topographical, climatic, etc. conditions which contribute to the alteration of sedimentary processes. 220 pages deal with products of sedimentation: minerals, limestones, dolomites, coals, etc. Vol. 2 continues the examination of products such as gypsum and saline residues, silica, strontium, manganese, etc. An extensive exposition of structures, textures and colors of sediments: stratification, cross-lamination, ripple mark, oolitic and pisolitic textures, etc. Chapters on environments or realms of sedimentation and field and laboratory techniques are also included. Indispensable to modern-day geologists and students. Index. List of authors cited. 1733-item bibliography. 121 diagrams. Total of xxxiii + 926pp. 5⅜ x 8½.
Vol. I: S950 Paperbound **$2.50**
Vol. II: S951 Paperbound **$2.50**
Two volume set Paperbound **$5.00**

THE EVOLUTION OF THE IGNEOUS ROCKS, N. L. Bowen. Invaluable serious introduction applies techniques of physics and chemistry to explain igneous rock diversity in terms of chemical composition and fractional crystallization. Discusses liquid immiscibility in silicate magmas, crystal sorting, liquid lines of descent, fractional resorption of complex minerals, petrogenesis, etc. Of prime importance to geologists & mining engineers, also to physicists, chemists working with high temperatures and pressures. "Most important," TIMES, London. 3 indexes. 263 bibliographic notes. 82 figures. xviii + 334pp. 5⅜ x 8. S311 Paperbound **$2.25**

INTERNAL CONSTITUTION OF THE EARTH, edited by Beno Gutenberg. Completely revised. Brought up-to-date, reset. Prepared for the National Research Council this is a complete & thorough coverage of such topics as earth origins, continent formation, nature & behavior of the earth's core, petrology of the crust, cooling forces in the core, seismic & earthquake material, gravity, elastic constants, strain characteristics and similar topics. "One is filled with admiration . . . a high standard . . . there is no reader who will not learn something from this book," London, Edinburgh, Dublin, Philosophic Magazine. Largest bibliography in print: 1127 classified items. Indexes. Tables of constants. 43 diagrams. 439pp. 6⅛ x 9¼.
S414 Paperbound **$3.00**

HYDROLOGY, edited by Oscar E. Meinzer. Prepared for the National Research Council. Detailed complete reference library on precipitation, evaporation, snow, snow surveying, glaciers, lakes, infiltration, soil moisture, ground water, runoff, drought, physical changes produced by water, hydrology of limestone terranes, etc. Practical in application, especially valuable for engineers. 24 experts have created "the most up-to-date, most complete treatment of the subject," AM. ASSOC. of PETROLEUM GEOLOGISTS. Bibliography. Index. 165 illustrations. xi + 712pp. 6⅛ x 9¼. S191 Paperbound **$3.50**

SNOW CRYSTALS, W. A. Bentley and W. J. Humphreys. Over 200 pages of Bentley's famous microphotographs of snow flakes—the product of painstaking, methodical work at his Jericho, Vermont studio. The pictures, which also include plates of frost, glaze and dew on vegetation, spider webs, windowpanes; sleet; graupel or soft hail, were chosen both for their scientific interest and their aesthetic qualities. The wonder of nature's diversity is exhibited in the intricate, beautiful patterns of the snow flakes. Introductory text by W. J. Humphreys. Selected bibliography. 2,453 illustrations. 224pp. 8 x 10¼. T287 Paperbound **$2.95**

PHYSICS OF THE AIR, W. J. Humphreys. A very thorough coverage of classical materials and theories in meteorology . . . written by one of this century's most highly respected physical meteorologists. Contains the standard account in English of atmospheric optics. 5 main sections: Mechanics and Thermodynamics of the Atmosphere, Atmospheric Electricity and Auroras, Meteorological Acoustics, Atmospheric Optics, and Factors of Climatic Control. Under these headings, topics covered are: theoretical relations between temperature, pressure, and volume in the atmosphere; composition, pressure, and density; circulation; evaporation and condensation; fog, clouds, thunderstorms, lightning; aurora polaris; principal ice-age theories; etc. New preface by Prof. Julius London. 226 illustrations. Index. xviii + 676pp. 5⅜ x 8½. S1044 Paperbound **$3.00**

MATHEMATICS, HISTORIES AND CLASSICS

HISTORY OF MATHEMATICS, D. E. Smith. Most comprehensive non-technical history of math in English. Discusses lives and works of over a thousand major and minor figures, with footnotes supplying technical information outside the book's scheme, and indicating disputed matters. Vol I: A chronological examination, from primitive concepts through Egypt, Babylonia, Greece, the Orient, Rome, the Middle Ages, the Renaissance, and up to 1900. Vol 2: The development of ideas in specific fields and problems, up through elementary calculus. Two volumes, total of 510 illustrations, 1355pp. 5⅜ x 8. Set boxed in attractive container.   T429, 430 Paperbound, the set **$6.00**

A SHORT ACCOUNT OF THE HISTORY OF MATHEMATICS, W. W. R. Ball. Most readable non-technical history of mathematics treats lives, discoveries of every important figure from Egyptian, Phoenician mathematicians to late 19th century. Discusses schools of Ionia, Pythagoras, Athens, Cyzicus, Alexandria, Byzantium, systems of numeration; primitive arithmetic; Middle Ages, Renaissance, including Arabs, Bacon, Regiomontanus, Tartaglia, Cardan, Stevinus, Galileo, Kepler; modern mathematics of Descartes, Pascal, Wallis, Huygens, Newton, Leibnitz, d'Alembert, Euler, Lambert, Laplace, Legendre, Gauss, Hermite, Weierstrass, scores more. Index. 25 figures. 546pp. 5⅜ x 8.   S630 Paperbound **$2.25**

A HISTORY OF GEOMETRICAL METHODS, J. L. Coolidge. Full, authoritative history of the techniques which men have employed in dealing with geometric questions . . . from ancient times to the modern development of projective geometry. Critical analyses of the original works. Contents: Synthetic Geometry—the early beginnings, Greek mathematics, non-Euclidean geometries, projective and descriptive geometry; Algebraic Geometry—extension of the system of linear coordinates, other systems of point coordinates, enumerative and birational geometry, etc.; and Differential Geometry—intrinsic geometry and moving axes, Gauss and the classical theory of surfaces, and projective and absolute differential geometry. The work of scores of geometers analyzed: Pythagoras, Archimedes, Newton, Descartes, Leibniz, Lobachevski, Riemann, Hilbert, Bernoulli, Schubert, Grassman, Klein, Cauchy, and many, many others. Extensive (24-page) bibliography. Index. 13 figures. xviii + 451pp. 5⅜ x 8½.   S1006 Paperbound **$2.25**

THE MATHEMATICS OF GREAT AMATEURS, Julian Lowell Coolidge. Enlightening, often surprising, accounts of what can result from a non-professional preoccupation with mathematics. Chapters on Plato, Omar Khayyam and his work with cubic equations, Piero della Francesca, Albrecht Dürer, as the true discoverer of descriptive geometry, Leonardo da Vinci and his varied mathematical interests, John Napier, Baron of Merchiston, inventor of logarithms, Pascal, Diderot, l'Hospital, and seven others known primarily for contributions in other fields. Bibliography. 56 figures. viii + 211pp. 5⅜ x 8½.   S1009 Paperbound **$1.50**

ART AND GEOMETRY, Wm. M. Ivins, Jr. A controversial study which propounds the view that the ideas of Greek philosophy and culture served not to stimulate, but to stifle the development of Western thought. Through an examination of Greek art and geometrical inquiries and Renaissance experiments, this book offers a concise history of the evolution of mathematical perspective and projective geometry. Discusses the work of Alberti, Dürer, Pelerin, Nicholas of Cusa, Kepler, Desargues, etc. in a wholly readable text of interest to the art historian, philosopher, mathematician, historian of science, and others. x + 113pp. 5⅜ x 8⅜.   T941 Paperbound **$1.25**

A SOURCE BOOK IN MATHEMATICS, D. E. Smith. Great discoveries in math, from Renaissance to end of 19th century, in English translation. Read announcements by Dedekind, Gauss, Delamain, Pascal, Fermat, Newton, Abel, Lobachevsky, Bolyai, Riemann, De Moivre, Legendre, Laplace, others of discoveries about imaginary numbers, number congruence, slide rule, equations, symbolism, cubic algebraic equations, non-Euclidean forms of geometry, calculus, function theory, quaternions, etc. Succinct selections from 125 different treatises, articles, most unavailable elsewhere in English. Each article preceded by biographical, historical introduction. Vol. I: Fields of Number, Algebra. Index. 32 illus. 338pp. 5⅜ x 8. Vol. II: Fields of Geometry, Probability, Calculus, Functions, Quaternions. 83 illus. 432pp. 5⅜ x 8.
Vol. 1: S552 Paperbound **$2.00**
Vol. 2: S553 Paperbound **$2.00**
2 vol. set, **$4.00**

A COLLECTION OF MODERN MATHEMATICAL CLASSICS, edited by R. Bellman. 13 classic papers, complete in their original languages, by Hermite, Hardy and Littlewood, Tchebychef, Fejér, Fredholm, Fuchs, Hurwitz, Weyl, van der Pol, Birkhoff, Kellogg, von Neumann, and Hilbert. Each of these papers, collected here for the first time, triggered a burst of mathematical activity, providing useful new generalizations or stimulating fresh investigations. Topics discussed include classical analysis, periodic and almost periodic functions, analysis and number theory, integral equations, theory of approximation, non-linear differential equations, and functional analysis. Brief introductions and bibliographies to each paper. xii + 292pp. 6 x 9.   S730 Paperbound **$2.00**

THE WORKS OF ARCHIMEDES, edited by T. L. Heath. All the known works of the great Greek mathematician are contained in this one volume, including the recently discovered Method of Archimedes. Contains: On Sphere & Cylinder, Measurement of a Circle, Spirals, Conoids, Spheroids, etc. This is the definitive edition of the greatest mathematical intellect of the ancient world. 186-page study by Heath discusses Archimedes and the history of Greek mathematics. Bibliography. 563pp. 5⅜ x 8.   S9 Paperbound **$2.45**

MATHEMATICAL PUZZLES AND RECREATIONS

AMUSEMENTS IN MATHEMATICS, Henry Ernest Dudeney. The foremost British originator of mathematical puzzles is always intriguing, witty, and paradoxical in this classic, one of the largest collections of mathematical amusements. More than 430 puzzles, problems, and paradoxes. Mazes and games, problems on number manipulation, unicursal and other route problems, puzzles on measuring, weighing, packing, age, kinship, chessboards, joining, crossing river, plane figure dissection, and many others. Solutions. More than 450 illustrations. vii + 258pp. 5⅜ x 8. T473 Paperbound **$1.25**

SYMBOLIC LOGIC and THE GAME OF LOGIC, Lewis Carroll. "Symbolic Logic" is not concerned with modern symbolic logic, but is instead a collection of over 380 problems posed with charm and imagination, using the syllogism, and a fascinating diagrammatic method of drawing conclusions. In "The Game of Logic," Carroll's whimsical imagination devises a logical game played with 2 diagrams and counters (included) to manipulate hundreds of tricky syllogisms. The final section, "Hit or Miss" is a lagniappe of 101 additional puzzles in the delightful Carroll manner. Until this reprint edition, both of these books were rarities costing up to $15 each. Symbolic Logic: Index, xxxi + 199pp. The Game of Logic: 96pp. Two vols. bound as one. 5⅜ x 8. T492 Paperbound **$1.50**

MAZES AND LABYRINTHS: A BOOK OF PUZZLES, W. Shepherd. Mazes, formerly associated with mystery and ritual, are still among the most intriguing of intellectual puzzles. This is a novel and different collection of 50 amusements that embody the principle of the maze: mazes in the classical tradition; 3-dimensional, ribbon, and Möbius-strip mazes; hidden messages; spatial arrangements; etc.—almost all built on amusing story situations. 84 illustrations. Essay on maze psychology. Solutions. xv + 122pp. 5⅜ x 8. T731 Paperbound **$1.00**

MATHEMATICAL RECREATIONS, M. Kraitchik. Some 250 puzzles, problems, demonstrations of recreational mathematics for beginners & advanced mathematicians. Unusual historical problems from Greek, Medieval, Arabic, Hindu sources; modern problems based on "mathematics without numbers," geometry, topology, arithmetic, etc. Pastimes derived from figurative numbers, Mersenne numbers, Fermat numbers; fairy chess, latruncles, reversi, many topics. Full solutions. Excellent for insights into special fields of math. 181 illustrations. 330pp. 5⅜ x 8. T163 Paperbound **$1.75**

MATHEMATICAL PUZZLES OF SAM LOYD, Vol. I, selected and edited by M. Gardner. Puzzles by the greatest puzzle creator and innovator. Selected from his famous "Cyclopedia of Puzzles," they retain the unique style and historical flavor of the originals. There are posers based on arithmetic, algebra, probability, game theory, route tracing, topology, counter, sliding block, operations research, geometrical dissection. Includes his famous "14-15" puzzle which was a national craze, and his "Horse of a Different Color" which sold millions of copies. 117 of his most ingenious puzzles in all, 120 line drawings and diagrams. Solutions. Selected references. xx + 167pp. 5⅜ x 8. T498 Paperbound **$1.00**

MY BEST PUZZLES IN MATHEMATICS, Hubert Phillips ("Caliban"). Caliban is generally considered the best of the modern problemists. Here are 100 of his best and wittiest puzzles, selected by the author himself from such publications as the London Daily Telegraph, and each puzzle is guaranteed to put even the sharpest puzzle detective through his paces. Perfect for the development of clear thinking and a logical mind. Complete solutions are provided for every puzzle. x + 107pp. 5⅜ x 8½. T91 Paperbound **$1.00**

MY BEST PUZZLES IN LOGIC AND REASONING, H. Phillips ("Caliban"). 100 choice, hitherto unavailable puzzles by England's best-known problemist. No special knowledge needed to solve these logical or inferential problems, just an unclouded mind, nerves of steel, and fast reflexes. Data presented are both necessary and just sufficient to allow one unambiguous answer. More than 30 different types of puzzles, all ingenious and varied, many one of a kind, that will challenge the expert, please the beginner. Original publication. 100 puzzles, full solutions. x + 107pp. 5⅜ x 8½. T119 Paperbound **$1.00**

MATHEMATICAL PUZZLES FOR BEGINNERS AND ENTHUSIASTS, G. Mott-Smith. 188 mathematical puzzles to test mental agility. Inference, interpretation, algebra, dissection of plane figures, geometry, properties of numbers, decimation, permutations, probability, all enter these delightful problems. Puzzles like the Odic Force, How to Draw an Ellipse, Spider's Cousin, more than 180 others. Detailed solutions. Appendix with square roots, triangular numbers, primes, etc. 135 illustrations. 2nd revised edition. 248pp. 5⅜ x 8. T198 Paperbound **$1.00**

MATHEMATICS, MAGIC AND MYSTERY, Martin Gardner. Card tricks, feats of mental mathematics, stage mind-reading, other "magic" explained as applications of probability, sets, theory of numbers, topology, various branches of mathematics. Creative examination of laws and their applications with scores of new tricks and insights. 115 sections discuss tricks wtih cards, dice, coins; geometrical vanishing tricks, dozens of others. No sleight of hand needed; mathematics guarantees success. 115 illustrations. xii + 174pp. 5⅜ x 8. T335 Paperbound **$1.00**

RECREATIONS IN THE THEORY OF NUMBERS: THE QUEEN OF MATHEMATICS ENTERTAINS, Albert H. Beiler. The theory of numbers is often referred to as the "Queen of Mathematics." In this book Mr. Beiler has compiled the first English volume to deal exclusively with the recreational aspects of number theory, an inherently recreational branch of mathematics. The author's clear style makes for enjoyable reading as he deals with such topics as: perfect numbers, amicable numbers, Fermat's theorem, Wilson's theorem, interesting properties of digits, methods of factoring, primitive roots, Euler's function, polygonal and figurate numbers, Mersenne numbers, congruence, repeating decimals, etc. Countless puzzle problems, with full answers and explanations. For mathematicians and mathematically-inclined laymen, etc. New publication. 28 figures. 9 illustrations. 103 tables. Bibliography at chapter ends. vi + 247pp. 5⅜ x 8½. **T1096 Paperbound $1.85**

PAPER FOLDING FOR BEGINNERS, W. D. Murray and F. J. Rigney. A delightful introduction to the varied and entertaining Japanese art of origami (paper folding), with a full crystal-clear text that anticipates every difficulty; over 275 clearly labeled diagrams of all important stages in creation. You get results at each stage, since complex figures are logically developed from simpler ones. 43 different pieces are explained: place mats, drinking cups, bonbon boxes, sailboats, frogs, roosters, etc. 6 photographic plates. 279 diagrams. 95pp. 5⅝ x 8⅜. **T713 Paperbound $1.00**

1800 RIDDLES, ENIGMAS AND CONUNDRUMS, Darwin A. Hindman. Entertaining collection ranging from hilarious gags to outrageous puns to sheer nonsense—a welcome respite from sophisticated humor. Children, toastmasters, and practically anyone with a funny bone will find these zany riddles tickling and eminently repeatable. Sample: "Why does Santa Claus always go down the chimney?" "Because it soots him." Some old, some new—covering a wide variety of subjects. New publication. iii + 154pp. 5⅜ x 8½. **T1059 Paperbound $1.00**

EASY-TO-DO ENTERTAINMENTS AND DIVERSIONS WITH CARDS, STRING, COINS, PAPER AND MATCHES, R. M. Abraham. Over 300 entertaining games, tricks, puzzles, and pastimes for children and adults. Invaluable to anyone in charge of groups of youngsters, for party givers, etc. Contains sections on card tricks and games, making things by paperfolding—toys, decorations, and the like; tricks with coins, matches, and pieces of string; descriptions of games; toys that can be made from common household objects; mathematical recreations; word games; and 50 miscellaneous entertainments. Formerly "Winter Nights Entertainments." Introduction by Lord Baden Powell. 329 illustrations. v + 186pp. 5⅜ x 8. **T921 Paperbound $1.00**

DIVERSIONS AND PASTIMES WITH CARDS, STRING, PAPER AND MATCHES, R. M. Abraham. Another collection of amusements and diversion for game and puzzle fans of all ages. Many new paperfolding ideas and tricks, an extensive section on amusements with knots and splices, two chapters of easy and not-so-easy problems, coin and match tricks, and lots of other parlor pastimes from the agile mind of the late British problemist and gamester. Corrected and revised version. Illustrations. 160pp. 5⅜ x 8½. **T1127 Paperbound $1.00**

STRING FIGURES AND HOW TO MAKE THEM: A STUDY OF CAT'S-CRADLE IN MANY LANDS, Caroline Furness Jayne. In a simple and easy-to-follow manner, this book describes how to make 107 different string figures. Not only is looping and crossing string between the fingers a common youthful diversion, but it is an ancient form of amusement practiced in all parts of the globe, especially popular among primitive tribes. These games are fun for all ages and offer an excellent means for developing manual dexterity and coordination. Much insight also for the anthropological observer on games and diversions in many different cultures. Index. Bibliography. Introduction by A. C. Haddon, Cambridge University. 17 full-page plates. 950 illustrations. xxiii + 407pp. 5⅜ x 8½. **T152 Paperbound $2.00**

CRYPTANALYSIS, Helen F. Gaines. (Formerly ELEMENTARY CRYPTANALYSIS.) A standard elementary and intermediate text for serious students. It does not confine itself to old material, but contains much that is not generally known, except to experts. Concealment, Transposition, Substitution ciphers; Vigenere, Kasiski, Playfair, multafid, dozens of other techniques. Appendix with sequence charts, letter frequencies in English, 5 other languages, English word frequencies. Bibliography. 167 codes. New to this edition: solution to codes. vi + 230pp. 5⅜ x 8. **T97 Paperbound $2.00**

MAGIC SQUARES AND CUBES, W. S. Andrews. Only book-length treatment in English, a thorough non-technical description and analysis. Here are nasik, overlapping, pandiagonal, serrated squares; magic circles, cubes, spheres, rhombuses. Try your hand at 4-dimensional magical figures! Much unusual folklore and tradition included. High school algebra is sufficient. 754 diagrams and illustrations. viii + 419pp. 5⅜ x 8. **T658 Paperbound $1.85**

CALIBAN'S PROBLEM BOOK: MATHEMATICAL, INFERENTIAL, AND CRYPTOGRAPHIC PUZZLES, H. Phillips ("Caliban"), S. T. Shovelton, G. S. Marshall. 105 ingenious problems by the greatest living creator of puzzles based on logic and inference. Rigorous, modern, piquant, and reflecting their author's unusual personality, these intermediate and advanced puzzles all involve the ability to reason clearly through complex situations; some call for mathematical knowledge, ranging from algebra to number theory. Solutions. xi + 180pp. 5⅜ x 8. **T736 Paperbound $1.25**

FICTION

THE LAND THAT TIME FORGOT and THE MOON MAID, Edgar Rice Burroughs. In the opinion of many, Burroughs' best work. The first concerns a strange island where evolution is individual rather than phylogenetic. Speechless anthropoids develop into intelligent human beings within a single generation. The second projects the reader far into the future and describes the first voyage to the Moon (in the year 2025), the conquest of the Earth by the Moon, and years of violence and adventure as the enslaved Earthmen try to regain possession of their planet. "An imaginative tour de force that keeps the reader keyed up and expectant," NEW YORK TIMES. Complete, unabridged text of the original two novels (three parts in each). 5 illustrations by J. Allen St. John. vi + 552pp. 5⅜ x 8½.
T1020 Clothbound **$3.75**
T358 Paperbound **$2.00**

AT THE EARTH'S CORE, PELLUCIDAR, TANAR OF PELLUCIDAR: THREE SCIENCE FICTION NOVELS BY EDGAR RICE BURROUGHS. Complete, unabridged texts of the first three Pellucidar novels. Tales of derring-do by the famous master of science fiction. The locale for these three related stories is the inner surface of the hollow Earth where we discover the world of Pellucidar, complete with all types of bizarre, menacing creatures, strange peoples, and alluring maidens—guaranteed to delight all Burroughs fans and a wide circle of adventure lovers. Illustrated by J. Allen St. John and P. F. Berdanier. vi + 433pp. 5⅜ x 8½.
T1051 Paperbound **$2.00**

THE PIRATES OF VENUS and LOST ON VENUS: TWO VENUS NOVELS BY EDGAR RICE BURROUGHS. Two related novels, complete and unabridged. Exciting adventure on the planet Venus with Earthman Carson Napier broken-field running through one dangerous episode after another. All lovers of swashbuckling science fiction will enjoy these two stories set in a world of fascinating societies, fierce beasts, 5000-ft. trees, lush vegetation, and wide seas. Illustrations by Fortunino Matania. Total of vi + 340pp. 5⅜ x 8½. T1053 Paperbound **$1.75**

A PRINCESS OF MARS and A FIGHTING MAN OF MARS: TWO MARTIAN NOVELS BY EDGAR RICE BURROUGHS. "Princess of Mars" is the very first of the great Martian novels written by Burroughs, and it is probably the best of them all; it set the pattern for all of his later fantasy novels and contains a thrilling cast of strange peoples and creatures and the formula of Olympian heroism amidst ever-fluctuating fortunes which Burroughs carries off so successfully. "Fighting Man" returns to the same scenes and cities—many years later. A mad scientist, a degenerate dictator, and an indomitable defender of the right clash—with the fate of the Red Planet at stake! Complete, unabridged reprinting of original editions. Illustrations by F. E. Schoonover and Hugh Hutton. v + 356pp. 5⅜ x 8½.
T1140 Paperbound **$1.75**

THREE MARTIAN NOVELS, Edgar Rice Burroughs. Contains: Thuvia, Maid of Mars; The Chessmen of Mars; and The Master Mind of Mars. High adventure set in an imaginative and intricate conception of the Red Planet. Mars is peopled with an intelligent, heroic human race which lives in densely populated cities and with fierce barbarians who inhabit dead sea bottoms. Other exciting creatures abound amidst an inventive framework of Martian history and geography. Complete unabridged reprintings of the first edition. 16 illustrations by J. Allen St. John. vi + 499pp. 5⅜ x 8½.
T39 Paperbound **$1.85**

THREE PROPHETIC NOVELS BY H. G. WELLS, edited by E. F. Bleiler. Complete texts of "When the Sleeper Wakes" (1st book printing in 50 years), "A Story of the Days to Come," "The Time Machine" (1st complete printing in book form). Exciting adventures in the future are as enjoyable today as 50 years ago when first printed. Predict TV, movies, intercontinental airplanes, prefabricated houses, air-conditioned cities, etc. First important author to foresee problems of mind control, technological dictatorships. "Absolute best of imaginative fiction," N. Y. Times. Introduction. 335pp. 5⅜ x 8.
T605 Paperbound **$1.50**

28 SCIENCE FICTION STORIES OF H. G. WELLS. Two full unabridged novels, MEN LIKE GODS and STAR BEGOTTEN, plus 26 short stories by the master science-fiction writer of all time. Stories of space, time, invention, exploration, future adventure—an indispensable part of the library of everyone interested in science and adventure. PARTIAL CONTENTS: Men Like Gods, The Country of the Blind, In the Abyss, The Crystal Egg, The Man Who Could Work Miracles, A Story of the Days to Come, The Valley of Spiders, and 21 more! 928pp. 5⅜ x 8.
T265 Clothbound **$4.50**

THE WAR IN THE AIR, IN THE DAYS OF THE COMET, THE FOOD OF THE GODS: THREE SCIENCE FICTION NOVELS BY H. G. WELLS. Three exciting Wells offerings bearing on vital social and philosophical issues of his and our own day. Here are tales of air power, strategic bombing, East vs. West, the potential miracles of science, the potential disasters from outer space, the relationship between scientific advancement and moral progress, etc. First reprinting of "War in the Air" in almost 50 years. An excellent sampling of Wells at his storytelling best. Complete, unabridged reprintings. 16 illustrations. 645pp. 5⅜ x 8½.
T1135 Paperbound **$2.00**

SEVEN SCIENCE FICTION NOVELS, H. G. Wells. Full unabridged texts of 7 science-fiction novels of the master. Ranging from biology, physics, chemistry, astronomy to sociology and other studies, Mr. Wells extrapolates whole worlds of strange and intriguing character. "One will have to go far to match this for entertainment, excitement, and sheer pleasure . . . ," NEW YORK TIMES. Contents: The Time Machine, The Island of Dr. Moreau, First Men in the Moon, The Invisible Man, The War of the Worlds, The Food of the Gods, In the Days of the Comet. 1015pp. 5⅜ x 8.   T264 Clothbound **$4.50**

BEST GHOST STORIES OF J. S. LE FANU, Selected and introduced by E. F. Bleiler. LeFanu is deemed the greatest name in Victorian supernatural fiction. Here are 16 of his best horror stories, including 2 novelles: "Carmilla," a classic vampire tale couched in a perverse eroticism, and "The Haunted Baronet." Also: "Sir Toby's Will," "Green Tea," "Schalken the Painter," "Ultor de Lacy," "The Familiar," etc. The first American publication of about half of this material: a long-overdue opportunity to get a choice sampling of LeFanu's work. New selection (1964). 8 illustrations. 5⅜ x 8⅜.   T415 Paperbound **$1.85**

THE WONDERFUL WIZARD OF OZ, L. F. Baum. Only edition in print with all the original W. W. Denslow illustrations in full color—as much a part of "The Wizard" as Tenniel's drawings are for "Alice in Wonderland." "The Wizard" is still America's best-loved fairy tale, in which, as the author expresses it, "The wonderment and joy are retained and the heartaches and nightmares left out." Now today's young readers can enjoy every word and wonderful picture of the original book. New introduction by Martin Gardner. A Baum bibliography. 23 full-page color plates. viii + 268pp. 5⅜ x 8.   T691 Paperbound **$1.50**

GHOST AND HORROR STORIES OF AMBROSE BIERCE, Selected and introduced by E. F. Bleiler. 24 morbid, eerie tales—the cream of Bierce's fiction output. Contains such memorable pieces as "The Moonlit Road," "The Damned Thing," "An Inhabitant of Carcosa," "The Eyes of the Panther," "The Famous Gilson Bequest," "The Middle Toe of the Right Foot," and other chilling stories, plus the essay, "Visions of the Night" in which Bierce gives us a kind of rationale for his aesthetic of horror. New collection (1964). xxii + 199pp. 5⅜ x 8⅜.   T767 Paperbound **$1.00**

HUMOR

MR. DOOLEY ON IVRYTHING AND IVRYBODY, Finley Peter Dunne. Since the time of his appearance in 1893, "Mr. Dooley," the fictitious Chicago bartender, has been recognized as America's most humorous social and political commentator. Collected in this volume are 102 of the best Dooley pieces—all written around the turn of the century, the height of his popularity. Mr. Dooley's Irish brogue is employed wittily and penetratingly on subjects which are just as fresh and relevant today as they were then: corruption and hypocrisy of politicans, war preparations and chauvinism, automation, Latin American affairs, superbombs, etc. Other articles range from Rudyard Kipling to football. Selected with an introduction by Robert Hutchinson. xii + 244pp. 5⅜ x 8½.   T626 Paperbound **$1.00**

RUTHLESS RHYMES FOR HEARTLESS HOMES and MORE RUTHLESS RHYMES FOR HEARTLESS HOMES, Harry Graham ("Col. D. Streamer"). A collection of Little Willy and 48 other poetic "disasters." Graham's funniest and most disrespectful verse, accompanied by original illustrations. Nonsensical, wry humor which employs stern parents, careless nurses, uninhibited children, practical jokers, single-minded golfers, Scottish lairds, etc. in the leading roles. A precursor of the "sick joke" school of today. This volume contains, bound together for the first time, two of the most perennially popular books of humor in England and America. Index. vi + 69pp. 5⅜ x 8.   T930 Paperbound **75¢**

A WHIMSEY ANTHOLOGY, Collected by Carolyn Wells. 250 of the most amusing rhymes ever written. Acrostics, anagrams, palindromes, alphabetical jingles, tongue twisters, echo verses, alliterative verses, riddles, mnemonic rhymes, interior rhymes, over 40 limericks, etc. by Lewis Carroll, Edward Lear, Joseph Addison, W. S. Gilbert, Christina Rossetti, Chas. Lamb, James Boswell, Hood, Dickens, Swinburne, Leigh Hunt, Harry Graham, Poe, Eugene Field, and many others. xiv + 221pp. 5⅜ x 8½.   T195 Paperbound **$1.25**

MY PIOUS FRIENDS AND DRUNKEN COMPANIONS and MORE PIOUS FRIENDS AND DRUNKEN COMPANIONS, Songs and ballads of Conviviality Collected by Frank Shay. Magnificently illuminated by John Held, Jr. 132 ballads, blues, vaudeville numbers, drinking songs, cowboy songs, sea chanties, comedy songs, etc. of the Naughty Nineties and early 20th century. Over a third are reprinted with music. Many perennial favorites such as: The Band Played On, Frankie and Johnnie, The Old Grey Mare, The Face on the Bar-room Floor, etc. Many others unlocatable elsewhere: The Dog-Catcher's Child, The Cannibal Maiden, Don't Go in the Lion's Cage Tonight, Mother, etc. Complete verses and introductions to songs. Unabridged republication of first editions, 2 Indexes (song titles and first lines and choruses). Introduction by Frank Shay. 2 volumes bounds as 1. Total of xvi + 235pp. 5⅜ x 8½.   T946 Paperbound **$1.25**

MAX AND MORITZ, Wilhelm Busch. Edited and annotated by H. Arthur Klein. Translated by H. Arthur Klein, M. C. Klein, and others. The mischievous high jinks of Max and Moritz, Peter and Paul, Ker and Plunk, etc. are delightfully captured in sketch and rhyme. (Companion volume to "Hypocritical Helena.") In addition to the title piece, it contians: Ker and Plunk; Two Dogs and Two Boys; The Egghead and the Two Cut-ups of Corinth; Deceitful Henry; The Boys and the Pipe; Cat and Mouse; and others. (Original German text with accompanying English translations.) Afterword by H. A. Klein. vi + 216pp. 5⅜ x 8½.
T181 Paperbound **$1.15**

THROUGH THE ALIMENTARY CANAL WITH GUN AND CAMERA: A FASCINATING TRIP TO THE INTERIOR, Personally Conducted by George S. Chappell. In mock-travelogue style, the amusing account of an imaginative journey down the alimentary canal. The "explorers" enter the esophagus, round the Adam's Apple, narrowly escape from a fierce Amoeba, struggle through the impenetrable Nerve Forests of the Lumbar Region, etc. Illustrated by the famous cartoonist, Otto Soglow, the book is as much a brilliant satire of academic pomposity and professional travel literature as it is a clever use of the facts of physiology for supremely comic purposes. Preface by Robert Benchley. Author's Foreword. 1 Photograph. 17 illustrations by O. Soglow. xii + 114pp. 5⅜ x 8½.
T376 Paperbound **$1.00**

THE BAD CHILD'S BOOK OF BEASTS, MORE BEASTS FOR WORSE CHILDREN, and A MORAL ALPHABET, H. Belloc. Hardly an anthology of humorous verse has appeared in the last 50 years without at least a couple of these famous nonsense verses. But one must see the entire volumes—with all the delightful original illustrations by Sir Basil Blackwood—to appreciate fully Belloc's charming and witty verses that play so subacidly on the platitudes of life and morals that beset his day—and ours. A great humor classic. Three books in one. Total of 157pp. 5⅜ x 8.
T749 Paperbound **$1.00**

THE DEVIL'S DICTIONARY, Ambrose Bierce. Sardonic and irreverent barbs puncturing the pomposities and absurdities of American politics, business, religion, literature, and arts, by the country's greatest satirist in the classic tradition. Epigrammatic as Shaw, piercing as Swift, American as Mark Twain, Will Rogers, and Fred Allen. Bierce will always remain the favorite of a small coterie of enthusiasts, and of writers and speakers whom he supplies with "some of the most gorgeous witticisms of the English language." (H. L. Mencken) Over 1000 entries in alphabetical order. 144pp. 5⅜ x 8.
T487 Paperbound **$1.00**

THE COMPLETE NONSENSE OF EDWARD LEAR. This is the only complete edition of this master of gentle madness available at a popular price. A BOOK OF NONSENSE, NONSENSE SONGS, MORE NONSENSE SONGS AND STORIES in their entirety with all the old favorites that have delighted children and adults for years. The Dong With A Luminous Nose, The Jumblies, The Owl and the Pussycat, and hundreds of other bits of wonderful nonsense. 214 limericks, 3 sets of Nonsense Botany, 5 Nonsense Alphabets. 546 drawings by Lear himself, and much more. 320pp. 5⅜ x 8.
T167 Paperbound **$1.00**

SINGULAR TRAVELS, CAMPAIGNS, AND ADVENTURES OF BARON MUNCHAUSEN, R. E. Raspe, with 90 illustrations by Gustave Doré. The first edition in over 150 years to reestablish the deeds of the Prince of Liars exactly as Raspe first recorded them in 1785—the genuine Baron Munchausen, one of the most popular personalities in English literature. Included also are the best of the many sequels, written by other hands. Introduction on Raspe by J. Carswell. Bibliography of early editions. xliv + 192pp. 5⅜ x 8. T698 Paperbound **$1.00**

HOW TO TELL THE BIRDS FROM THE FLOWERS, R. W. Wood. How not to confuse a carrot with a parrot, a grape with an ape, a puffin with nuffin. Delightful drawings, clever puns, absurd little poems point out farfetched resemblances in nature. The author was a leading physicist. Introduction by Margaret Wood White. 106 illus. 60pp. 5⅜ x 8.
T523 Paperbound **75¢**

JOE MILLER'S JESTS OR, THE WITS VADE-MECUM. The original Joe Miller jest book. Gives a keen and pungent impression of life in 18th-century England. Many are somewhat on the bawdy side and they are still capable of provoking amusement and good fun. This volume is a facsimile of the original "Joe Miller" first published in 1739. It remains the most popular and influential humor book of all time. New introduction by Robert Hutchinson. xxi + 70pp. 5⅜ x 8½.
T423 Paperbound **$1.00**

Prices subject to change without notice.

Dover publishes books on art, music, philosophy, literature, languages, history, social sciences, psychology, handcrafts, orientalia, puzzles and entertainments, chess, pets and gardens, books explaining science, intermediate and higher mathematics, mathematical physics, engineering, biological sciences, earth sciences, classics of science, etc. Write to:

Dept. catrr.
Dover Publications, Inc.
180 Varick Street, N.Y. 14, N.Y.